AF410825

Artificial Intelligence Techniques in Hydrology and Water Resources Management

Artificial Intelligence Techniques in Hydrology and Water Resources Management

Editors

Fi-John Chang
Li-Chiu Chang
Jui-Fa Chen

MDPI • Basel • Beijing • Wuhan • Barcelona • Belgrade • Manchester • Tokyo • Cluj • Tianjin

Editors

Fi-John Chang
Department of
Bioenvironmental Systems
Engineering
National Taiwan University
Taipei
Taiwan

Li-Chiu Chang
Department of Water
Resources and Environmental
Engineering
Tamkang University
New Taipei City
Taiwan

Jui-Fa Chen
Department of Computer
Science and Information
Engineering
Tamkang University
New Taipei City
Taiwan

Editorial Office
MDPI
St. Alban-Anlage 66
4052 Basel, Switzerland

This is a reprint of articles from the Special Issue published online in the open access journal *Water* (ISSN 2073-4441) (available at: www.mdpi.com/journal/water/special_issues/AI_hydrology).

For citation purposes, cite each article independently as indicated on the article page online and as indicated below:

LastName, A.A.; LastName, B.B.; LastName, C.C. Article Title. *Journal Name* **Year**, *Volume Number*, Page Range.

ISBN 978-3-0365-7785-2 (Hbk)
ISBN 978-3-0365-7784-5 (PDF)

Contents

About the Editors

Fi-John Chang

Fi-John Chang holds the position of Distinguished Professor at National Taiwan University. Additionally, he serves as the Founding President of the Taiwan Hydro-Informatics Society and holds the role of National Correspondent for the International Association of Hydrological Sciences (IAHS). His primary research focus lies in the field of artificial intelligence (AI) technology applied to smart water resources management and environmental sciences. Over the past several years, he has expanded his research scope to include the water–energy–food nexus. Through his innovative theoretical advancements and the development of novel methodologies tailored to Taiwan's specific water resources and environmental management needs, he has carried out numerous successful case studies. These contributions have significantly enhanced hydro-informatics technology, strengthened smart water resources management practices, and generated practical applications. In recognition of his outstanding research contributions, he has received the Outstanding Research Award from the National Science and Technology Council twice (in 2010 and 2018) and the International Award from the PAWEES in 2014. His research findings have been published in leading SCI journals in the fields of hydrology, environmental engineering, and computer science, amounting to over 169 papers with more than 7,100 citations to date. His h-index stands at 49. Prof. Chang has effectively raised Taiwan's profile and influence within the international academic community.

Li-Chiu Chang

Li-Chiu Chang, a Professor at Tamkang University, possesses over two decades of experience in utilizing artificial neural network models, fuzzy inference systems, and data mining to develop innovative approaches for hydro-informatics systems. She has successfully coordinated more than 60 government projects focused on the development and design of real-time engines for reservoir flood operations, as well as integration platforms for intelligent urban flood warning systems in Taiwan and Malaysia. With her expertise, she has contributed to over 60 technical and scientific papers published in peer-reviewed journals and has co-authored a textbook titled "Introduction to Artificial Neural Networks: Principles and Applications". Her current research interests encompass various areas, including agricultural water management, reservoir operations, artificial intelligence (particularly artificial neural networks, genetic algorithms, optimization, and fuzzy sets), flood forecasting and early warning, urban inundation forecasting, and big data mining. Li-Chiu Chang has achieved an h-index of 35, showcasing the impact of her research.

Jui-Fa Chen

Jui-Fa Chen, an Associate Professor at Tamkang University, specializes in several domains, including computer networks, wireless sensor networks, medical engineering, and artificial intelligence (AI). His research interests have also expanded to include wearable devices. Notably, his patented technology, the "Electronic device for assessing workers' overwork risk", received the prestigious Platinum Award at the 2019 Taiwan Innotech Expo. In recognition of his innovative work, he was invited to showcase his sensing device, which integrates big data, Internet of Things (IoT) technologies, and applications, at the Sustainable Development Pavilion of the 2020 Taiwan Innotech Expo. This demonstration received significant attention. Recently, he has been engaged in transdisciplinary research encompassing the computer science, engineering, and medical fields. His aim is to invent more wearable devices that contribute to enhancing the quality of human life.

Preface to "Artificial Intelligence Techniques in Hydrology and Water Resources Management"

The effective management of water cycles is of utmost importance in the context of climate change and global warming. This entails overseeing the global, regional, and local water cycles, as well as the water cycles within urban, agricultural, and industrial settings, with the aim of preserving water resources and their interactions with energy, food production, microclimates, biodiversity, ecosystem functioning, and human activities. Hydrological modeling plays an important role in achieving this objective because it is crucial for water resources management and the mitigation of natural disasters. In recent years, the integration of artificial intelligence (AI) techniques in hydrology and water resources management has seen significant progress. When confronted with uncertainties related to hydrology, geology, and meteorology, AI approaches have proven to be powerful tools for accurately simulating intricate and nonlinear hydrological processes, as well as effectively leveraging various digital and imaging data sources such as ground gauges, remote sensing tools, and in situ Internet of Things (IoT) devices. The collection of thirteen research papers featured in this Special Issue has made noteworthy contributions to both long- and short-term hydrological modeling and water resources management in the face of changing environments, utilizing AI techniques in combination with diverse analytical tools. These contributions encompass areas such as hydrological forecasting, microclimate regulation, and climate adaptation, and have the potential to enhance hydrology research and guide policy-making efforts toward sustainable and integrated water resources management.

Fi-John Chang, Li-Chiu Chang, and Jui-Fa Chen
Editors

Editorial

Artificial Intelligence Techniques in Hydrology and Water Resources Management

Fi-John Chang [1,*] , Li-Chiu Chang [2] and Jui-Fa Chen [3]

[1] Department of Bioenvironmental Systems Engineering, National Taiwan University, Taipei 10617, Taiwan
[2] Department of Water Resources and Environmental Engineering, Tamkang University, New Taipei City 25137, Taiwan; changlc@mail.tku.edu.tw
[3] Department of Computer Science and Information Engineering, Tamkang University, New Taipei City 25137, Taiwan; alpha@mail.tku.edu.tw
* Correspondence: changfj@ntu.edu.tw

Abstract: The sustainable management of water cycles is crucial in the context of climate change and global warming. It involves managing global, regional, and local water cycles—as well as urban, agricultural, and industrial water cycles—to conserve water resources and their relationships with energy, food, microclimates, biodiversity, ecosystem functioning, and anthropogenic activities. Hydrological modeling is indispensable for achieving this goal, as it is essential for water resources management and mitigation of natural disasters. In recent decades, the application of artificial intelligence (AI) techniques in hydrology and water resources management has made notable advances. In the face of hydro-geo-meteorological uncertainty, AI approaches have proven to be powerful tools for accurately modeling complex, non-linear hydrological processes and effectively utilizing various digital and imaging data sources, such as ground gauges, remote sensing tools, and in situ Internet of Things (IoTs). The thirteen research papers published in this Special Issue make significant contributions to long- and short-term hydrological modeling and water resources management under changing environments using AI techniques coupled with various analytics tools. These contributions, which cover hydrological forecasting, microclimate control, and climate adaptation, can promote hydrology research and direct policy making toward sustainable and integrated water resources management.

Keywords: machine learning; deep learning; hydroinformatics; hydrological modeling; early warning; uncertainty; sustainability

Citation: Chang, F.-J.; Chang, L.-C.; Chen, J.-F. Artificial Intelligence Techniques in Hydrology and Water Resources Management. *Water* **2023**, *15*, 1846. https://doi.org/10.3390/w15101846

Received: 9 May 2023
Accepted: 11 May 2023
Published: 12 May 2023

1. Introduction

Artificial intelligence (AI) encompasses a broad range of computer-related disciplines that focus on creating intelligent models to conduct work previously carried out by humans [1,2]. AI enables computers to model or even surpass human cognitive abilities, thereby rapidly rationalizing and taking steps to achieve specific objectives, such as several-steps-ahead predictions and pattern recognition. AI is also recognized for its ability to manage massive amounts of data and sophisticated models with ease [3]. Since the mid-20th century, the use of AI techniques has grown in a wide variety of engineering and scientific disciplines [4,5]. With numerous interdisciplinary scientific approaches, recent advancements in AI have triggered a paradigm shift in almost every field, including engineering, hydrology, technology, and medical imaging [6–9].

Over the past two decades, AI approaches have rapidly emerged as a solution to overcome the challenges presented by the high complexity, dynamics, non-linearity and non-stationarity observed in hydrological processes [10,11]. The increase in severe natural disasters resulting from climate change and global warming has posed a significant threat to sustainable hydrology and water resources management. As a result, there has been a notable surge in exploring AI models to characterize and predict hydrological variability under growing hydro-geo-meteorological uncertainty [12–14]. AI techniques offer a promising alternative or supplement to conventional physical-based or statistical approaches to hydrological modeling [15–19]. By utilizing data from various sources, including micro-sensing, imaging, in situ, and remote sensing devices, AI techniques are currently enabling the creation of reliable and robust hydrological models at finer spatio-temporal resolutions of interest, which is crucial for addressing highly nonlinear hydro-meteorological processes [20–24]. Therefore, there is a need to explore innovative AI models to better allocate, regulate, and conserve water resources, which will significantly contribute to sustainable their management.

In 2021 and 2022, Water (MDPI) published thirteen research papers in a Special Issue entitled "Artificial Intelligence Techniques in Hydrology and Water Resources Management".

The objectives of the current Special Issue are as follows:

- To advance the use of AI techniques in hydrological modeling and water resources management.
- To develop innovative solutions for hydrological forecasting and problem-solving in watershed hydrology under changing environments.
- To improve water and environmental systems.
- To promote urban water–energy–food nexus synergies.
- To quantify uncertainty of hydrological modeling.

This editorial provides an overview of the Special Issue, offering insights and suggestions for future research.

2. Highlights of the Articles in the Special Issue

The thirteen articles presented in the Special Issue have made substantial contributions across five main research areas:

- The application of machine learning and deep learning techniques in hydro-meteorological forecasting, classification and time series generation under changing environmental conditions;
- The use of AI techniques for smart microclimate control;
- The current and future roles of Geospatial Artificial Intelligence (GeoAI) in hydrological and fluvial systems;
- Adaptation strategies for extreme hydrological events to mitigate hazards;
- The utilization of AI for processing hydro-geo-meteorological data.

These articles are grouped and highlighted as follows.

2.1. Smart Microclimate Control System Using AI

The prediction of a short-term microclimate is a challenging task due to the rapid changes and strong interconnections among meteorological variables. To address this issue, Chen et al. introduced a water-centric smart microclimate control system (SMCS) that incorporates system dynamics and machine learning techniques, which can regulate the micro-environment within a greenhouse canopy to induce environmental cooling while improving resource-use efficiency [25]. The proposed SMCS demonstrates the practicality of machine-learning-enabled greenhouse automation that enhances crop productivity and resource-use efficiency, thereby contributing to the mitigation of carbon emissions and a sustainable water–energy–food nexus.

2.2. Weather Typing for Smart Urban Agriculture Using AI

In outdoor agricultural production, weather is a crucial factor that affects crop growth. Climate information can be utilized to help farmers plan their planting and production schedules, especially for urban agriculture. Huang and Chang used a self-organizing map (SOM) to investigate the spatiotemporal weather features of Taipei City by analyzing the observed data of six key weather factors from five weather stations in Northern Taiwan between 2014 and 2018 [26]. The results provide practical references for anticipating upcoming weather types and features within designated time frames, arranging potential cultivation tasks or making necessary adjustments, and efficiently utilizing water and energy resources to achieve sustainable production in smart urban agriculture.

2.3. AI-Driven Forecasting

2.3.1. Precipitation Forecasting

Abnormal changes in precipitation and temperature caused by climate change have increased the risks of climate disasters and rainfall damage. Despite quantitative rainfall estimates from weather forecasts, it remains difficult to estimate the damage caused by rainfall. To address this issue, Chu et al. employed various methods, such as support vector machine (SVM), random forest, and eXtreme Gradient Boosting (XGBoost), finding that XGBoost has the best performance [27]. Using XGBoost, the threshold rainfall of ungauged watersheds was calculated and verified using past rainfall events and damage cases, enabling the accurate prediction of flooding-induced rainfall and preparation for vulnerable areas. Alternatively, Pakdaman et al. proposed a learning approach based on an artificial neural network (ANN) and random forest algorithms to provide multi-model ensemble forecasting of monthly precipitation in Southwest Asia [28]. The approach employed four forecasting models from the North American multi-model ensemble (NMME) project, including GEM-NEMO, NASA-GEOSS2S, CanCM4i, and COLA-RSMAS-CCSM4, and used the ERA5 reanalysis dataset to train the models. The results show that the ANN and random forest post-processing both performed better than individual NMME models, with random forest outperforming ANN for all lead times and months of the year.

2.3.2. Temperature Forecasting

Temperature is a crucial weather variable required for various studies. Changes in temperature and precipitation can have a significant impact on river basins. Hernández-Bedolla et al. developed a stochastic model for daily precipitation occurrence and its effect on maximum and minimum temperatures [29]. The study employed a Markov model to identify the daily occurrence of rainfall and a multisite multivariate autoregressive model (MASCV) to represent the short-term memory of daily temperature. The research was conducted on the Jucar River Basin in Spain, where the proposed model could accurately represent both the occurrence of rainfall and the maximum and minimum temperature using a two-state and a lag-one multivariate stochastic model.

2.3.3. Streamflow Forecasting

Ghobadi and Kang proposed a probabilistic forecasting model for multi-step-ahead daily streamflow forecasting, which uses Bayesian sampling in a long short-term memory (BLSTM) neural network to address the subproblem of univariate time series models and quantify epistemic and aleatory uncertainty [30]. The proposed method was validated by three case studies in the USA, and three forecasting horizons demonstrate that BLSTM outperformed the other models in terms of forecasting reliability, accuracy, and overall performance. Moreover, BLSTM can handle data with higher variation and peaks, particularly for long-term multi-step-ahead streamflow forecasting, compared to other models. Alternatively, Forghanparast and Mohammadi compared the performance of three deep learning algorithms, including convolutional neural networks (CNN), long short-term memory (LSTM), and self-attention LSTM models, against a baseline extreme learning machine (ELM) model for monthly streamflow prediction in the headwaters of the Colorado River in Texas [31]. The LSTM model was identified as a more appropriate, effective, and parsimonious streamflow prediction tool for the headwaters of the Colorado River in Texas, with better evaluation metrics than the ELM and CNN algorithms and a more competitive performance then the SA-LSTM model.

2.3.4. Dam Inflow Prediction

Kim et al. illustrated the process and methodology for selecting the most suitable deep learning model using 16 design scenarios to predict dam inflow using hydrologic data from the past two decades [32]. The study focused on Andong Dam and Imha Dam, located upstream of the Nakdong River in South Korea. The optimal recurrent-neural-network-based models demonstrated a better prediction of observed inflow than the storage function model (SFM), which is currently used by both dams. Most deep learning models provided more accurate predictions than the SFM under various typhoon conditions. Therefore, it is crucial to make an informed decision by comparing the inflow predictions of both the SFM and deep learning models for efficient dam operation and management.

2.3.5. Real-Time Inundation Depths Estimation

Wu et al. developed a stochastic model (SM_EID_IOT) for estimating the inundation depths and associated 95% confidence intervals at the specific locations of the roadside water-level gauges (IoTs) sensors under observed water levels/rainfalls and the precipitation forecasts [33]. The goal was to improve the accuracy and reliability of inundation depth estimations at the IoT sensors. The model was tested in the Nankon catchment in northern Taiwan, and the results show that the SM_EID_IOT model was capable of estimating inundation depths at various lead times with a high reliability and accuracy, as validated by the datasets. The corrected inundation depth estimates also exhibited a good agreement with the validated data over time, with an acceptable bias.

2.3.6. Rainfall Time Series Generation

Nguyen and Chen employed a Monte Carlo simulation, a bivariate copula, and a modified Huff curve method to create a stochastic rainfall generator to produce continuous rainfall time series at a high temporal resolution of 10 min [34]. The created rainfall generator was then applied to duplicate rainfall time series for the Yilan River Basin in Taiwan, with statistical indices staying close to those of the observed rainfall time series. The results suggest the need and appropriateness of the newly generated rainfall type for rainfall type classification. In summary, the developed stochastic rainfall generator is capable of adequately reproducing continuous rainfall time series at a 10 min resolution.

2.4. Review of Geospatial Artificial Intelligence (GeoAI)

Gonzales-Inca et al. conducted a review of the current applications of GeoAI and machine learning in various hydrological and hydraulic modeling fields [35]. GeoAI is an effective tool for handling vast amounts of spatial and non-spatial data. GeoAI demonstrates advantages in non-linear modeling, computational efficiency, integration of multiple data sources, high prediction accuracy, and revealing new hydrological patterns and processes. However, a significant drawback of most GeoAI models is the lack of physical interpretability, explainability, and model generalization due to inadequate model settings. Recent GeoAI research has focused on integrating physical-based models with GeoAI methods and developing autonomous prediction and forecasting systems.

2.5. Data Processing Using AI

Measuring water levels in rivers is crucial for producing early warnings and detecting risks. However, data collected by devices installed in remote locations may contain errors due to malfunctions, which can result in missed or false alarms. Khampuengson and Wang investigated deep reinforcement learning (DRL) due to its ability to automatically detect anomalies. They found that this approach lacked consistency despite achieving a higher accuracy than some machine learning models [36]. Thus, an ensemble approach combining multiple DRL models was proposed and achieved higher consistency and accuracy than other models such as multilayer perceptrons (MLP) and LSTM. On the other hand, Papailiou et al. proposed a methodology using ensembles of ANNs to estimate the missing data of daily precipitation in Chania, Greece [37]. The methodology aimed to generate precipitation time series based on observed data from neighboring stations. The results indicate that ANNs achieved more accurate results but were more time-consuming compared to multiple linear regression (MLR) models.

3. Conclusions

In recent decades, the field of hydrology and water resources management has witnessed significant advances in the use of AI techniques. This Special Issue includes 12 research articles and 1 review article that propose innovative AI-based solutions for addressing the critical challenges associated with hydrology and water resources, with promising outcomes.

As AI techniques continue to rapidly evolve across the globe, future research should focus on developing AI techniques and methodologies and integrating advanced hydrological monitoring devices with varying spatial and temporal scales to conduct comprehensive analyses of complex nonlinear hydrological processes in light of scientific and socio-economic considerations. Furthermore, AI-powered solutions can also incorporate low-carbon pathways to support hydrological and engineering sectors in achieving the net zero goal by 2050.

The foundations of Earth and environmental studies lie in the modeling of dynamic geophysical phenomena. While the geoscientific community has conventionally depended on physically based models, the emergence of big Earth data and the widespread success of AI tools suggest a more in-depth adoption of AI. A new grand vision for geoscience involves the fusion of physically based mechanisms and AI techniques to generate hybrid models, but the question of how to implement these approaches remains open.

We would like to extend our sincere gratitude to the authors, reviewers, and editorial staff of *Water* for their valuable contributions to this Special Issue.

Author Contributions: Conceptualization, L.-C.C. and F.-J.C.; Resources, F.-J.C., L.-C.C. and J.-F.C.; Supervision, F.-J.C.; Writing—Review and Editing, F.-J.C., L.-C.C. and J.-F.C. All authors have read and agreed to the published version of the manuscript.

Funding: The authors would like to thank the National Science and Technology Council, Taiwan (111-2625-M-002-014 and 110-2313-B-002-034-MY3).

Conflicts of Interest: The authors declare no conflict of interest.

References

1. Sun, Z.; Sandoval, L.; Crystal-Ornelas, R.; Mousavi, S.M.; Wang, J.; Lin, C.; Cristea, N.; Tong, D.; Carande, W.H.; Ma, X.; et al. A review of Earth Artificial Intelligence. *Comput. Geosci.* **2022**, *159*, 105034. [CrossRef]
2. Sharma, P.; Singh, S.; Sharma, S.D. Artificial neural network approach for hydrologic river flow time series forecasting. *Agric. Res.* **2022**, *11*, 465–476. [CrossRef]
3. Kao, I.F.; Liou, J.Y.; Lee, M.H.; Chang, F.J. Fusing stacked autoencoder and long short-term memory for regional multistep-ahead flood inundation forecasts. *J. Hydrol.* **2021**, *598*, 126371. [CrossRef]
4. Chang, F.J.; Chang, K.Y.; Chang, L.C. Counterpropagation fuzzy-neural network for city flood control system. *J. Hydrol.* **2008**, *358*, 24–34. [CrossRef]
5. Kasiviswanathan, K.S.; Sudheer, K.P. Methods used for quantifying the prediction uncertainty of artificial neural network based hydrologic models. *Stoch. Environ. Res. Risk Assess.* **2017**, *31*, 1659–1670. [CrossRef]
6. Blake, R.B.; Mathew, R.; George, A.; Papakostas, N. Impact of artificial intelligence on engineering: Past, present and future. *Procedia CIRP* **2021**, *104*, 1728–1733. [CrossRef]
7. Shi, F.; Wang, J.; Shi, J.; Wu, Z.; Wang, Q.; Tang, Z.; He, K.; Shi, Y.; Shen, D. Review of artificial intelligence techniques in imaging data acquisition segmentation and diagnosis for COVID-19. *IEEE Rev. Biomed. Eng.* **2020**, *14*, 4–15. [CrossRef] [PubMed]
8. Xie, H.; Chu, H.C.; Hwang, G.J.; Wang, C.C. Trends and development in technology-enhanced adaptive/personalized learning: A systematic review of journal publications from 2007 to 2017. *Comput. Educ.* **2019**, *140*, 103599. [CrossRef]
9. Zhou, Y.; Guo, S.; Hong, X.; Chang, F.J. Systematic impact assessment on inter-basin water transfer projects of the Hanjiang River Basin in China. *J. Hydrol.* **2017**, *553*, 584–595. [CrossRef]
10. Apaydin, H.; Sattari, M.T.; Falsafian, K.; Prasad, R. Artificial intelligence modelling integrated with Singular Spectral analysis and Seasonal-Trend decomposition using Loess approaches for streamflow predictions. *J. Hydrol.* **2021**, *600*, 126506. [CrossRef]
11. Nourani, V.; Baghanam, A.H.; Adamowski, J.; Kisi, O. Applications of hybrid wavelet–artificial intelligence models in hydrology: A review. *J. Hydrol.* **2014**, *514*, 358–377. [CrossRef]

12. Chang, L.C.; Chang, F.J.; Yang, S.N.; Tsai, F.H.; Chang, T.H.; Herricks, E.E. Self-organizing maps of typhoon tracks allow for flood forecasts up to two days in advance. *Nat. Commun.* **2020**, *11*, 1983. [CrossRef] [PubMed]

13. Mohseni, U.; Agnihotri, P.G.; Pande, C.B.; Durin, B. Understanding the Climate Change and Land Use Impact on Streamflow in the Present and Future under CMIP6 Climate Scenarios for the Parvara Mula Basin, India. *Water* **2023**, *15*, 1753. [CrossRef]

14. Visweshwaran, R.; Ramsankaran, R.; Eldho, T.I.; Jha, M.K. Hydrological Impact Assessment of Future Climate Change on a Complex River Basin of Western Ghats, India. *Water* **2022**, *14*, 3571. [CrossRef]

15. Chang, F.J.; Wang, Y.C.; Tsai, W.P. Modelling intelligent water resources allocation for multi-users. *Water Resour. Manag.* **2016**, *30*, 1395–1413. [CrossRef]

16. Chang, L.C.; Chang, F.J.; Yang, S.N.; Kao, I.F.; Ku, Y.Y.; Kuo, C.L.; Amin, I.M.Z.M. Building an intelligent hydroinformatics integration platform for regional flood inundation warning systems. *Water* **2018**, *11*, 9. [CrossRef]

17. Jiang, S.; Zheng, Y.; Solomatine, D. Improving AI system awareness of geoscience knowledge: Symbiotic integration of physical approaches and deep learning. *Geophys. Res. Lett.* **2020**, *47*, e2020GL088229. [CrossRef]

18. Tsai, W.P.; Chang, F.J.; Chang, L.C.; Herricks, E.E. AI techniques for optimizing multi-objective reservoir operation upon human and riverine ecosystem demands. *J. Hydrol.* **2015**, *530*, 634–644. [CrossRef]

19. Zhou, Y.; Guo, S.; Chang, F.J. Explore an evolutionary recurrent ANFIS for modelling multi-step-ahead flood forecasts. *J. Hydrol.* **2019**, *570*, 343–355. [CrossRef]

20. Afan, H.A.; El-shafie, A.; Mohtar, W.H.M.W.; Yaseen, Z.M. Past, present and prospect of an Artificial Intelligence (AI) based model for sediment transport prediction. *J. Hydrol.* **2016**, *541*, 902–913. [CrossRef]

21. Sihag, P.; Singh, V.P.; Angelaki, A.; Kumar, V.; Sepahvand, A.; Golia, E. Modelling of infiltration using artificial intelligence techniques in semi-arid Iran. *Hydrol. Sci. J.* **2019**, *64*, 1647–1658. [CrossRef]

22. Singh, B.; Sihag, P.; Parsaie, A.; Angelaki, A. Comparative analysis of artificial intelligence techniques for the prediction of infiltration process. *Geol. Ecol. Landsc.* **2021**, *5*, 109–118. [CrossRef]

23. Xiong, M.; Liu, P.; Cheng, L.; Deng, C.; Gui, Z.; Zhang, X.; Liu, Y. Identifying time-varying hydrological model parameters to improve simulation efficiency by the ensemble Kalman filter: A joint assimilation of streamflow and actual evapotranspiration. *J. Hydrol.* **2019**, *568*, 758–768. [CrossRef]

24. Yaseen, Z.M.; El-Shafie, A.; Jaafar, O.; Afan, H.A.; Sayl, K.N. Artificial intelligence based models for stream-flow forecasting: 2000–2015. *J. Hydrol.* **2015**, *530*, 829–844. [CrossRef]

25. Chen, T.-H.; Lee, M.-H.; Hsia, I.-W.; Hsu, C.-H.; Yao, M.-H.; Chang, F.-J. Develop a Smart Microclimate Control System for Greenhouses through System Dynamics and Machine Learning Techniques. *Water* **2022**, *14*, 3941. [CrossRef]

26. Huang, A.; Chang, F.-J. Using a Self-Organizing Map to Explore Local Weather Features for Smart Urban Agriculture in Northern Taiwan. *Water* **2021**, *13*, 3457. [CrossRef]

27. Chu, K.-S.; Oh, C.-H.; Choi, J.-R.; Kim, B.-S. Estimation of Threshold Rainfall in Ungauged Areas Using Machine Learning. *Water* **2022**, *14*, 859. [CrossRef]

28. Pakdaman, M.; Babaeian, I.; Bouwer, L.M. Improved Monthly and Seasonal Multi-Model Ensemble Precipitation Forecasts in Southwest Asia Using Machine Learning Algorithms. *Water* **2022**, *14*, 2632. [CrossRef]

29. Hernández-Bedolla, J.; Solera, A.; Paredes-Arquiola, J.; Sanchez-Quispe, S.T.; Domínguez-Sánchez, C. A Continuous Multisite Multivariate Generator for Daily Temperature Conditioned by Precipitation Occurrence. *Water* **2022**, *14*, 3494. [CrossRef]

30. Ghobadi, F.; Kang, D. Multi-Step Ahead Probabilistic Forecasting of Daily Streamflow Using Bayesian Deep Learning: A Multiple Case Study. *Water* **2022**, *14*, 3672. [CrossRef]

31. Forghanparast, F.; Mohammadi, G. Using Deep Learning Algorithms for Intermittent Streamflow Prediction in the Headwaters of the Colorado River, Texas. *Water* **2022**, *14*, 2972. [CrossRef]

32. Kim, B.-J.; Lee, Y.-T.; Kim, B.-H. A Study on the Optimal Deep Learning Model for Dam Inflow Prediction. *Water* **2022**, *14*, 2766. [CrossRef]

33. Wu, S.-J.; Hsu, C.-T.; Chang, C.-H. Stochastic Modeling for Estimating Real-Time Inundation Depths at Roadside IoT Sensors Using the ANN-Derived Model. *Water* **2021**, *13*, 3128. [CrossRef]

34. Nguyen, D.T.; Chen, S.T. Generating Continuous Rainfall Time Series with High Temporal Resolution by Using a Stochastic Rainfall Generator with a Copula and Modified Huff Rainfall Curves. *Water* **2022**, *14*, 2123. [CrossRef]

35. Gonzales-Inca, C.; Calle, M.; Croghan, D.; Torabi Haghighi, A.; Marttila, H.; Silander, J.; Alho, P. Geospatial Artificial Intelligence (GeoAI) in the Integrated Hydrological and Fluvial Systems Modeling: Review of Current Applications and Trends. *Water* **2022**, *14*, 2211. [CrossRef]

36. Khampuengson, T.; Wang, W. Deep Reinforcement Learning Ensemble for Detecting Anomaly in Telemetry Water Level Data. *Water* **2022**, *14*, 2492. [CrossRef]

37. Papailiou, I.; Spyropoulos, F.; Trichakis, I.; Karatzas, G.P. Artificial Neural Networks and Multiple Linear Regression for Filling in Missing Daily Rainfall Data. *Water* **2022**, *14*, 2892. [CrossRef]

Article

Develop a Smart Microclimate Control System for Greenhouses through System Dynamics and Machine Learning Techniques

Ting-Hsuan Chen [1], Meng-Hsin Lee [1], I-Wen Hsia [1], Chia-Hui Hsu [1], Ming-Hwi Yao [2,*] and Fi-John Chang [1,*]

[1] Department of Bioenvironmental Systems Engineering, National Taiwan University, Taipei 10617, Taiwan
[2] Taiwan Agricultural Research Institute, Taichung City 41362, Taiwan
* Correspondence: mhyao@tari.gov.tw (M.-H.Y.); changfj@ntu.edu.tw (F.-J.C.)

Abstract: Agriculture is extremely vulnerable to climate change. Greenhouse farming is recognized as a promising measure against climate change. Nevertheless, greenhouse farming frequently encounters environmental adversity, especially greenhouses built to protect against typhoons. Short-term microclimate prediction is challenging because meteorological variables are strongly interconnected and change rapidly. Therefore, this study proposes a water-centric smart microclimate-control system (SMCS) that fuses system dynamics and machine-learning techniques in consideration of the internal hydro-meteorological process to regulate the greenhouse micro-environment within the canopy for environmental cooling with improved resource-use efficiency. SMCS was assessed by in situ data collected from a tomato greenhouse in Taiwan. The results demonstrate that the proposed SMCS could save 66.8% of water and energy (electricity) used for early spraying during the entire cultivation period compared to the traditional greenhouse-spraying system based mainly on operators' experiences. The proposed SMCS suggests a practicability niche in machine-learning-enabled greenhouse automation with improved crop productivity and resource-use efficiency. This will increase agricultural resilience to hydro-climate uncertainty and promote resource preservation, which offers a pathway towards carbon-emission mitigation and a sustainable water–energy–food nexus.

Keywords: smart microclimate-control system (SMCS); machine learning; system dynamics; water–energy–food nexus; agricultural resilience

Citation: Chen, T.-H.; Lee, M.-H.; Hsia, I.-W.; Hsu, C.-H.; Yao, M.-H.; Chang, F.-J. Develop a Smart Microclimate Control System for Greenhouses through System Dynamics and Machine Learning Techniques. *Water* **2022**, *14*, 3941. https://doi.org/10.3390/w14233941

Academic Editor: Maria Mimikou

Received: 17 August 2022
Accepted: 29 November 2022
Published: 3 December 2022

Publisher's Note: MDPI stays neutral with regard to jurisdictional claims in published maps and institutional affiliations.

1. Introduction

The Sustainable Development Goals (SDGs) call for imperative action to ensure food security while preserving natural resources and maintaining environmental sustainability, especially in the era of climate change [1]. Significant changes in Earth's climate have fostered more extreme weather events in recent decades and therefore have increasingly impacted global agriculture by deeply implicating the fate of food systems and directly affecting the future of "eating" for humans. For instance, Taiwan suffered from 15 extreme weather events in 2016, including 4 typhoons, 3 torrential rains, 4 severe rains, and 4 cold snaps. The huge agricultural loss caused by these extreme weather events accounted for 10.3% of the total value of agricultural production, resulting in severe fluctuations in food prices and disturbance in social equilibrium. Besides, changes in temperature and precipitation patterns may increase crop failures and production declines [2].

Agricultural systems are vulnerable to changes not only in climate but also in other evolving factors like farming practices and technology. The impacts of climate change on agricultural systems globally have been investigated in recent decades [3–6]. Greenhouses are an expensive and technological solution for the challenges climate change poses to agriculture. However, they are not a universal tool that will solve all problems since it is infeasible to grow all crops indoors. For specific, high-value crops, this makes sense. Climate-smart agriculture is an integrated approach that seeks to manage landscapes by assessing interlinked food security and climate change to simultaneously improve crop productivity as well as reduce agricultural vulnerability to pests and climate-related risks [7].

Greenhouse cultivation that creates a controllable and stable environment facilitating crop growth and yield could be a climate-smart practice [8–10]. Hemming et al. [11] indicated that the opportunities and challenges for the future implementation of sensor systems in greenhouses could be explored by using artificial-intelligence techniques. Greenhouse farming is recognized as a promising measure to cope with climate change because this physical practice can promote crop growth and productivity by adequately controlling a microclimate to increase food security [12–14]. Due to the high agricultural loss induced by extreme weather events in 2016, the Council of Agriculture in Taiwan launched a five-year funding program in December of 2016 to encourage greenhouse construction or upgrades (2000 ha expected) for mitigating agricultural losses and maintaining stable food prices in the future. Among limited managerial tools, spraying plays a pivotal role in greenhouse control of environmental cooling, especially for places like Taiwan with hot and humid weather, where environmental adversity can occur in greenhouses. For instance, Bwambale et al. [15] conducted a review of smart irrigation-monitoring and control strategies that aimed to improve water-use efficiency in precision agriculture. Tona et al. [16] conducted a technical–economic analysis on spraying equipment for specialty crops and indicated that the purchase price would make the robotic platform profitable. Spraying systems are evidently one of the key environmental-control strategies for greenhouse cultivation. Nevertheless, most of the previous research related to spraying for environmental cooling focused mainly on cooling effects [17], without considering resource consumption. For resource preservation, it is required to consider the resource-use efficiency of spraying for environmental cooling.

Greenhouse cultivation by nature substantially depends on environmental controls to stabilize crop productivity [18,19]. Accurate prediction or simulation of a greenhouse internal environment is needed to evaluate environmental-control strategies for crop growth [20–25]. Besides, short-term microclimate prediction is challenging because meteorological variables are strongly interconnected with values changing rapidly during an event. With the motivation to fill the research gap and support the above-mentioned governmental greenhouse policy to achieve SDGs #2 (Zero Hunger), #12 (Responsible Consumption and Production), and #13 (Climate Action), this study developed a water-centric smart microclimate-control system (SMCS) for greenhouse cultivation in response to climatic variation. The SMCS was designed to automatically activate early spraying for environmental cooling while consuming less water and energy. The SMCS seamlessly integrates a system-dynamics (SD) model coupled with a physically based (i.e., a hydro-meteorological process) estimation model, a machine-learning prediction model, and a spray mechanism. A traditional greenhouse-spraying system based on the physically based estimation model and the spray mechanism coupled with operators' experience served as a benchmark for exploring the usefulness and applicability of the proposed SMCS. A tomato greenhouse located in Changhua County of Taiwan formed the case study, where the in situ datasets for use in this study were collected by Internet of Things (IoT) devices. The SMCS is expected to increase greenhouse automation and reinforce the efficiency of resource utilization, which can pave the way to reducing carbon emissions and promoting water–energy–food-nexus synergies in greenhouse farming.

2. Materials and Methods

This study proposes a water-centric SMCS that fuses system-dynamics and machine-learning techniques to regulate the greenhouse micro-environment within the canopy, with improved resource-use efficiency. The research flow chart is shown in Figure 1. We first collected the historical IoT monitoring data of the investigative greenhouse. Based on the IoT data, the SD model simulated the greenhouse microclimate within the canopy before and after spraying for environmental cooling. The back-propagation neural-network (BPNN) model predicted one-hour-ahead greenhouse internal temperature and relative humidity, where the initial inputs were the IoT data. Based on the prediction results, a spray mechanism was designed to determine the necessity of early spraying for environmental cooling. Consequently, the impacts of spraying on the internal environment and resource consumption were investigated. This study further compared the spray effects between the SMCS and the traditional greenhouse-spraying system (a benchmark), with the main focus on the resource consumption of spraying for environmental cooling. In the end, the potential of the SMCS for agricultural-loss mitigation in the perspective of water–energy–food-nexus synergies was discussed. It was noted that both traditional and machine-learning-based systems were constructed based on the IoT data collected from the same trial during 20 May and 20 July 2019.

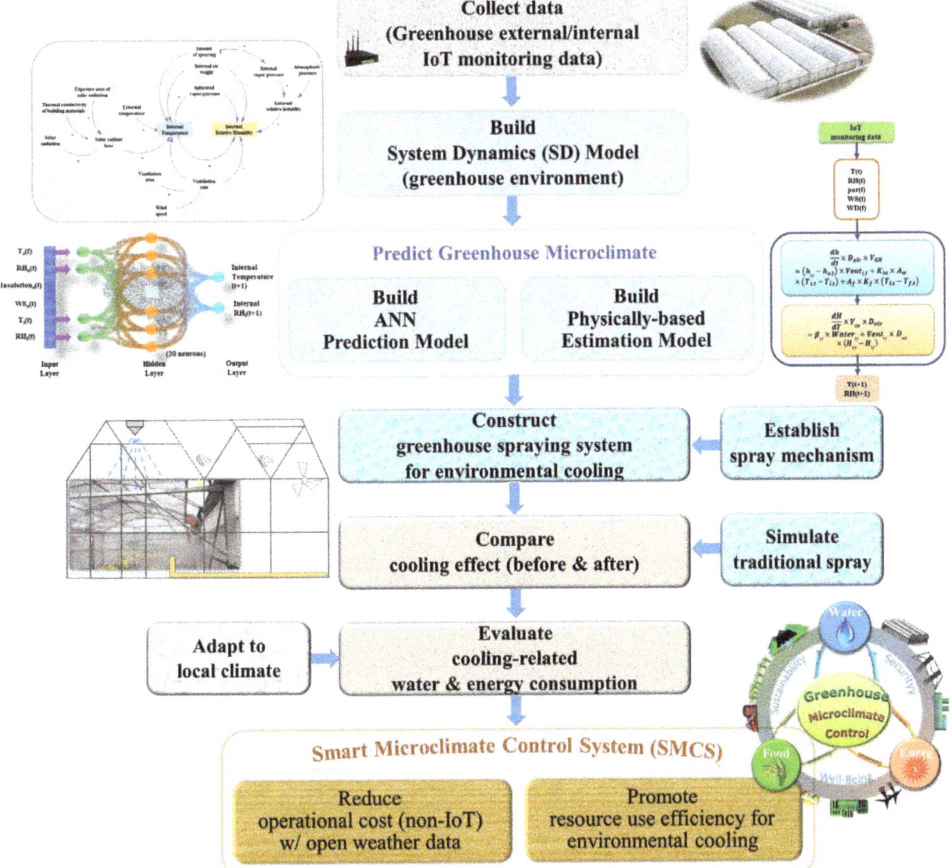

Figure 1. Research flow chart of this study.

2.1. Study Area and Materials

In this study, a total of 1488 hourly meteorological datasets related to tomato cultivation were collected on 20 May and 20 July 2019 by IoT devices installed inside and outside a privately owned greenhouse located in Changhua County of Taiwan (Figure 2). The IoT devices (Figure 2) installed in the greenhouse were developed by the Taiwan Agricultural Research Institute. The size of the greenhouse is about 52 m × 30 m × 6 m (length × width × height), indicating that the land area of the greenhouse is about 1560 m². Monitoring items consisted of internal/external temperature, internal/external relative humidity, external insolation, wind speed, and wind direction (Table 1). It is noted that this study adopted IoT datasets for model-construction and evaluation purposes only.

Figure 2. Location and structure of the greenhouse investigated in this study.

Table 1. IoT monitoring data collected in this study for model-validation purposes (20 May–20 July 2019 at a 10 min scale).

Item	Notation	SI Unit
External temperature	T_o	°C
External relative humidity	RH_o	%
External insolation	par_o	W/m^2
Wind speed	WS	m/s
Wind direction	WD	°
Internal temperature	T_i	°C
Internal relative humidity	RH_i	%

2.2. System Dynamics (SD) for Simulating Greenhouse Environment

SD is a set of process-oriented research methods specializing in the causal-feedback relationship among many variables and high-order non-linear systems [26–28]. It also specializes in explaining the results of system behavior through structural reasons behind the behavior [29]. SD has been widely used for simulating the non-linear behaviors in complex systems over time in various fields, including greenhouse management, forecasting and experimentation [30–32], rooftop farming [33], and the water–food–energy nexus [34,35].

This study explored the causal loops of SD for greenhouse cultivation by consideration the spray effect (Figure 3a). It is noted that the SMCS was constructed to reduce internal temperature and increase internal relative humidity by raising the partial pressure of water vapor to achieve the effect of cooling and humidification. A physically based model was constructed based on the SD model to estimate the greenhouse internal temperature and relative humidity before and after spraying. The framework of the SD model coupled with the physically based estimation model is shown in Figure 3.

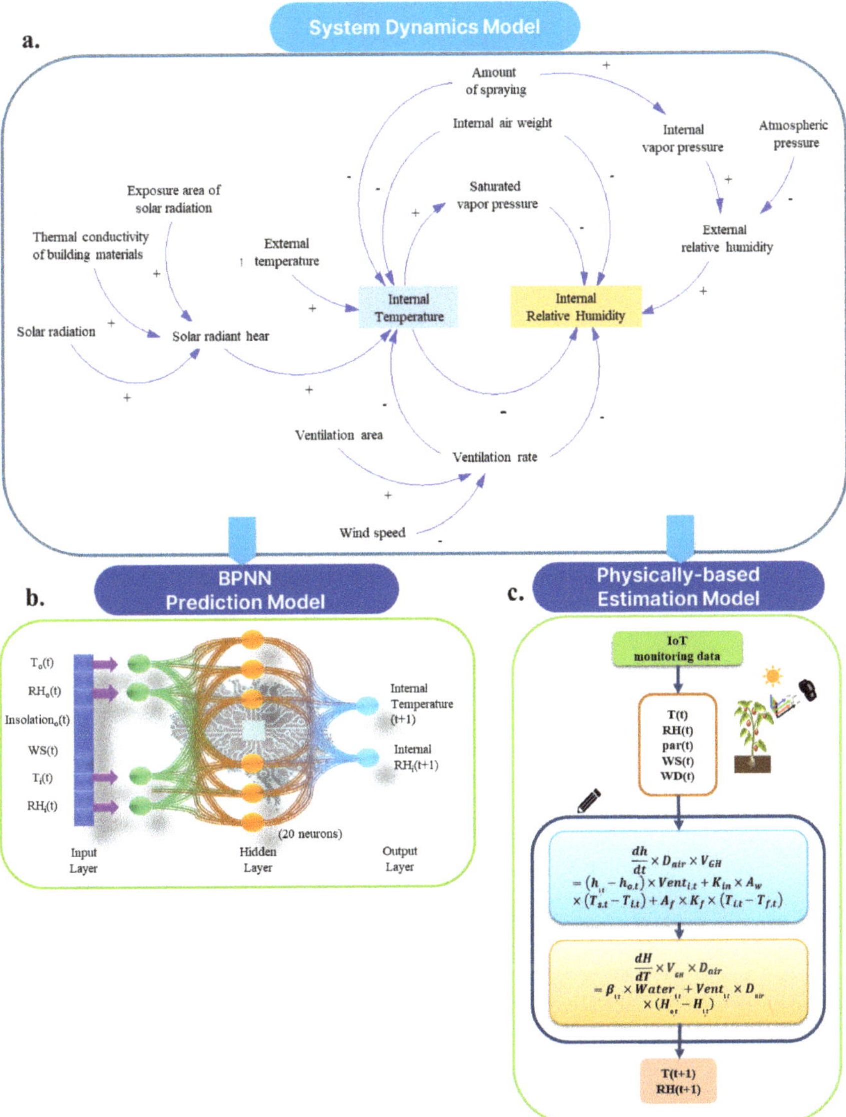

Figure 3. Model construction of the proposed smart microclimate-control system (SMCS) for greenhouse cultivation in consideration of the spray effect. (**a**). SD model. (**b**). BPNN prediction model. (**c**). Physically based estimation model.

Referring to Lee et al. [17], greenhouse internal relative humidity and temperature were considered to be a function of the conservation of mass and the conservation of energy, which consisted of two parts. Part 1 estimated the internal relative humidity by calculating enthalpy and heat conduction. Part 2 estimated the internal temperature by calculating the variation in moisture in the air. The formulation of greenhouse internal relative humidity and temperature is briefly introduced below.

2.2.1. Formulation of Greenhouse Internal Relative Humidity

The physically based estimation model of internal relative humidity was constructed by the equations of the conservation of mass and the conservation of energy (Equation (1)).

$$\frac{dH}{dt} \times V_{GH} \times D_{air} = \beta_{i,t} \times Water_{i,t} + Vent_{i,t} \times D_{air} \times (H_{o,t} - H_{i,t}) \tag{1}$$

where $\frac{dH}{dt}$ is the indoor absolute humidity change rate in a time period (kg/m^3 h), $\beta_{i,t}$ is the spray efficiency (%), $Water_{i,t}$ denotes the amount of spray (kg), $Vent_{i,t}$ denotes the indoor ventilation (kg/h), and $H_{i,t}$ ($H_{o,t}$) denote the internal (external) absolute humidity (kg/m^3) at t. V_{GH} denotes the total capacity of the greenhouse (m^3), and D_{air} denotes the air density (1.2 kg/m^3).

$$H_{i,t} = 0.62198 \times \frac{RH_{i,t} \times esi_{i,t}}{(P_{atm} - RH_{i,t} \times esi_{i,t})} \tag{2}$$

$$H_{o,t} = 0.62198 \times \frac{RH_{o,t} \times esi_{o,t}}{(P_{atm} - RH_{o,t} \times esi_{o,t})} \tag{3}$$

where $RH_{i,t}$ ($RH_{o,t}$) denotes the indoor (external) relative humidity (%) at t, $esi_{i,t}$ ($esi_{o,t}$) denotes the indoor (external) saturated vapor pressure (kpa) at t, and P_{atm} denotes the atmospheric pressure (101 kpa).

$$esi_{i,t} = 0.6178 \times e^{\frac{17.2694 \times T_{i,t}}{(T_{i,t} + 237.3)}} \tag{4}$$

$$esi_{o,t} = 0.6178 \times e^{\frac{17.2694 \times T_{o,t}}{(T_{o,t} + 237.3)}} \tag{5}$$

where $T_{i,t}$ ($T_{o,t}$) denotes the indoor (external) temperature (°C) at t.

$$\beta_{i,t} = 1.1906 - 0.09077 \times RH_{i,t} \tag{6}$$

$$Vent_{i,t} = C_{i,t} \times WS_t \times A_{GH} \tag{7}$$

where $C_{i,t}$ is the ventilation utilization factor at t, A_{GH} is the ventilation area of the greenhouse (m^2), and WS_t denotes the wind speed (m/h) at t.

$$H_{i,t+1} = H_{i,t} + \frac{dH}{dt} \tag{8}$$

where $H_{i,t+1}$ and $H_{i,t}$ denote the indoor absolute humidity at t + 1 and t (kg/m^3), respectively.

$$ei_{i,t+1} = \frac{H_{i,t+1} \times P_{atm}}{H_{i,t+1} + 0.62198} \tag{9}$$

where $ei_{i,t+1}$ denotes the indoor partial pressure of water vapor (kpa) at t + 1.

Consequently, the internal relative humidity ($RH_{i,t+1}$) at t + 1 could be calculated by Equation (10).

$$RH_{i,t+1} = \frac{ei_{i,t+1}}{esi_{i,t+1}} \tag{10}$$

2.2.2. Formulation of Greenhouse Internal Temperature

The internal temperature was also constructed by the equations of the conservation of mass and the conservation of energy (Equation (11)).

$$\frac{dh}{dt} \times D_{air} \times V_{GH} = (h_{i,t} - h_{o,t}) \times Vent_{i,t} + K_{in} \times A_w \times (T_{s,t} - T_{i,t}) + A_f \times K_f \times (T_{i,t} - T_{f,t}) \tag{11}$$

where $\frac{dh}{dt}$ denotes the indoor change rate of enthalpy in a time period (kj/kg h); $h_{i,t}$ and $h_{o,t}$ denote the indoor and external enthalpies (kj/kg) in the air at t, respectively; $Vent_{i,t}$ denotes the ventilation rate (m^3/h) at t; V_{GH} denotes the total capacity of the greenhouse (m^3); D_{air} denotes the air density (1.2 kg/m^3); K_{in} denotes the indoor coating material's heat-convection parameter in the air (6.4 W/m^2 °C); A_w denotes the area of the coating material (m^2); $T_{s,t}$, $T_{i,t}$, and $T_{f,t}$ denote the indoor temperature (°C) of the coating material, the indoor temperature (°C), and the indoor ground temperature (°C) at t, respectively; A_f denotes the total ground area of the greenhouse (m^2); and K_f denotes the indoor ground-to-air heat convection parameter (4.65 W/m^2 °C).

$$h_{i,t} = 1.006 \times T_{i,t} + H_{i,t} \times (2501 + 1.085 \times T_{i,t}) \tag{12}$$

where $H_{i,t}$ denotes the indoor absolute humidity (kg/m^3) at t.

$$h_{o,t} = 1.006 \times T_{o,t} + H_{o,t} \times (2501 + 1.085 \times T_{o,t}) \tag{13}$$

where $T_{o,t}$ denotes the external temperature (°C) at t, and $H_{o,t}$ denotes the external absolute humidity (kg/m^3) at t.

$$T_{s,t} = T_{o,t} + a \times \left(\frac{Rn_{o,t}}{K_{out}} \right) \tag{14}$$

where a is the solar-absorption rate on the surface of the material (0.65%), $Rn_{o,t}$ denotes the external solar radiation (W/m^2) at t, and K_{out} denotes the thermal conductivity on the surface of the material (6.3 W/m^2 °C).

$$Rn_{o,t} = (1 - ref) \times par_{o,t} + Rn_{lon} \tag{15}$$

where ref denotes the ground reflectivity (0.2), $par_{o,t}$ denotes the external insolation at t (W/m^2), and Rn_{lon} denotes the atmospheric long-wave radiation (343 W/m^2).

$$T_{f,t} = T_{o,t} + \frac{Rn_{o,t} - B \times (T_{o,t} + 273.15)^4}{\left(4 \times B \times (T_{o,t} + 273.15)^3 \right)} \tag{16}$$

where B is the Boltzmann constant (5.67 $\times$ 10^{-8} Wm^{-2}K^{-4}).

Because this study considered spray to be a means of humidification and cooling, it required calculating the internal heat moving away due to spray, as shown in Equation (17) (refer to [36]).

$$Q_t = \beta_{i,t} \times Water_{i,t} \times H_{fg} \tag{17}$$

where Q_t denotes the heat moving away due to spray (kj/h), $\beta_{i,t}$ denotes the indoor spray efficiency (%) at t, $Water_{i,t}$ denotes the indoor spray amount (kg/h) at t, and H_{fg} denotes the latent heat of water evaporation (2256.6 kj/kg).

$$dT = \frac{\frac{dh}{dt} \times V_{GH} \times D_{air} - Q_t}{4.186 \times C_p \times V_{GH} \times D_{air}} \tag{18}$$

where dT denotes the indoor temperature change in a time period (°C/h), and C_p denotes the specific heat of the air (1.0052 kj/kg °C).

Consequently, the internal temperature at t + 1 could be obtained from Equation (19).

$$T_{i,t+1} = T_{i,t} + dT \tag{19}$$

where $T_{i,t+1}$ and $T_{i,t}$ denote the indoor temperature (°C) at t + 1 and t, respectively.

Details of the formulation of greenhouse relative humidity (Part 1) and internal temperature (Part 2) can be found in the Supplementary Material.

2.3. Machine Learning for Predicting Greenhouse Internal Environment

Artificial neural networks (ANNs) in machine learning are a family of computation methods that imitate the operation and learning of the human nerve system. ANNs are broadly used to tackle diverse environmental issues, such as rainfall forecasts [37,38], evaporation prediction [39,40], flood forecasts [41–46], hydrological analysis [47–51], ecological-environment analysis [52,53], air-quality estimation [54], agricultural automation [55], and greenhouse environmental control [22,56,57].

The BPNN is one of the most widely used ANNs. This study utilized the BPNN to predict one-hour-ahead greenhouse internal temperature ($T_i(t + 1)$) and relative humidity ($RH_i(t + 1)$) based on current information on six meteorological factors, including external temperature (T_o), external relative humidity (RH_o), external insolation (par_o) and wind speed (WS), internal temperature (T_i), and internal relative humidity (RH_i) (Figure 3b). The construction of the BPNN prediction model was based on a total of 1488 hourly IoT data, where 64, 16, and 20% of the data were shuffled and randomly allocated into training, validation, and testing stages, respectively. The architecture of the BPNN model constructed in this study is illustrated in Figure 3b. The parameter setting of the BPNN model is shown in Table 2, where the number of neurons in the hidden layer and the batch size were determined to be 20 and 64, respectively, through trial-and-error processes. The relevant trial-and-error results are presented in Tables 3 and 4.

Table 2. Parameter setting of the BPNN model.

Item	BPNN
Number of hidden neurons	10, 20, 40
Number of epochs	200
Early stopping	20
Batch size	8, 16, 32, 64
Learning rate	0.001
Activation function	Scaled exponential linear unit (SELU)
Optimizer	Adam

Table 3. Trial-and-error results of the number of hidden neurons in the BPNN model.

Number of Hidden Neurons	Temperature		Relative Humidity	
	R^2	RMSE	R^2	RMSE
10	0.80	1.61 °C	0.87	4.45%
20 [1]	0.82	1.55 °C	0.88	4.19%
40	0.81	2.42 °C	0.87	4.40%

Note: [1] The number of hidden neurons that was determined for constructing the BPNN model in consideration of the model complexity and the values of the evaluation indicators.

Table 4. Trial-and-error results of the batch number in the BPNN model.

Batch Number	Temperature		Relative Humidity	
	R^2	RMSE	R^2	RMSE
8	0.83	2.08 °C	0.88	4.28%
16	0.81	1.56 °C	0.87	4.53%
32	0.82	1.67 °C	0.88	4.35%
64 [1]	0.83	1.55 °C	0.88	4.19%

Note: [1] The batch number that was determined for constructing the BPNN model in consideration of the values of the evaluation indicators.

2.4. Construction of the Spray Mechanism

Figure 4 presents the spray-simulation flow chart of the SMCS. According to the one-hour-ahead predictions ($t + 1$) of greenhouse internal temperature and relative humidity obtained from the BPNN model, a spray mechanism with spraying criteria was designed to determine the time to spray, which is introduced as follows.

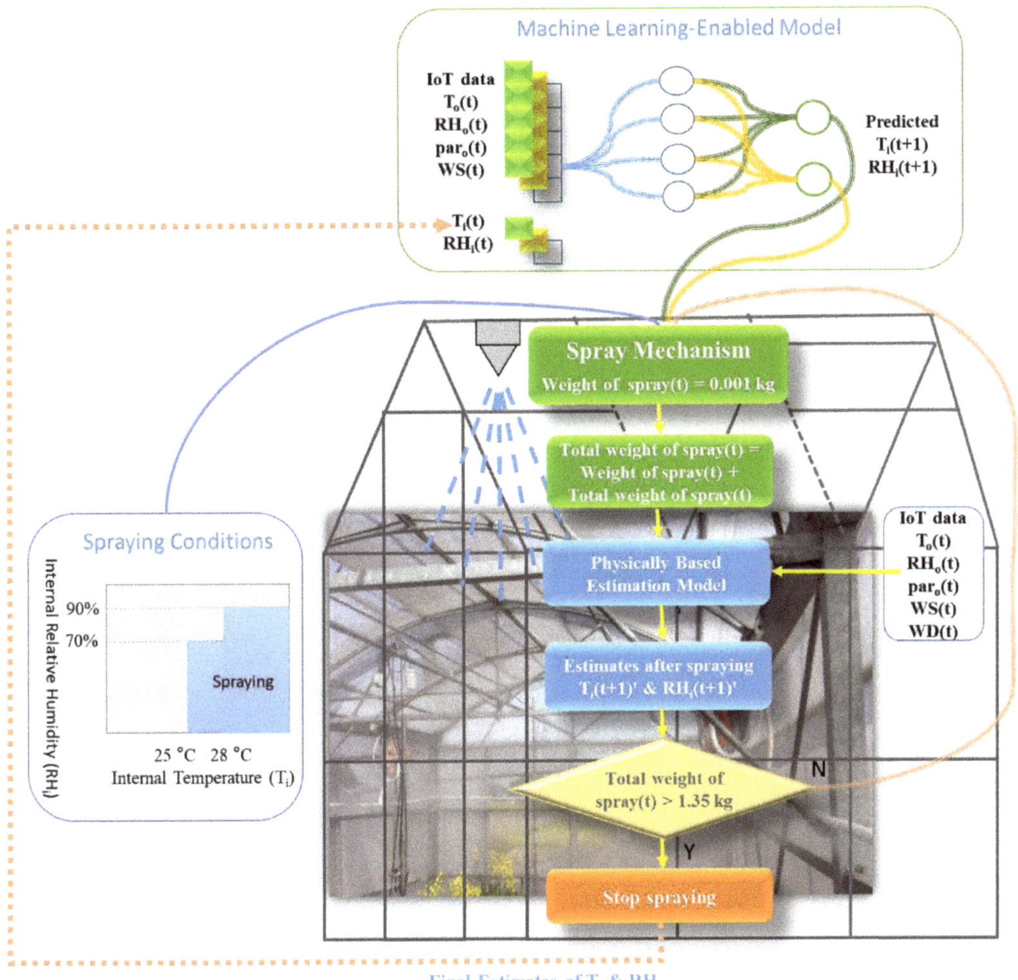

Figure 4. Spray-simulation flow chart of the SMCS that integrates the SD model, the BPNN prediction model, the physically based estimation model, and the spray mechanism.

According to Xue et al. [58] on greenhouse cultivation, the net photosynthetic rate and cumulative photosynthesis of tomato leaves could be significantly improved when the internal relative humidity reached 70%. Liou et al. [59] indicated that the formation of lycopene in tomatoes would be reduced if the greenhouse internal temperature exceeded 28 °C. Therefore, this study managed to activate sprayers for environmental cooling under two conditions: when the internal relative humidity fell below 70%, and when the internal relative humidity and the internal temperature exceeded 90% and 28 °C, respectively.

To avoid resource over-consumption, sprayers would not be activated if the internal relative humidity exceeded 90% or the internal temperature fell below 25 °C. Besides, the switching on/off of the sprayers would be carried out based on the predicted values of internal relative humidity and temperature. Therefore, the spray mechanism would activate sprayers for environmental cooling subject to two criteria: (1) the one-hour-ahead prediction of internal relative humidity would be less than 70% and the one-hour-ahead prediction of internal temperature would be higher than 25 °C, and (2) the one-hour-ahead prediction of internal relative humidity would be less than 90% and the one-hour-ahead prediction of internal temperature would be higher than 28 °C. Spraying would terminate either when the internal temperature and relative humidity met the environmental suitability for tomato growth or when the total amount of spray exceeded the maximal spray volume within one hour (i.e., 1.35 kg).

In the case of no spraying being required for environmental cooling, the one-hour-ahead predictions of internal temperature and relative humidity obtained from the BPNN model would be fed back to the system and serve as the initial input values of the BPNN model at the next time-step (the orange dotted line in Figure 4). If either above-mentioned activation criterion for spraying was met, a spray of 0.001 kg would be carried out, leading to a re-calculation of the internal temperature and relative humidity after spraying by using the physically based estimation model. The spraying process would repeat until reaching the stop criteria. It is noted that a sprayer would not be activated if the required amount of spray was less than its minimal spray volume (=the minimal duration of spray × the rate of spray). When spraying terminates, the final one-hour-ahead estimates (t + 1) of internal temperature and relative humidity obtained from the physically based model would be fed back to the system and serve as the initial input values of the BPNN model at the next time-step (the orange dotted line in Figure 4). For the greenhouse investigated and the sprayer selected for use in this study, it would require three sprayers to cover the entire greenhouse farm (1560 m^2). The weight of spray each time would be 0.001 kg per sprayer, and the total weight of spray per hour would be 1.35 kg for three sprayers. Therefore, the control loop would be evaluated at a rate of 8 s.

2.5. Evaluation of Model Performances

To explore the spray effect of the SMCS on greenhouse farming, the above-mentioned spraying process for environmental cooling was implemented on all 1488 IoT data collected in this study. For comparison purposes, a traditional greenhouse-spraying system was established by integrating the physically based estimation model with the spray mechanism only, whereas the physically based model was responsible for estimating one-hour-ahead greenhouse internal temperature and relative humidity before and after spraying.

This study used the root-mean-square error (RMSE) and the coefficient of determination (R^2) as the statistical indicators to evaluate model performance. Their mathematical formulas refer to Equations (20) and (21).

Root-Mean-Square Error

$$\text{RMSE} = \sqrt{\frac{1}{N} \sum_{i=1}^{N} (y_i - o_i)^2} \tag{20}$$

Coefficient of Determination

$$R^2 = \left[\frac{\sum_{i=1}^{N} (y_i - \bar{y})(o_i - \bar{o})}{\sqrt{\sum_{i=1}^{N} (y_i - \bar{y})^2} \sqrt{\sum_{i=1}^{N} (o_i - \bar{o})^2}} \right]^2 \tag{21}$$

where N is the total number of data, y_i is the output value of the model, o_i is the observation value, and $\bar{y}$ and $\bar{o}$ are the average of the output value and the observation value, respectively.

According to the definitions of the two indicators, it is obvious that a model is considered to perform well if it produces a higher R^2 value but a lower RMSE value than the comparative model(s).

3. Results

This study developed a water-centric SMCS dedicated to greenhouse farming and the spray effect on greenhouse microclimate for environmental cooling with the relevant resource consumption being investigated. The operation of the SMCS was composed of four main phases: to simulate greenhouse environmental dynamics in consideration of the spray effect (by the SD model), to predict one-hour-ahead internal temperature and relative humidity (by the BPNN model), to determine the necessity of spraying for environmental cooling (by the spray mechanism), and to estimate the required amount of spray to manage a microclimate suitable for tomato growth in the coming hour (by the physically based model). The SMCS was applied to the 1488 in situ data collected from a greenhouse on 20 May 2019 and 20 July 2019. The modeling results are presented and discussed as follows.

3.1. Comparison of Model Accuracy and Reliability between the Physically Based and ANN Models

Table 5 shows the performance of the physically based estimation model and the BPNN prediction model with respect to greenhouse internal temperature and relative humidity based on test datasets. For the physically based estimation model, the R^2 and RMSE values of the internal temperature were 0.80 and 1.89 °C, respectively, whereas those of the internal relative humidity were 0.79 and 8.17%, respectively. The results demonstrate the accuracy and reliability of the physically based model. As for the BPNN prediction model, its R^2 and RMSE values of the internal temperature were 0.83 and 1.37 °C, respectively, whereas those of the internal relative humidity were 0.88 and 3.9%, respectively. The results also demonstrate the accuracy and reliability of the BPNN model. It appears that the BPNN model is superior to the physically based model in terms of higher R^2 and lower RMSE values.

Table 5. Performance of the physically based estimation model and the BPNN prediction model with respect to greenhouse internal temperature and relative humidity based on test datasets.

Indicators	Temperature		Relative Humidity	
	Physically Based	**BPNN**	**Physically Based**	**BPNN**
R^2	0.80	0.83	0.79	0.88
RMSE	1.89 °C	1.37 °C	8.17%	3.9%

Figures 5 and 6 show the errors and error distributions of internal-temperature and relative-humidity estimates obtained from the physically based model and the BPNN model, respectively. In both error plots, positive values indicate overestimation whereas negative values indicate underestimation. Regarding the physically based estimation model, it can be seen in Figure 5a that the errors of the internal temperature mostly fell within 1 and 2 °C (overestimated), with an overestimation occurrence frequency (1098 times) much higher than the underestimation one (387 times). According to Figure 5b, the errors in the internal relative humidity were mostly concentrated within −3% and −6% (underestimated), with an underestimation occurrence frequency (787 times) higher than the overestimation one (699 times).

Regarding the BPNN prediction model, the results of Figure 6a indicate that the errors of the internal temperature mainly fell within −1 and 0 °C (under prediction), where underprediction (1176 times) occurred more frequently than overprediction (302 times). According to Figure 6b, the errors in the internal relative humidity were mainly concentrated within −3% and 0%, where underprediction (959 times) also occurred more frequently than overprediction (517 times). It also appears that the BPNN model performed better than the physically based model in terms of smaller error ranges and error distributions centering at zero.

Furthermore, the results shown in Table 5 and Figures 5 and 6 are quite consistent, which shows that the overall performance of the BPNN model was slightly better than that of the physically based model. This recommended the incorporation of the BPNN model into the SMCS to predict one-hour-ahead internal temperature and relative humidity in this study.

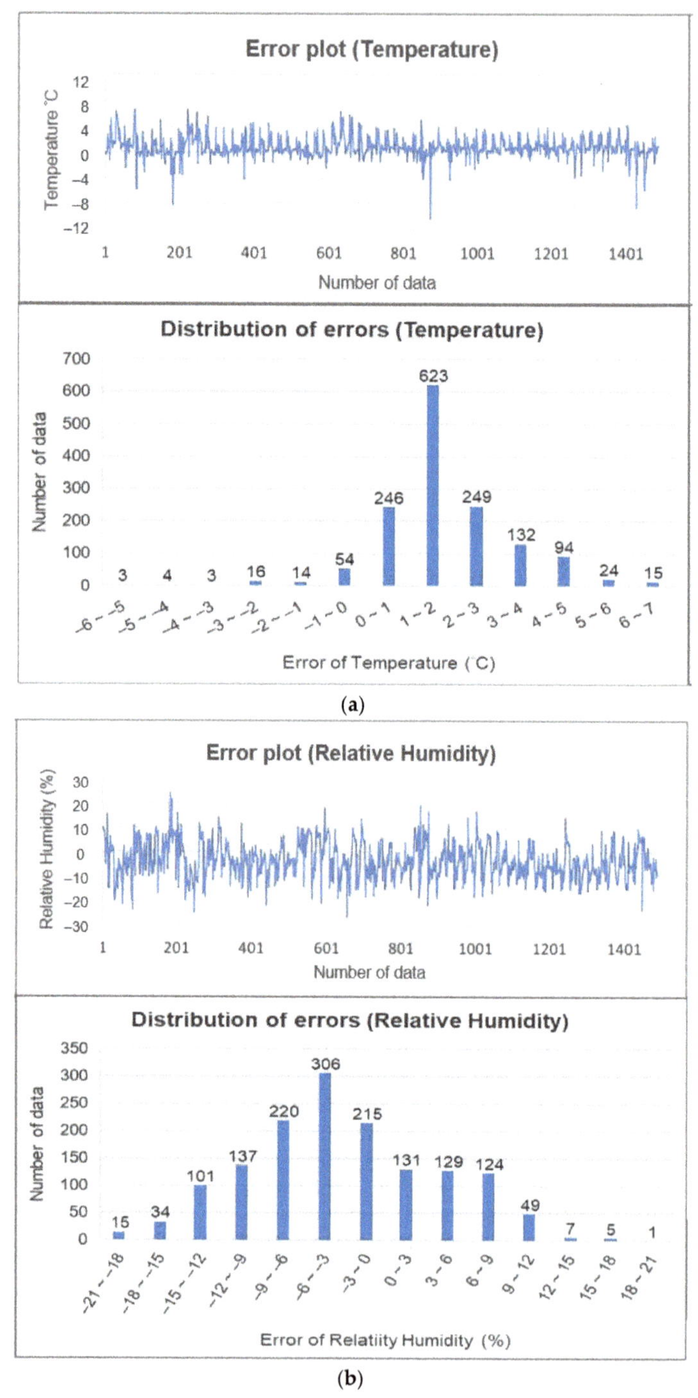

Figure 5. Errors and error distributions of greenhouse microclimate estimates from the physically based model (20 May 2019–20 July 2019). (**a**). Internal temperature. (**b**). Internal relative humidity.

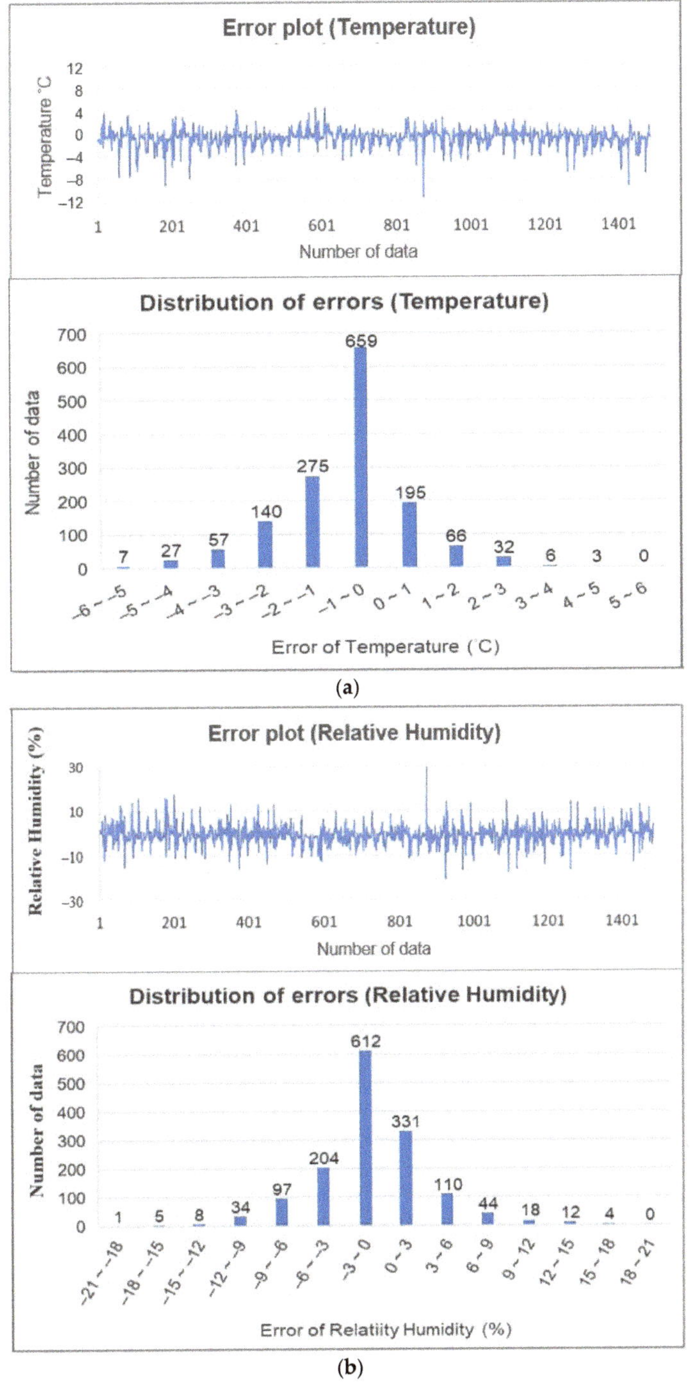

Figure 6. Errors and error distributions of greenhouse microclimate predictions from the BPNN model (20 May 2019–20 July 2019). (**a**). Internal temperature. (**b**). Internal relative humidity.

3.2. Comparison of the Spray Effect of Traditional and Smart Control Systems on Greenhouse Internal Environment

Tables 6 and 7 show the results of internal temperature and relative humidity before and after spraying by the traditional spraying system and the proposed SMCS, respectively.

Table 6. Results of greenhouse environmental control on internal temperature and relative humidity before and after spraying for environmental cooling by the traditional spraying system (20 May 2019–20 July 2019).

Indicators	Temperature		Relative Humidity	
	Before	**After**	**Before**	**After**
Max	38.8 °C	28.1 °C	100%	100%
Min	23.8 °C	23.8 °C	37%	56%
Average	29.6 °C	27.0 °C	72%	86%
Standard deviation	3.6 °C	1.3 °C	16%	7%

Table 7. Results of greenhouse environmental control on internal temperature and relative humidity before and after spraying for environmental cooling by the SMCS (20 May 2019–20 July 2019).

Indicators	Temperature		Relative Humidity	
	Before	**After**	**Before**	**After**
Max	34.3 °C	32.9 °C	91%	100%
Min	21.3 °C	22.1 °C	48%	69%
Average	28.0 °C	26.6 °C	74%	89%
Standard deviation	2.9 °C	1.5 °C	12%	4%

The results of Table 6 indicate that the average and standard deviation of the internal temperature after spraying decreased by 2.6 and 2.3 °C, respectively. For the internal relative humidity after spraying, the average value increased from 72% to 86%, whereas the standard deviation dropped from 16% to 7%.

The results of Table 7 show that both the average and standard deviation of the internal temperature after spraying decreased by 1.4 °C. For the internal relative humidity after spraying, the average value increased from 74% to 89%, whereas the standard deviation dropped from 12% to 4%. These results demonstrate that the SMCS could more effectively reduce the internal temperature while increasing the internal relative humidity after spraying than the traditional one, which supports the practicability of the proposed SMCS on greenhouse farms.

3.3. Comparison of Resource Consumption between Traditional and Smart Microclimate-Control Systems

Concerning spray-related resources utilization for greenhouse environmental control over the entire investigative period, water consumption could be obtained directly from summing up the amount of spray at each time-step while power consumption would be converted from horsepower and the total operating hours of the sprayers. For spray-simulation purposes, this study adopted the "FH-09 power spray motor" sprayer launched by the Fog Century Environmental Protection and Energy Saving Enterprise Co. Ltd., located in Taichung City, Taiwan. The main specifications of the sprayer are a horsepower of 1.125 kW, a water absorption of 0.15 kg/h, and an applicable area of about 400 to 600 m^2. Considering the greenhouse investigated in this study occupies an area of 1560 m^2, it would require three sprayers to cover the entire greenhouse farm.

Table 8 compares the traditional and the proposed control systems regarding the resource consumption of spraying for environmental cooling.

Table 8. Comparison between traditional and smart microclimate-control systems regarding the resource consumption of spraying for environmental cooling.

Item	Water (kg)	Electric Power (kWh)	Number of On/Off Switch of Sprayers
Traditional spraying system	129,478	90.0	736/1488
Smart spraying system	42,962	29.8	726/1488
Resource-saving amount [1]	86,516	60.2	10/-
Resource-saving rate [2]	66.8%	66.8%	1.4%/-
Traditional spraying system	129,478	90.0	736/1488

Notes: [1] Amount of the traditional spraying system—amount of the SMCS. [2] Resource saving amount/amount of the traditional spraying system.

It is noted that the numbers of the on/off switches of the sprayers associated with the two comparative systems differed slightly (736 times for the traditional system vs. 726 times for the smart system). Therefore, the difference of the two systems in power consumption enabling the switching on/off of the sprayers could be ignored. Under this assumption, the traditional system consumed about 129,478 kg of water and 90 kWh of electric power for greenhouse environmental control during the entire tomato-cultivation period. In contrast, the SMCS only consumed about 42,962 kg of water and 29.8 kWh of electric power. The results demonstrate that the SMCS consumed far fewer resources for spraying than the traditional system, with water- and power-saving rates reaching 66.8%. It was further noticed that early spraying for environmental cooling suggested by the SMCS allowed the wind to blow away excess internal water vapor one hour ahead, leading to a decrease in the internal relative humidity. Spray efficiency is known to be inversely proportional to the internal relative humidity. Therefore, the amount of spray could be reduced due to early spraying.

4. Discussion

4.1. Evaluation of Hazard Mitigation by the SMCS

This study further evaluated the potential contribution of the proposed SMCS to the governmental greenhouse policy launched in 2016 regarding the construction of 2000 ha of reinforced greenhouses within five years. Taking the agricultural loss in 2020 released by the Council of Agriculture in Taiwan as an example, under the scenario that all 2000 ha of greenhouses could be equipped with the SMCS, the agricultural loss caused by extreme weather events would be significantly reduced by 22% (=2000 (greenhouse area in ha)/9097 (total damaged area in ha)) on average. Besides, resource saving in water and energy would achieve 1,109,918 tons (=((86,516 kg/1560 m^2) × 10,000) × 2000 ha/1000) and 771,795 kWh (=((60.2 kWh/1560 m^2) × 10,000) × 2000 ha), respectively (Table 8). This suggests the smart greenhouse microclimate-control practice bears high potential for tackling climate change and can significantly promote the nexus synergies among water, energy, and food, especially when encountering extreme weather events.

4.2. Conributions of the SMCS

The proposed SMCS makes two main contributions. Firstly, for maintaining an environment suitable for crop growth, the traditional greenhouse-spraying system requires monitoring sensors like IoT devices to detect the internal temperature or relative humidity for switching sprayers on/off. Nevertheless, this may impose the risk of an unsuitable environment on greenhouse farming between two time-steps. For example, the operational time interval was one hour in this study. Even if the greenhouse environment complies with the suitability conditions of crop growth at the current minute, it may violate the suitability conditions in the next minute. In contrast, the SMCS can predict the greenhouse microclimate for the next hour well, thereby spraying in advance to prevent an unsuitable environment for crop growth. Besides, the SMCS avoids using IoT sensors because the extra hardware and maintenance costs of the monitoring devices also place a heavy burden on greenhouse owners. Secondly, the SMCS consumes fewer resources of water and energy (electricity) when spraying for environmental cooling than the traditional method, indicating that the SMCS can mitigate greenhouse-gas emissions. Low resource consumption also represents cost-effectiveness and relatively high profits, leading to more commercial value that can be achieved by the SMCS.

The greenhouse-management practice developed can be applied to crops and areas of interest with adequate modification of the environmental suitability for crop growth. Similar methodology for developing the SMCS can also be applied to different greenhouse types. Future research can consider incorporating crop evapotranspiration, soil-moisture content, nutrients, and fertilization into the SMCS to increase the prediction accuracy of the greenhouse environment and promote crop productivity and quality. Ventilation is also a major factor in the control of greenhouse temperature. In future research, ventilation will be considered by incorporating the greenhouse-control factors (e.g., skylight, roller shades on each wall, and inner shade net) into the proposed water-centric smart microclimate-control system (SMCS) to increase its operational efficiency and effectiveness.

5. Conclusions

This study proposed a water-centric smart microclimate-control system (SMCS) for greenhouse farming, with a mission to manage the microclimate through efficient spraying for environmental cooling. The SMCS can maintain stable crop productivity when extreme weather events occur. The SMCS can determine the necessity of spraying for environmental cooling according to the predictions of greenhouse internal temperature and relative humidity. The results demonstrate that the SMCS could achieve the same environmental-control effect as the traditional one while consuming far fewer resources for spraying, which makes greenhouse farming move towards carbon-emission mitigation and sustainable management of the water–energy–food nexus. There are four main findings drawn from this study, shown below.

Firstly, the cost of sensor installation is a major concern for farmers in Taiwan, especially concerning device investment and maintenance issues. The BPNN model could (Figure 1) predict greenhouse microclimate based on external climate conditions with less water and energy. After the BPNN model is constructed, this science-based management practice requires no in situ monitoring sensors, which favorably lessens greenhouse owners' investment in environmental control and makes a positive contribution to the overall cost–benefit ratio of greenhouse farming. The physically based model engaging the internal hydro-meteorological process could produce satisfactory accuracy and reliability in estimating greenhouse microclimate, despite it performing slightly worse than the BPNN prediction model.

Secondly, the SMCS could predict the greenhouse internal environment well one hour ahead and spray in advance when needed for environmental cooling, which prevents crops from being exposed to an unsuitable cultivation environment.

Thirdly, the SMCS could achieve savings as high as 66.8% of water and energy compared to the traditional method. Therefore, the SMCS gains more commercial value than the traditional method because low resource consumption means low production cost and relatively high profits.

Fourthly, the reduction in agricultural loss caused by extreme weather events in 2020 would reach 22% if the SMCS could be implemented in 2000 ha of greenhouses (the goal of the governmental greenhouse policy launched in 2016 in Taiwan). This would lead to effective resource saving in water and energy of 1,109,918 tons and 771,795 kWh per year, respectively. This greenhouse-control strategy significantly contributes to environmental sustainability and greenhouse-gas-emission mitigation.

This study suggests a practicability niche in machine-learning-enabled greenhouse automation with improved crop productivity and resource-use efficiency. The proposed SMCS substantially moves greenhouse farming towards the SDGs in the perspectives of food security, natural-resource preservation, and environmental sustainability.

Supplementary Materials: The following supporting information can be downloaded at: https://www.mdpi.com/article/10.3390/w14233941/s1, Formulation of the physically based estimation model.

Author Contributions: Conceptualization, T.-H.C. and F.-J.C.; methodology, T.-H.C. and C.-H.H.; software, T.-H.C. and C.-H.H.; validation, M.-H.L., I.-W.H. and M.-H.Y.; formal analysis, T.-H.C. and C.-H.H.; investigation, M.-H.L. and I.-W.H.; resources, M.-H.Y. and F.-J.C.; data curation, T.-H.C. and M.-H.Y.; writing—original draft preparation, T.-H.C.; writing—review and editing, F.-J.C.; visualization, T.-H.C., M.-H.L. and I.-W.H.; supervision, F.-J.C.; project administration, F.-J.C.; funding acquisition, F.-J.C. and M.-H.Y. All authors have read and agreed to the published version of the manuscript.

Funding: This research was funded by the National Science and Technology Council, Taiwan, grant numbers 110-2313-B-002-034-MY3 and 110-2321-B-055-001-.

Data Availability Statement: The data presented in this study are available on request from the corresponding author.

Acknowledgments: The datasets provided by the Central Weather Bureau in Taiwan and the Taiwan Agricultural Research Institute are acknowledged. The authors would like to thank the editors and anonymous reviewers for their constructive comments that greatly enriched the manuscript.

Conflicts of Interest: The authors declare no conflict of interest.

References

1. United Nations. *The Sustainable Development Goals Report 2022*; United Nations Publications: New York, NY, USA, 2022; pp. 6–25.
2. Huang, A.; Chang, F.J. Using a self-organizing map to explore local weather features for smart urban agriculture in northern Taiwan. *Water* **2021**, *13*, 3457. [CrossRef]
3. Walters, S.A.; Gajewski, C.; Sadeghpour, A.; Groninger, J.W. Mitigation of climate change for urban agriculture: Water management of culinary herbs grown in an extensive green roof environment. *Climate* **2022**, *10*, 180. [CrossRef]
4. Holzkämper, A. Adapting agricultural production systems to climate change—What's the use of models? *Agriculture* **2017**, *7*, 86. [CrossRef]
5. Salpina, D.; Pagliacci, F. Are we adapting to climate change? Evidence from the high-quality agri-food sector in the Veneto region. *Sustainability* **2022**, *14*, 11482. [CrossRef]
6. Shayanmehr, S.; Porhajašová, J.I.; Babošová, M.; Sabouhi Sabouni, M.; Mohammadi, H.; Rastegari Henneberry, S.; Shahnoushi Foroushani, N. The impacts of climate change on water resources and crop production in an arid region. *Agriculture* **2022**, *12*, 1056. [CrossRef]
7. Xin, Y.; Tao, F. Developing climate-smart agricultural systems in the North China Plain. *Agric. Ecosyst. Environ.* **2020**, *291*, 106791. [CrossRef]
8. Esmaili, M.; Aliniaeifard, S.; Mashal, M.; Vakilian, K.A.; Ghorbanzadeh, P.; Azadegan, B.; Seif, M.; Didaran, F. Assessment of adaptive neuro-fuzzy inference system (ANFIS) to predict production and water productivity of lettuce in response to different light intensities and CO_2 concentrations. *Agric. Water Manag.* **2021**, *258*, 107201. [CrossRef]
9. Kalkhajeh, Y.K.; Huang, B.; Hu, W.; Ma, C.; Gao, H.; Thompson, M.L.; Hansen, H.C.B. Environmental soil quality and vegetable safety under current greenhouse vegetable production management in China. *Agric. Ecosyst. Environ.* **2021**, *307*, 107230. [CrossRef]
10. Li, M.; Chen, S.; Liu, F.; Zhao, L.; Xue, Q.; Wang, H.; Chen, M.; Lei, P.; Wen, D.; Sanchez-Molina, J.A.; et al. A risk management system for meteorological disasters of solar greenhouse vegetables. *Precis. Agric.* **2017**, *18*, 997–1010. [CrossRef]
11. Hemming, S.; Zwart, F.D.; Elings, A.; Petropoulou, A.; Righini, I. Cherry tomato production in intelligent greenhouses—Sensors and AI for control of climate, irrigation, crop yield, and quality. *Sensors* **2020**, *20*, 6430. [CrossRef]
12. Huang, Y.I. Study of fog and fan system for plastic greenhouse cooling in Taiwan. *J. Agric. Mach.* **1999**, *8*, 17–27.
13. Joudi, K.A.; Farhan, A.A. A dynamic model and an experimental study for the internal air and soil temperatures in an innovative greenhouse. *Energy Convers. Manag.* **2015**, *91*, 76–82. [CrossRef]
14. Pawlowski, A.; Sánchez-Molina, J.A.; Guzmán, J.L.; Rodríguez, F.; Dormido, S. Evaluation of event-based irrigation system control scheme for tomato crops in greenhouses. *Agric. Water Manag.* **2017**, *183*, 16–25. [CrossRef]
15. Bwambale, E.; Abagale, F.K.; Anornu, G.J. Smart irrigation monitoring and control strategies for improving water use efficiency in precision agriculture: A review. *Agric. Water Manag.* **2022**, *260*, 107324. [CrossRef]
16. Tona, E.; Calcante, A.; Oberti, R. The profitability of precision spraying on specialty crops: A technical–economic analysis of protection equipment at increasing technological levels. *Precis. Agric.* **2018**, *19*, 606–629. [CrossRef]
17. Lee, S.C.; Chang, C.Y.; Lee, W.S. The Study on Greenhouse Cooling Effect on Different Control Strategies for Fogging System. *J. Agric. Mach.* **2006**, *15*, 23–36.

18. Chen, S.; Cai, L.; Zhang, H.; Zhang, Q.; Song, J.; Zhang, Z.; Deng, Y.; Liu, Y.; Wang, X.; Fang, H. Deposition distribution, metabolism characteristics, and reduced application dose of difenoconazole in the open field and greenhouse pepper ecosystem. *Agric. Ecosyst. Environ.* **2021**, *313*, 107370. [CrossRef]

19. Hu, J.; Gettel, G.; Fan, Z.; Lv, H.; Zhao, Y.; Yu, Y.; Wang, J.; Butterbach-Bahl, K.; Li, G.; Lin, S. Drip fertigation promotes water and nitrogen use efficiency and yield stability through improved root growth for tomatoes in plastic greenhouse production. *Agric. Ecosyst. Environ.* **2021**, *313*, 107379. [CrossRef]

20. Ding, J.T.; Tu, H.Y.; Zang, Z.L.; Huang, M.; Zhou, S.J. Precise control and prediction of the greenhouse growth environment of Dendrobium candidum. *Comput. Electron. Agric.* **2018**, *151*, 453–459. [CrossRef]

21. Hamrani, A.; Akbarzadeh, A.; Madramootoo, C.A. Machine learning for predicting greenhouse gas emissions from agricultural soils. *Sci. Total Environ.* **2020**, *741*, 140338. [CrossRef]

22. Jung, D.H.; Kim, H.S.; Jhin, C.; Kim, H.J.; Park, S.H. Time-serial analysis of deep neural network models for prediction of climatic conditions inside a greenhouse. *Comput. Electron. Agric.* **2020**, *173*, 105402. [CrossRef]

23. Katzin, D.; van Henten, E.J.; van Mourik, S. Process-based greenhouse climate models: Genealogy, current status, and future directions. *Agric. Syst.* **2022**, *198*, 103388. [CrossRef]

24. Rincón, V.J.; Grella, M.; Marucco, P.; Alcatrão0, A.E.; Sanchez-Hermosilla, J.; Balsari, P. Spray performance assessment of a remote-controlled vehicle prototype for pesticide application in greenhouse tomato crops. *Sci. Total Environ.* **2020**, *726*, 138509. [CrossRef] [PubMed]

25. Rodriguez-Ortega, W.M.; Martinez, V.; Rivero, R.M.; Camara-Zapata, J.M.; Mestre, T.; Garcia-Sanchez, F. Use of a smart irrigation system to study the effects of irrigation management on the agronomic and physiological responses of tomato plants grown under different temperatures regimes. *Agric. Water Manag.* **2017**, *183*, 158–168. [CrossRef]

26. Astegiano, P.; Fermi, F.; Martino, A. Investigating the impact of e-bikes on modal share and greenhouse emissions: A system dynamic approach. *Transp. Res. Procedia* **2019**, *37*, 163–170. [CrossRef]

27. Forrester, J.W. System dynamics and the lessons of 35 years. In *A Systems-Based Approach to Policymaking*; Springer: Boston, MA, USA, 1993; pp. 199–240.

28. Li, J.W.; Cao, Y.C.; Zhu, Y.Q.; Xu, C.; Wang, L.X. System dynamic analysis of greenhouse effect based on carbon cycle and prediction of carbon emissions. *Appl. Ecol. Environ. Res.* **2019**, *17*, 5067–5080. [CrossRef]

29. Forrester, J.W. Industrial dynamics. *J. Oper. Res. Soc.* **1997**, *48*, 1037–1041. [CrossRef]

30. Château, P.A.; Wunderlich, R.F.; Wang, T.W.; Lai, H.T.; Chen, C.C.; Chang, F.J. Mathematical modeling suggests high potential for the deployment of floating photovoltaic on fish ponds. *Sci. Total Environ.* **2019**, *687*, 654–666. [CrossRef] [PubMed]

31. Lu, D.; Iqbal, A.; Zan, F.; Liu, X.; Chen, G. Life-cycle-based rgeenhouse gas, energy, and economic analysis of municipal solid wastemanagement using system dynamics model. *Sustainability* **2021**, *13*, 1641. [CrossRef]

32. Stasinopoulos, P.; Shiwakoti, N.; Beining, M. Use-stage life cycle greenhouse gas emissions of the transition to an autonomous vehicle fleet: A system dynamics approach. *J. Clean. Prod.* **2021**, *278*, 123447. [CrossRef]

33. Huang, A.; Chang, F.J. Prospects for rooftop farming system dynamics: An action to stimulate water-energy-food nexus synergies toward green cities of tomorrow. *Sustainability* **2021**, *13*, 9042. [CrossRef]

34. Amadei, B. *A Systems Approach to Modeling the Water-Energy-Land-Food Nexus: System Dynamics MODELING and dynamic Scenario Planning*, 1st ed.; Momentum Press: New York, NY, USA, 2019; Volume I, II.

35. Gary, I.E.; Grigg, N.; Reagan, W. Dynamic behavior of the water-food-energy nexus: Focus on crop production and consumption. *Irrig. Drain.* **2017**, *66*, 19–33.

36. Fang, W. Quantitative measures of the effectiveness of evaporative cooling systems in greenhouse. *J. Agric. Mach.* **1995**, *4*, 15–25.

37. Chang, F.J.; Tsai, M.J. A nonlinear spatio-temporal lumping of radar rainfall for modeling multi-step-ahead inflow forecasts by data-driven techniques. *J. Hydrol.* **2016**, *535*, 256–269. [CrossRef]

38. Mirabbasi, R.; Kisi, O.; Sanikhani, H.; Gajbhiye Meshram, S. Monthly long-term rainfall estimation in Central India using M5Tree, MARS, LSSVR, ANN and GEP models. *Neural Comput. Appl.* **2019**, *31*, 6843–6862. [CrossRef]

39. Arya Azar, N.; Kardan, N.; Ghordoyee Milan, S. Developing the artificial neural network–evolutionary algorithms hybrid models (ANN–EA) to predict the daily evaporation from dam reservoirs. *Eng. Comput.* **2021**, *37*, 1–19. [CrossRef]

40. Chang, F.J.; Chang, L.C.; Kao, H.S.; Wu, G.R. Assessing the effort of meteorological variables for evaporation estimation by self-organizing map neural network. *J. Hydrol.* **2010**, *384*, 118–129. [CrossRef]

41. Chang, L.C.; Amin, M.; Yang, S.N.; Chang, F.J. Building ANN-based regional multi-step-ahead flood inundation forecast models. *Water* **2018**, *10*, 1283. [CrossRef]

42. Chang, L.C.; Chang, F.J.; Yang, S.N.; Tsai, F.H.; Chang, T.H.; Herricks, E.E. Self-organizing maps of typhoon tracks allow for flood forecasts up to two days in advance. *Nat. Commun.* **2020**, *11*, 1–13. [CrossRef]

43. Chang, L.C.; Liou, C.Y.; Chang, F.J. Explore training self-organizing map methods for clustering high-dimensional flood inundation maps. *J. Hydrol.* **2022**, *595*, 125655. [CrossRef]

44. Kao, I.F.; Zhou, Y.; Chang, L.C.; Chang, F.J. Exploring a long short-term memory based encoder-decoder framework for multi-step-ahead flood forecasting. *J. Hydrol.* **2020**, *583*, 124631. [CrossRef]

45. Zhou, Y.; Chang, L.C.; Uen, T.S.; Guo, S.; Xu, C.Y.; Chang, F.J. Prospect for small-hydropower installation settled upon optimal water allocation: An action to stimulate synergies of water-food-energy nexus. *Appl. Energy* **2019**, *238*, 668–682. [CrossRef]

46. Zhou, Y.; Guo, S.; Xu, C.Y.; Chang, F.J.; Yin, J. Improving the reliability of probabilistic multi-step-ahead flood forecasting by fusing unscented Kalman filter with recurrent neural network. *Water* **2020**, *12*, 578. [CrossRef]
47. Chang, F.J.; Guo, S. Advances in hydrologic forecasts and water resources management. *Water* **2020**, *12*, 1819. [CrossRef]
48. Bai, T.; Tsai, W.P.; Chiang, Y.M.; Chang, F.J.; Chang, W.Y.; Chang, L.C.; Chang, K.C. Modeling and investigating the mechanisms of groundwater level variation in the Jhuoshui River Basin of Central Taiwan. *Water* **2019**, *11*, 1554. [CrossRef]
49. Chen, I.T.; Chang, L.C.; Chang, F.J. Exploring the spatio-temporal interrelation between groundwater and surface water by using the self-organizing maps. *J. Hydrol.* **2018**, *556*, 131–142. [CrossRef]
50. Ghimire, S.; Deo, R.C.; Downs, N.J.; Raj, N. Global solar radiation prediction by ANN integrated with European Centre for medium range weather forecast fields in solar rich cities of Queensland Australia. *J. Clean. Prod.* **2019**, *216*, 288–310. [CrossRef]
51. Pradhan, P.; Tingsanchali, T.; Shrestha, S. Evaluation of soil and water assessment tool and artificial neural network models for hydrologic simulation in different climatic regions of Asia. *Sci. Total Environ.* **2020**, *701*, 134308. [CrossRef]
52. Cheng, S.T.; Tsai, W.P.; Yu, T.C.; Herricks, E.E.; Chang, F.J. Signals of stream fish homogenization revealed by AI-based clusters. *Sci. Rep.* **2018**, *8*, 15960. [CrossRef]
53. Hu, J.H.; Tsai, W.P.; Cheng, S.T.; Chang, F.J. Explore the relationship between fish community and environmental factors by machine learning techniques. *Environ. Res.* **2020**, *184*, 109262. [CrossRef]
54. Kow, P.Y.; Wang, Y.S.; Zhou, Y.; Kao, I.F.; Issermann, M.; Chang, L.C.; Chang, F.J. Seamless integration of convolutional and back-propagation neural networks for regional multi-step-ahead PM2.5 forecasting. *J. Clean. Prod.* **2020**, *261*, 121285. [CrossRef]
55. Saleem, M.H.; Potgieter, J.; Arif, K.M. Automation in agriculture by machine and deep learning techniques: A review of recent developments. *Precis. Agric.* **2021**, *22*, 2053–2091. [CrossRef]
56. Nicolosi, G.; Volpe, R.; Messineo, A. An innovative adaptive control system to regulate microclimatic conditions in a greenhouse. *Energies* **2017**, *10*, 722. [CrossRef]
57. Riahi, J.; Vergura, S.; Mezghani, D.; Mami, A. Intelligent control of the microclimate of an agricultural greenhouse powered by a supporting PV system. *Appl. Sci.* **2020**, *10*, 1350. [CrossRef]
58. Xue, Y.X.; Li, Y.L.; Wen, X.Z. Effects of air humidity on the photosynthesis and fruit-set of yomato under high Temperature. *Acta Hortic. Sin.* **2010**, *37*, 397–404.
59. Liou, Y.C.; Huang, R.J.; Cai, M.L.; Huang, S.W. Facility cultivation and health management techniques of grape tomato. *Tech. Issue Tainan Dist. Agric. Res. Ext. Stn.* **2016**, *164*, 3–47. (In Chinese)

 *water*

Article

Multi-Step Ahead Probabilistic Forecasting of Daily Streamflow Using Bayesian Deep Learning: A Multiple Case Study

Fatemeh Ghobadi and Doosun Kang *

Department of Civil Engineering, Kyung Hee University, 1732 Deogyeong-daero, Giheung-gu, Yongin-si 17104, Korea
* Correspondence: doosunkang@khu.ac.kr; Tel.: +82-31-201-2513

Abstract: In recent decades, natural calamities such as drought and flood have caused widespread economic and social damage. Climate change and rapid urbanization contribute to the occurrence of natural disasters. In addition, their destructive impact has been altered, posing significant challenges to the efficiency, equity, and sustainability of water resources allocation and management. Uncertainty estimation in hydrology is essential for water resources management. By quantifying the associated uncertainty of reliable hydrological forecasting, an efficient water resources management plan is obtained. Moreover, reliable forecasting provides significant future information to assist risk assessment. Currently, the majority of hydrological forecasts utilize deterministic approaches. Nevertheless, deterministic forecasting models cannot account for the intrinsic uncertainty of forecasted values. Using the Bayesian deep learning approach, this study developed a probabilistic forecasting model that covers the pertinent subproblem of univariate time series models for multi-step ahead daily streamflow forecasting to quantify epistemic and aleatory uncertainty. The new model implements Bayesian sampling in the Long short-term memory (LSTM) neural network by using variational inference to approximate the posterior distribution. The proposed method is verified with three case studies in the USA and three forecasting horizons. LSTM as a point forecasting neural network model and three probabilistic forecasting models, such as LSTM-BNN, BNN, and LSTM with Monte Carlo (MC) dropout (LSTM-MC), were applied for comparison with the proposed model. The results show that the proposed Bayesian long short-term memory (BLSTM) outperforms the other models in terms of forecasting reliability, sharpness, and overall performance. The results reveal that all probabilistic forecasting models outperformed the deterministic model with a lower RMSE value. Furthermore, the uncertainty estimation results show that BLSTM can handle data with higher variation and peak, particularly for long-term multi-step ahead streamflow forecasting, compared to other models.

Keywords: Bayesian neural network; forecasting uncertainty; multi-step ahead forecasting; probabilistic streamflow forecasting; variational inference

Citation: Ghobadi, F.; Kang, D. Multi-Step Ahead Probabilistic Forecasting of Daily Streamflow Using Bayesian Deep Learning: A Multiple Case Study. *Water* **2022**, *14*, 3672. https://doi.org/10.3390/w14223672

Academic Editors: Fi-John Chang, Li-Chiu Chang and Jui-Fa Chen

Received: 17 October 2022
Accepted: 11 November 2022
Published: 14 November 2022

Publisher's Note: MDPI stays neutral with regard to jurisdictional claims in published maps and institutional affiliations.

1. Introduction

Sustainable water resource management is an essential requirement worldwide, and streamflow forecasting is an essential component of an effective water resource management plan [1]. Accurate streamflow forecasting plays a critical role in many decision-making scenarios related to water resource management such as flood/drought control and mitigation, reservoir management, hydropower generation, sediment transport, and irrigation management [2]. Owing to the complex and nonlinear characteristics associated with streamflow [3], forecasting is challenging. Sustainable water resource management plans are used to meet the requirements of people today and in the future. To support risk-aware decision-making in water resource management, current streamflow forecasting approaches should be improved to estimate forecasting uncertainties and leverage large volumes of data with complex dependencies [4].

Deep learning (DL), a sophisticated and mathematically complex evolution of machine learning (ML) algorithms, has recently received huge attention from researchers and has gradually become the most widely used forecasting approach in hydrology [5–9]. The advantage of DL is its flexibility to learn massive data and the ease of incorporating exogenous covariates [10]. The advantages of DL techniques over traditional ML algorithms for streamflow forecasting have been discussed in many studies [1,11]. However, DL has not been extensively explored in forecasting uncertainty.

Only a few studies have been conducted on probabilistic approaches to streamflow forecasting. Thus, the existing uncertainty was not addressed by most DL approaches. However, deterministic approaches may not be as efficient as probabilistic methods and exhibit suboptimal performance. In general, deterministic forecasting is widely used in hydrology as an input for various water resource management plans. The transition from deterministic to probabilistic forecasting methods with uncertainty quantification is strongly favored in academia and industry. The primary issue in streamflow forecasting is the complex uncertainty rooted in the stochastic characteristics of streamflow time series. Furthermore, probabilistic forecasting has emerged to overcome the shortcomings of conventional deterministic methods and to deal with uncertainty more effectively. The probabilistic approach has recently gained importance because it can extract more valuable information from historical data and quantify the uncertainty of the future by forming a probability distribution over possible outcomes. The probabilistic approach extends beyond single-point forecasting for each time step and can provide a band of likely forecasting intervals above and below the mean forecasted value. Existing deterministic methods report the mean of possible outcomes, and they are unable to reflect the inherent uncertainty that exists in the real world.

Despite the fact that hydrological prediction can be most helpful when given in probabilistic form [12], the use of probabilistic modeling is still a relatively new concept in the field of hydrology [13]. Moreover, the probabilistic approach is crucial to optimal decision-making that reveals the upper and lower bounds between which the uncertain actual future values may exist. Occasionally, decision-making requires more than single-point forecasting; this is where distribution would be beneficial. To make reliable forecasts and to conduct a comprehensive performance evaluation, a probabilistic approach should be considered in streamflow forecasting. Most existing streamflow forecasting methods focus on deterministic forecasting. The application of various machine learning algorithms in deterministic prediction has been investigated in many studies [14–19]. Limited research has been conducted on multistep-ahead streamflow predictions [1,20–23]. Even though considerable efforts have been made to improve the performance of streamflow forecasting models from short- to long-term [24,25] and from single- to multi-step ahead [9,26,27], they are still limited by uncertainties [28–31].

An effective way to perform probability forecasting in the field of hydrology is to apply the Bayesian approach due to the benefits of uncertainty representation, understanding generalization, and reliable prediction through the lens of probability theory. The Bayesian approach can be classified into four primary groups: Bayesian model averaging (BMA), Bayesian model updating (BMU), Bayesian networks (BN), and Bayesian neural networks (BNN) [32]. A BNN is a type of stochastic artificial neural network that uses a BMU for training and updating the probabilistic distributions of network parameters. Furthermore, BMU and BN have become prevalent, and they have been implemented in various fields such as computer vision, natural language processing, medical diagnostics, autonomous driving, and flood hazard analysis [32]. Han and Coulibaly [33] presented a comprehensive review of Bayesian approaches applied to flood forecasting from 1999 to 2015. The results reveal that probabilistic flood forecasting can reduce uncertainty and provide more accurate and reliable forecasting. Moreover, they mentioned that only a limited number of river basins have been studied from the Bayesian perspective to date. Furthermore, we should determine if the Bayesian approaches are suitable for different watersheds with different sizes and physical and climatic characteristics. Costa and Fernandes [34] developed a

Bayesian framework to estimate the extreme flood quantile from a rainfall-runoff model of a dam in California. Xu et al. [35] developed a real-time probabilistic channel flood forecasting model by combining a hydraulic model with the Bayesian approach in the upstream reaches of the Three Gorges Dam on the Yangtze River, China. A state-of-the-art review was provided by Huang et al. [36] to summarize the application of Bayesian inference in system identification and damage assessment for civil infrastructure. Goodarzi et al. [37] proposed a decision-making model using BN to predict heavy precipitation in the Kan Basin. Bayesian neural networks are yet to be applied to probabilistic streamflow forecasting, as aforementioned.

Recent studies on probabilistic prediction in the field of hydrology are summarized in Table 1. As shown in Table 1, a few researchers trained a deterministic model and used the obtained deterministic result to obtain a probabilistic forecasting result to estimate the uncertainty [38]. However, in a few studies, a deterministic layer has been coupled with a probabilistic layer to achieve forecasting uncertainty [39]. Conversely, a few studies have focused on developing a probabilistic model by introducing stochastic components into the network by giving the network either stochastic activation or weights [40–42].

Table 1. Overview of recent probabilistic prediction studies in the field of hydrology.

Field	Probabilistic Method	Base Models	Posterior Approximation *		Evaluation Metrics		Ref.
			VI	MCM	Deterministic	Probabilistic	
Streamflow	LSTM-HetGP	ANN, HetGP, GLM, LSTM	-	-	NSE, RMSE, MRE, MSLE	percentage of coverage (POC) and the average interval width (AIW)	[39]
Flood	LSTM	ARIMA	-	-	RMSE, MAE		[40]
Streamflow	LSTM with multiparameter ensemble and dropout ensemble	–	-	✓	PBIAS, MARE, RMSE, NSE, KGE	POC, average width (AW), average interval score (AIS)	[41]
Streamflow	Variational Bayesian Long Short-Term Memory network (VB-LSTM)	Bayesian model Averaging (BMA)	✓	-	MAE	CRPS	[42]
Runoff	XGBoost (XGB) and Gaussian process regression (GPR) with Bayesian optimization algorithm (BOA)	GBR, LGB, CNN, LSTM, ANN, SVR, QR, GPR—combined with GPR	-	-	RMSE, MAPE, R^2	Coverage probability, Mean width percentage, Suitability metric, CRPS	[38]
Runoff	B-spline quantile regression model combined with kernel density estimation	QR, QRNN	-	-	RMSE, R^2, Q_r	PICP, PINAW, CRPS	[43]
Streamflow	Bayesian LSTM model	physics-based hydrologic model (Precipitation-Runoff Modeling System)	-	✓	NSE, RMSE-observations standard deviation ratio (RSR)		[44]

* The tick mark (✓) denotes the application of Posterior Approximation.

The application of probabilistic DL showed superior performance in various fields, including residential net load forecasting [45,46], short-term scheduling in power markets [47], photovoltaic power [48], load forecasting for buildings [49], and electricity consumption [50]. This indicates the wide range of the potential applicability of the probabilistic DL approaches. Univariate streamflow forecasting using conventional data-driven models has been investigated in the previous studies [51–54]. To the best of the authors' knowledge, the application of BLSTM in multi-step ahead probabilistic prediction using a retrospective univariate time series has not been applied to streamflow prediction yet. To address the aforementioned research gaps, this study proposed a framework for transforming a deterministic model into a probabilistic model with improved performance.

This study developed a Bayesian deep neural network framework to characterize the prognostic uncertainties for probabilistic streamflow forecasting, which investigated both epistemic and aleatoric uncertainties. The motivation of the framework was to transform existing deterministic prediction models into their probabilistic counterparts for better performance in water resources management and decision-making, and to cover newly emerged challenges that humankind encountered primarily due to climate change.

The primary contributions of the study are as follows: For the first time, in streamflow prediction, we introduced the Bayesian LSTM network's application for multi-step ahead probabilistic forecasting in water resource management. Bayesian theory and LSTM networks were combined to generate probabilistic streamflow forecasts to capture both epistemic and aleatoric uncertainties. This is the first study to exploit Bayesian deep learning for streamflow prediction. Moreover, a comprehensive comparison with a series of state-of-the-art probabilistic prediction methods is conducted. The superior performance of the proposed scheme was demonstrated with respect to both the deterministic and probabilistic forecasting results. Moreover, to demonstrate the superiority of probabilistic forecasting, particularly for water resource management, a comparative analysis was conducted for three case studies with different forecasting horizons and timescales.

The paper is organized as follows: In Section 2, the materials and methods are presented in subsections on Bayesian long-short-term memory (BLSTM) (2.1), experimental setup (2.2), and performance evaluation (2.3). In Section 3, the case study, study area (3.1), and experimental setup (3.2) are detailed. The results are presented in Section 4, with two subsections focusing on the probabilistic forecasting performance assessment (4.1) and the impact of the forecast horizon on probabilistic forecasting performance (4.2). Furthermore, the concluding remarks of this study with directions for future research are discussed in Section 5.

2. Materials and Methods

The proposed Bayesian deep-learning approach for probabilistic streamflow forecasting is presented in detail in the following sections.

2.1. Bayesian Long Short-Term Memory (BLSTM)

In this study, the Bayesian approach is employed, which is a well-established and thorough approach to fit probabilistic models that capture and distinguish different sources of uncertainties [55]. The BNN is a stochastic artificial neural network (ANN) trained using the Bayesian approach. Probability is defined in terms of the degree of belief in the Bayesian approach; the more likely an outcome is, the higher its degree of belief. The primary idea of the Bayesian approach in deep learning is to replace each weight with a distribution [56]. An LSTM network overcomes the long-term dependency issue of conventional RNNs through additional interactions in its various unit cells. Additionally, LSTM cells (memory cells) are composed of three gates (input, forget, and output) for short-term memory selection and a state vector transmission responsible for long-term memory. Information can be selectively passed during the learning procedure by manipulating the gate settings. The LSTM network is mathematically represented as follows [57]:

$$i_t = \sigma(W_i.[h_{t-1}, x_t] + b_i)f_t = \sigma(W_f.[h_{t-1}, x_t] + b_f), \tag{1}$$

$$o_t = \sigma(W_o.[h_{t-1}, x_t] + b_o), \tag{2}$$

$$h_t = o_t \times \tanh(C_t), \tag{3}$$

$$\tilde{C} = \tanh(W_c.[h_{t-1}, x_t] + b_c), \tag{4}$$

$$C_t = f_t \circ C_{t-1} + i_t \circ \tilde{C}, \tag{5}$$

$$\sigma(x) = sigmoid(x) = \frac{1}{1 + e^{-x}}, \tag{6}$$

$$\tanh(x) = \frac{e^x - e^{-x}}{e^x + e^{-x}}, \tag{7}$$

where at time step t, x_t is the input vector, h_t is the LSTM output and hidden state (short-term memory), and i_t, f_t, and o_t are the input, forget, and output gates, respectively. W and b are the weight matrix and bias, respectively. C_t is the current cell state (long-term memory), and $\tilde{C}$ is the candidate cell state value. σ is a sigmoid activation function that uses h_{t-1} and x_t to make decisions regarding the input, forget, and output gates [57].

Given the input data, $X_{train} = [x_1, \ldots, x_{Train}]$ and their corresponding output labels $Y_{train} = [y_1, \ldots, y_{Train}]$. The primary goal of the Bayesian approach is to identify the parameter W of a function $y = f^W(x)$ that probably generates the desired output [58,59]. In this approach, a prior distribution that represents the prior belief about the neural network parameter distribution before observing the inputs is employed over W to capture epistemic uncertainty. The Bayesian neural network structure is illustrated in Figure 1. Rather than a single network, this method trains a set of networks in which the weight of each network is derived from a shared learning probability distribution [59].

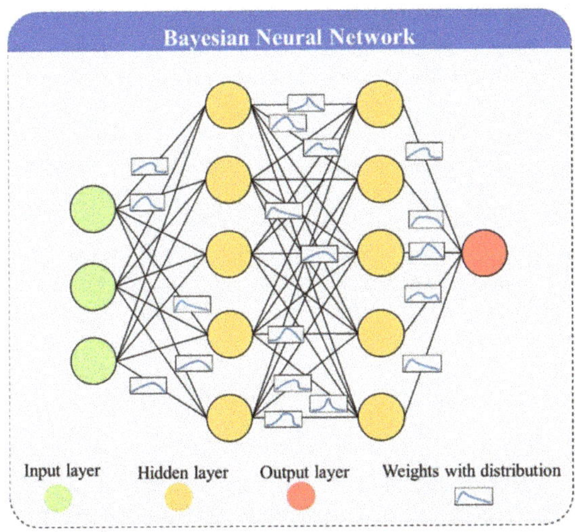

Figure 1. Structure of the Bayesian neural network.

Setting a standard normal distribution as a prior with zero mean, which can bring the benefit of regularization, has been demonstrated as one of the most effective solutions when the prior distribution is difficult to identify. After training the Bayesian deep neural network and observing data, the model likelihood distribution $p(Y_{Train}|f^W)$ should be defined as a normal distribution $\mathcal{N}(f^W(X_{Train}), \sigma^2)$ and observation noise σ to capture roughly suitable parameters. Based on the Bayesian rule, the posterior $p(W|X_{Train}, Y_{Train})$ is employed over the weights to generate samples of predictions rather than the prior distribution. The posterior is calculated as follows [59]:

$$p(W|X_{Train}, Y_{Train}) = \frac{p(Y_{Train}|X_{Train}, W).p(W)}{p(Y_{Train}|X_{Train})} \tag{8}$$

where $p(Y_{Train}|X_{Train})$ is the marginal likelihood probability that cannot be estimated, thereby, the posterior is not tractable without a variational inference to approximate it. With this distribution, suitable parameters given by the input data can be captured, and the output y can be predicted for a new input x by integration [58]:

$$p(y|x, X_{Train}, Y_{Train}) = \int p(y|x, W)p(W|X_{Train}, Y_{Train})dW. \tag{9}$$

To evaluate the true posterior $p(W|X_{Train}, Y_{Train})$, an approximation variational distribution $q_\theta(W)$, which is parameterized by θ, is required to ensure that the optimal distribution $\widetilde{q}_\theta(W)$ can represent the posterior by minimizing the Kullback–Leibler (KL) divergence [60] between the approximation variational and posterior distributions [61]:

$$KL(q_\theta(W) \parallel p(W|X_{Train}, Y_{Train})) = \int q_\theta(W) \log \frac{q_\theta(W)}{p(W|X_{Train}, Y_{Train})} dW. \tag{10}$$

Generally, two methods are available to approximate the posterior distribution: variational inference (VI) and Monte Carlo (MC) dropout [56,61,62]. The study employed the VI to solve the optimization issue analytically. Interested readers can refer to [62] for detailed information on the approximation method. The predictive distribution can be approximated by:

$$p(y|x, X_{Train}, Y_{Train}) = \int p(y|x, W)\widetilde{q}_\theta(W) dW = \widetilde{q}_\theta(y|x). \tag{11}$$

2.1.1. Epistemic Uncertainty

Mathematically, by simulating the model based on input x, the predictive mean can be estimated with an unbiased estimator, as follows [63]:

$$\widetilde{\mathbb{E}}[y] := \frac{1}{T} \sum_{t=1}^{T} f^{\hat{W}_t}(x), \tag{12}$$

where $\widetilde{\mathbb{E}}[y]$ is the predictive mean, $f^{\hat{W}_t}$ is the stochastic output of the prediction model, $\hat{W}_t$ represents the sample weights, and T denotes the number of samples at time t. Similar to the estimation of the predictive mean, given that $\hat{W}_t \sim \widetilde{q}_\theta(W)$ and $p(y|f^W(x)) = \mathcal{N}(y; f^w(x), \sigma^2)$ for $\sigma > 0$, the predictive variance can be estimated by an unbiased estimator as follows [63]:

$$\widetilde{\mathbb{E}}[y^T y] := \frac{1}{T} \sum_{t=1}^{T} f^{\hat{W}_t}(x)^T f^{\hat{W}_t}(x) + \sigma^2. \tag{13}$$

The term σ^2 corresponds to inherent noise in the input data. Afterward, the epistemic uncertainty, which represents the uncertainty of the model about its prediction outputs, is captured by the predictive variance that can be approximated as [63]:

$$\widetilde{Var}[y] = \widetilde{\mathbb{E}}[y^T y] - \widetilde{\mathbb{E}}[y]^T \widetilde{\mathbb{E}}[y]. \tag{14}$$

2.1.2. Aleatoric Uncertainty

Aleatoric uncertainty can be divided into homoscedastic and heteroscedastic uncertainties. To capture the aleatoric uncertainty, parameter σ should be tuned. For each input x, in the homoscedastic uncertainty, the observation noise σ is assumed to be constant. In contrast, heteroscedastic uncertainty assumes that observation noise varies with the input. Heteroscedastic models are data-dependent and can be expressed as:

$$\mathcal{L}(\theta) = \frac{1}{T_{train}} \sum_{i=1}^{T_{train}} \frac{1}{2\sigma(x_i)^2} \parallel y_i - f(x_i)^2 \parallel + \frac{1}{2} \log(x_i)^2. \tag{15}$$

Because the maximum posterior is performed to find a single value for parameter θ, this approach does not capture the epistemic uncertainty since it is a property of the model, not the input data.

2.1.3. Combining Aleatoric and Epistemic Uncertainty

Abdar et al. [64] explained that an effective way to combine both uncertainties in a single model is to transform the heteroscedastic model into a Bayesian model by placing a distribution over its weight and bias parameters. Thus, both the predictive mean and variance were derived from the developed prediction model.

$$\left[\hat{y}, \hat{\sigma}^2\right] = f_M^W(X),\qquad(16)$$

where f_M^W is the prediction model (BLSTM) used in this study, parameterized by the model weight W [55,56]. The Gaussian likelihood is used to model the aleatoric uncertainty, and the final loss function of the prediction model can be expressed as [58]:

$$\mathcal{L}_M(\theta) = \frac{1}{T_{train}} \sum_{i=1}^{T_{train}} \frac{1}{2\sigma(x_i)^2} \parallel y_i - \hat{y}_i \parallel^2 + \frac{1}{2}\log\hat{\sigma}_i^2.\qquad(17)$$

Finally, the predictive uncertainty of the prediction model, consisting of both aleatoric and epistemic uncertainties, can be approximated as

$$\widetilde{\mathrm{Var}}[y] = \frac{1}{T_{sample}} \sum_{t=1}^{T_{sample}} \hat{y}_t^2 - \left(\frac{1}{T_{sample}} \sum_{t=1}^{T_{sample}} \hat{y}_t\right)^2 + \frac{1}{T_{sample}} \sum_{t=1}^{T_{sample}} \hat{\sigma}_t^2,\qquad(18)$$

where T_{sample} denotes the number of training samples. An example of the Bayesian LSTM cell of the proposed BLSTM network is shown in Figure 2, with a zoomed-in plot of the forget gate at time step t in the first layer.

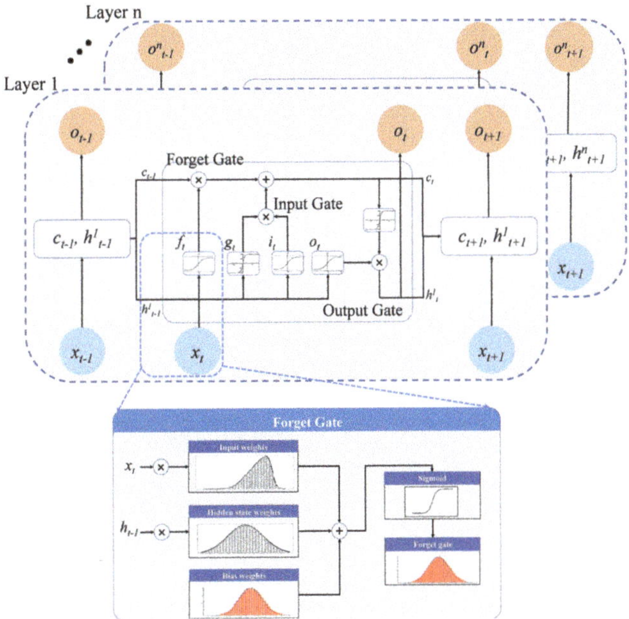

Figure 2. Example of the proposed BLSTM network with a zoomed-in plot of the forget gate at time step t in the first layer.

2.2. Performance Evaluation Metrics

To assess the performance of the prediction models, this study adopted the root mean square error (*RMSE*) metric for deterministic prediction and three metrics for probabilistic prediction: continuous ranked probability score (CRPS), prediction interval coverage probability (PICP), and mean prediction interval width (MPIW), which are formulated as follows.

1. Metric for Deterministic Forecasting

To evaluate the accuracy of the deterministic forecasting results, the root-mean-square error (*RMSE*) was selected as a commonly used hydrological evaluation indicator. The *RMSE* is defined as follows:

$$RMSE\sqrt{\frac{1}{n}\sum_{i=1}^{n}(Y_i - \hat{Y}_i)^2},\tag{19}$$

where $\hat{Y}_i$ is the predicted variable, Y_i is the observed value, and n is the number of samples.

2. Metrics for Probabilistic Forecasting

A useful metric to assess the accuracy of probabilistic prediction models is the CRPS. The CRPS expresses the distance between the probabilistic forecast p and the observed value Y_i and is defined as

$$\text{CRPS} = \int_{-\infty}^{+\infty}\left[P(\hat{Y}_i) - H(\hat{Y}_i - Y_i)\right]^2 d\hat{Y}_i,\tag{20}$$

$$P(\hat{Y}_i) = \int_{-\infty}^{\hat{Y}_i} p(x)dx,\tag{21}$$

$$H(\hat{Y}_i - Y_i) = \begin{cases} 0\ for\ \hat{Y}_i < Y_i \\ 1\ for\ \hat{Y}_i \geq Y_i \end{cases},\tag{22}$$

where $p(x)$ is the probability density function (PDF), $P(\hat{Y}_i)$ is the prediction cumulative distribution function (CDF), and H is the Heaviside step function, which equals 0 if $\hat{Y}_i < Y_i$ and equals 1 otherwise.

The mean prediction interval width (MPIW) is an effective representation of sharpness in probabilistic predictions. This metric is defined as

$$\text{MPIW} = \frac{1}{n}\sum_{i=1}^{n}(\hat{Y}_i^{\,u} - \hat{Y}_i^{\,l}),\tag{23}$$

where n is the size of the test set, and $\hat{Y}_i^{\,u}$ and $\hat{Y}_i^{\,l}$ denote the upper and lower bounds of the 95% prediction interval, respectively.

Prediction interval coverage probability (PICP) or (PI) is the probability that the target lies within the interval provided by the prediction model. PICP is defined as:

$$\text{PICP} = \frac{1}{n}\sum_{i=1}^{n} c_i,\ c_i = \begin{cases} 1,\ if\ Y_i \in \left[\hat{Y}_i^{\,l}, \hat{Y}_i^{\,u}\right] \\ 0,\ if\ Y_i \notin \left[\hat{Y}_i^{\,l}, \hat{Y}_i^{\,u}\right] \end{cases}.\tag{24}$$

Thus, PICP indicates the frequency with which the prediction interval (PI) captures the observed value, ranging from 0 if all predicted values lie outside the PI and 1 if all predicted values lie inside the PI.

3. Case Study

To evaluate the performance of the probabilistic data-driven models under different conditions, three basins in the United States with different hydroclimatic conditions and drainage areas were selected as study areas, as described in the following section.

3.1. Study Area

The study basins were located in different climate regions of three states across the United States, i.e., IN (Indiana), MN (Minnesota), and CA (California). Figure 3 shows the locations of the three basins. The first case study was conducted in Bartholomew County, IN, the second was conducted in Koochiching County, MN, and third was conducted in Shasta County, CA. The drainage area of the river basins was approximately 1560–4420 km^2.

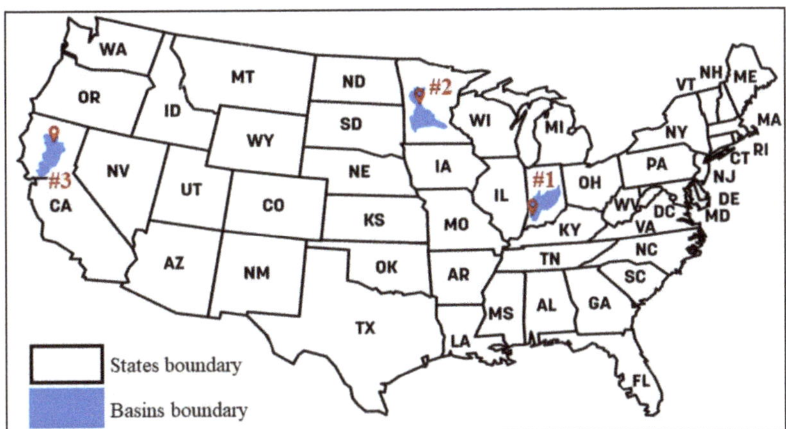

Figure 3. Location of 3 study basins in different climate regions across the United States.

Based on the USGS statewide streamflow–water year 2021 report, the annual mean streamflow was ranked by state from 1 to 92, indicating the maximum and minimum annual flow for all years analyzed. Streamflow rankings were grouped into categories of much below normal, below normal, normal, above normal, and much above normal based on percentiles of flow (<10%, 10–24%, 25–75%, 76–90%, and >90%, respectively) [65]. Much-below-normal streamflow with a rank 84–91 is reported in CA and below-normal streamflow with a rank 70–83 is reported in MN. The annual mean streamflow rank for IN was reported to be normal, with a rank 24–50. Daily historical streamflow data for the three selected case studies were obtained from the United States Geological Survey (USGS) website, (https://waterdata.usgs.gov/nwis) (accessed on 1 February 2022).

The descriptive information of the daily streamflow in the three case studies is presented in Table 2. Details of the case studies, including gauge ID, gauge name, and drainage area, are presented in Table 3. For all catchments, streamflow with a 30-day lag was considered owing to the cross-correlation results.

Table 2. Descriptive information of daily historical streamflow data for three case studies.

Criteria	Case Study 1	Case Study 2	Case Study 3
No. Samples	27,146	34,205	22,645
Mean (m^3/s)	58	30	31
Std (m^3/s)	90	51	42
Min (m^3/s)	2.5	0.6	3
25% (m^3/s)	13	4	9
50% (m^3/s)	31	11	20
75% (m^3/s)	64	32	36
Max (m^3/s)	1654	702	1274

Table 3. Details of the selected case studies.

Case Study. No.	Station ID	G-Name	Elev. (m)	Drainage Area (km^2)	Lon. (°)	Lat. (°)	Period
1	03364000	EAST FORK WHITE RIVER AT COLUMBUS, IN	183.8	4421	85°55′32″	39°12′00″	1948–2022
2	05131500	LITTLE FORK RIVER AT LITTLEFORK, MN	330.3	4403	93°32′57″	48°23′45″	1928–2022
3	11368000	MCCLOUD R AB SHASTA LK CA	335.3	1564	122°13′07″	40°57′30″	1945–2007

3.2. Experiment Setup

Before using the data to train the model, data preprocessing began with min-max normalization and log transfer as the initial phase of model development. The input time step was then derived from an autocorrelation analysis of the transformed-streamflow time series. Using a threshold of more than 0.5, which represents a moderate relationship, the past 30 days were selected as input. The autocorrelation analysis results for three case studies are given in the Supplementary File. The datasets for the three case studies were split into three sets: the first set accounted for 60% of the data, and it was used for model training; the second set was used for model validation (20%), and the remaining 20% was used for test purposes. Subsequently, the sliding window technique with a window size of 30 days was used. To demonstrate the superior performance of the Bayesian forecasting approach, probabilistic methods that have been widely used in the literature were employed for comparison. More specifically, LSTM-BNN [66], LSTM Monte-Carlo Dropout (LSTM-MC) [62,67], BNN [68], and deterministic LSTM were implemented in this study. Monte Carlo dropout is a straightforward epistemic uncertainty extension to the neural network. In general, dropout is a technique used to avoid overfitting by randomly dropping units during training. This can be considered the application of random noise in training. When this dropout was performed multiple times, multiple results were obtained. The distribution of the samples represents the uncertainty of the prediction model. The structure of the prediction models along with their graphical scheme are given in the Supplementary File.

The prediction models were developed in Python 3.6.9 with the Keras [69], Tensor-Flow [70], and PyTorch [71] libraries. The prediction model was implemented by an NVIDIA® GeForce® RTX 2070 SUPER and an Intel® Core i9-10920X central processing unit at 3.5 GHz utilizing 128 GB random access memory. For a fair comparison among the prediction models, a grid search for hyperparameter tuning was used to ensure identical evaluation.

4. Result and Discussion

To better clarify the forecasting performance of BLSTM method, the study compares the BLSTM to the LSTM-BNN, BNN, and LSTM-MC in terms of prediction interval uncertainty, sharpness, prediction reliability, and multi-step ahead probabilistic prediction performance. Moreover, LSTM is used as a deterministic model to evaluate the performance of all probabilistic prediction models against the deterministic model. In this section, the predictive ability of the four probabilistic models for 1 day (Scenario I), 7 days (Scenario II), and 30 days (Scenario III) ahead of streamflow prediction is investigated.

4.1. Probabilistic Prediction Performance Assessment

The PICP, MPIW, and CRPS values for the four models in the three case studies during the test period are listed in Table 4. The length of the bar represents the value of the evaluation metrics. In terms of PICP, the higher the value, the better and longer the bar, and vice versa for the other measures. Three major aspects must be considered simultaneously to evaluate probabilistic forecasting performance. PICP refers to the reliability of a model, MPIW refers to the model's sharpness, and CRPS indicates overall performance. In Scenario I, case study I was considered as an example because the models used the same mechanism to quantify the forecast uncertainty, and the PICP values of the four models were relatively close. Note that the larger the PICP and the smaller the MPIW and CRPS, the better the model performance. We observed that BLSTM showed better performance in handling datasets with high Std and peak streamflow. Case study III had 22,645 samples, which was ~17% and ~34% less than that of case studies I and II, with 27,146 and 34,205 samples, respectively. This difference did not lead to a particular change in the prediction performance of all the models for the first scenario. In this case study, from Scenarios I to III in BLSTM models, PICP decreased ~2% and 4%, respectively. While for case study I, PICP decreased ~25% and 50%, respectively, and for case study II, PICP decreased ~1% and 2%, respectively. Therefore, from the obtained results, we inferred that for single-step ahead prediction, the results were promising for all case studies, and the number of samples and peak streamflow did not affect the prediction performance. This made BNNs extremely data-efficient because they could learn from even a small dataset without overfitting. Furthermore, we predict that more uncertainty is associated with the results, particularly for the case study with a higher peak of streamflow, leading to wider prediction intervals and lower coverage. As expected, all models exhibited better predictive performance during shorter lead times (1–7 days) than during the longer horizon (30 days). Therefore, from the obtained results, we can infer that the probabilistic forecasting model can lead to higher uncertainty and lower accuracy over a longer forecasting horizon.

Table 4. Summary of prediction performance results for three case studies and three scenarios.

Model	Metric	Forecast Horizon = 1			Forecast Horizon = 7			Forecast Horizon = 30		
		Case I	Case II	Case III	Case I	Case II	Case III	Case I	Case II	Case III
BLSTM	PICP	0.950	0.964	0.956	0.709	0.958	0.941	0.477	0.943	0.921
	MPIW	0.021	0.006	0.016	0.023	0.024	0.023	0.032	0.042	0.028
	CRPS	0.087	0.035	0.066	0.375	0.212	0.214	0.576	0.437	0.337
LSTM-BNN	PICP	0.957	0.967	0.971	0.591	0.962	0.956	0.371	0.953	0.942
	MPIW	0.023	0.008	0.018	0.035	0.037	0.032	0.039	0.076	0.040
	CRPS	0.086	0.034	0.070	0.400	0.226	0.237	0.615	0.457	0.367
BNN	PICP	0.955	0.953	0.961	0.496	0.779	0.870	0.281	0.630	0.865
	MPIW	0.027	0.009	0.022	0.045	0.050	0.040	0.053	0.141	0.047
	CRPS	0.101	0.041	0.083	0.412	0.240	0.262	0.634	0.461	0.371
LSTM-MC	PICP	0.973	0.994	0.981	0.454	0.948	0.913	0.268	0.945	0.909
	MPIW	0.046	0.040	0.027	0.049	0.050	0.032	0.057	0.076	0.039
	CRPS	0.109	0.071	0.106	0.414	0.251	0.248	0.636	0.545	0.376

To further evaluate the results of the four probabilistic models for all scenarios in the three case studies, LSTM as a well-known deterministic model was used to make a comparison in terms of *RMSE*, as shown in Figure 4. All probabilistic models in all horizons performed well and provided more accurate prediction performance in terms of *RMSE* than the deterministic LSTM, indicating the superiority of all probabilistic models in comparison with the conventional deterministic model. The range of *RMSE* indicated that all models were fairly trained, and they showed promising predictability performance in terms of *RMSE*.

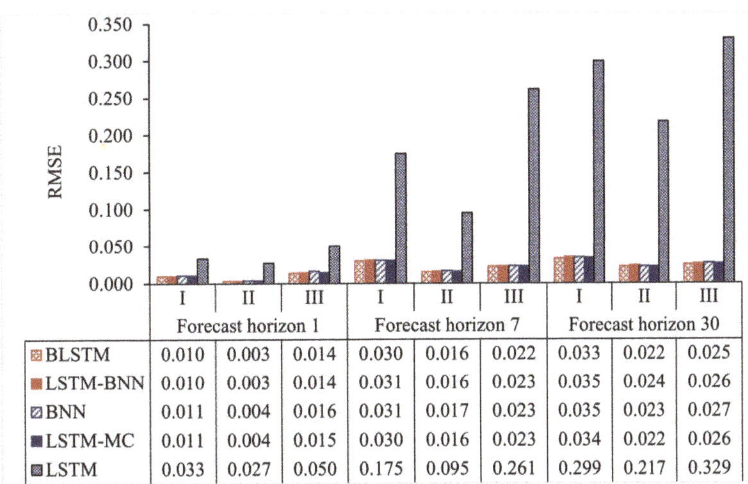

	I	II	III	I	II	III	I	II	III
		Forecast horizon 1			Forecast horizon 7			Forecast horizon 30	
BLSTM	0.010	0.003	0.014	0.030	0.016	0.022	0.033	0.022	0.025
LSTM-BNN	0.010	0.003	0.014	0.031	0.016	0.023	0.035	0.024	0.026
BNN	0.011	0.004	0.016	0.031	0.017	0.023	0.035	0.023	0.027
LSTM-MC	0.011	0.004	0.015	0.030	0.016	0.023	0.034	0.022	0.026
LSTM	0.033	0.027	0.050	0.175	0.095	0.261	0.299	0.217	0.329

Figure 4. Comparison among prediction models in terms of RMSE.

As shown in Figure 5, the probability that the prediction lies within the prediction interval by the LSTM-MC model is higher, followed by the LSTM-BNN, BNN, and BLSTM in the first scenario. The values of PICP indicate the percentage of the observed streamflow data lies within their 95% predictive intervals. In Scenario I, LSTM-MC outperformed BNN in terms of PICP. Moreover, for a longer horizon (7 days and 30 days ahead), due to the gradient vanishing of LSTM and BNNs as non-sequential models, both showed the lowest coverage and the points falling within the interval decreased by increasing the uncertainty in comparison with BLSTM and LSTM-BNN.

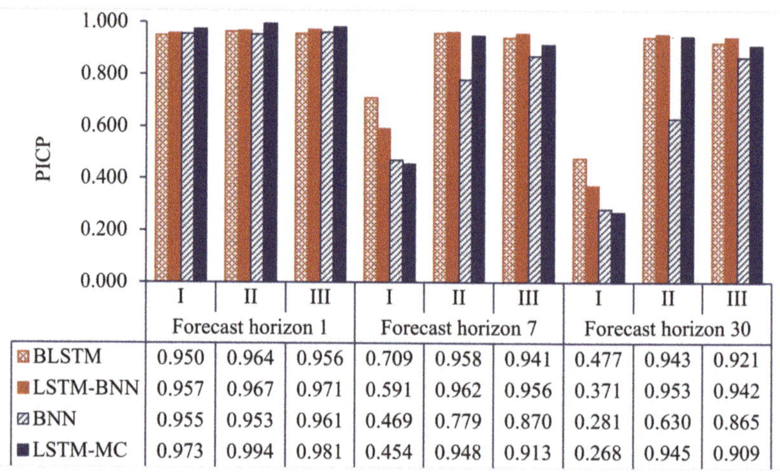

	| I |	II	| III |	| I |	II	| III |	| I |	II	| III |
	Forecast horizon 1			Forecast horizon 7			Forecast horizon 30		
▨ BLSTM	0.950	0.964	0.956	0.709	0.958	0.941	0.477	0.943	0.921
■ LSTM-BNN	0.957	0.967	0.971	0.591	0.962	0.956	0.371	0.953	0.942
▨ BNN	0.955	0.953	0.961	0.469	0.779	0.870	0.281	0.630	0.865
■ LSTM-MC	0.973	0.994	0.981	0.454	0.948	0.913	0.268	0.945	0.909

Figure 5. Comparison among prediction models in terms of PICP.

The MPIW values of the four models during the test period are shown in Figure 6a. The MPIW was considered an effective representation of sharpness in probabilistic predictions, and referred to the concentration of the predictive distributions. The more concentrated the predictive distributions, the lower the MPIW, the sharper the prediction, and consequently the better the predictive performance. As shown in Figure 6a, BLSTM has the lowest MPIW, which indicates that it is the sharpest predictive model among the other models in all scenarios. The stand-alone BNN and LSTM-MC were slightly different, whereas the LSTM-MC obtained the highest MPIW value among the other models, particularly by increasing the horizon. Compared with the BLSTM model with the narrowest MPIW, LSTM-MC had the worst prediction sharpness over all horizons. The fact that BLSTM presents the best predictive capability indicates the significance of capturing both epistemic and aleatoric uncertainties.

To comprehensively evaluate the probability prediction accuracy and reliability, a comparison among all prediction models in terms of the CRPS is shown in Figure 6b. Overall, BLSTM and LSTM-BNN competed with each other in all case studies and scenarios. However, in the longer horizon, BLSTM outperformed other models and proved its superiority by keeping more points falling within its forecasting interval while keeping the interval as narrow as possible while also increasing the uncertainty for a longer horizon. Therefore, the proposed BLSTM model outperformed the other models in terms of *RMSE*, MPIW, and CRPS, demonstrating the forecasting accuracy, sharpness, and overall performance of the model.

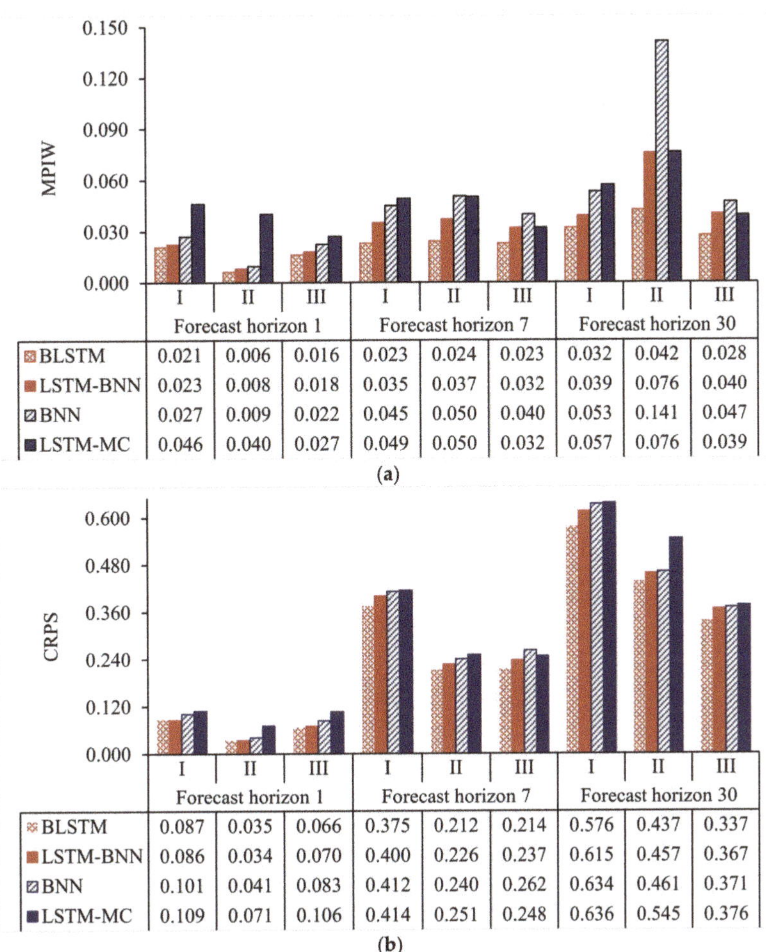

	I	II	III	I	II	III	I	II	III
	Forecast horizon 1			Forecast horizon 7			Forecast horizon 30		
BLSTM	0.021	0.006	0.016	0.023	0.024	0.023	0.032	0.042	0.028
LSTM-BNN	0.023	0.008	0.018	0.035	0.037	0.032	0.039	0.076	0.040
BNN	0.027	0.009	0.022	0.045	0.050	0.040	0.053	0.141	0.047
LSTM-MC	0.046	0.040	0.027	0.049	0.050	0.032	0.057	0.076	0.039

(**a**)

	I	II	III	I	II	III	I	II	III
	Forecast horizon 1			Forecast horizon 7			Forecast horizon 30		
BLSTM	0.087	0.035	0.066	0.375	0.212	0.214	0.576	0.437	0.337
LSTM-BNN	0.086	0.034	0.070	0.400	0.226	0.237	0.615	0.457	0.367
BNN	0.101	0.041	0.083	0.412	0.240	0.262	0.634	0.461	0.371
LSTM-MC	0.109	0.071	0.106	0.414	0.251	0.248	0.636	0.545	0.376

(**b**)

Figure 6. Comparison among prediction models in terms of (**a**) MPIW and (**b**) CRPS.

4.2. Impact of Forecast Horizon in Probabilistic Prediction Performance

The prediction results of all models are compared graphically in Figures 7–10 in the form of time series for the entire test set for case study II for all scenarios (forecasting horizon). Considering space limitations, we have avoided adding the results of all case studies and scenarios in the main text, and only the results of case study II are presented. The actual streamflow and forecast value for the test period are represented by black and red curves, respectively. The red band represents the prediction interval, with a 95% confidence level. The probabilistic forecasts generated with the BLSTM model presented the benefits of high prediction coverage of observed streamflow data (PICP) with a tighter prediction width (MPIW) and better overall performance (CRPS), corresponding to reliability, sharpness, and resolution. Furthermore, accurate peak prediction, which is a crucial factor for disaster prevention and water resources management, can be predicted with reasonable magnitudes with the proposed BLSTM. Additionally, with increasing the forecast horizon, BLSTM still showed reliable performance, while other models were incapable of handling such a situation. In forecast horizon 30, massive fluctuations in the prediction results occurred for all the models and case studies. However, most of the prediction results were covered by the 95% interval in Scenario I, followed by Scenario II.

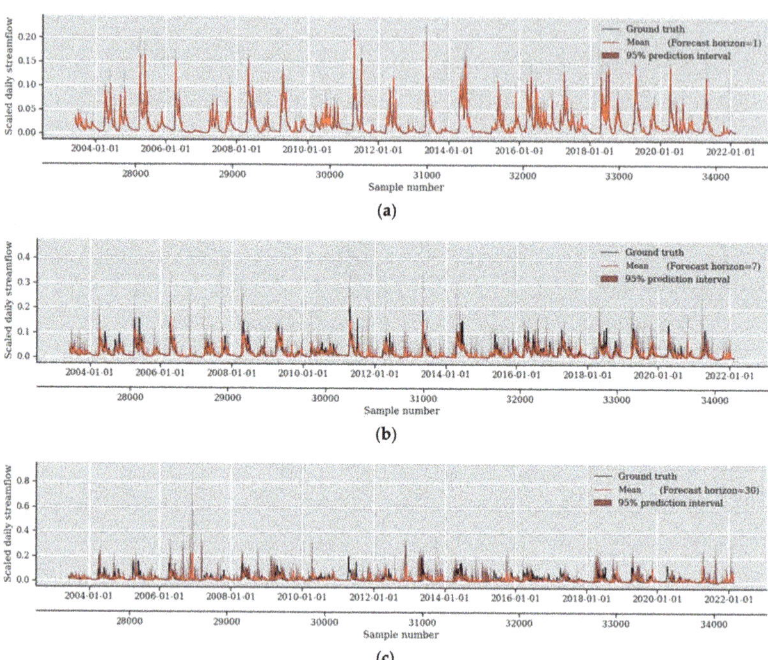

Figure 7. Probabilistic streamflow forecasting results obtained by BLSTM for case study II for (**a**) forecast horizon 1, (**b**) forecast horizon 7, and (**c**) forecast horizon 30.

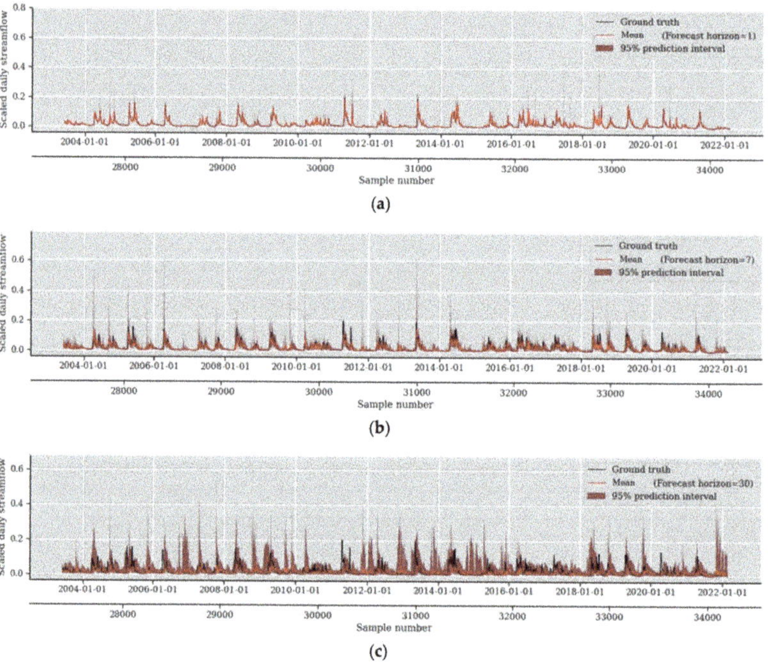

Figure 8. Probabilistic streamflow forecasting results obtained by LSTM-BNN for case study II for (**a**) forecast horizon 1, (**b**) forecast horizon 7, and (**c**) forecast horizon 30.

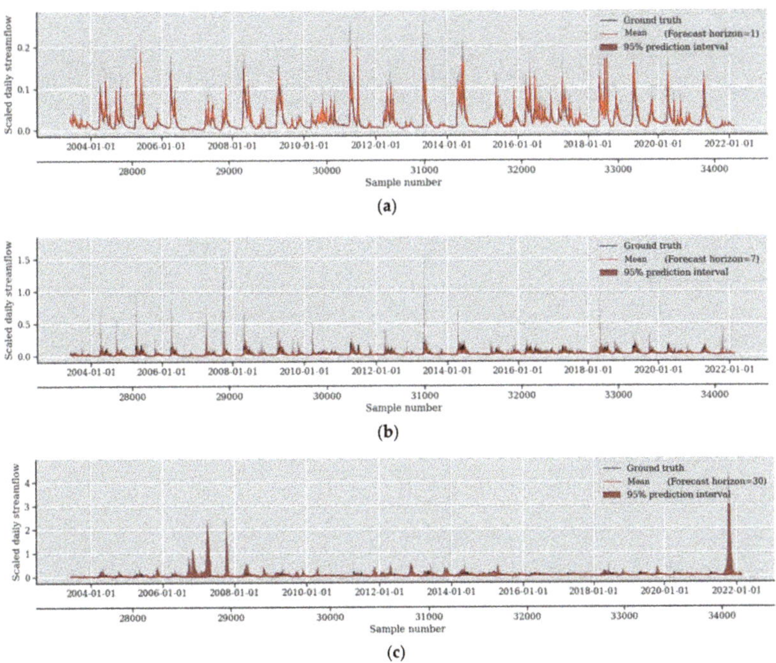

Figure 9. Probabilistic streamflow forecasting results obtained by BNN for case study II for (**a**) forecast horizon 1, (**b**) forecast horizon 7, and (**c**) forecast horizon 30.

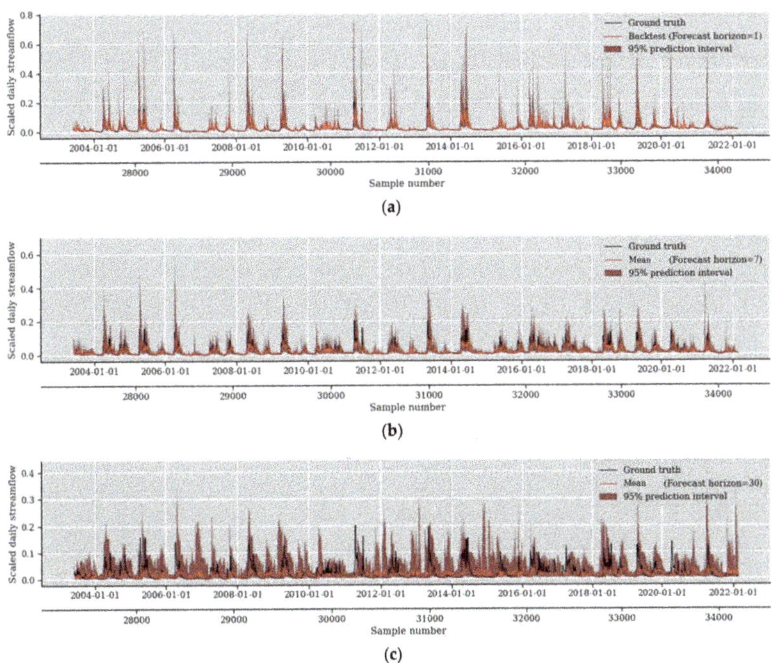

Figure 10. Probabilistic streamflow forecasting results obtained by LSTM-MC for case study II (**a**) forecast horizon 1, (**b**) forecast horizon 7, and (**c**) forecast horizon 30.

As shown in Figure 11, with an increase in the forecasting horizon from 1 to 30 days, the MPIW and CRPS values increase, and the PICP decreases. This indicates that the accuracy of the prediction models decreases with an increase in the forecasting horizon. The prediction accuracy over longer horizons decreased mainly as a result of the accumulative error issue in multi-step ahead recursive models and the gradient vanishing issue in long sequence time-series forecasting. Nevertheless, the predictive mean values of probabilistic streamflow from the BLSTM model matched the observations better than the other three models. The performance of all models gradually worsened with increasing lead times for the three case studies. As shown in Figure 11, when the overall prediction accuracy was low, MPIW was smaller. The interval width of the forecasting with LSTM-MC increased rapidly with the prediction horizon. The average interval width of the proposed BLSTM was much smaller than that of the other models. Simultaneously, the overall performance in terms of CRPS was higher, proving the superiority of the proposed method for the probability forecasting of daily streamflow, particularly for longer prediction horizons.

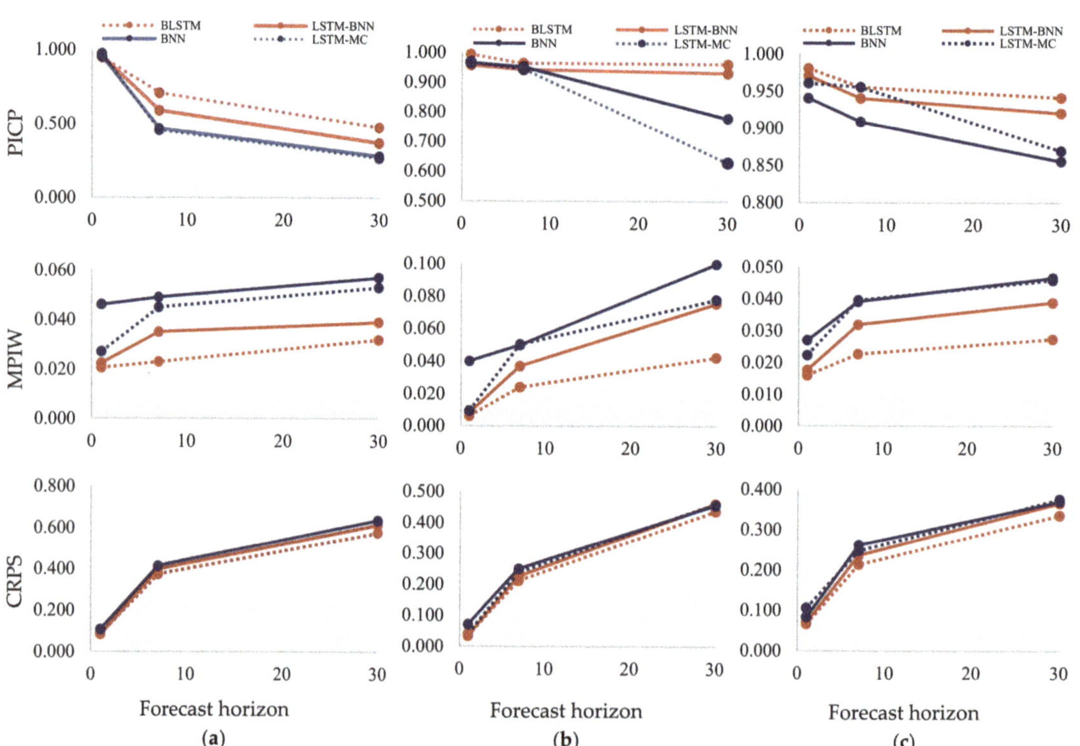

Figure 11. Change in probabilistic streamflow forecasting results by increasing horizon for (**a**) case study I, (**b**) case study II, and (**c**) case study III.

For a better and more vivid comparison of all model performances, the time series of all models for all case studies in the three scenarios are shown in Figure 12a–c for the first year of the test period (365 days). We observed that BNN and LSTM-MC underestimated the peak flows with a misleading trend in the first 365 days. In the first scenario, a 95% PI was relatively narrow and constant for all models, indicating that models captured both low and high flow values appropriately with low uncertainty. However, longer horizons in Scenarios II and III were associated with a wider 95% PI, indicating greater model uncertainty.

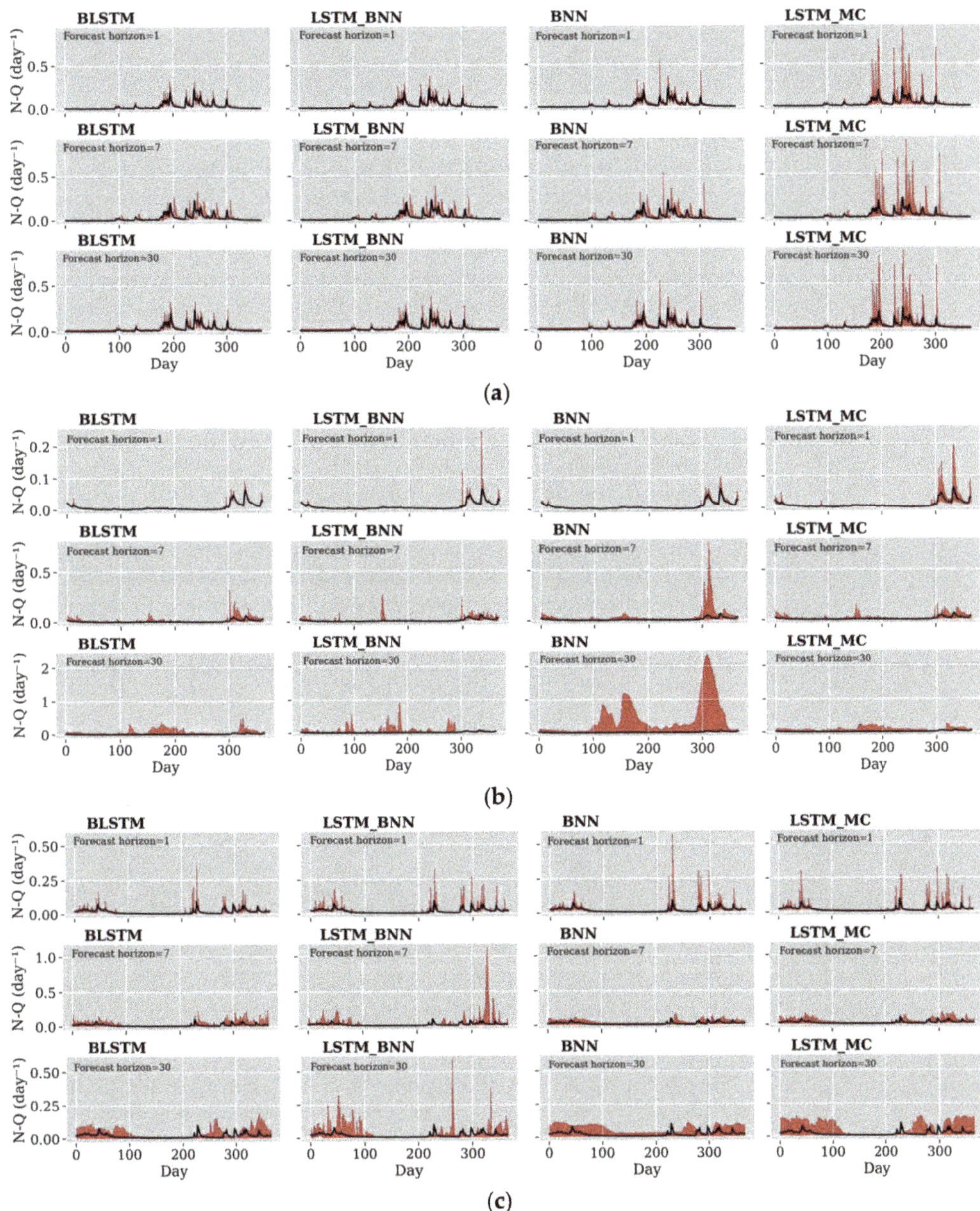

Figure 12. Prediction results of all models with 1, 7, and 30 days ahead forecasting for (**a**) case study I, (**b**) case study II, and (**c**) case study III.

Furthermore, we observed that all case studies can be effectively covered by the PI. Furthermore, for case study II, which had the lowest peak, BLSTM achieved the best results in all three scenarios. In contrast, case study I, with the highest peak at 1654 m^3/s and Std. of 90 m^3/s, achieved the worst prediction results for all scenarios. From Scenarios I to II in case study I for BLSTM, LSTM-BNN, BNN, and LSTM-MC, PICP decreased by approximately 25, 38, 48, and 53%, respectively, indicating the best performance of BLSTM in maintaining its predictability in case study I, with the highest peak and Std in the extended horizon prediction. Moreover, by increasing the horizon, prediction performance for case study I dramatically decreased, whereas, in terms of PICP for BLSTM, case studies II and III from Scenarios I to III decreased by approximately 1–2% and 2–4%, respectively. Furthermore, LSTM-MC and BNN achieved the worst overall prediction performance for all the scenarios. The catchment area of case study I was relatively large, and heavy rain was the primary source of streamflow. These two characteristics cause the seasonal and annual variations in streamflow to be greater than those in the other two case studies. In this case study, the streamflow was very stable and small during the dry season, whereas in the rainy season, the streamflow increased steeply and then decreased. This made forecasting challenging and resulted in a higher uncertainty than that in the other case studies. Therefore, for this type of catchment, using more in-situ meteorological predictors, such as precipitation and temperature, along with available high-resolution large-scale hydroclimate data, can improve forecasting accuracy.

The kernel density estimation plots of the daily streamflow prediction of all models for case study II are displayed in Figure 13a–c. As depicted, the kernel density estimation plots are on the top with boxplots, and the data points of the prediction are underneath. The boxes represent the inner quartiles, the vertical lines within the box indicate the median, and the diamonds represent the outliers in each model. As shown in Figure 13, the prediction variance of BLSTM is lower in comparison with the other models, in particular for the Scenario III which is forecasting horizon 30. Moreover, the inter-quartile range of BLSTM is smaller which indicates that the BLSTM prediction results has less dispersion while LSTM-MC has the highest dispersion. The results of this study indicate that BLSTM shows the best overall probabilistic prediction performance.

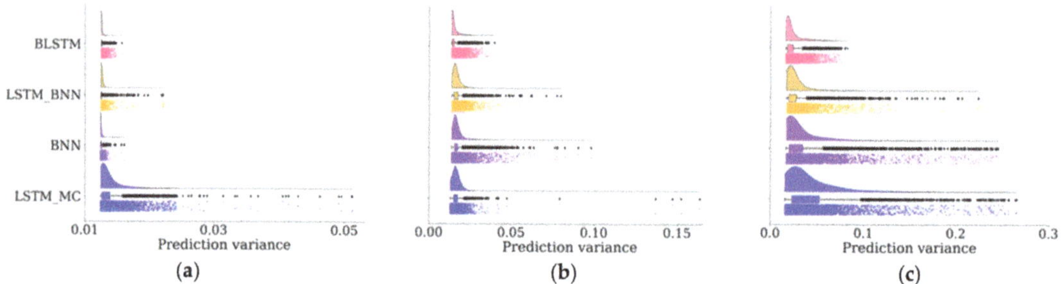

Figure 13. Kernel density estimation plots of daily streamflow prediction of all models in case study II for (**a**) forecast horizon 1, (**b**) forecast horizon 7, and (**c**) forecast horizon 30.

5. Conclusions

This study proposes BLSTM as a probabilistic prediction model to estimate streamflow uncertainty. For comparison, three probabilistic models and one deterministic model, including LSTM-BNN, BNN, LSTM-MC, and LSTM, are developed under three scenarios: 1 day, 7 days, and 30 days ahead daily streamflow forecasting. The results are compared in terms of reliability, sharpness, and overall performance for three different case studies in the USA. Reliability is measured by computing the PICP, sharpness is measured by computing the MPIW, and overall performance is measured by CRPS. The results show that all probabilistic models outperformed the deterministic model (LSTM). Moreover, among the probabilistic models, BLSTM is superior. The Bayesian LSTM achieves better results with less computing time and is easier to train than those of LSTM-BNN and BNN. The results reveal that the BLSTM network with variational inference achieves the highest accuracy. The fact that BLSTM shows the best predictive performance indicates the significance of capturing temporal dependencies by considering both uncertainties. Moreover, taking advantage of the long- and short-term dependencies and capturing the inherent uncertainty that is inevitable in hydrology provides better prediction results. For longer forecast horizons, models such as the BNN and LSTM-MC perform poorly due to the fact that the former is not an autoregressive model, and both have the gradient-vanishing problem in the long sequence time series. In addition, the issue of cumulative error in multi-step ahead recursive model is inevitable. Future research will investigate an enhanced network structure with a large prediction capacity, such as attention-based and parallel-feed architectures, to handle the long sequence time-series forecasting. In addition to the recursive models, other multi-step ahead prediction strategies, such as direct and hybrid, can be studied to minimize the accumulated error issue for longer horizons forecasting. Moreover, in the relevant future work, meteorological parameters such as precipitation, temperature, and humidity will be included as input to allow the model to detect the complexity necessary to enhance the accuracy of prediction, particularly for the longer horizon, and to evaluate the effect of multivariate input on model uncertainty.

Supplementary Materials: The following supporting information can be downloaded at: https://www.mdpi.com/article/10.3390/w14223672/s1, Figure S1: Autocorrelation function plots of transformed-streamflow timeseries; Figure S2: The visualizations of network execution graphs and traces for BLSTM models' outputs; Figure S3: The visualizations of network execution graphs and traces for LSTM-BNN models' outputs; Figure S4: The visualizations of network execution graphs and traces for BNN models' outputs; Figure S5: The visualizations of network execution graphs and traces for LSTM-MC models' outputs; Figure S6: Prediction results of all models for case study II; Table S1: General structures of deep neural networks.

Author Contributions: F.G.: conceptualization, methodology, investigation, software, validation, formal analysis, data curation, writing—original draft, writing—review and editing, visualization. D.K.: supervision, validation, writing—review and editing, resources, and funding acquisition. All authors have read and agreed to the published version of the manuscript.

Funding: This research was funded by (1) the Korea Ministry of Environment (MOE) as "Graduate School specialized in Climate Change" and (2) the Korea Institute of Energy Technology Evaluation and Planning (KETEP) and the Korea Ministry of Trade, Industry & Energy (MOTIE) grant number [20224000000260].

Data Availability Statement: Data will be made available on request. Daily historical streamflow data for the three selected case studies were obtained from the United States Geological Survey (USGS) website, accessed on 1 February 2022 (https://waterdata.usgs.gov/nwis).

Conflicts of Interest: The authors declare no conflict of interest.

References

1. Ghobadi, F.; Kang, D. Improving Long-Term Streamflow Prediction in a Poorly Gauged Basin Using Geo-Spatiotemporal Mesoscale Data and Attention-Based Deep Learning: A Comparative Study. *J. Hydrol.* **2022**, *615*, 128608. [CrossRef]
2. Wang, Y.; Liu, J.; Li, R.; Suo, X.; Lu, E.H. Medium and Long-Term Precipitation Prediction Using Wavelet Decomposition-Prediction-Reconstruction Model. *Water Resour. Manag.* **2022**, *36*, 971–987. [CrossRef]
3. Dikshit, A.; Pradhan, B.; Santosh, M. Artificial Neural Networks in Drought Prediction in the 21st Century–A Scientometric Analysis. *Appl. Soft Comput.* **2022**, *114*, 108080. [CrossRef]
4. Van den Hurk, B.J.J.M.; Bouwer, L.M.; Buontempo, C.; Döscher, R.; Ercin, E.; Hananel, C.; Hunink, J.E.; Kjellström, E.; Klein, B.; Manez, M.; et al. Improving Predictions and Management of Hydrological Extremes through Climate Services: Www.Imprex.Eu. *Clim. Serv.* **2016**, *1*, 6–11. [CrossRef]
5. Lange, H.; Sippel, S. *Machine Learning Applications in Hydrology BT—Forest-Water Interactions*; Levia, D.F., Carlyle-Moses, D.E., Iida, S., Michalzik, B., Nanko, K., Tischer, A., Eds.; Springer International Publishing: Cham, Switzerland, 2020; pp. 233–257. ISBN 978-3-030-26086-6.
6. Lin, Y.; Wang, D.; Wang, G.; Qiu, J.; Long, K.; Du, Y.; Xie, H.; Wei, Z.; Shangguan, W.; Dai, Y. A Hybrid Deep Learning Algorithm and Its Application to Streamflow Prediction. *J. Hydrol.* **2021**, *601*, 126636. [CrossRef]
7. Hagen, J.S.; Leblois, E.; Lawrence, D.; Solomatine, D.; Sorteberg, A. Identifying Major Drivers of Daily Streamflow from Large-Scale Atmospheric Circulation with Machine Learning. *J. Hydrol.* **2021**, *596*, 126086. [CrossRef]
8. Ren, K.; Wang, X.; Shi, X.; Qu, J.; Fang, W. Examination and Comparison of Binary Metaheuristic Wrapper-Based Input Variable Selection for Local and Global Climate Information-Driven One-Step Monthly Streamflow Forecasting. *J. Hydrol.* **2021**, *597*, 126152. [CrossRef]

9. Tayerani Charmchi, A.S.; Ifaei, P.; Yoo, C.K. Smart Supply-Side Management of Optimal Hydro Reservoirs Using the Water/Energy Nexus Concept: A Hydropower Pinch Analysis. *Appl. Energy* **2021**, *281*, 116136. [CrossRef]
10. Papacharalampous, G.; Tyralis, H. A Review of Machine Learning Concepts and Methods for Addressing Challenges in Probabilistic Hydrological Post-Processing and Forecasting. *arXiv* **2022**. [CrossRef]
11. Ghimire, S.; Yaseen, Z.M.; Farooque, A.A.; Deo, R.C.; Zhang, J.; Tao, X. Streamflow Prediction Using an Integrated Methodology Based on Convolutional Neural Network and Long Short-Term Memory Networks. *Sci. Rep.* **2021**, *11*, 17497. [CrossRef]
12. Klotz, D.; Kratzert, F.; Gauch, M.; Keefe Sampson, A.; Brandstetter, J.; Klambauer, G.; Hochreiter, S.; Nearing, G. Uncertainty Estimation with Deep Learning for Rainfall-Runoff Modeling. *Hydrol. Earth Syst. Sci.* **2022**, *26*, 1673–1693. [CrossRef]
13. Papacharalampous, G.; Tyralis, H.; Langousis, A.; Jayawardena, A.W.; Sivakumar, B.; Mamassis, N.; Montanari, A.; Koutsoyiannis, D. Probabilistic Hydrological Post-Processing at Scale: Why and How to Apply Machine-Learning Quantile Regression Algorithms. *Water* **2019**, *11*, 2126. [CrossRef]
14. Adnan, R.M.; Liang, Z.; Parmar, K.S.; Soni, K.; Kisi, O. Modeling Monthly Streamflow in Mountainous Basin by MARS, GMDH-NN and DENFIS Using Hydroclimatic Data. *Neural Comput. Appl.* **2021**, *33*, 2853–2871. [CrossRef]
15. Apaydin, H.; Taghi Sattari, M.; Falsafian, K.; Prasad, R. Artificial Intelligence Modelling Integrated with Singular Spectral Analysis and Seasonal-Trend Decomposition Using Loess Approaches for Streamflow Predictions. *J. Hydrol.* **2021**, *600*, 126506. [CrossRef]
16. Mehdizadeh, S.; Fathian, F.; Safari, M.J.S.; Adamowski, J.F. Comparative Assessment of Time Series and Artificial Intelligence Models to Estimate Monthly Streamflow: A Local and External Data Analysis Approach. *J. Hydrol.* **2019**, *579*, 124225. [CrossRef]
17. Nanda, T.; Sahoo, B.; Chatterjee, C. Enhancing Real-Time Streamflow Forecasts with Wavelet-Neural Network Based Error-Updating Schemes and ECMWF Meteorological Predictions in Variable Infiltration Capacity Model. *J. Hydrol.* **2019**, *575*, 890–910. [CrossRef]
18. Khosravi, K.; Golkarian, A.; Tiefenbacher, J.P. Using Optimized Deep Learning to Predict Daily Streamflow: A Comparison to Common Machine Learning Algorithms. *Water Resour. Manag.* **2022**, *36*, 699–716. [CrossRef]
19. Xu, W.; Chen, J.; Zhang, X.J. Scale Effects of the Monthly Streamflow Prediction Using a State-of-the-Art Deep Learning Model. *Water Resour. Manag.* **2022**, *36*, 3609–3625. [CrossRef]
20. Cheng, M.; Fang, F.; Kinouchi, T.; Navon, I.M.; Pain, C.C. Long Lead-Time Daily and Monthly Streamflow Forecasting Using Machine Learning Methods. *J. Hydrol.* **2020**, *590*, 125376. [CrossRef]
21. Cui, Z.; Zhou, Y.; Guo, S.; Wang, J.; Xu, C.Y. Effective Improvement of Multi-Step-Ahead Flood Forecasting Accuracy through Encoder-Decoder with an Exogenous Input Structure. *J. Hydrol.* **2022**, *609*, 127764. [CrossRef]
22. Yin, H.; Zhang, X.; Wang, F.; Zhang, Y.; Xia, R.; Jin, J. Rainfall-Runoff Modeling Using LSTM-Based Multi-State-Vector Sequence-to-Sequence Model. *J. Hydrol.* **2021**, *598*, 126378. [CrossRef]
23. Kao, I.F.; Zhou, Y.; Chang, L.C.; Chang, F.J. Exploring a Long Short-Term Memory Based Encoder-Decoder Framework for Multi-Step-Ahead Flood Forecasting. *J. Hydrol.* **2020**, *583*, 124631. [CrossRef]
24. Babaeian, E.; Paheding, S.; Siddique, N.; Devabhaktuni, V.K.; Tuller, M. Short- and Mid-Term Forecasts of Actual Evapotranspiration with Deep Learning. *J. Hydrol.* **2022**, *612*, 128078. [CrossRef]
25. Xiang, Z.; Yan, J.; Demir, I. A Rainfall-Runoff Model With LSTM-Based Sequence-to-Sequence Learning. *Water Resour. Res.* **2020**, *56*, e2019WR025326. [CrossRef]
26. Ferreira, L.B.; da Cunha, F.F. Multi-Step Ahead Forecasting of Daily Reference Evapotranspiration Using Deep Learning. *Comput. Electron. Agric.* **2020**, *178*, 105728. [CrossRef]
27. Granata, F.; Di Nunno, F.; de Marinis, G. Stacked Machine Learning Algorithms and Bidirectional Long Short-Term Memory Networks for Multi-Step Ahead Streamflow Forecasting: A Comparative Study. *J. Hydrol.* **2022**, *613*, 128431. [CrossRef]
28. Masrur Ahmed, A.A.; Deo, R.C.; Feng, Q.; Ghahramani, A.; Raj, N.; Yin, Z.; Yang, L. Deep Learning Hybrid Model with Boruta-Random Forest Optimiser Algorithm for Streamflow Forecasting with Climate Mode Indices, Rainfall, and Periodicity. *J. Hydrol.* **2021**, *599*, 126350. [CrossRef]
29. Rahimzad, M.; Moghaddam Nia, A.; Zolfonoon, H.; Soltani, J.; Danandeh Mehr, A.; Kwon, H.H. Performance Comparison of an LSTM-Based Deep Learning Model versus Conventional Machine Learning Algorithms for Streamflow Forecasting. *Water Resour. Manag.* **2021**, *35*, 4167–4187. [CrossRef]
30. Barzegar, R.; Aalami, M.T.; Adamowski, J. Coupling a Hybrid CNN-LSTM Deep Learning Model with a Boundary Corrected Maximal Overlap Discrete Wavelet Transform for Multiscale Lake Water Level Forecasting. *J. Hydrol.* **2021**, *598*, 126196. [CrossRef]
31. Granata, F.; Di Nunno, F. Forecasting Evapotranspiration in Different Climates Using Ensembles of Recurrent Neural Networks. *Agric. Water Manag.* **2021**, *255*, 107040. [CrossRef]
32. Zheng, Y.; Xie, Y.; Long, X. A Comprehensive Review of Bayesian Statistics in Natural Hazards Engineering. *Nat. Hazards* **2021**, *108*, 63–91. [CrossRef]
33. Han, S.; Coulibaly, P. Bayesian Flood Forecasting Methods: A Review. *J. Hydrol.* **2017**, *551*, 340–351. [CrossRef]
34. Costa, V.; Fernandes, W. Bayesian Estimation of Extreme Flood Quantiles Using a Rainfall-Runoff Model and a Stochastic Daily Rainfall Generator. *J. Hydrol.* **2017**, *554*, 137–154. [CrossRef]
35. Xu, X.; Zhang, X.; Fang, H.; Lai, R.; Zhang, Y.; Huang, L.; Liu, X. A Real-Time Probabilistic Channel Flood-Forecasting Model Based on the Bayesian Particle Filter Approach. *Environ. Model. Softw.* **2017**, *88*, 151–167. [CrossRef]

36. Huang, Y.; Shao, C.; Wu, B.; Beck, J.L.; Li, H. State-of-the-Art Review on Bayesian Inference in Structural System Identification and Damage Assessment. *Adv. Struct. Eng.* **2019**, *22*, 1329–1351. [CrossRef]
37. Goodarzi, L.; Banihabib, M.E.; Roozbahani, A.; Dietrich, J. Bayesian Network Model for Flood Forecasting Based on Atmospheric Ensemble Forecasts. *Nat. Hazards Earth Syst. Sci.* **2019**, *19*, 2513–2524. [CrossRef]
38. Bai, H.; Li, G.; Liu, C.; Li, B.; Zhang, Z.; Qin, H. Hydrological Probabilistic Forecasting Based on Deep Learning and Bayesian Optimization Algorithm. *Hydrol. Res.* **2021**, *52*, 927–943. [CrossRef]
39. Zhu, S.; Luo, X.; Yuan, X.; Xu, Z. An Improved Long Short-Term Memory Network for Streamflow Forecasting in the Upper Yangtze River. *Stoch. Environ. Res. Risk Assess.* **2020**, *34*, 1313–1329. [CrossRef]
40. Gude, V.; Corns, S.; Long, S. Flood Prediction and Uncertainty Estimation Using Deep Learning. *Water* **2020**, *12*, 884. [CrossRef]
41. Althoff, D.; Rodrigues, L.N.; Bazame, H.C. Uncertainty Quantification for Hydrological Models Based on Neural Networks: The Dropout Ensemble. *Stoch. Environ. Res. Risk Assess.* **2021**, *35*, 1051–1067. [CrossRef]
42. Li, D.; Marshall, L.; Liang, Z.; Sharma, A. Hydrologic Multi-Model Ensemble Predictions Using Variational Bayesian Deep Learning. *J. Hydrol.* **2022**, *604*, 127221. [CrossRef]
43. He, Y.; Fan, H.; Lei, X.; Wan, J. A Runoff Probability Density Prediction Method Based on B-Spline Quantile Regression and Kernel Density Estimation. *Appl. Math. Model.* **2021**, *93*, 852–867. [CrossRef]
44. Lu, D.; Konapala, G.; Painter, S.L.; Kao, S.C.; Gangrade, S. Streamflow Simulation in Data-Scarce Basins Using Bayesian and Physics-Informed Machine Learning Models. *J. Hydrometeorol.* **2021**, *22*, 1421–1438. [CrossRef]
45. Sun, M.; Zhang, T.; Wang, Y.; Strbac, G.; Kang, C. Using Bayesian Deep Learning to Capture Uncertainty for Residential Net Load Forecasting. *IEEE Trans. Power Syst.* **2020**, *35*, 188–201. [CrossRef]
46. Wang, Y.; Gan, D.; Sun, M.; Zhang, N.; Lu, Z.; Kang, C. Probabilistic Individual Load Forecasting Using Pinball Loss Guided LSTM. *Appl. Energy* **2019**, *235*, 10–20. [CrossRef]
47. Toubeau, J.F.; Bottieau, J.; Vallee, F.; De Greve, Z. Deep Learning-Based Multivariate Probabilistic Forecasting for Short-Term Scheduling in Power Markets. *IEEE Trans. Power Syst.* **2019**, *34*, 1203–1215. [CrossRef]
48. Wang, H.; Yi, H.; Peng, J.; Wang, G.; Liu, Y.; Jiang, H.; Liu, W. Deterministic and Probabilistic Forecasting of Photovoltaic Power Based on Deep Convolutional Neural Network. *Energy Convers. Manag.* **2017**, *153*, 409–422. [CrossRef]
49. Xu, L.; Hu, M.; Fan, C. Probabilistic Electrical Load Forecasting for Buildings Using Bayesian Deep Neural Networks. *J. Build. Eng.* **2022**, *46*, 103853. [CrossRef]
50. Van der Meer, D.W.; Shepero, M.; Svensson, A.; Widén, J.; Munkhammar, J. Probabilistic Forecasting of Electricity Consumption, Photovoltaic Power Generation and Net Demand of an Individual Building Using Gaussian Processes. *Appl. Energy* **2018**, *213*, 195–207. [CrossRef]
51. Zhang, Z.; Zhang, Q.; Singh, V.P. Univariate Streamflow Forecasting Using Commonly Used Data-Driven Models: Literature Review and Case Study. *Hydrol. Sci. J.* **2018**, *63*, 1091–1111. [CrossRef]
52. Lange, H.; Sippel, S. Machine Learning Applications in Hydrology. In *Forest-Water Interactions*; Ecological Studies, Volume 240; Springer: Cham, Switzerland, 2020; pp. 233–257.
53. Abdul Kareem, B.; Zubaidi, S.L.; Ridha, H.M.; Al-Ansari, N.; Al-Bdairi, N.S.S. Applicability of ANN Model and CPSOCGSA Algorithm for Multi-Time Step Ahead River Streamflow Forecasting. *Hydrology* **2022**, *9*, 171. [CrossRef]
54. Wegayehu, E.B.; Muluneh, F.B. Short-Term Daily Univariate Streamflow Forecasting Using Deep Learning Models. *Adv. Meteorol.* **2022**, *2022*, 1860460. [CrossRef]
55. Kendall, A.; Gal, Y. What Uncertainties Do We Need in Bayesian Deep Learning for Computer Vision? *Adv. Neural Inf. Process. Syst.* **2017**, *2017*, 5575–5585. [CrossRef]
56. Blundell, C.; Cornebise, J.; Kavukcuoglu, K.; Wierstra, D. Weight Uncertainty in Neural Networks. *arXiv* **2015**. [CrossRef]
57. Hochreiter, S.; Schmidhuber, J. Long Short-Term Memory. *Neural Comput.* **1997**, *9*, 1735–1780. [CrossRef] [PubMed]
58. Li, G.; Yang, L.; Lee, C.G.; Wang, X.; Rong, M. A Bayesian Deep Learning RUL Framework Integrating Epistemic and Aleatoric Uncertainties. *IEEE Trans. Ind. Electron.* **2021**, *68*, 8829–8841. [CrossRef]
59. Bernardo, J.M.; Smith, A.F.M. *Bayesian Theory*; Wiley Blackwell: Hoboken, NJ, USA, 2008; ISBN 9780470316870.
60. Runnalls, A.R. Kullback-Leibler Approach to Gaussian Mixture Reduction. *IEEE Trans. Aerosp. Electron. Syst.* **2007**, *43*, 989–999. [CrossRef]
61. Jospin, L.V.; Buntine, W.; Boussaid, F.; Laga, H.; Bennamoun, M. Hands-on Bayesian Neural Networks–A Tutorial for Deep Learning Users. *IEEE Comput. Intell. Mag.* **2020**, *17*, 29–48. [CrossRef]
62. Gal, Y.; Ghahramani, Z. Dropout as a Bayesian Approximation: Representing Model Uncertainty in Deep Learning. *33rd Int. Conf. Mach. Learn. ICML* **2016**, *3*, 1651–1660. [CrossRef]
63. Gal, Y. Uncertainty in Deep Learning. Ph.D. Thesis, University of Cambridge, Cambridge, UK, 2016.
64. Abdar, M.; Pourpanah, F.; Hussain, S.; Rezazadegan, D.; Liu, L.; Ghavamzadeh, M.; Fieguth, P.; Cao, X.; Khosravi, A.; Acharya, U.R.; et al. A Review of Uncertainty Quantification in Deep Learning: Techniques, Applications and Challenges. *Inf. Fusion* **2021**, *76*, 243–297. [CrossRef]
65. Jian, X.; Wolock, D.M.; Lins, H.F.; Henderson, R.J.; Brady, S.J. *Streamflow—Water Year 2021: U.S. Geological Survey Fact Sheet 2022–3072*; USGS: Reston, VA, USA, 2022.

66. Chen, R.; Cao, J.; Zhang, D. Probabilistic Prediction of Photovoltaic Power Using Bayesian Neural Network-LSTM Model. In Proceedings of the 2021 IEEE 4th International Conference on Renewable Energy and Power Engineering (REPE), Beijing, China, 9–11 October 2021; pp. 294–299. [CrossRef]

67. Srivastava, N.; Hinton, G.; Krizhevsky, A.; Salakhutdinov, R. Dropout: A Simple Way to Prevent Neural Networks from Overfitting. *J. Mach. Learn. Res.* **2014**, *15*, 1929–1958.

68. Fortunato, M.; Blundell, C.; Vinyals, O. Bayesian Recurrent Neural Networks. *arXiv* **2019**. [CrossRef]

69. Ketkar, N. Introduction to Keras. In *Deep Learning with Python*; Ketkar, N., Ed.; Apress: Berkeley, CA, USA, 2017; pp. 97–111. ISBN 978-1-4842-2766-4.

70. Abadi, M.; Barham, P.; Chen, J.; Chen, Z.; Davis, A.; Dean, J.; Devin, M.; Ghemawat, S.; Irving, G.; Isard, M.; et al. TensorFlow: A System for Large-Scale Machine Learning. In Proceedings of the 12th USENIX Symposium on Operating Systems Design and Implementation (OSDI 16), Savannah, GA, USA, 2–4 November 2016; pp. 265–283.

71. PyTorch Documentation—PyTorch 1.13 Documentation. Available online: https://pytorch.org/docs/stable/index.html (accessed on 7 November 2022).

 *water*

Article

A Continuous Multisite Multivariate Generator for Daily Temperature Conditioned by Precipitation Occurrence

Joel Hernández-Bedolla [1], Abel Solera [2], Javier Paredes-Arquiola [2], Sonia Tatiana Sanchez-Quispe [1,*] and Constantino Domínguez-Sánchez [1]

[1] Faculty of Civil Engineering, Michoacan University of Saint Nicolas of Hidalgo, Morelia 58030, Mexico
[2] Research Institute of Water and Environmental Engineering (IIAMA), Universitat Politècnica de València, 46022 Valencia, Spain
* Correspondence: quispe@umich.mx; Tel.: +52-443-246-6616

Abstract: Temperature is one of the most influential weather variables necessary for numerous studies, such as climate change, integrated water resources management, and water scarcity, among others. The temperature and precipitation are relevant in river basins because they may be particularly affected by modifications in the variability, for example, due to climate change. We developed a stochastic model for daily precipitation occurrences and their influence on maximum and minimum temperatures with a straightforward approach. The Markov model has been used to determine everyday occurrences of rainfall. Moreover, we developed a multisite multivariate autoregressive model to represent the short-term memory of daily temperature, called MASCV. The reduction of parameters is an essential factor addressed in this approach. For this reason, the normalization of the temperatures was performed through different nonparametric transformations. The case study is the Jucar River Basin in Spain. The multisite multivariate stochastic model of two states and a lag-one accurately represents both occurrences as well as maximum and minimum temperature. The simulation and generation of occurrences and temperature is considered a continuous multivariate stochastic process. Additionally, time series of multiple correlated climate variables are completed. Therefore, we simplify the complexity and reduce the computational time for the simulation.

Keywords: multivariate stochastic model; autoregressive model; Markov model; daily temperature; temperature generator

Citation: Hernández-Bedolla, J.; Solera, A.; Paredes-Arquiola, J.; Sanchez-Quispe, S.T.; Domínguez-Sánchez, C. A Continuous Multisite Multivariate Generator for Daily Temperature Conditioned by Precipitation Occurrence. *Water* **2022**, *14*, 3494. https://doi.org/10.3390/w14213494

Academic Editors: Fi-John Chang, Li-Chiu Chang and Jui-Fa Chen

Received: 11 September 2022
Accepted: 24 October 2022
Published: 1 November 2022

1. Introduction

The stochastic modeling approach has been widely used for hydrologic time series analysis, including modeling weather variables [1,2] and flood prediction [3]. Therefore, it is essential to build accurate forecast models in the hydrologic process [4]. The stochastic modeling of temperature conditioned by precipitation is proposed when a day is wet or dry. It is necessary to estimate the temperature for both dry and wet days.

The most common stochastic model is the first-order Markov model with two states. It was introduced by Gabriel and Newman [5] and has since been used and modified by many authors [3–19]. The rainfall occurrence process focuses on representing the dry and wet days. Critical probability (p_c) depends on transition probabilities of dry–wet (p_{01}) and wet–wet days (p_{11}) [8]. Thus, the occurrence is a bivariate function (0,1) that relies on the uniform random process (u) and critical probabilities [12].

The temperature displays a near-normal distribution. It is common to consider the normal distribution to minimize the skewness coefficient of observed data. In other cases, root transformation is used [20].

The statistical characteristics of the series change throughout the whole year. The mean, variance, and standard deviation are periodic—these statistical changes are yearly, monthly, daily, and even hourly. This periodicity is commonly modeled by the finite Fourier series [9,16,21,22]. The periodic component is necessary for standardization, which is a process that converts into a series with a mean of zero and a standard deviation of one. Standardization is usually applied to series with no follow-up normal distributions [7,23–25]. Moreover, standardization helps remove the seasonal effects [26]. Standardized series are used to calculate the autoregressive parameters of the stochastic model [27].

It is desirable to generate synthetic temperature based on weather data [9]. The process should be capable of representing different statistics from historical data [18]. The basic statistics of the series are the sample mean, standard deviation, skewness coefficient, cross-correlation coefficients, and probability distribution. We must fit a probability distribution function (two or more parameters) or use a nonparametric transformation. Many authors have proposed stochastic weather generators (WG) for different purposes, a few of which are downscaling, prediction, or simulation [28]. Some developed models are Weather GENerator (WGEN) [29], CLImate GENerator (CLIGEN) [30], Agriculture and Agri-Food Canada Weather Generator (AAA-FC) [27], CLIMate GENerator (CLIMGEN) [31], Long Ashton Research Station-Weather Generator (LARS-WG) [21], the weather generator CLIMA [32], the closed skew-normal weather generator (WACS-Gen) [26], École de technologie supérieure Weather Generator (WeaGETS) [22], and the Multi-site Rainfall Simulator (MRS) [18]. We propose the multivariate autoregressive model of climate variables (MASCV).

For precipitation, weather generators analyze the occurrence similarly. For example, CLIMA [32], CLIGEN [30], and WGEN [9] model the occurrence process based on transition probabilities of lag-one and two states for each month (two parameters). LARS-WG [21] is based on an empirical function of wet and dry days for each month (21 parameters). CLIMGEN [31] models the occurrence process through transition probabilities of a second-order Markov model at a monthly scale (four parameters). WeaGETS [22] models the transition probabilities of a third-order Markov model with two states on a biweekly scale (eight parameters). WASC-Gen uses a lag-one Markov model with different states and parameters derived from the Bayesian information criterion [26]. AA-FC-WG models the occurrence process with a second-order Markov model at a monthly scale using two parameters [27]. Other stochastic models have proposed modifications or new methods, such as the Markov renewal model [33], Copula-based models [34], the Hidden Markov model [35,36], the Semi-Markov model [37], generalized chain-dependent process models [38], the autoregressive model [39], and artificial intelligence [40].

Artificial intelligence (AI) has been used to predict hydrological data, including rainfall, rainfall runoff, and temperature [4]. Various machine learning methods have been applied for rainfall and temperature prediction, such as the support vector machine (SVM) classifier [41], ANN [42], long short-term memory (LSTM), statistical and multiple linear regression (MLR) [43,44], and classification and regression trees [45].

Stochastic weather generators are used to represent different variables such as precipitation and temperature. Long temperature series are commonly required, but historical data are sometimes short in length. Therefore, it is necessary to apply stochastic weather generators for hydrological design [12]. The integrated water resources management (IWRM) depends on water availability [46]. For the hydric balance and various decisions at a river basin scale, the rainfall is fundamental [10], primarily to generate long synthetic series.

The issue with synthetic series is preserving the low-frequency variability in temperature and rainfall [47]. Daily temperature series have a strong correlation [48]. To preserve low-frequency variability, different authors have proposed the fast Fourier transform (FFT) algorithm [49], using crop variables [50], annual and monthly autoregressive models [51], k-nearest neighbor bootstrap [52], and the copulas approach [53].

The objective of this paper was to develop an adequate stochastic model for daily maximum and minimum temperature with a straightforward approach. We propose a multisite multivariate stochastic model for precipitation occurrences and temperature capable of reproducing spatial and temporal dependence. We corrected the daily stochastic model using an annual multisite multivariate autoregressive model. For precipitation, we applied a multivariate multisite lag-one Markov model with two states (wet and dry) and few parameters. A normal distribution was used to define the precipitation occurrence process. For the maximum and minimum temperature, we implemented the first-order multisite multivariate autoregressive model (MAR(p)) for daily and annual temperature. Our approach simplifies the temperature simulation (continuous and nonparametric), which implies a considerable advantage and versatility concerning other stochastic generators (parametric distribution function for each month or biweekly).

The normalized mean and variance periodicity was modeled as a continuous daily scale through the Fourier series. The model validation was performed by synthetic series of precipitation, which accurately represent the main statistics of observed data for different climates. MASCV was programmed in MATLAB and is capable of simulating with various parameters for both occurrences and temperature. Moreover, we proposed a multisite multivariate stochastic model in which different wet thresholds and nonparametric transformations can be applied. A relevant advantage is that we modeled continuously for the whole series. The method developed in this paper generates constant day-to-day rainfall occurrences and temperature.

2. Materials and Methods

The purpose of generating long series of temperatures is to evaluate the effects of hydrologic changes [4,9] and analyze different scenarios, such as environmental, agricultural [28], or climate change. We focused on developing long multivariate synthetic series of maximum and minimum temperature. The lag-one Markov model has been applied for plenty of weather modelers due to its accurate representation of dependence [12]. Said model is based on terms of occurrence given the previous day. We modeled the rainfall occurrence and the temperature separately (Figure 1), so we defined the model according to Equation (1):

$$Y_t = T_t X_t \tag{1}$$

where T_t is the temperature model, X_t is the precipitation occurrence model, and Y_t is the whole stochastic process.

2.1. Multivariate Precipitation Occurrence (Dry–Wet)

First, we must determine whether a day is wet or dry according to the day prior. We used a bivariate function (X_t), and if the precipitation is greater than a given threshold, the t day is wet ($X_t = 1$); otherwise, the t day is dry ($X_t = 0$). The first-order Markovian approach only depends on the previous day, whether it was wet or dry. The high-order Markov model has been notably studied [54,55] and is recommended for vast persistence [10]. The two- and three-order Markov models significantly improve the fit [22,56,57]. In other cases, the results are nearly equivalent to the first-order Markov model [54]. One disadvantage of the high-order Markov model is the increase in the number of parameters [28]. The threshold for precipitation occurrence could change according to different data. This depends on the minimum value of precipitation amount that the station in the study can report. Common threshold values are 0 [9,21,29,30], 0. [22,26,58], 0.2 [25], 0.254, and 0.3 mm [18]. We used the wet thresholds (0.001, 0.01, 0.1, and 0.25) and identified the ones that provided the best results. A disadvantage in the case of stochastic modeling is the limited data from the historical period.

For the occurrence process, we used the Wilks approach [12]. The conditional probabilities for a one-order Markov model are a dry day followed by a wet day, p_{01}, a wet day followed by a wet day, p_{11}, a wet day followed by a dry day, p_{10}, and a dry day followed by a dry day, p_{00}. The conditional probabilities have a complementary relationship between them.

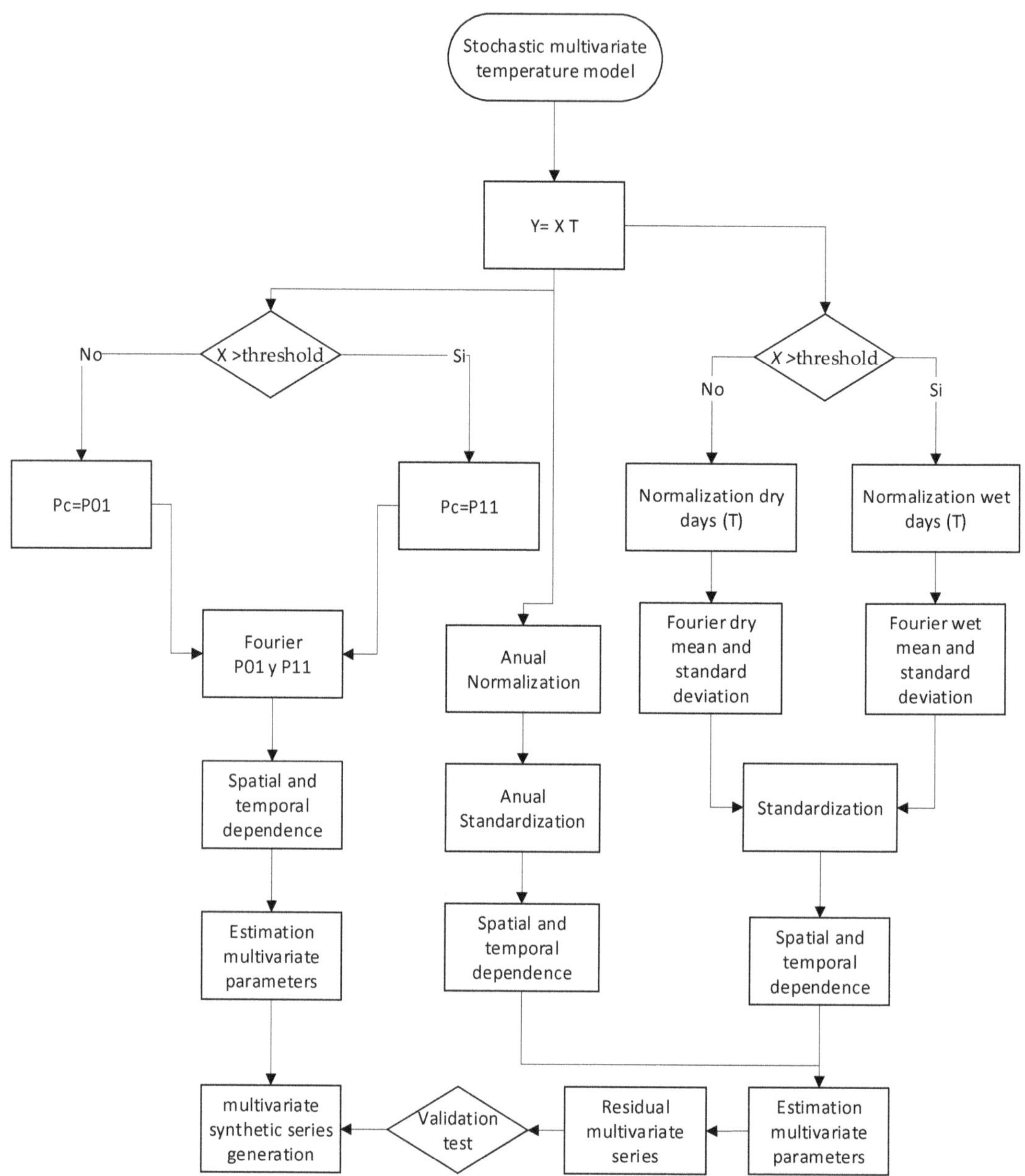

Figure 1. Proposed methodology for multisite multivariate precipitation occurrence, daily and annual temperatures.

The normal critical probability (p_c) depends on the previous day (X_{t-1}); if it is dry, $p_c = p_{01}$, otherwise $p_c = p_{11}$. The boundary for the precipitation occurrence process is determined for a random spatially correlated normal matrix $n_t = \varnothing^{-1} u_t\ [0,1]$. Only one day is computed if the random variable is equal to or less than a critical probability.

$$if \quad n_t \leq p_c \therefore X_t = 1; \qquad n_t > p_c \therefore X_t = 0 \tag{2}$$

The transition probability vectors vary on a monthly and daily scale. Therefore, the occurrence process was simulated through different parameters. The lag-one Markov model with two states enables the interaction between wet and dry days. Moreover, it allows the residence time and periodicity in both the first (dry) and second (wet) states.

For the daily occurrence process, one of the disadvantages is the limited number of data, primarily for dry seasons and arid river basins. These data limit the quality of results, and it is common to model the transition probabilities monthly or biweekly [9,22,26,27,30,31].

Our approach was to simulate the transition probabilities on a daily scale, similar to Woolhiser and Pegram [59]. It is imperative to preserve most of the characteristics for days, months, and years. The number of parameters increases from a monthly to a daily scale; therefore, it is essential to determine the most efficient model. In this study, we focused on the relevance of evaluating different analyses, which depend on the number of parameters, $p_{01\tau} = P_{01}|_{\tau}$ and $p_{01\tau} = P_{11}|_{\tau}$.

Due to the distinct temporal scales of analysis, the number of parameters will increase according to the variability of the occurrence process. The transition probabilities follow a uniform distribution applicable in the case of univariate modeling. However, in the case of multivariate modeling, a transformation to the normal distribution is more efficient [15]. A transformation was performed from uniform to normal, $p_{n11\tau} = \varnothing^{-1}(p_{11\tau})$ and $p_{n01\tau} = \varnothing^{-1}(p_{01\tau})$. The spatial correlation between probability vectors with lag-k is shown in Equation (3):

$$r_k^{(i,j)}(p) = \left(p_{n11}^{(i,j)} - p_{n01}^{(i,j)}\right)^k i \neq j \tag{3}$$

where $r_k^{(i,j)}$ is the spatial correlation with lag-k, $p_{n11}^{(i,j)}$ is the normalized probability vector of two consecutive days of rainfall occurrence, and $p_{n01}^{(i,j)}$ is the normalized probability vector of a dry day followed by a wet day. Based on these vectors, a spatially correlated matrix was proposed in Equation (4):

$$M_k = \begin{bmatrix} r_k^{(1,1)}(p) & \cdots & r_k^{(1,n)}(p) \\ \vdots & \ddots & \vdots \\ r_k^{(n,1)}(p) & \cdots & r_k^{(n,n)}(p) \end{bmatrix} \tag{4}$$

where M_k is the cross-correlation matrix. The Cholesky factorization was used to determine the spatial dependence of the series [15,19,60]. For a positive definite matrix, the Cholesky factorization is a lower triangular matrix, $D = [M][M]'$. Multiplying this lower triangular matrix by a random normal matrix results in a random spatially correlated normal matrix, n, in Equation (5):

$$n = [M]'[N] = \begin{bmatrix} M'^{(1,1)} & \cdots & 0 \\ \vdots & \ddots & \vdots \\ M'^{(n,1)} & \cdots & M'^{(n,n)} \end{bmatrix} x \begin{bmatrix} N^{(1,1)} & \cdots & N^{(1,m)} \\ \vdots & \ddots & \vdots \\ N^{(n,1)} & \cdots & N^{(n,m)} \end{bmatrix} \tag{5}$$

where n is the random spatially correlated normal matrix, $M'^{(n,n)}$ is the lower triangular matrix for n series, and $N^{(n,m)}$ is the random normal matrix for n series and m days. We used this matrix to generate multivariate precipitation occurrences. In addition, they were used to determine synthetic wet and dry days.

2.2. Multivariate Maximum and Minimum Temperature

The maximum and minimum temperatures were modeled using the original maximum temperature data and the difference between the maximum and minimum temperature, Trange in Equation (6). The target of modeling the temperature range is to avoid negative values of the synthetic series [61].

$$Trange = Tmax - Tmin \tag{6}$$

The temperature process is commonly modeled considering normal distribution. However, according to our experience, in some cases, the maximum temperature and temperature range follow a normal distribution. We focused on the nonparametric transformations. These are practical solutions to model the temperature [25,56,62].

Seasonal variability is one of the most significant characteristics of the stochastic temperature process. The daily temperature shows recurrent changes within the year. Parameters such as the mean and standard deviation are periodic components [1]. The periodicity was analyzed through the Fourier series [59]. The number of parameters is reduced when these are periodic or seasonal. The Fourier series [23,59] is present in Equation (7). Several harmonics were used to represent 90% of the explicative variance of the observed data.

$$v = \overline{u} + \sum_{j=1}^{h} \left[A_j \cos\left(\frac{2\pi\tau}{w}\right) + B_j \sin\left(\frac{2\pi\tau}{w}\right) \right] \tag{7}$$

where $\overline{u}$ is the normalized temperature mean for wet ($\mu_{1\tau}$) and dry days ($\mu_{0\tau}$) and standard deviation for wet ($s_{1\tau}$) and dry days ($s_{0\tau}$). A and B are the Fourier coefficient vectors, j is the harmonic, and h is the total number of harmonics, equal to $(w-1)/2$ for even numbers and equal to $w/2$ for uneven numbers. For example, in a daily simulation, we have 365 days, and the maximum number of harmonics is 182. Important harmonics were selected according to the accumulated period-gram, defined as the ratio of mean standard deviation (MSD) of the harmonics to the observed series. We accepted 90% of the original data representation to select the significant harmonics applied at the mean, standard deviation, and transition probabilities ($\mu_{1\tau}$, $\mu_{0\tau}$, $s_{1\tau}$, $s_{0\tau}$, $p_{01\tau}$, $p_{11\tau}$).

Once the periodic component was modeled, we standardized both temperatures (maximum and range), allowing the analysis of temporal dependence. In addition, a standardized series served to generate residual series. Standardization removes the periodicity of the series based on the mean ($\mu_{1\tau}$, $\mu_{0\tau}$) and standard deviation ($s_{1\tau}$, $s_{0\tau}$). We determined the standardized series (z_τ) at a daily scale, canceling the mean and standard deviation of the normalized series according to Equation (8):

$$z_\tau = \frac{y_\tau - \mu_{0\tau}}{s_{0\tau}}; \quad y_\tau = 0; \quad z_\tau = \frac{y_\tau - \mu_{1\tau}}{s_{1\tau}} \quad y_\tau > 0; \tag{8}$$

A multivariate autoregressive model, MAR(1), with constant parameters was applied, in which the best fit needs to represent the conditions of temporal and spatial dependence. The first order of the multivariate autoregressive model was determined based on the Cholesky factorization. In the same way, temporal dependence was conditioned by precipitation occurrence in Equation (9):

$$[\phi]_1 = M_0 M_1; \quad [\phi]_0 [\phi]_0^T = M_0 - [\phi]_1 M_1^T; \quad [\phi]_0 [\phi]_0^T = D \tag{9}$$

where $[\phi]_1$ is the lag 1 autoregressive coefficient matrix, $[\phi]_0$ is the lag 0 autoregressive coefficient matrix, $[\phi]_0^T$ is the transposed matrix, M_0 and M_1 are the cross-correlation matrices, and D is the positive definite matrix. Finally, we determined the white noise based on the multivariate autoregressive coefficient matrix using Equation (10):

$$\varepsilon_\tau = [\phi]_0^{-1}\left(\{z\}_\tau - [\phi]_1\{z\}_{\tau-1}\right) \tag{10}$$

2.3. Evidence for the Goodness of Fit

The residual series must satisfy the residual normality ($\varepsilon \cong 0$, $s_\varepsilon \cong 1$), which is neither correlated ($r_k(\varepsilon) \cong 0$) nor has a biased judgment ($g_\varepsilon \cong 0$). The probability of the residual series is verified by the confidence interval limits as well as for the mean, skewness coefficient, standard deviation, and correlations to corroborate the residual series normality. It must satisfy the mean and correlation of the entire series within 95% of the confidence limits [63]. In the case of the standard deviation of the chi-squared test, it should comply with a 95% confidence for a normal distribution [64]. The skewness coefficient of the residual series must be within the confidence limits [63]. Another evaluated aspect in the stochastic process is the Akaike information criterion (AIC) [65].

2.4. Generation of Multivariate Synthetic Series

For the generation of synthetic series, the stochastic model is divided into two states, the precipitation occurrence (X_t) and the maximum temperature and temperature range (T_t). The precipitation occurrence will appear if the transition vector (p_c) is a random normally distributed number greater than the random normal distribution (n_t); in other words, if $n_t \leq p_c$, it reaches the wet state. The stochastic temperature process starts by generating a random number with a normal distribution (ε) and then obtains the coupled standardized series (z_t) and corrected using the annual low-frequency multivariate stochastic model. Finally, the inverse normalization (y_τ^{-1}) was used.

The obtainment of synthetic series through the stochastic process allows for validating the developed model. The multivariate series were generated with the same characteristics as the observed data. The statistical parameters of both series were assessed and should not be significantly different; consequently, the developed model can be validated. The main metrics were determined. Mean absolute error (MAE), root mean square error (RMSE), and percent error estimate (PE) were defined.

2.5. Study Area

The Jucar River Basin is part of the Jucar Basin Agency (JBA), located in the eastern portion of the Iberian Peninsula, Spain. The basin covers an area of approximately 22,291 km^2. Information regarding the zone was obtained from the official website of the confederation (www.chj.es). The most relevant surface runoff is the Jucar River, which captures the surface runoff of all sub-basins [66]. The most significant reservoirs are Alarcon (1088 hm^3) and Contreras (852 hm^3). The river rises from the Tragacete (1600 ms.n.m) and subsequently arrives at reservoirs Toba, Alarcon, Molinar, and Tous. The study area's limit ends where the Mediterranean Sea is reached (Figure 2). Rainfall in the Jucar River Basin has decreased since 1980 [67,68]. Temporal and spatial variation characteristics of meteorological elements in the Jucar River Basin are presented in Appendix A (Figures A1–A4).

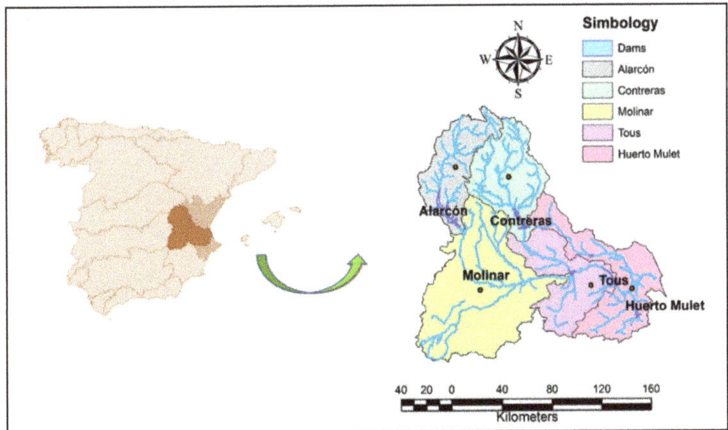

Figure 2. Location of Jucar River Basin.

The Jucar River Basin is divided into five sub-basins: Alarcon, Contreras, Molinar, Tous, and Huerto Mulet. The precipitation data for the study area were obtained from the Spain 02 database [69], a regular grid (20 × 20 km). The historical data have information from

1950 to 2015. The observed data were interpolated using the inverse distance weighting (IDW) method to generate a rainfall and temperature series for each sub-basin. The IDW method is one of the most common interpolation techniques [46,70–72].

3. Results

According to the information of the Jucar River Basin, the average annual temperature was between 17.5 °C (Contreras) and 21 °C for the years 1950–2015. In the present study, we defined four wet day thresholds (0.001, 0.01, 0.1, and 0.25 mm). The range of rainy days from October to May was between 7.3 and 12.17, and for June to September, the average precipitation occurrences were between 2.4 and 8.72 days. The months with few precipitation occurrences were between June and September, an important factor because there were little data for the stochastic modeling process. In the case of the months from October to May, the information was essential for the stochastic modeling to perform with better confidence.

3.1. Multivariate Occurrence Synthetic Series

The occurrence process was developed through a Markov model of two states. The transition probability vectors were identified, and the daily noise level can be observed. The transition probabilities for all sub-basins $p_{01\tau}$ were in a range of 0.05 (minimum) to 0.45 (maximum). On the other hand, the transition probabilities $p_{11\tau}$ were between 0.3 and 0.95 (according to Figure 3).

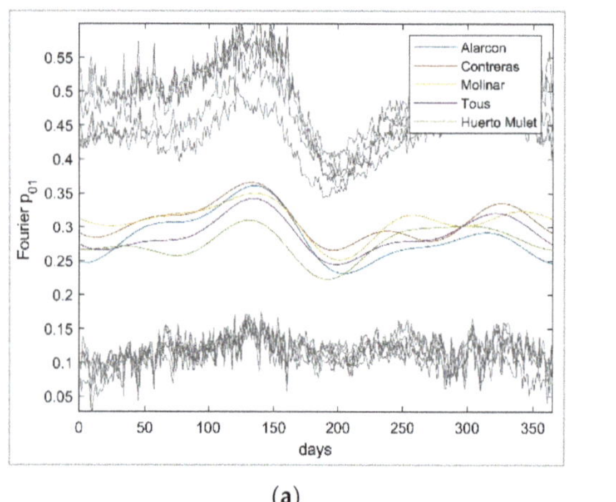
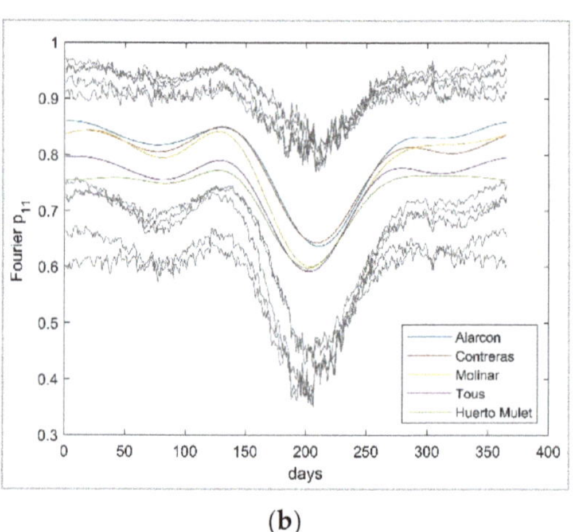

(a)   (b)

Figure 3. Fourier series and confidence limits for Alarcon: (**a**) p_{01} (**b**) p_{11} with four parameters. Confidence limits at 95% (lower and upper limits).

The transition vectors were evaluated with four parameters for the Fourier series. The objective of analyzing the transition probability vectors was to select the best representation of the wet–dry event. Simulations were carried out from two harmonics in the Fourier series to reach approximately 90% of the explicative variance. However, the process of rainfall occurrence can be represented by only a few parameters [9,22,29,73]; therefore, using few harmonics is acceptable. The frequency of the precipitation for four parameters represents a smoothed probability of occurrences, and with these few parameters, a good approximation of $p_{11\tau}$ and $p_{01\tau}$ can be obtained (Figure 3). On the other hand, the confidence limits for the vectors $p_{11\tau}$ and $p_{01\tau}$ were determined from the approximation to the t-distribution with 95% confidence. For the Fourier probability $p_{01\tau}$, four main patterns were observed, predominantly the Fourier probability increase in March, April, and May, after a decrease in June and July, an increase again in August and September, and finally, from October to February, the smoothed Fourier probabilities were nearly constant. A similar pattern was present for the Fourier probability $p_{11\tau}$, which had no considerable fluctuations between October and May, decreases in June and July, and increases once more in August and September. These Fourier series are sufficient for proper stochastic performance of rainfall occurrence.

3.2. Stochastic Multisite Multivariate Temperature Series

In the case of the maximum temperature and temperature range, the skewness coefficient of the historical series was near normal. For this reason, we did not consider normalization. Normality was assessed based on the skewness coefficient test for the 95% confidence limit. The daily temperature skewness coefficient was near the normal distribution, according to the confidence limits (Figure 4).

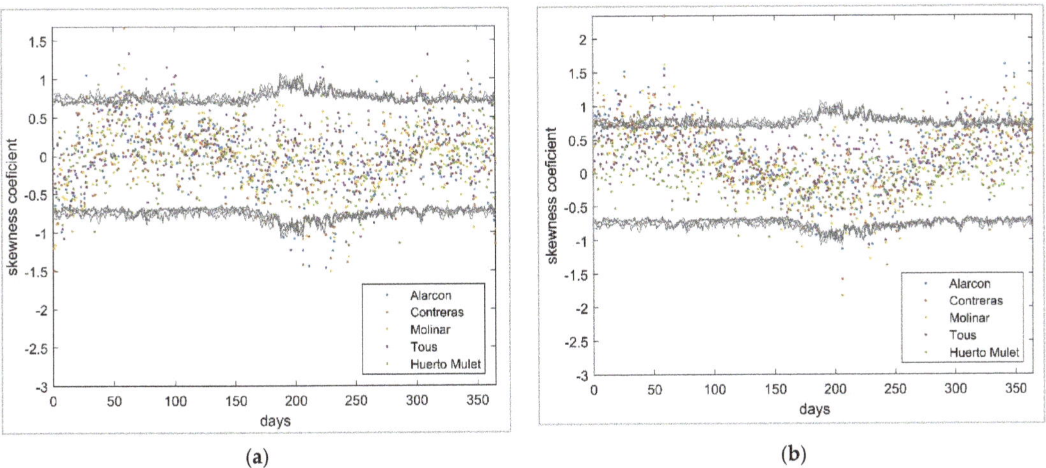

Figure 4. Skewness coefficient (daily average) of normalized series (66 years): (**a**) maximum temperature and (**b**) temperature range. Anderson confidence limits (95%).

Fourier series was applied to the mean and standard deviation (μ_τ, s_τ), and the objective was to reduce the number of parameters (Figure 5). The series were carried out with four parameters, which provide a good fit for the model for the Jucar River Basin. In Figure 5, we can observe the mean, standard deviation, and the Fourier series for the stochastic models for wet and dry days with a 95% confidence interval. The results from fitting Fourier series for the model in Figure 5a,b reflect the smoothed mean for wet and dry days. In case of the standard deviation presented in Figure 6a,b, the fitted curve includes the original data noise.

We calculated the standardization based on the normal series, mean, and standard deviation of the fitted series. The objective of the standardization is to remove the series periodicities and to obtain a mean of zero and variance of one. On the other hand, the standardized series were determined by the multivariate autoregressive model. For the Jucar River Basin, we defined the autoregressive parameters and residual series for different wet thresholds.

We selected the wet threshold of 0.001 for determining the temperature multisite multivariate stochastic generator. First, we evaluated the normality of the residual series. Mean, deviation, skewness coefficient, and lag-one autocorrelation were calculated (Table 1).

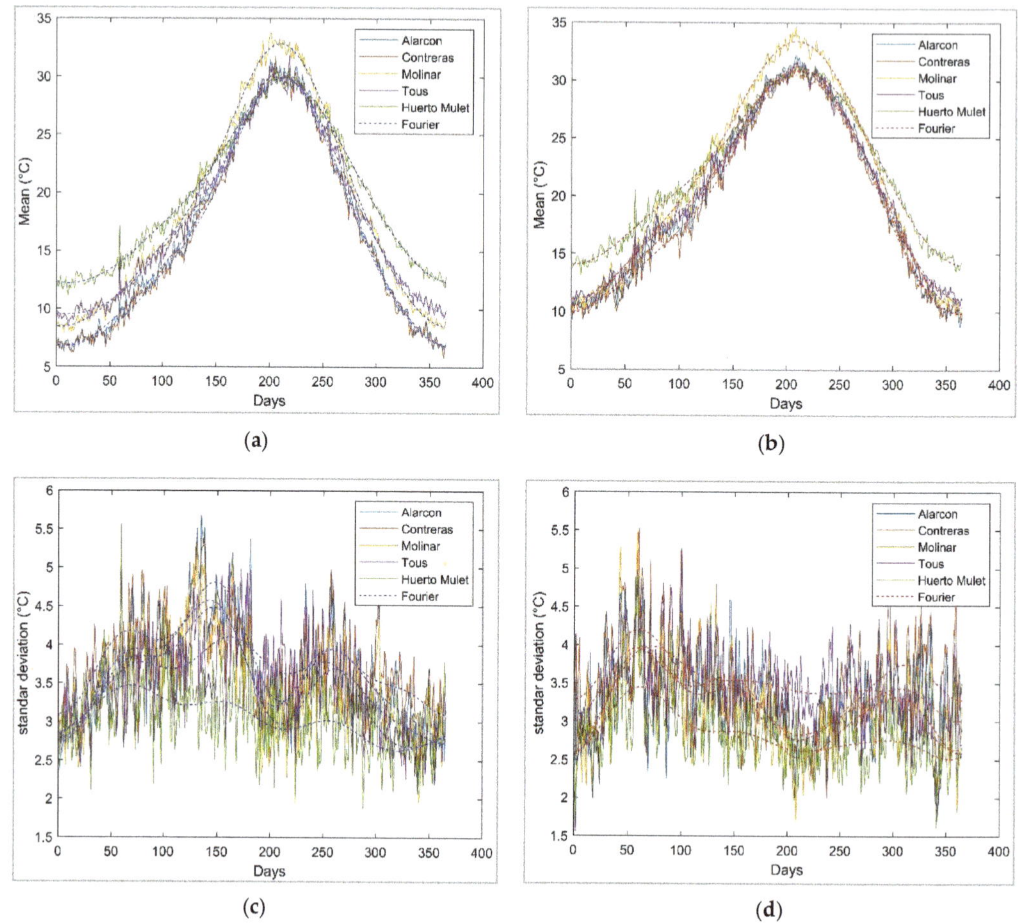

Figure 5. Fourier series (daily average) of maximum temperature series (66 years): (**a**) mean wet days, (**b**) mean dry days, (**c**) standard deviation wet days, and (**d**) standard deviation dry days.

Table 1. AIC for different wet thresholds and stochastic occurrence model.

Sub-Basin	Wet Day Threshold (mm)			
	0.001 *	0.01	0.10	0.25
Alarcon	−590.2	−515.3	−362.2	−310.9
Contreras	−681.5	−551.2	−459.5	−421.3
Molinar	−562.3	−420.7	−261.2	−215.1
Tous	−587.4	−463.6	−380.5	−340.0
Huerto Mulet	−610.5	−554.7	−467.3	−427.3

* Best performance.

The autocorrelation for this series was also determined within the 95% confidence limits for both models' maximum temperature and temperature range. In addition, we applied different tests to confirm that the residual series can be considered a normal distribution with a mean of zero, variance of one, and skewness coefficient of zero (Figure 7). The residual series of the two developed stochastic models were very similar.

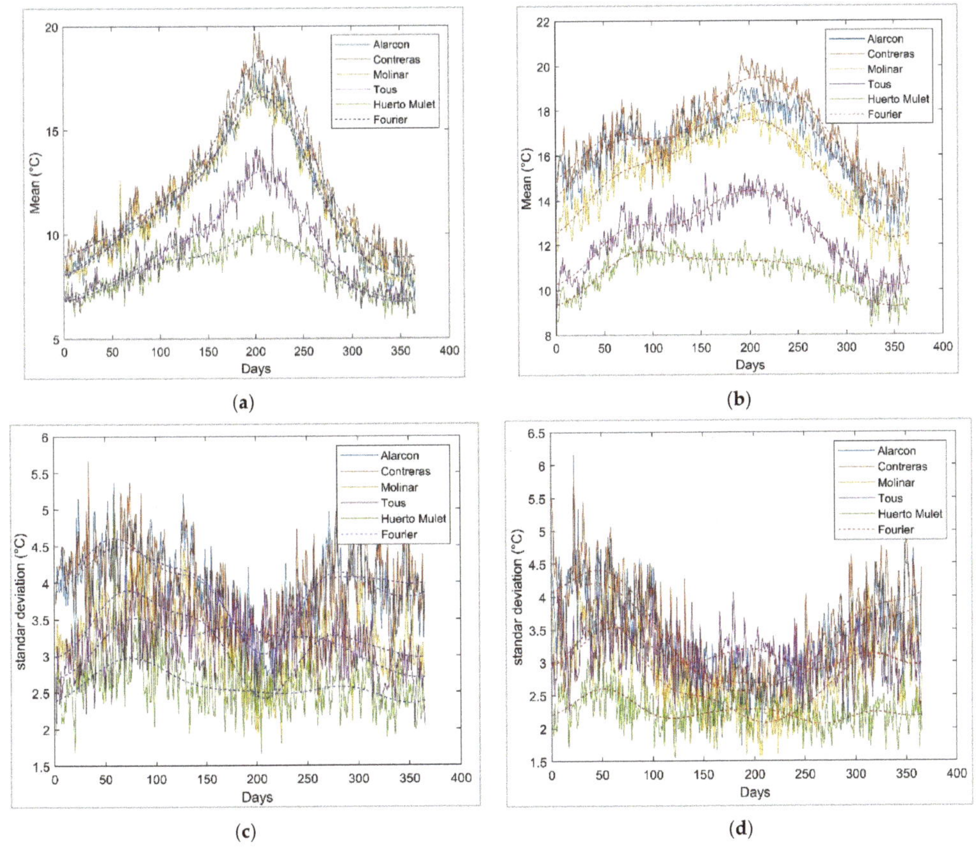

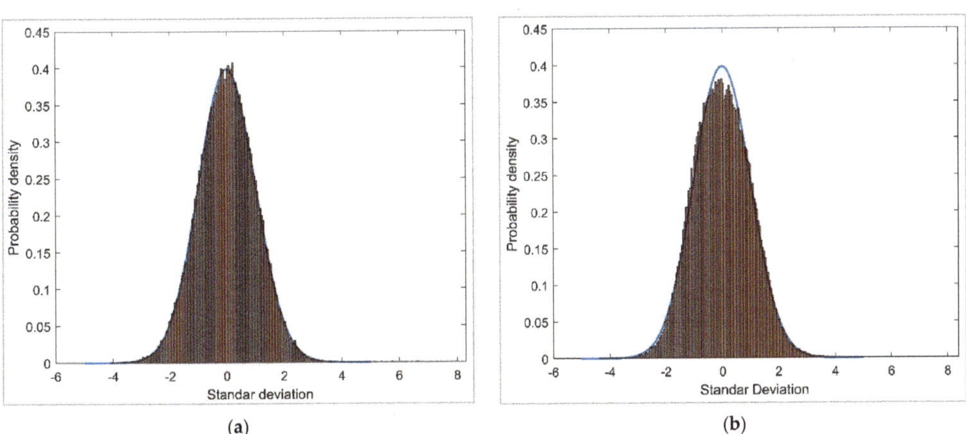

Figure 6. Fourier series (daily average) of temperature range series (66 years): (**a**) mean wet days, (**b**) mean dry days, (**c**) standard deviation wet days, and (**d**) standard deviation dry days.

Figure 7. Theoretical normal distribution (blue) and histogram for residual series for all sub-basins: (**a**) Model 1 (maximum temperature) and (**b**) Model 2 (temperature range).

The residual series of the MAR(1) for the maximum temperature had a mean near zero, the variance was around 0.85, the skewness coefficient was between -0.245 and 0.026, and the lag-one autocorrelation was around ± 0.01. The skewness coefficient and autocorrelation are within the confidence limits, the average is within 99% of the confidence limits, and therefore we assumed the normality of the residual series (Table 2).

Table 2. Normality analysis for residual series for M1 and M2 (wet threshold 0.001).

Model	Statistical/Sub-Basin	Alarcon	Contreras	Molinar	Tous	Huerto Mulet
1 *	Mean	-6.7×10^{-5}	-2.0×10^{-4}	7.5×10^{-5}	-2.6×10^{-4}	5.6×10^{-5}
	Deviation	0.842	0.821	0.834	0.812	0.850
	Skewness coefficient	-0.185	-0.242	-0.245	-0.089	0.026
	Lag-one autocorrelation	0.005	0.003	0.009	-0.041	-0.042
	AIC	-8369	-9508	-8814	$-10,152$	-7958
2 **	Mean	-3.09×10^{-4}	-5.98×10^{-4}	-6.83×10^{-5}	-2.70×10^{-4}	-2.65×10^{-4}
	Deviation	0.937	0.920	0.942	0.900	0.942
	Skewness coefficient	-0.009	-0.038	-0.044	0.112	-0.028
	Lag-one autocorrelation	0.036	0.043	-0.030	-0.082	-0.039
	AIC	-6471	-8233	-5957	$-10,294$	-5896

* Maximum temperature. ** Temperature range.

On the other hand, the series was also considered stationary since it complied with $\varnothing_1 < 1$ for all sub-basins. The AIC value was $-10,152$ with the examined parameters.

Similar results were obtained for the stochastic model for temperature range: the residual series had a mean of around -0.0005, variance of at least 0.92, skewness coefficient between -0.044 and 0.112, and lag-one autocorrelation of about ± 0.08 (Figure A5). For this stochastic model, we assumed normality and stationarity of the residual series as well as for stochastic Model 1 (Table 2). According to the AIC, the stochastic maximum temperature model (M1) was similar to the stochastic temperature range (M2) for all sub-basins.

3.3. Generation of Multivariate Synthetic Temperature Series

For the process of generating synthetic series, 1000 series were created considering the same length as the sample (66 years). The statistical sum, mean, standard deviation, and skewness coefficient were determined for both the synthetic and historical series. The occurrence depends on the correlated multivariate precipitation probabilities, critical probability, and normal distribution. Multivariate synthetic series of rainfall occurrences were generated for the five sub-basins using the stochastic process. The same occurrence series was applied for both Model 1 (M1) and Model 2 (M2) to avoid the bias from the MAR(1) model.

For the historical series, the sum of the number of rainy days in 66 years was calculated. In the same way, the monthly occurrences of the synthetic and the historical series were determined. Several statistical tests were performed to validate that the generated series are not substantially different from the historical period. The tests applied to the precipitation occurrence were the k-s test to verify that the results come from the same distribution, the t-test for equality of means, and the Wilcoxon test for equality of medians. The tests were applied considering a 95% reliability, and it was concluded that there is insufficient evidence to refute that the generated precipitation occurrences and historical series, both monthly and daily, are significant.

The scatter plots of daily mean occurrences (66 years) of historical versus simulated data can be seen in Figure 8 for the five sub-basins. The daily rainfall occurrences varied ± 5 days from the 1:1 line (Figure 8a). The monthly rainfall occurrences varied by ± 0.2 days. Monthly occurrences provide better results than daily occurrences due to the number of parameters applied (four in total). We used the same simulated rainfall occurrence (four parameters) to generate the maximum temperature and temperature range.

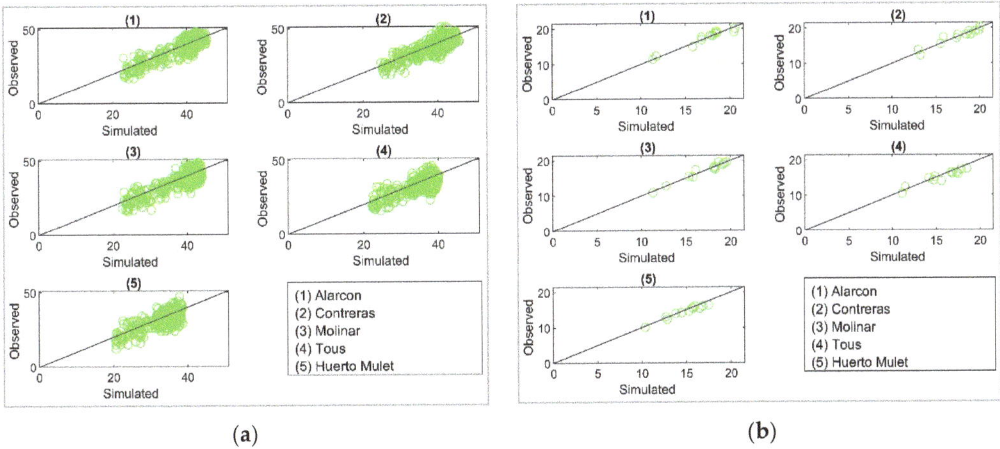

Figure 8. Scatter plots for rainfall occurrence (mean 66 years observed and 1000 simulated series) for the five sub-basins: (**a**) daily mean for each calendar day (green) and (**b**) monthly mean for each calendar moth (green) for M1 and M2.

The statistical mean, standard deviation, and skewness coefficient of daily data were computed. The MAR(1) for mean daily temperatures had a deviation of ± 1 °C, which is more accurate than the stochastic Model 2 for the daily average temperature range (± 1.5 °C). Model 1 achieved better results on both a daily and a monthly scale (Figure 9). The observed and generated series were not significantly different according to the k-s test, which indicates that they originate from the same distribution, in addition to sharing the same average according to the t-test and the same median according to the Wilcoxon test. These tests were applied to the daily average temperatures and the monthly averages. The best approximations were provided for the monthly data with higher reliability. Moreover, the RMSE, MAE, and PE (Table 3) display the adequate performance of Model 1 (M1) and Model 2 (M2).

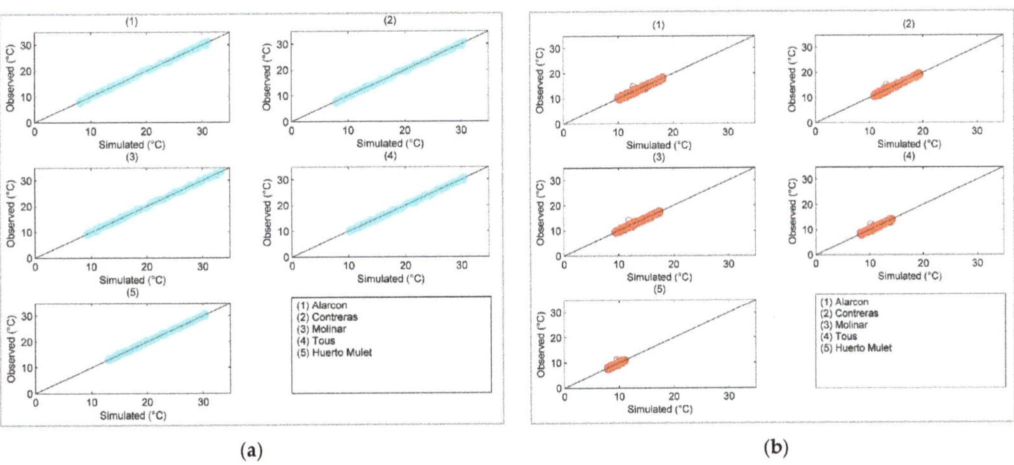

Figure 9. Scatter plots for observed mean versus generated temperature (mean 66 years observed and 1000 simulated series) for two models: (**a**) maximum temperature for each calendar day (blue) and (**b**) temperature range for each calendar day (red).

Table 3. Performance analysis for Model 1 and Model 2.

Parameter	Model 1 (M1)					Model 2 (M2)				
	1	2	3	4	5	1	2	3	4	5
RMSE (°C/day)	1.881	1.107	1.392	1.156	0.767	1.786	1.019	1.290	1.078	0.780
MAE (°C/day)	1.503	0.820	1.092	0.896	0.572	1.455	0.773	1.021	0.852	0.582
PE (%)	0.043	0.027	0.017	0.031	0.021	0.035	0.049	0.040	0.025	0.012

The f-test was used for the standard deviation (Figure 10), which indicates the equality of deviations for both models and showed minor differences. Both the maximum temperature and temperature range performed well in the daily standard deviation. It is worth mentioning that the deviations were overestimated concerning the observed data. For a monthly scale, both models effectively reproduced the standard deviation.

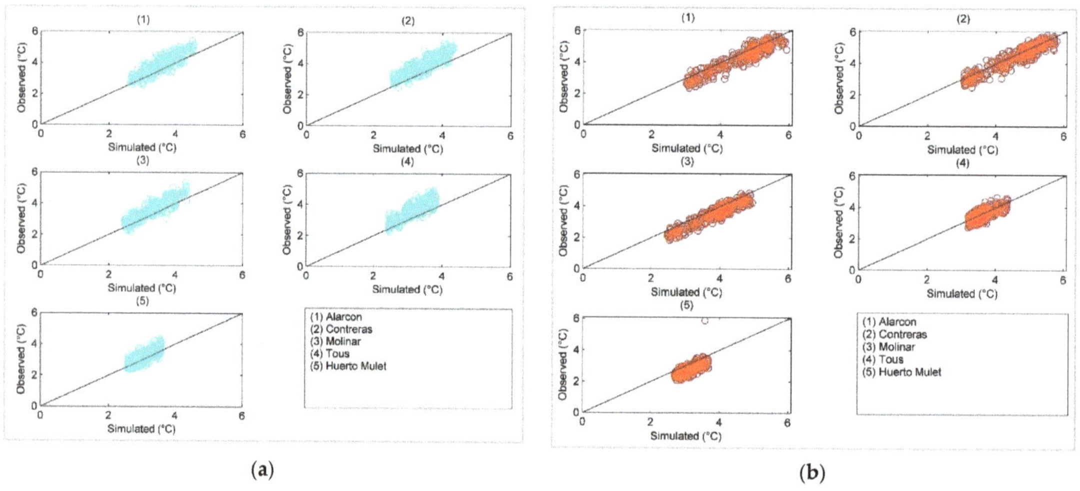

Figure 10. Scatter plots for observed standard deviations versus generated (mean 66 years observed and 1000 simulated series) for two models: (**a**) maximum temperature for each calendar day (blue) and (**b**) temperature range for each calendar day (red).

Regarding the skewness coefficient, MAR(1) was underestimated for the observed data. The same dispersion was present for the two models (M1 and M2) and for the daily and monthly averages (Figure 11).

Even though the normalization's function can be adjusted on average for the confidence limits, this offers a variation for the observed skewness. The skewness of observed daily data was between −1.2 and 1.2, and the simulated was between −0.5 and 0.5. Therefore, the multivariate stochastic model underestimated the skewness coefficient by less than −0.5 and more than 0.5. On a monthly scale, the skewness of the simulated precipitation distribution was underestimated similarly to the daily scale.

For monthly temperature, the multisite multivariate model preserved the main statistics. Figure 12 presents all months for the 5 sub-basins and 66 years, in which the temporal and spatial dependence was adequately performed. For maximum temperature, the values were between 5.5 and 35 °C, with variability of ±2 °C (Figure 12a). In the case of temperature range, the values were between 5 and 25 °C with the same variability. For the monthly mean of all sub-basins, the error was only ±0.1 °C.

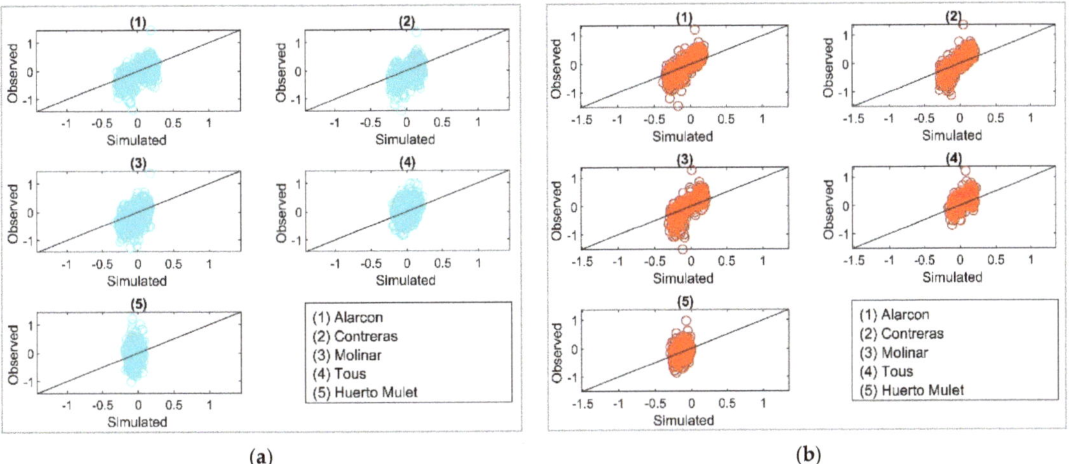

Figure 11. Scatter plots for observed skewness coefficient versus generated (mean 66 years observed and 1000 simulated series) for two models: (**a**) maximum temperature for each calendar day (blue) and (**b**) temperature range for each calendar day (red).

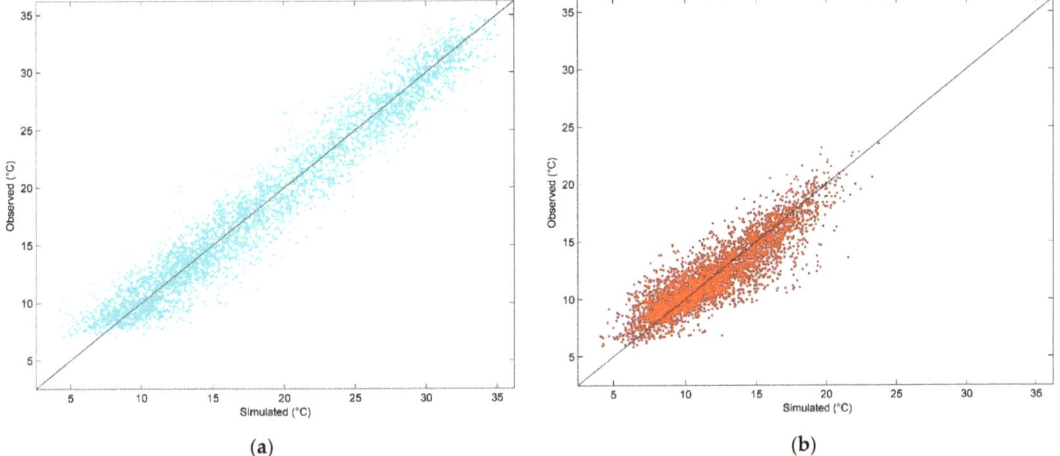

Figure 12. Monthly temperature for all sub-basins (mean 66 years observed and 1000 simulated series) observed and simulated for (**a**) maximum temperature for each month (blue) and (**b**) temperature range for each month (red), for 66 years, 5 sub-basins, and all months.

The multisite multivariate stochastic autoregressive Model 1 was selected to compare temperature years with regard to the observed data. The stochastic model can represent the temporal tendency of the results, providing an adequate indication of the yearly temperatures (Figure 13). Due to the design of the multisite multivariate stochastic model corrected by the annual model, we can simulate low frequency. The interannual variability represents the autocorrelation and cross-correlations of observed data (Table 4). The results indicate that variability was well-reproduced for the stochastic process. Moreover, the maximum and minimum values were performed adequately. For annual temperature, the stochastic model can produce the variability of both temperatures. The variability expressed for the model (sim) was greater than the observed data. Accordingly, the MAR(1) can define the maximum and minimum temperature values.

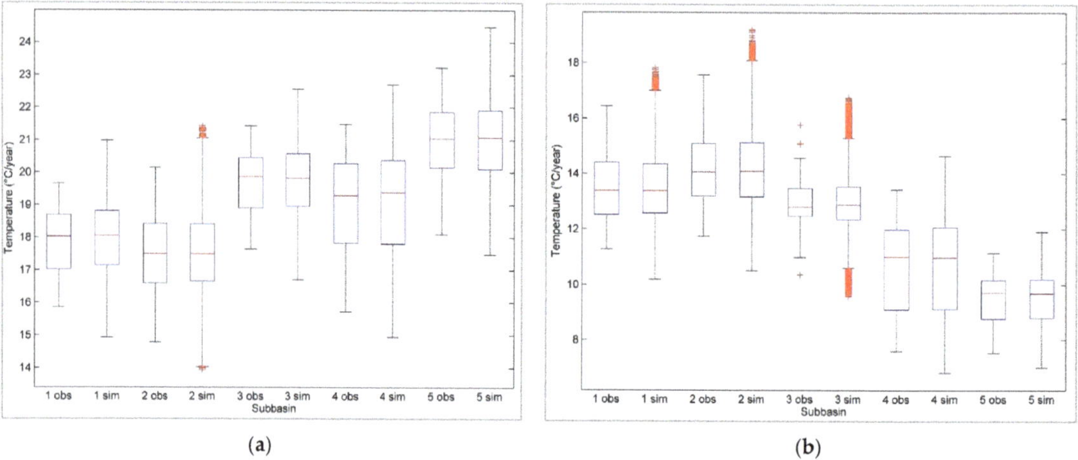

(a) **(b)**

Figure 13. Yearly temperature for all sub-basins observed (obs) and simulated (sim) for (**a**) maximum temperature and (**b**) temperature range. (1) Alarcon, (2) Contreras, (3) Molinar, (4) Tous, and (5) Huerto Mulet. The outliers are plotted individually using the '+' marker symbol.

Table 4. Annual cross-correlation matrix for all sub-basins.

Maximum Temperature Cross-Correlation										
Sub-Basin	Alarcon	Contreras	Molinar	Tous	Huerto M	* Alarcon	* Contreras	* Molinar	* Tous	* Huerto M
Alarcon	1.000					1.000				
Contreras	0.833	1.000				0.834	1.000			
Molinar	0.760	0.797	1.000			0.765	0.819	1.000		
Tous	0.239	0.492	0.598	1.000		0.243	0.489	0.610	1.000	
Huerto M	0.223	0.371	0.390	0.802	1.000	0.230	0.377	0.398	0.800	1.000
Temperature Range Cross-Correlation										
Sub-Basin	Alarcon	Contreras	Molinar	Tous	Huerto M	* Alarcon	* Contreras	* Molinar	* Tous	* Huerto M
Alarcon	1.000					1.000				
Contreras	0.719	1.000				0.718	1.000			
Molinar	0.891	0.592	1.000			0.895	0.590	1.000		
Tous	0.563	0.494	0.754	1.000		0.565	0.499	0.759	1.000	
Huerto M	0.504	0.331	0.710	0.918	1.000	0.502	0.333	0.708	0.917	1.000

* Simulated cross-correlation.

4. Discussion

The multisite multivariate autoregressive stochastic model (MASCV) was developed using MATLAB and was verified for different stations within the same basin with similar results. A Markov model of two states for the multivariate precipitation occurrence and the conditioned multivariate stochastic model for temperature can represent spatial and temporal parameters to study on-site conditions on daily, monthly, and annual scales. The Markov model of two states with few parameters has been able to depict the precipitation occurrence process. On the other hand, the temperature was accomplished considering normal distribution. Therefore, a reduction of parameters was achieved in generating the temperature. This represents a critical simplification in obtaining the daily temperature, which according to the performed parameterization, provides acceptable results for the Jucar River Basin. Moreover, it simplifies the complexity and reduces computational time.

The stochastic multivariate autoregressive model with few parameters adequately reproduced the daily and monthly temperatures. The tests showed insignificant differences between the observed and generated temperatures.

The primary objective of this stochastic model is to determine the monthly runoff and incorporate it into the integrated water resources management. It is noteworthy that this validation is performed on a monthly scale. Therefore, MASCV must be able to represent the statistics of the precipitation occurrence and temperatures in different timescales. The developed stochastic MAR(1) adequately reproduced the main statistics.

5. Conclusions

The multivariate autoregressive model of climate variables (MASCV) is a daily stochastic weather generator, programmed in MATLAB, with several user advantages, i.e., wet day selection, number of harmonics for each case, normalization type, and synthetic series to generate and automatize graphs. Moreover, MASCV can generate extreme temperatures over extended periods of time. This is unique in stochastic weather generators because only a few can reproduce extreme events. Furthermore, MASVC can be used for bias correction for climate change studies, in this case, perturbing parameters according to climate models. Finally, the results of MASCV can be incorporated for environment analysis.

MASCV presents the completion of a multivariate model for precipitation occurrences, i.e., a Markov model of two states and the dependence of temperature with rainfall occurrence process. This multisite multivariate stochastic model is meant to become a beneficial tool in a simplified manner, which may allow the incorporation of different climatic and hydrological variables.

A Markov model of first order can reflect the time dependence of the precipitation occurrence, preserving comparative statistical data with the historical series. Moreover, the spatial and temporal structure using a stochastic multisite multivariate model and coupling daily and annual temperature correction reproduced adequately in the different timescales.

Models M1 and M2 were suitably performed for different temperatures. For wet and dry days, the multisite multivariate stochastic model can adjust to real dependence with precipitation occurrence and spatial and temporal dependence of daily, monthly, and annual temperatures.

This approach greatly simplifies the process of simulating precipitation, which implies a considerable advantage and versatility over other stochastic generators. The reduction of parameters is an important factor addressed in this approach for determining the temperature and considering continuous modeling for days, months, and years.

Author Contributions: Conceptualization, J.H.-B. and A.S.; methodology, A.S.; software, J.H.-B.; validation, A.S., J.P.-A. and S.T.S.-Q.; formal analysis, A.S.; investigation, J.H.-B.; resources, A.S. and C.D.-S.; data curation, J.H.-B.; writing—original draft preparation, J.H.-B.; writing—review and editing, A.S., J.P.-A., C.D.-S. and S.T.S.-Q.; visualization, A.S.; supervision, J.P.-A.; project administration, A.S. and C.D.-S. All authors have read and agreed to the published version of the manuscript.

Funding: This research received no external funding.

Data Availability Statement: Not applicable.

Acknowledgments: We thank the anonymous reviewers and the editor for their constructive comments on the manuscript.

Conflicts of Interest: The authors declare no conflict of interest.

Appendix A

The spatial and temporal variation of the characteristics of meteorological elements in the Jucar River Basin are presented in Figures A1–A4, showing the spatial distribution of annual elements based on coordinates and 16 meteorological stations in Jucar River Basin. The grid interpolation uses inverse distance weight interpolation (IDW). Figure A5 shows the daily lag correlation for residual series.

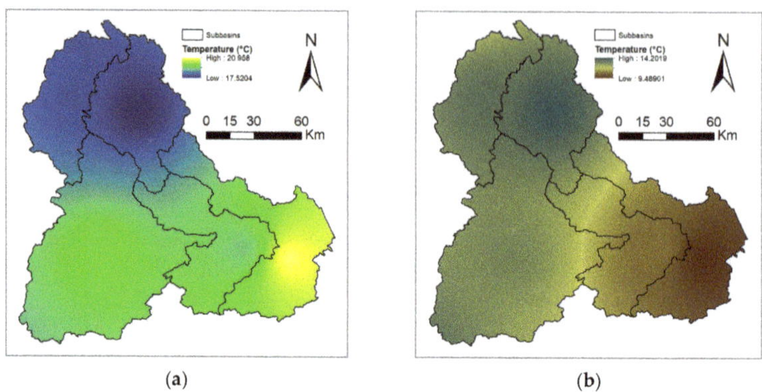

Figure A1. Spatial distribution of annual (**a**) maximum temperature (°C) and (**b**) temperature range (°C).

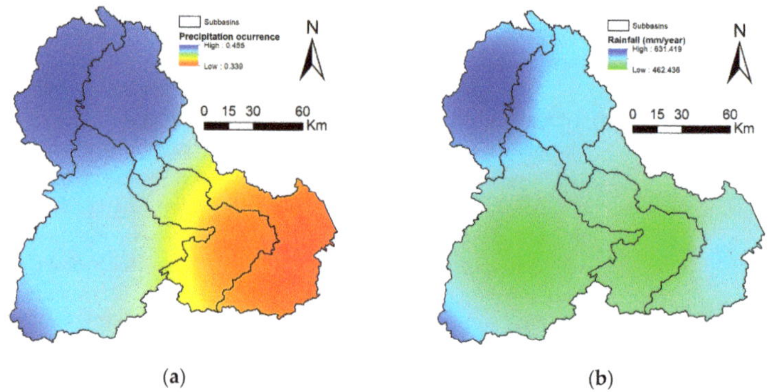

Figure A2. Spatial distribution of annual (**a**) precipitation occurrence (wet days/year) and (**b**) rainfall (mm/year).

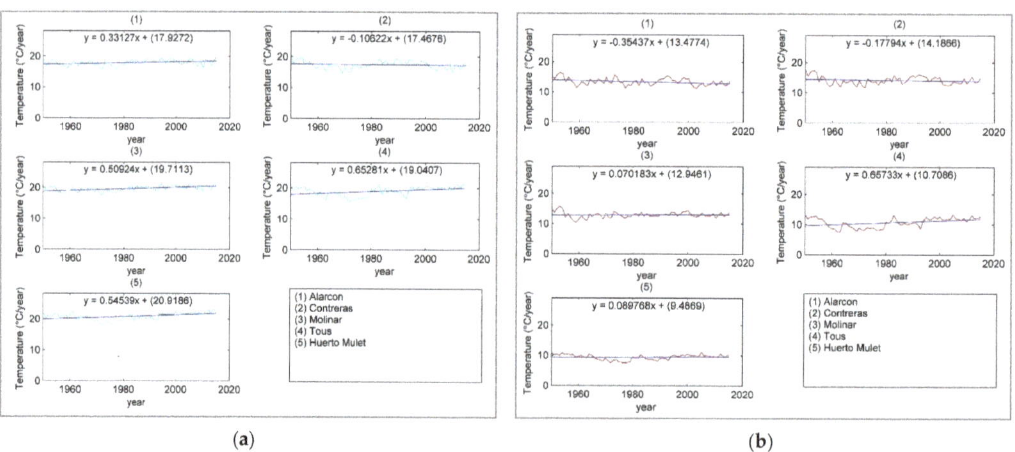

Figure A3. Interannual variation trends of the average (**a**) maximum temperature (°C) and (**b**) temperature range (°C).

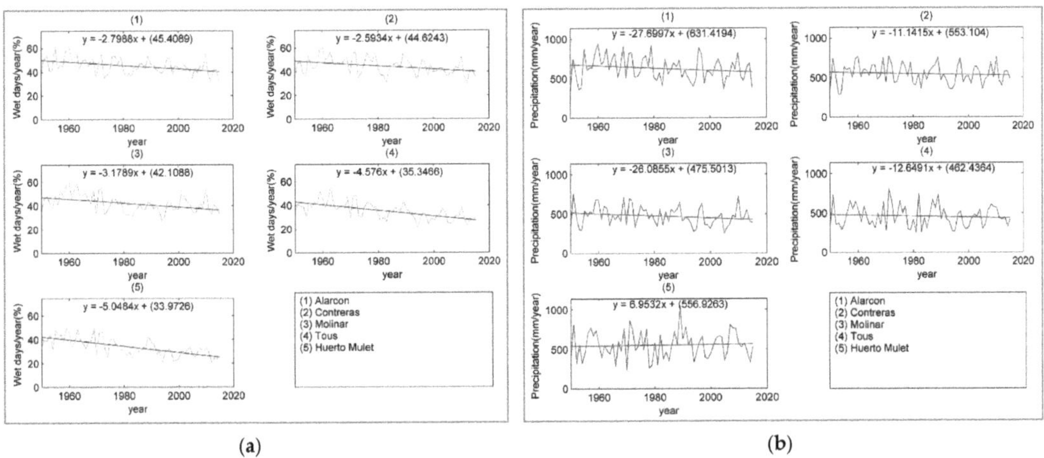

Figure A4. Interannual variation trends of the average (**a**) precipitation occurrence (wet days year^{-1}) and (**b**) rainfall (mm year^{-1}).

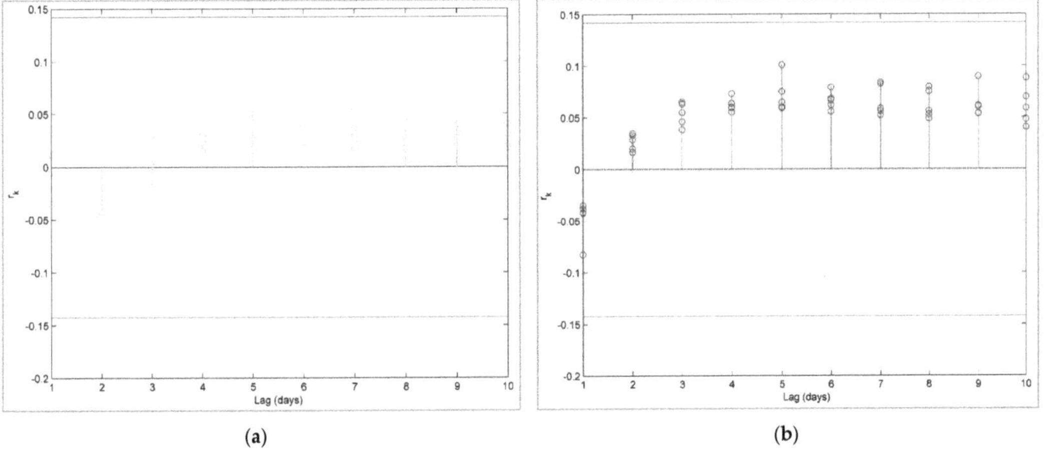

Figure A5. Daily correlation function for residual series considering the ten lag days: (**a**) Model 1 (M1) and (**b**) Model 2 (M2).

References

1. Sivakumar, B. *Chaos in Hydrology: Bridging Determinism and Stochasticity*, 1st ed.; Springer Science: Dordrecht, The Netherlands, 2017; pp. 63–111.
2. Beneyto, C.; Aranda, J.Á.; Benito, G.; Francés, F. New Approach to Estimate Extreme Flooding Using Continuous Synthetic Simulation Supported by Regional Precipitation and Non-Systematic Flood Data. *Water* **2020**, *12*, 3174. [CrossRef]
3. Chang, F.; Hsu, K.; Chang, L. Flood Forecasting Using Machine Learning Methods. *Water* **2019**, *2*, 14–53. [CrossRef]
4. Chang, F.J.; Guo, S. Advances in Hydrologic Forecasts and Water Resources Management. *Water* **2020**, *12*, 1819. [CrossRef]
5. Gabriel, K.R.; Neumann, J. A Markov Chain model for daily rainfall occurrence at tel aviv. *Q. J. R. Meteorol. Soc.* **1962**, *88*, 90–95. [CrossRef]
6. Caskey, E. A Markov Chain model for the probability of precipitation occurrence in intervals of various length. *Mon. Weather Rev.* **1963**, *91*, 298–301. [CrossRef]
7. Matalas, N.C. *Time Series Analysis*, 4th ed.; John Wiley & Sons, Inc.: Hoboken, NJ, USA, 1967; Volume 3. [CrossRef]
8. Todorovic, P.; Woolhiser, D.A. A Stochastic Model of n -Day Precipitation. *J. Appl. Meteorol.* **1975**, *14*, 17–24. [CrossRef]

9. Richardson, C.W. Stochastic Simulation of Daily Precipitation, Temperature, and Solar Radiation. *Water Resour. Res.* **1981**, *17*, 182–190. [CrossRef]
10. Roldán, J. Tendencias Actuales En El Modelado de La Precipitación Diaria. *Ing. Agua* **1994**, *I*, 89–100. [CrossRef]
11. Rajagopalan, B.; Lall, U.; Tarboton, D.G. Nonhomogeneous Markov Model for Daily Precipitation. *J. Hydrol. Eng.* **1996**, *1*, 33–40. [CrossRef]
12. Wilks, D.S. Multisite Generalization of a Daily Stochastic Precipitation Generation Model. *J. Hydrol.* **1998**, *210*, 178–191. [CrossRef]
13. Wilks, D.S.S.; Wilby, R.L.L. The Weather Generation Game: A review of stochastic weather models. *Prog. Phys. Geogr.* **1999**, *23*, 329–357. [CrossRef]
14. Harrold, T.I. A Nonparametric Model for Stochastic Generation of Daily Rainfall Amounts. *Water Resour. Res.* **2003**, *39*, 1–12. [CrossRef]
15. Brissette, F.P.; Khalili, M.; Leconte, R. Efficient Stochastic Generation of Multi-Site Synthetic Precipitation Data. *J. Hydrol.* **2007**, *345*, 121–133. [CrossRef]
16. Liu, Y.; Zhang, W.; Shao, Y.; Zhang, K. A comparison of four precipitation distribution models used in daily stochastic models. *Adv. Atmos. Sci.* **2011**, *28*, 809–820. [CrossRef]
17. Li, C.; Singh, V.P.; Mishra, A.K. Simulation of the entire range of daily precipitation using a hybrid probability distribution. *Water Resour. Res.* **2012**, *48*, 1–17. [CrossRef]
18. Mehrotra, R.; Li, J.; Westra, S.; Sharma, A. A programming tool to generate multi-site daily rainfall using a two-stage semi parametric model. *Environ. Model. Softw.* **2015**, *63*, 230–239. [CrossRef]
19. So, B.J.; Kwon, H.H.; Kim, D.; Lee, S.O. Modeling of daily rainfall sequence and extremes based on a semiparametric pareto tail approach at multiple locations. *J. Hydrol.* **2015**, *529*, 1442–1450. [CrossRef]
20. Wilks, D.S. Simultaneous stochastic simulation of daily precipitation, temperature and solar radiation at multiple sites in complex terrain. *Agric. For. Meteorol.* **1999**, *96*, 85–101. [CrossRef]
21. Semenov, M.A.; Barrow, E.M. LARS-WG A Stochastic Weather Generator for Use in Climate Impact Studies LARS-WG: Stochastic Weather Generator Contents, Harpenden, Hertfordshire, United Kigdom. Available online: http://resources.rothamsted.ac.uk/sites/default/files/groups/mas-models/download/LARS-WG-Manual.pdf (accessed on 10 December 2021).
22. Chen, J.; Brissette, F.P.; Leconte, R. WeaGETS—A Matlab-Based Daily Scale Weather Generator for Generating Precipitation and Temperature. *Procedia Environ. Sci.* **2012**, *13*, 2222–2235. [CrossRef]
23. Salas, J.J.D.; Delleur, J.W.; Yevjevich, V.M.; Lane, W.L. *Applied Modeling of Hydrologic Time Series*; Water Resources Publication: Littleton, CO, USA, 1980; ISBN 0-918334-37-3.
24. Srikanthan, R.; McMahon, T.A. Stochastic generation of annual, monthly and daily climate data: A review. *Hydrol. Earth Syst. Sci.* **2001**, *5*, 653–670. [CrossRef]
25. Qian, B.; Gameda, S.; Hayhoe, H.; De Jong, R.; Bootsma, A. Comparison of LARS-WG and AAFC-WG Stochastic Weather Generators for Diverse Canadian Climates. *Clim. Res.* **2004**, *26*, 175–191. [CrossRef]
26. Flecher, C.; Naveau, P.; Allard, D.; Brisson, N. A stochastic daily weather generator for skewed data. *Water Resour. Res.* **2010**, *46*, 1–15. [CrossRef]
27. Hayhoe, H.N. Improvements of stochastic weather data generators for diverse climates. *Clim. Res.* **2000**, *14*, 75–87. [CrossRef]
28. Ailliot, P.; Allard, D.; Monbet, V.; Naveau, P. Stochastic Weather Generators: An Overview of Weather Type Models. *J. Société Française Stat.* **2015**, *156*, 101–113.
29. Richardson, C.W.; Wright, D.A.; Nofziger, D.L.; Hornsby, A.G. WGEN: A Model for Generating Daily Weather Variables. *ARS* **1984**, *8*, 1–83.
30. Carter, T.; Posch, M.; Tuomenvirta, H. SILMUSCEN and CLIGEN User's Guide: Guidelines for the Construction of Climatic Scenarios and Use of a Stochastic Weather Generator in the Finnish. Available online: https://www.osti.gov/etdeweb/biblio/458148 (accessed on 5 November 2021).
31. Stöckle, C.O.; Nelson, R.; Donatelli, M.; Castellvì, F. ClimGen: A flexible weather generation program. In Proceedings of the 2nd International Symposium Modelling Cropping Systems, Florence, Italy, 17 July 2001.
32. Marcello, D.; Gianni, B.; Ephrem, H.; Simone, B.; Roberto, C.; Bettina, B. CLIMA: A Weather Generator Framework. In Proceedings of the 18th World IMACS/MODSIM Congress, Carins, Australia, 13–17 July 2009; Available online: https://core.ac.uk/download/pdf/38616113.pdf (accessed on 17 July 2009).
33. Foufoula-georgiou, E.; Lettenmaier, D.P. A Markov renewal model for rainfall occurrences. *Water Resour. Res.* **1987**, *23*, 875–884. [CrossRef]
34. Bárdossy, A.; Pegram, G.G.S. Copula Based Multisite Model for Daily Precipitation Simulation. *Hydrol. Earth Syst. Sci.* **2009**, *13*, 2299–2314. [CrossRef]
35. Sansom, J. A Hidden Markov Model for Rainfall Using Breakpoint Data. *J. Clim.* **1998**, *11*, 42–53. [CrossRef]
36. Ailliot, P.; Thompson, C.; Thomson, P. Space-Time Modelling of Precipitation by Using a Hidden Markov Model and Censored Gaussian Distributions. *J. R. Stat. Soc. Ser. C Appl. Stat.* **2009**, *58*, 405–426. [CrossRef]
37. Racsko, P.; Szeidl, L.; Semenov, M. A serial approach to local stochastic weather models. *Ecol. Model.* **1991**, *57*, 27–41. [CrossRef]
38. Zheng, X.; Katz, R.W. Mixture Model of Generalized Chain-Dependent Processes and Its Application to Simulation of Interannual Variability of Daily Rainfall. *J. Hydrol.* **2008**, *349*, 191–199. [CrossRef]

39. Hannachi, A. Intermittency, Autoregression and Censoring: A First-Order AR Model for Daily Precipitation. *Meteorol. Appl.* **2014**, *21*, 384–397. [CrossRef]
40. Khan, R.S.; Abul, M.; Bhuiyan, E.; Khan, R.S.; Bhuiyan, M.A.E.; García-Ortega, E.; Rigo, T. Artificial Intelligence-Based Techniques for Rainfall Estimation Integrating Multisource Precipitation Datasets. *Atmosphere* **2021**, *12*, 1239. [CrossRef]
41. Chiang, Y.M.; Chang, F.J.; Jou, B.J.D.; Lin, P.F. Dynamic ANN for Precipitation Estimation and Forecasting from Radar Observations. *J. Hydrol.* **2007**, *334*, 250–261. [CrossRef]
42. Cachim, P. ANN Prediction of Fire Temperature in Timber. *J. Struct. Fire Eng.* **2019**, *10*, 233–244. [CrossRef]
43. Li, X.; Li, Z.; Huang, W.; Zhou, P. Performance of Statistical and Machine Learning Ensembles for Daily Temperature Downscaling. *Theor. Appl. Climatol.* **2020**, *140*, 571–588. [CrossRef]
44. Bochenek, B.; Ustrnul, Z. Machine Learning in Weather Prediction and Climate Analyses—Applications and Perspectives. *Atmosphere* **2022**, *13*, 180. [CrossRef]
45. Oses, N.; Azpiroz, I.; Marchi, S.; Guidotti, D.; Quartulli, M.; Olaizola, I.G. Analysis of Copernicus' Era5 Climate Reanalysis Data as a Replacement for Weather Station Temperature Measurements in Machine Learning Models for Olive Phenology Phase Prediction. *Sensors* **2020**, *20*, 6381. [CrossRef]
46. Hernández-Bedolla, J.; Solera, A.; Paredes-Arquiola, J.; Pedro-Monzonís, M.; Andreu, J.; Sánchez-Quispe, S. The Assessment of Sustainability Indexes and Climate Change Impacts on Integrated Water Resource Management. *Water* **2017**, *9*, 213. [CrossRef]
47. Tang, Y.; Zeng, G.; Yang, X.; Iyakaremye, V.; Li, Z. Intraseasonal Oscillation of Summer Extreme High Temperature in Northeast China and Associated Atmospheric Circulation Anomalies. *Atmosphere* **2022**, *13*, 387. [CrossRef]
48. Yang, Q. Extended-Range Forecast for the Low-Frequency Oscillation of Temperature and Low-Temperature Weather over the Lower Reaches of the Yangtze River in Winter. *Chin. J. Atmos. Sci.* **2021**, *45*, 21–36. [CrossRef]
49. Chen, J.; Brissette, F.P.; Leconte, R. A Daily Stochastic Weather Generator for Preserving Low-Frequency of Climate Variability. *J. Hydrol.* **2010**, *388*, 480–490. [CrossRef]
50. Hansen, J.W.; Mavromatis, T. Correcting Low-Frequency Variability Bias in Stochastic Weather Generators. *Agric. For. Meteorol.* **2001**, *109*, 297–310. [CrossRef]
51. Chen, J.; Arsenault, R.; Brissette, F.P.; Côté, P.; Su, T. Coupling Annual, Monthly and Daily Weather Generators to Simulate Multisite and Multivariate Climate Variables with Low—Frequency Variability for Hydrological Modelling. *Clim. Dyn.* **2019**, *53*, 3841–3860. [CrossRef]
52. Apipattanavis, S.; Podestá, G.; Rajagopalan, B.; Katz, R.W. A Semiparametric Multivariate and Multisite Weather Generator. *Water Resour. Res.* **2007**, *43*, W11401. [CrossRef]
53. Li, X.; Babovic, V. A New Scheme for Multivariate, Multisite Weather Generator with Inter-Variable, Inter-Site Dependence and Inter-Annual Variability Based on Empirical Copula Approach. *Clim. Dyn.* **2019**, *52*, 2247–2267. [CrossRef]
54. Ghosh Dastidar, A.; Ghosh, D.; Dasgupta, S.; De, U.K. Higher Order Markov Chain Models for Monsoon Rainfall over West Bengal, India. *Indian J. Radio Space Phys.* **2010**, *39*, 39–44.
55. Hosseini, R.; Le, N.; Zidek, J. Selecting a Binary Markov Model for a Precipitation Process. *Environ. Ecol. Stat.* **2011**, *18*, 795–820. [CrossRef]
56. Lennartsson, J.; Baxevani, A.; Chen, D. Modelling Precipitation in Sweden Using Multiple Step Markov Chains and a Composite Model. *J. Hydrol.* **2008**, *363*, 42–59. [CrossRef]
57. Otienoongála, J.; Ster, D.; Stern, R. Extending Genstat Capability to Analyze Rainfall Data Using a Markov Chain Model. *Eur. Sci. J. August Ed.* **2012**, *8*, 1857–7881.
58. Chen, J.; Brissette, F.P. Stochastic Generation of Daily Precipitation Amounts: Review and Evaluation of Different Models. *Clim. Res.* **2014**, *59*, 189–206. [CrossRef]
59. Woolhiser, D.A.; Pegram, G.G.S. Maximum Likelihood Estimation of Fourier Coefficients to Describe Seasonal Variations of Parameters in Stochastic Daily Precipitation Models. *J. Appl. Meteorol.* **1979**, *18*, 34–42. [CrossRef]
60. Keller, D.E.; Fischer, A.M.; Frei, C.; Liniger, M.A.; Appenzeller, C.; Knutti, R. Implementation and Validation of a Wilks-Type Multi-Site Daily Precipitation Generator over a Typical Alpine River Catchment. *Hydrol. Earth Syst. Sci.* **2015**, *19*, 2163–2177. [CrossRef]
61. Chen, J.; Brissette, F.P. Comparison of Five Stochastic Weather Generators in Simulating Daily Precipitation and Temperature for the Loess Plateau of China. *Int. J. Climatol.* **2014**, *34*, 3089–3105. [CrossRef]
62. Sakia, R.M. The Box-Cox Transformation Technique: A review. *J. R. Stat. Soc.* **1992**, *41*, 169–178. [CrossRef]
63. Anderson, R.L. Distribution of the Serial Correlation Coefficient. *Ann. Math. Stat.* **1942**, *13*, 1–13. [CrossRef]
64. Moors, D. Stubblebine Chi-Square Tests for multivariate normality with application to common Stock prices. *Comun. Stat. -Theory Methods* **1981**, *10*, 713–738. [CrossRef]
65. Hu, S. Akaike Information Criterion Statistics. *Math. Comput. Simul.* **1987**, *29*, 452. [CrossRef]
66. Pedro-Monzonís, M.; Ferrer, J.; Solera, A.; Estrela, T.; Paredes-Arquiola, J. Key Issues for Determining the Exploitable Water Resources in a Mediterranean River Basin. *Sci. Total Environ.* **2015**, *503–504*, 319–328. [CrossRef]
67. Pérez-Martín, M.A.; Thurston, W.; Estrela, T.; del Amo, P. Cambio En Las Series Hidrológicas de Los Últimos 30 Años y Sus Causas. El Efecto 80. *III Jorn. Ing. Agua (JIA 2013). La Protección Contra Los Riesgos Hídricos* **2013**, *2*, 527–534.

68. CHJ Plan Hidrológico de La Demarcación Hidrográfica Del Júcar, Memoria-Anejo 2. CJH, Valencia, España. 2015. Available online: https://www.chj.es/es-es/medioambiente/planificacionhidrologica/Paginas/PHC-2015-2021-Plan-Hidrologico-cuenca.aspx (accessed on 10 December 2021).
69. Herrera, S.; Fernández, J.; Gutiérrez, J.M. Update of the Spain02 Gridded Observational Dataset for EURO-CORDEX Evaluation: Assessing the Effect of the Interpolation Methodology. *Int. J. Climatol.* **2016**, *36*, 900–908. [CrossRef]
70. Daly, C. Guidelines for Assessing the Suitability of Spatial Climate Data Sets. *Int. J. Climatol.* **2006**, *26*, 707–721. [CrossRef]
71. Pérez-Martín, M.A.; Estrela, T.; Andreu, J.; Ferrer, J.; Pérez-Martín, M.A.; Andreu, J.; Estrela, T.; Ferrer, J. Modeling Water Resources and River-Aquifer Interaction in the Júcar River Basin, Spain. *Water Resour. Manag.* **2014**, *28*, 4337–4358. [CrossRef]
72. Melsen, L.; Teuling, A.; Torfs, P.; Zappa, M.; Mizukami, N.; Clark, M.; Uijlenhoet, R. Representation of Spatial and Temporal Variability in Large-Domain Hydrological Models: Case Study for a Mesoscale Pre-Alpine Basin. *Hydrol. Earth Syst. Sci.* **2016**, *20*, 2207–2226. [CrossRef]
73. Parlange, M.B.; Katz, R.W.; Parlange, M.B.; Katz, R.W. An Extended Version of the Richardson Model for Simulating Daily Weather Variables. *J. Appl. Meteorol.* **2000**, *39*, 610–622. [CrossRef]

 *water*

Article

Using Deep Learning Algorithms for Intermittent Streamflow Prediction in the Headwaters of the Colorado River, Texas

Farhang Forghanparast * and Ghazal Mohammadi

Department of Civil, Environmental and Construction Engineering, Texas Tech University, Lubbock, TX 79409, USA
* Correspondence: farhang.forghanparast@ttu.edu; Tel.: +1-210-639-3108

Abstract: Predicting streamflow in intermittent rivers and ephemeral streams (IRES), particularly those in climate hotspots such as the headwaters of the Colorado River in Texas, is a necessity for all planning and management endeavors associated with these ubiquitous and valuable surface water resources. In this study, the performance of three deep learning algorithms, namely Convolutional Neural Networks (CNN), Long Short-Term Memory (LSTM), and Self-Attention LSTM models, were evaluated and compared against a baseline Extreme Learning Machine (ELM) model for monthly streamflow prediction in the headwaters of the Texas Colorado River. The predictive performance of the models was assessed over the entire range of flow as well as for capturing the extreme hydrologic events (no-flow events and extreme floods) using a suite of model evaluation metrics. According to the results, the deep learning algorithms, especially the LSTM-based models, outperformed the ELM with respect to all evaluation metrics and offered overall higher accuracy and better stability (more robustness against overfitting). Unlike its deep learning counterparts, the simpler ELM model struggled to capture important components of the IRES flow time-series and failed to offer accurate estimates of the hydrologic extremes. The LSTM model (K.G.E. > 0.7, R^2 > 0.75, and r > 0.85), with better evaluation metrics than the ELM and CNN algorithm, and competitive performance to the SA–LSTM model, was identified as an appropriate, effective, and parsimonious streamflow prediction tool for the headwaters of the Colorado River in Texas.

Keywords: LSTM; CNN; ELM; temporary rivers; hydrological extremes

Citation: Forghanparast, F.; Mohammadi, G. Using Deep Learning Algorithms for Intermittent Streamflow Prediction in the Headwaters of the Colorado River, Texas. *Water* **2022**, *14*, 2972. https://doi.org/10.3390/w14192972

Academic Editors: Fi-John Chang, Li-Chiu Chang and Jui-Fa Chen

Received: 2 September 2022
Accepted: 19 September 2022
Published: 22 September 2022

Publisher's Note: MDPI stays neutral with regard to jurisdictional claims in published maps and institutional affiliations.

1. Introduction

The cessation of flow for at least a portion of a year is a defining characteristic of intermittent rivers and ephemeral streams (IRES) [1]. Different forms of IRES, from headwater streams to the tributaries of mountainous rivers and snow-fed streams, make up about 60% of the river network in the United States [2,3] and more than 50% of all streams globally [4]. These streams play a crucial role in their landscape's environmental and hydrological connectivity [5–7]. The transition between wet and dry states in IRES is an influential factor in promoting the peak biodiversity of riparian vegetation [8], controlling the kinetics of biogeochemical cycles [9], and channel geomorphology [10]. Additionally, IRES offer beneficial ecosystem services like forage, nesting sites, and transportation routes for both aquatic and terrestrial wildlife [11–15]. Further, there is significant interest in utilizing IRES systems to address anthropogenic water supply needs [16].

Flow in IRES is primarily influenced by soil and precipitation characteristics, both of which are heavily affected by changes in climatic patterns [4,17,18]. Many perennial streams are projected to become ephemeral as a result of climate change [19–21]. Hence, reliable models are required to capture the link between meteorological variables (e.g., precipitation, temperature) and streamflow in IRES. Accurate streamflow prediction in IRES settings is an essential step for including these increasingly necessary water resources in various planning and management endeavors, from floodplain design and ecosystem conservation efforts to long-term supply management and climate impact analysis.

The headwaters of the Colorado River in Texas, a vital source of water for the state's agricultural, municipal, and industrial sectors, is an intermittent stream. The Colorado River in Texas flows through several major reservoirs throughout the state and serves a variety of purposes, such as power plant operations, drought mitigation, and flood control [22]. Thus, accurate streamflow models are required to support the decision-making, planning, and management endeavors associated with this valuable water resource. Further, the Colorado River originates in the semi-arid lands of West Texas (Llano Estacado), known to be a climate hotspot that has already been and is projected to be heavily influenced by precipitation and temperature variability [23]. Therefore, the flow dynamics in the headwaters of the Colorado River are likely to change, and developing effective hydro-meteorological models that are capable of investigating the relationship between climatic variability and IRES streamflow is a necessary step for supporting the active management of this headwater stream. Despite the importance of the Colorado River in Texas, there have not been any attempts to develop appropriate models for predicting streamflow in its intermittent headwaters. This study seeks to fill this gap and answer some key questions that could provide valuable guidelines for the Colorado River, as well as for numerous other headwater streams with similar circumstances.

Accurate estimation of the two types of extreme hydrological events that bookend the IRES flow spectrum—extremely high flows (large floods) and no-flow events—is essential for modeling and understanding IRES flow dynamics. During flooding events, IRES transport significant amounts of water and materials; thus, forecasting these extreme high flows is critical for flood control and management applications [24]. Flow cessation (no-flow conditions) occurs when water in the stream channel becomes disconnected and exists in discontinuous pockets. The dryness promotes local biodiversity by providing habitat and food for semi-aquatic and terrestrial biota [11]. Ultimately, periodic flow intermittency helps improve biota resilience to drying and the development of new survival and adaptation mechanisms [25]. Further, the no-flow periods are essential from the water supply perspective, as they could serve as indicators of water stress and drought, particularly in headwater streams. Reliable streamflow prediction models for IRES must be capable of accurately capturing both extreme high flows and no-flow events.

Predicting streamflow in IRES is challenging; IRES flow often varies by several orders of magnitude [26]. Moreover, in arid and semi-arid regions, IRES flow shows considerable monthly variability, ranging from very high flowrates in one month to complete flow cessation in the next [27,28]. Due to these natural characteristics of IRES flow data and factors such as the paucity of gauging stations and long-term reliable flow records in numerous headwater and low-order streams, many common rainfall-runoff approaches may not be applicable for IRES streamflow prediction [29]. A variety of data-driven machine learning techniques have been proposed as alternatives to model IRES flow over the last two decades: Cheebane et al. [30] used a stochastic autoregressive approach to reproduce monthly intermittent streamflow. Aksoy and Bayazit [31] generated daily flowrates of an intermittent stream using a Markov chains-based model and reported that their model is capable of preserving flow characteristics (e.g., hydrograph ascension and recession, mean, serial correlation). For two stations in the European part of Turkey, Kisi [32] proposed that the use of a conjunction model of discrete wavelet transform and artificial neural networks (ANN) yields more accurate 1-day-ahead streamflow forecasts than a single ANN model. Makwana and Tiwari [33] also recommended the use of wavelet transformations to improve the predictive ability of ANNs, in forecasting daily intermittent streamflow data in Gujarat, India's semi-arid region, particularly over extreme values. Mehr [29] combined a genetic algorithm (GA) with gene expression programming (GEP) and reported that it outperformed a set of classic genetic programming-based models in modeling monthly IRES streamflow in Shavir Creek, Iran. Badrzadeh et al. [34] concluded that coupling wavelet pre-processing analysis with the adaptive neuro-fuzzy inference system (ANFIS) for modeling IRES flow series in Western Australia could significantly improve the performance of ANFIS models over daily, weekly, and monthly temporal scales.

Rahmani-Rezaeieh et al. [35] used an ensemble gene expression programming (EGEP) modeling approach for 1-day- and 2-day-ahead streamflow forecasting in Iran's Shahrchay River and reported competitive performance, sometimes with higher accuracy, compared to regular ANN. Mehr and Gandomi [36] developed MSPG-LASSO, a new multi-stage genetic programming technique coupled with multiple regression LASSO methods, for univariate streamflow forecasting in Turkey's Sedre River and found it superior to a series of models from the genetic programming variant family. Kisi et al. [37] investigated the predictive abilities of the Extreme Learning Machines (ELM) model coupled with Discrete Wavelet Transform for monthly intermittent streamflow forecasting and found it superior to regular ANN models. Li et al. [38] devised a staged error model that treats zero flows as censored data for hourly streamflow forecasting over 18 ephemeral streams in Australia. In one of the most recent researches on IRES flow, Alizadeh et al. [39] developed an attention-based Long-Short Term Memory (LSTM) cell deep learning (DL) model and examined it for one- to seven-day-ahead predictions of daily flows for four basins in different climatological regimes of the United States and reported accurate and promising results.

While the reported advancements have improved the forecasting abilities of IRES systems, capturing the high variability in most intermittent flow series and modeling their extreme hydrologic events is still challenging. The streamflow models introduced in the literature tend to over-predict the low flow events and under-predict the high flows. Further, in many of these models, lagged values of streamflow (i.e., the model's output at previous time-steps) are utilized as inputs, which severely restricts their application for long-term forecasting, as propagating the prediction errors using endogenous lags causes the accuracy of the prediction to deteriorate quickly [40]. Many water planning and flood management activities (e.g., damage mitigation, food production, environmental protection) associated with IRES, such as the headwaters of the Colorado River in Texas, depend upon accurate streamflow forecasts with the longest possible lead time [36,41]. Therefore, flexible and exogenous (inputs independent of output) hydro-meteorological models are required to model the flow dynamics in intermittent settings and deliver reliable long-term streamflow predictions.

Following the most recent recommendations on modeling IRES flow data, the main objective of this study is to investigate the application of deep learning algorithms for predicting intermittent streamflow in the headwaters of the Colorado River in Texas. Three models, namely a Convolutional Neural Network (CNN), a Long Short-Term Memory (LSTM), and a Self-Attention LSTM (SA–LSTM) model, were chosen to represent deep learning algorithms of different levels of complexity. An Extreme Learning Machine (ELM) model was developed as a baseline shallow learning model for better comparisons, and to highlight the impacts of the use of deep learning versus shallow learning models. Considering the importance of the Colorado River for long-term water planning for the state, the heavy influence of climate on flow generation in IRES, and the location of the Texas Colorado River headwaters in a climate hotspot, this study adopted a monthly timescale and focused on capturing the links between climate variables and streamflow. This research seeks to answer the following questions about the intermittent headwaters of the Colorado River in Texas: (a) What is the difference between the performance of the deep learning algorithms and that of the baseline ELM model in terms of capturing the hydrological extremes and the entire range of flowrates? (b) Are deep learning algorithms appropriate for intermittent streamflow prediction? (c) How much complexity is warranted for predicting intermittent streamflow using deep learning algorithms?

2. Materials and Methods

2.1. Study Area

As illustrated in Figure 1a, The Colorado River rises near the Texas–New Mexico border (south of Lubbock, TX, USA) and flows southeast for 1560 km into the Gulf of Mexico at Matagorda Bay, making it the longest and largest river by length and drainage area in Texas [42,43]. Its drainage area, which is about 16 percent of the total area of Texas, is stretched from the dryer west side of Texas, with higher elevations and lower precipitation rates, to the more humid and lesser-elevated southeast of the state (Figure 1b,c). The average annual runoff of the Texas Colorado River reaches a volume of more than 2 billion cubic meters near the Gulf of Mexico [44]. Several dams (e.g., the J.B. Thomas, E.V. Spence, and O.H. Ivie) and lakes (e.g., the Texas Highland Lakes) along the Colorado River serve water supply, flood mitigation, recreational, and energy production purposes. The headwaters of the Texas Colorado River are in the High Plains level III ecoregion at an elevation of 1195 m asl, where the annual average temperature is 13.9 °C, and the average yearly precipitation is 40 cm [45–47].

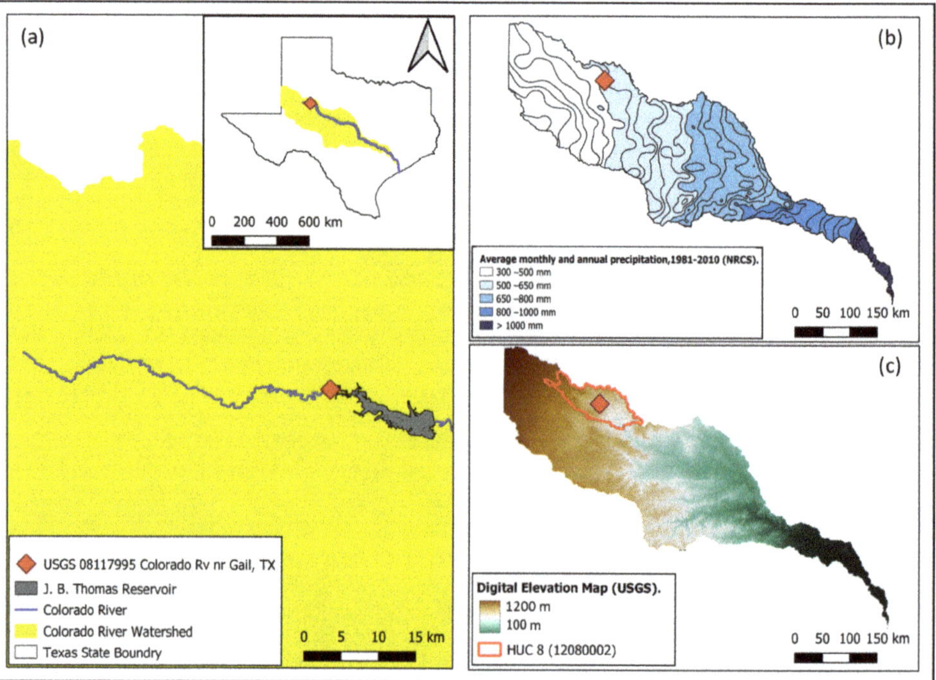

Figure 1. (**a**) Map of the study area and the relative location of the streamflow monitoring station, (**b**) Average precipitation, and (**c**) Digital elevation map of the Colorado River watershed in Texas.

2.2. Data

Streamflow data of the headwaters of the Colorado River in Texas were accessed from the closest USGS streamflow monitoring station to the point of origin (Station 08117995, located near Gail in Borden County, TX, USA) and used for this study. This USGS station has a drainage area of 1290 square kilometers and is located upstream of Lake J.B. Thomas, one of the three reservoirs operated by the Colorado River Municipal Water District, supplying water to the rapidly growing Midland–Odessa region in West Texas. Monthly streamflow records were obtained from March 1988 to May 2022 [48], with a total of 0.5% (the equivalent of 2 months) missing data. Kalman filtering was applied as an imputation technique to fill these missing records based on the available data [49,50].

The streamflow records indicate that Texas Colorado River headwaters were dry for 127 months during the study period, making its intermittency ratio (ratio of the duration of dry runs to the total duration of the study) equal 31%. The frequency of extreme high flow events has increased in the stream over the last decade because of changing climatic patterns. The station of interest has recorded three extreme flooding events (greater than 5 cubic meters per second) in the last six years, including the floods of September 2014

(14.29 cubic meters per second) and May 2015 (20.10 cubic meters per second) that were almost two and three times greater than the greatest previously recorded flood (7.45 cubic meters per second in May 1992), respectively.

Climate variables were required to develop appropriate hydro-meteorological stream-flow prediction models. Utilizing precipitation and evaporation data for modeling IRES systems is recommended in the literature [51–54]. Monthly precipitation and temperature data were extracted for the location of the streamflow monitoring site and the same 410-month period (March 1988–May 2022) from PRISM [55]. The Thronthwaite method was used to calculate potential evapotranspiration based on temperature records [56]. A summary of the collected data is provided in Table 1.

Table 1. Summary of the hydro-meteorological data used for this study.

	Minimum	Median	Mean	Maximum
Flowrate (m^3/s)	0	0.013	0.377	20.102
Precipitation (mm)	0	29.8	42.0	251.2
ET (mm)	2.1	64.8	80.6	221.33

PPT and PET autocorrelation function (ACF) plots (Figure 2) revealed that the first two lags correlated positively with observed precipitation and evapotranspiration fluxes. This finding implies that the previous two months' rainfall and evaporative fluxes influenced streamflow observations during any given month. Seasonality of rainfall and PET can be seen in these plots, but higher lags were not taken into account due to the parsimony principle [57].

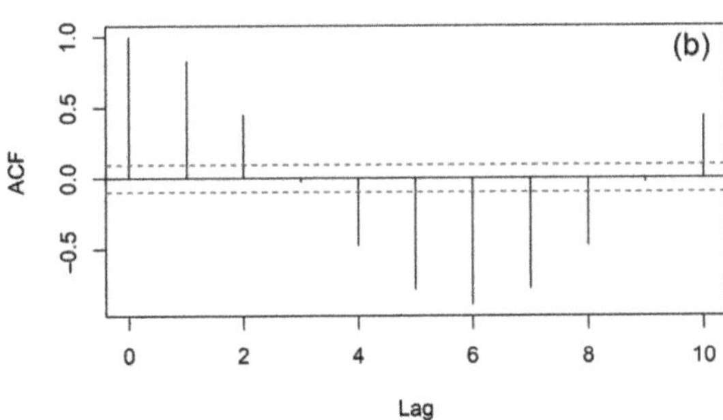

Figure 2. Autocorrelation Functions (ACF) plots for (**a**) PPT and (**b**) PET.

Correlation analysis between the climate variables (and their first two lags) and the streamflow data (Figure 3) indicated the flowrate at any given month shows a strong correlation with the observed precipitation in that month (r = 0.7) and moderate correlation with the PET (r = 0.27) and the first lag of precipitation (r = 0.31) and PET (r = 0.24). The second lags of precipitation and PET had weak correlations with streamflow (r < 0.2) and, therefore, were not included in the final set of the inputs. All parameters were standardized by subtracting the mean values and dividing them by the standard deviations so that the scale effects were removed, and the impacts of outliers were minimized.

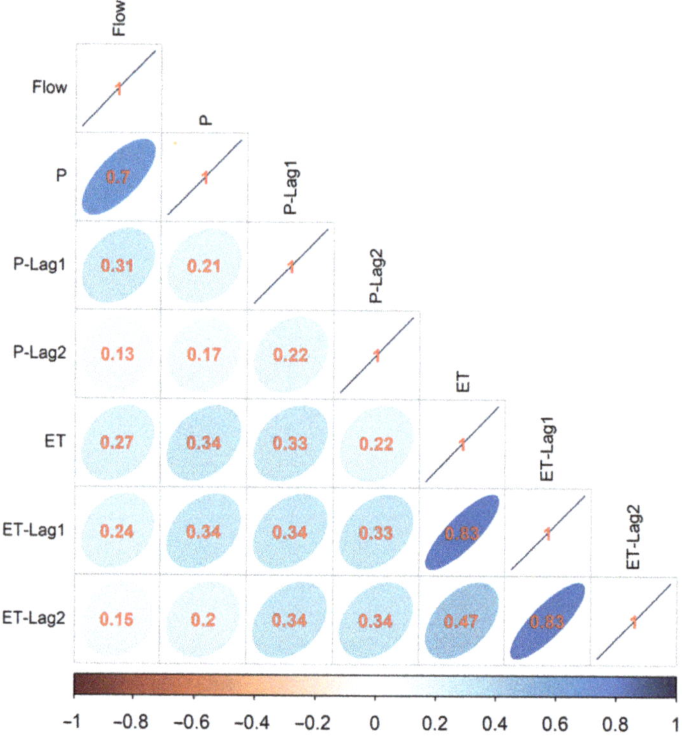

Figure 3. Correlation coefficients between the climate variables, Precipitation at each month (P), Precipitation of the last month (P-Lag1), Precipitation of the second last month (P-Lag2), Evapotranspiration at each month (ET), Evapotranspiration of the last month (ET-Lag1), Evapotranspiration of the second last month (ET-Lag2), and the streamflow data.

2.3. Methods

2.3.1. Extreme Learning Machine

An Extreme Learning Machine is composed of three main layers: input, hidden, and output layer, which employ various weights to convey information through the network (Figure 4a). Huang et al. [58,59] suggested the ELM method in which the weights from the input layer to the hidden layer are randomly assigned. ELM reduces the computational time and enhances the generalization ability of the single-layer Artificial Neural Network (ANN) model [58,60,61]. ELMs have gained popularity in the hydrologic literature [62–67] and are established as fast and effective streamflow forecasting models [37,68–73].

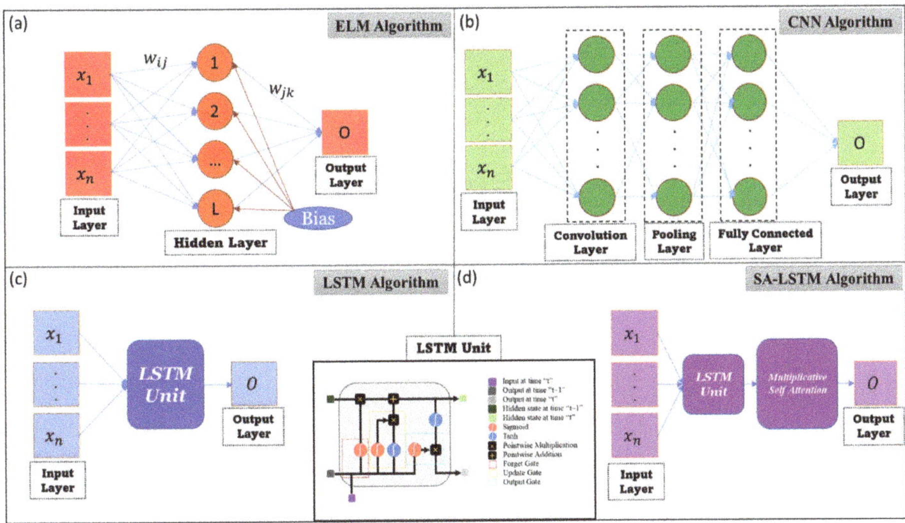

Figure 4. The model architectures for the investigated algorithms in this study.

An ANN architecture consisting of one input layer and L hidden neurons with an activation function called $g(x)$ and a bias term (B), is presented mathematically as Equation (1):

$$\sum_{i=1}^{L} \beta_i g_i(x_j) = \sum_{i=1}^{L} \beta_i g_i(w_i.x_j + B_i) = o_k \tag{1}$$

where w_i is the vector weights that connect ith hidden neuron to the input neurons, β_i is the vector of weights connecting ith hidden neuron to the output neurons, o_k is the kth output vector, B_i is the bias regarding ith hidden neuron, and g_i is the output of ith hidden neuron. Moreover, the equation is written based on the assumption of the network being trained using a dataset composed of N arbitrary patterns (X_i, Y_i). Equation (1) can be written as the following matrix [58,74,75]:

$$H\beta = T \tag{2}$$

where,

$$H = \begin{bmatrix} g_1(w_1.x_1 + b_1) & \cdots & g_L(w_L.x_1 + b_L) \\ \vdots & \vdots & \vdots \\ g_1(w_1.x_N + b_1) & \cdots & g_L(w_L.x_N + b_L) \end{bmatrix} \tag{3}$$

$$\beta = \begin{bmatrix} \beta_1^T \\ . \\ . \\ . \\ \beta_L^T \end{bmatrix} \tag{4}$$

$$T = \begin{bmatrix} t_1^T \\ . \\ . \\ . \\ t_L^T \end{bmatrix} \tag{5}$$

In which H is the hidden layer output of the network.

The following equation provides the output weights between the hidden layer and the output layer:

$$\hat{\beta} = H^+ T \tag{6}$$

where H^+ represents the Moore–Penrose generalized inverse of the hidden layer output matrix H [74]. Based on what is presented in Huang et al. [58,59], in an ELM model w_i weights and β_i bias are randomly assigned (based on a probability density function). Then, the H matrix is calculated based on Equation (3), and finally, H^+ can be calculated.

2.3.2. Convolutional Neural Networks

A Convolutional Neural Network (CNN) is a specific architecture of neural networks that is designed based on the weight sharing concept and employs convolution and pooling layers [76–78]. The family of CNN models include one-dimensional CNN (Conv1D), two-dimensional CNN (Conv2D), and three-dimensional CNN (Conv3D) models, and their primary difference is in the structure of the model inputs [79]. A standard CNN architecture consists of an input layer, an output layer, and a hidden layer that is composed of a convolution layer, a pooling layer, and a fully connected layer (Figure 4b). A unique feature of the CNN is that a given neuron is only connected to its nearby local neurons in the previous layer. While the neurons in the convolution layer are fully connected to the input layer neurons, they are not connected to all the neurons in the pooling layer.

Convolution and pooling layers, as the core building blocks in CNN, extract different features from the input layer and convert them to small dimensions by performing convolution operations on the input layer and merging neuron cluster outputs into a single neuron. The pooling mechanism significantly reduces the number of coefficients in the network and makes the training (learning) phase of the CNNs more efficient, easier, and faster than the regular ANN networks [79,80]. Following that, the fully connected layer flattens all feature maps in a feature vector and uses them as input variables to make predictions [81,82].

The application of CNN models for streamflow prediction has received more attention over the last few years, and they have been found to be relatively fast, accurate, and stable alternatives among the growing family of deep learning algorithms [78,83–85].

Streamflow is a one-dimensional data; thus, for this study, a Conv1D model is adopted. The application of a rectified linear unit (ReLU) activation function for the convolutional layer is recommended to enhance the model's ability to capture non-linearity [86]. The mean squared error (MSE) was used as the loss function for the fully connected layer.

2.3.3. Long Short-Term Memory

Hochreiter and Schmidhuber [87] introduced the Long Short-Term Memory (LSTM) model as a form of recurrent neural network. Contextual state cells are used in LSTM models as either long-term memory cells or short-term memory cells, making them suitable substitutes for representing sequential data [88,89]. The LSTM model's architecture (shown in Figure 4c) is made up of unique units (memory blocks) in the recurrent hidden layer. Self-connected memory cells and multiplicative units implemented in the memory blocks are utilized to store the network's temporal state. The input, output, and forget gates, which are multiplicative units, are in charge of managing the information flow. The following equations are used in different LSTM cells:

$$Input\ Gate: i_t = \sigma(W_i X_t + U_i h_{t-1} + b_i) = \sigma(W_i X_t + U_i h_{t-1} + b_i) \tag{7}$$

$$Forget\ Gate: f_t = \sigma\left(W_f X_t + U_f h_{t-1} + b_f\right) \tag{8}$$

$$Output\ Gate: O_t = \sigma(W_o X_t + U_o h_{t-1} + b_o) \tag{9}$$

$$Previous\ Cell\ State: \tilde{C}_t = \tanh(W_C X_t + U_C h_{t-1} + b_C) \tag{10}$$

$$Current\ Cell\ State: C_t = f_t \otimes C_{t-1} + i_{t-1} \otimes \tilde{C}_t \tag{11}$$

$$HiddenState: h_t = O_t \otimes tanh(C_{t-1}) \tag{12}$$

In which, i_t, f_t, and O_t represent the input gate, forget gate, and output gate, respectively. W_i, W_f, and W_o stand for the weights connecting the input, forget, and output gates with the input, respectively; U_i, U_f, and U_o are the weights from the input, forget, and output gates to the hidden layer, respectively; b_i, b_f, and b_o indicate the input, forget, and output gate bias vectors, respectively. C_t is the current state of the cell and $\widetilde{C}_t$ is the state of the cell at the previous time. Moreover, h_t refers to the output of the cell at the current time [90,91]. Additionally, the dropout mechanism was used to enhance the generalization of the model and avoid overfitting [92].

LSTM models have become extremely popular for modeling hydrological time-series [93–96], in particular, streamflow prediction [97–102], primarily because of their unique architectural design and abilities to model highly nonlinear sequential data.

2.3.4. Attention-Based Long Short-Term Memory

Attention is a deep learning strategy and can be viewed as essentially implementing a neural network within another neural network to weigh various portions of a sequence for relative feature importance [103,104]. Multiplicative self-attention, a special type of attention, is used for the SA–LSTM model in this study (Figure 4d), which is the mechanism of relating different positions of a single sequence in order to compute a representation of the same sequence [39,105].

$$h_t = tanh(W_x X_t + W_h h_{t-1} + b_h) \tag{13}$$

$$e_t = \sigma\left(X_t{}^T W_a X_{t-1} + b_t \right) \tag{14}$$

$$a_t = softmax(e_t) \tag{15}$$

The self-attention mechanism is known as an effective technique for improving LSTMs and enhancing the model's performance by "paying attention" and assigning attention scores (weights) to each observation [106,107]. Attention-based LSTM models are among the most recent developments in machine learning that have found application for streamflow prediction [39,108,109].

The first 75% of the period of study, from April 1988 until September 2013, was used for the training phase. 25% of these 307 months was used as the evaluation phase to find the best values for the hyperparameters of the models. The period from October 2013 until May 2022 was used as the independent testing dataset. For this study, the pre-processing (e.g., Kalman filtering, data standardization) and post-processing (e.g., model evaluation metric calculations, visualizations) were done in R [110]. The models were developed and run in Python [111]. 25% of the training dataset was used for the validation process, where the best values for the hyperparameters (e.g., the number of hidden neurons, dropout rate, learning rate, and number of epochs) were determined based on grid search.

2.4. Model Evaluation Metrics

Common guidelines for model evaluation (e.g., [112,113]) were utilized here to compare the closeness of model predictions to observations over a broad range of statistical measures. The following model evaluation metrics were used in this study:

- *The Mean Absolute Error (MAE) and Root Mean Square Error (RMSE):*

MAE and RMSE measure the errors associated with the low and high flowrates, respectively, and together they support model comparison with respect to accuracy. MAE and RMSE are calculated as:

$$\text{MAE} = \frac{1}{N} \sum_{i=1}^{N} |S_i - O_i| \tag{16}$$

$$\text{RMSE} = \sqrt{\frac{1}{N} \sum_{i=1}^{N} (S_i - O_i)^2} \tag{17}$$

where N is the number of observations, and S_i and O_i are the simulated and observed flowrates, respectively.

- *The Index of Agreement (d):*

Developed by Willmott [114], the index of agreement is a standardized measure between 0 and 1 and describes the degree of model prediction error. This index can identify additive and proportional differences between observed and simulated means and variances, but it should be noted that this index is extremely sensitive to extreme values [115]. The formula for the index of agreement is as follows:

$$d = 1 - \frac{\sum_{i=1}^{N}(O_i - S_i)^2}{\sum_{i=1}^{N}\left(\left|S_i - \overline{O}\right| + \left|O_i - \overline{O}\right|\right)^2} \tag{18}$$

where, $\overline{O}$ is the average of observed flowrates.

- *The Pearson's r (r):*

Pearson [116] developed the Pearson (Product–Moment) correlation (r), which was based on the work of others, including Galton [117], who first introduced the concept of correlation [118,119]. The r coefficient is considered the most common measure of association between variables and is widely used for describing linear relationships. Pearson's r is calculated as:

$$r = \frac{\sum_{i=1}^{N}(O_i - \overline{O})(S_i - \overline{S})}{\sqrt{\sum_{i=1}^{N}(O_i - \overline{O})^2 \sum_{i=1}^{n}(S_i - \overline{S})^2}} \tag{19}$$

where, $\overline{S}$ is the average of simulated flowrates. There are guidelines in the literature to interpret different ranges of r. According to Schober et al. [120], thresholds of 0.1, 0.39, 0.696, and 0.89 can be used to describe negligible, weak, moderate, strong, and very strong correlations, respectively. It should be noted that extraordinarily high outliers (extreme high floods in the case of this study) can have a huge effect on Pearson's r [119].

- *The Coefficient of Determination (R^2):*

The coefficient of determination describes how well observed outcomes are simulated by the model, based on the proportion of total variation of outcomes explained by the model [121].

$$R^2 = 1 - \frac{RSS}{TSS} \tag{20}$$

where RSS is the sum of squares of residuals and TSS is the total sum of squares.

- *The Nash–Sutcliffe Efficiency (NSE):*

The Nash–Sutcliffe efficiency (NSE) is a normalized statistic that measures the relative magnitude of residual variance ("noise") versus measured data variance [122]. The NSE is computed using the following equation:

$$NSE = 1 - \frac{\sum_{i=1}^{N}(S_i - O_i)^2}{\sum_{i=1}^{N}(O_i - \overline{O})^2} \tag{21}$$

The NSE varies between -infinity and 1, with NSE = 1 corresponding to an ideal match between the model estimation and the observed data. While positive values of NSE are generally considered "acceptable levels of performance," a negative NSE score suggests that the mean of the observed data is a better predictor than the model [113]. The NSE is a reliable and widely used model evaluation metric in the field of hydrology [123,124]. According to Moriasi et al. [113] thresholds of 0.5, 0.65, and 0.75 can be used to rate the model performance as "very good", "good", "satisfactory", and "unsatisfactory", respectively.

- *The Kling–Gupta Efficiency (KGE):*

The Kling–Gupta Efficiency (KGE) was developed initially by Gupta et al. [125] and later revised by Kling et al. [112] to decompose the Nash–Sutcliffe Efficiency metric into correlation (Pearson's r), bias (the ratio between the mean of the simulated values and the mean of the observed ones), and variability components to facilitate its use and providing more insight to model performance. Similar to the NSE, the KGE metric has been increasingly used as a model evaluation metric in the hydrologic literature [126–130].

The KGE is computed as:

$$KGE = 1 - ED \tag{22}$$

$$ED = \sqrt{(s[1] * (r - 1))^2 + (s[2] * (vr - 1))^2 + (s[3] * (\beta - 1))^2} \tag{23}$$

where r is Pearson's r, and s is a vector of length three which contains the scaling factor for correlation, variability, and bias. Variability (vr) and bias (β) can be calculated as:

$$\text{Variability Ratio (vr)} = \frac{\sigma_s/\mu_s}{\sigma_o/\mu_o} \tag{24}$$

$$Bias\ (\beta) = \frac{\sum_{i=1}^{N}(S_i - O_i)}{\sum_{i=1}^{N}(O_i)} \tag{25}$$

where σ_s and σ_o are the standard deviation of simulated and observed flowrates, and μ_s and μ_o are the average values of simulated and observed flowrates, repextively. Knoben et al. [131] showed that a KGE > −0.41 indicates that the model is more informative than the mean of the observed data.

3. Results and Discussion

3.1. Predictive Performance over the Entire Range of Flowrates

A summary of the model evaluation metrics during the training and testing periods is provided in Tables 2 and 3, respectively. The error indices, MAE, and RMSE, were significantly lower for the deep learning algorithms against the ELM model. With almost 20% lower Pearson's r value, 10% lower index of agreement, and almost 30% lower R^2, the baseline ELM model was outperformed by the more complex deep learning counterparts during the testing period. According to the NSE scores, the deep learning models achieved "very good" levels of performance (NSE > 0.75) against the unsatisfactory performance of the ELM model (NSE < 0.5). All models achieved "skillful" predictions (KGE > −0.41); however, the deep learners, particularly LSTM-based models offered better estimations with lower biases and better variability ratios and, thus, considerably better KGE scores in comparison to the ELM.

Table 2. Summary of model evaluation metrics during the training period.

	ELM	CNN	LSTM	SA–LSTM
MAE (m³/s)	0	0.02	0.04	0.04
RMSE (m³/s)	0	0.04	0.07	0.07
d	1	1	1	1
r	1	1	1	1
R²	1	1	0.99	0.99
NSE	1	1	0.99	0.99
KGE	1	0.98	0.99	0.99

A serious problem associated with utilizing artificial neural networks is overfitting or poor generalizability, which occurs when the model performs well with the training data and fails to maintain the same performance quality on the independent testing data [132–135]. A comparison of the metrics achieved by the four models during the training and testing periods indicated that the ELM model exhibited the worst performance decline from training to testing (substantially higher than the deep learning models). The ELM model achieved the best scores during training (almost perfect with respect to all metrics) and the worst scores during testing compared to the other models. Finding ELM prone to overfitting is consistent with the previous reports of its application in literature and is mostly due to the large number of hidden nodes required to capture complex non-linear relationships [74,136,137]. A variety of contributing factors to the problem of overfitting are stated in the literature: from architecture-related reasons (e.g., high model complexity, an extensive number of hidden units) [138–140] to data-related reasons (e.g., noisy training samples, under-sampled training data) [141–143]. Bejani and Ghatee [144] categorize methods for controlling overfitting as passive, active, and semi-active. The pooling mechanism built into CNN [79] and the dropout mechanism [92] utilized with the LSTM and SA–LSTM models belong to the category of active (regularization) and semi-active (dynamic architecture), respectively. The results indicated that the applied overfitting control mechanisms and architectural advancement of the deep learning models granted them an enhanced ability to learn the information (input-output relationships) and distinguish the noise in data during the learning (training) phase.

Table 3. Summary of model evaluation metrics during the testing period (the best performing model is highlighted with respect to each metric).

	ELM	**CNN**	**LSTM**	**SA–LSTM**
MAE (m^3/s)	1.15	0.51	0.49	0.47
RMSE (m^3/s)	1.85	1.28	1.26	1.24
d	0.81	0.9	0.92	0.92
r	0.7	0.9	0.88	0.88
R^2	0.49	0.82	0.78	0.77
NSE	0.47	0.75	0.76	0.76
KGE	0.48	0.63	0.7	0.73

While the metrics of the three deep learning algorithms were close (within a 10% difference), SA–LSTM achieved slightly less errors and higher correlations, as reflected in higher KGE scores. To further explore the streamflow predictions by the four algorithms, their estimated flowrates are plotted against the observed values during the testing period in Figure 5.

During the testing period, the streamflow monitoring station at the headwaters of the Texas Colorado River recorded 35 dry months (no-flow), which is equivalent of 34% of the testing period. Additionally, three relatively large flooding events (greater than 5 cubic meters per second) were recorded that were greater than the highest previously recorded flood at the station of interest. Given the limitation of data-driven models in predicting beyond the range of training data, the first two large flooding events were expected to be more challenging for all models. According to the results, the baseline ELM model predicted a considerable number of physically unrealistic negative streamflow predictions, mostly when the stream was dry (Figure 5a). Further, ELM severely underpredicted the largest flood events. While the overall better performance of the deep learning algorithms, previously discussed using the evaluation metrics, can also be seen with the time-series plot, Figure 5b–d provide more insight into the differences between the CNN and LSTM counterparts. Compared to the ELM model, the extent of negative flowrate estimates is less severe with the deep learning models. Additionally, the more complex models provided more accurate predictions of the extremely high flows. Based on the results, the LSTM and SA–LSTM models were superior among the investigated algorithms as they captured the extreme hydrologic events more accurately.

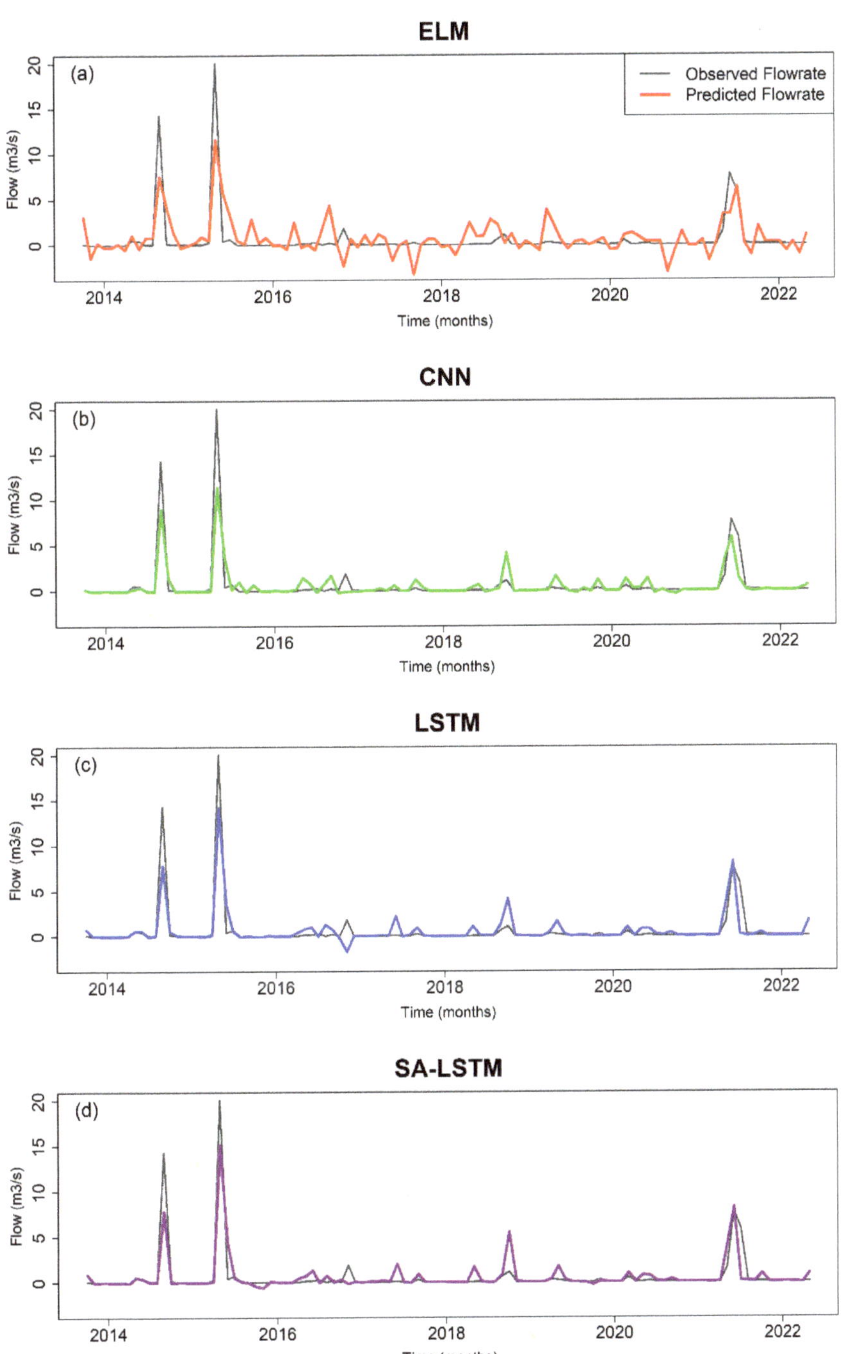

Figure 5. Observed vs. simulated flowrates by the investigated models during the testing period.

3.2. Predictive Performance for the No-Flow Events

As discussed earlier, the distinguishing characteristic of IRES is the presence of no-flow events (zero flowrate entries) in their flow time-series, and capturing these events is both important and challenging. Figure 6 illustrates the flowrate estimations of the four models for the cases when a no-flow event was recorded. While ideally, all the model predictions should be on the horizontal line of 0, according to the results, none of the investigated models in this study estimated an absolute zero flowrate. When predicting a true zero flowrate, the models either exhibited over-estimation and predicted a positive flowrate, or under-estimated the no-flow event and predicted physically unrealistic negative flowrates. The inability to capture the no-flow events leads to an uncertainty concern for the application of these streamflow prediction algorithms in IRES settings, and the extent of over- and under-estimation errors (as measured by the MAE) can be viewed as a measure of this uncertainty. The deep learning algorithms achieved closer values to zero when predicting no-flow events with lower errors (MAE < 0.1 m^3/s) in comparison to the ELM model (MAE = 0.67 m^3/s).

Table 4 summarizes the percentage of negative flowrate estimations by each model during the testing period. Despite achieving a relatively low MAE, the CNN model had the highest percentage of negative predictions, followed by the baseline ELM model. However, the advanced architecture of the LSTM models, and particularly, utilizing the attention unit, considerably reduced the extent of negative flowrate estimations. There are a number of factors contributing to the limited predictive ability of these models in estimating no-flow events.

From a hydrological perspective, the flow in headwater IRES systems tends to be seasonal and is largely controlled by overland flows following rainfall or snow-melt event. In arid and semi-arid regions, water tables tend to be deep, and the river systems are hydraulically disconnected from the water-bearing subsurface system [145]. Flow cessation (i.e., no-flow conditions) begins as water in the stream channel becomes disconnected and is present in discontinuous pockets. In the absence of precipitation, slow-moving water paths (e.g., groundwater discharges), or anthropogenic discharges (e.g., wastewater discharge from municipalities), the IRES dries up. Eventually, the intermittent headwater stream translates to ephemeral flow conditions, and the disconnected parcels of water and the exposed soils will continue to undergo evaporation until the complete flow cessation happens, resulting in no-flow conditions. As the mechanisms of flow and no-flow regimes are controlled by different hydrological processes, the assumption that they arise from a single underlying distribution is perhaps the main limitation of current data-driven models.

Further, as the streamflow prediction models try to match both zero and extremely high flows using a limited set of calibration parameters, they underestimate the high flows and overestimate the no-flow events. Therefore, the results from such models must be post-processed to induce intermittency. A cut-off threshold is often subtracted from the predicted flows to simulate intermittent flow conditions [146]. This approach, while pragmatic, is also subjective, and requires a careful assessment by experts and a detailed understanding of the surface and subsurface hydrological and geological conditions, which may not be known with certainty, even in well-characterized streams.

Table 4. Summary of the percentage of negative flowrate estimations by each model during the testing period.

	ELM	CNN	LSTM	SA–LSTM
% of negative flowrates	36%	45%	30%	21%

Thus, even though the deep learning models, particularly the LSTM models, outperformed the baseline ELM model and estimated considerably fewer negative flowrate estimations, there is still room to improve the performance of IRES streamflow prediction models and develop algorithms capable of accurately capturing no-flow conditions.

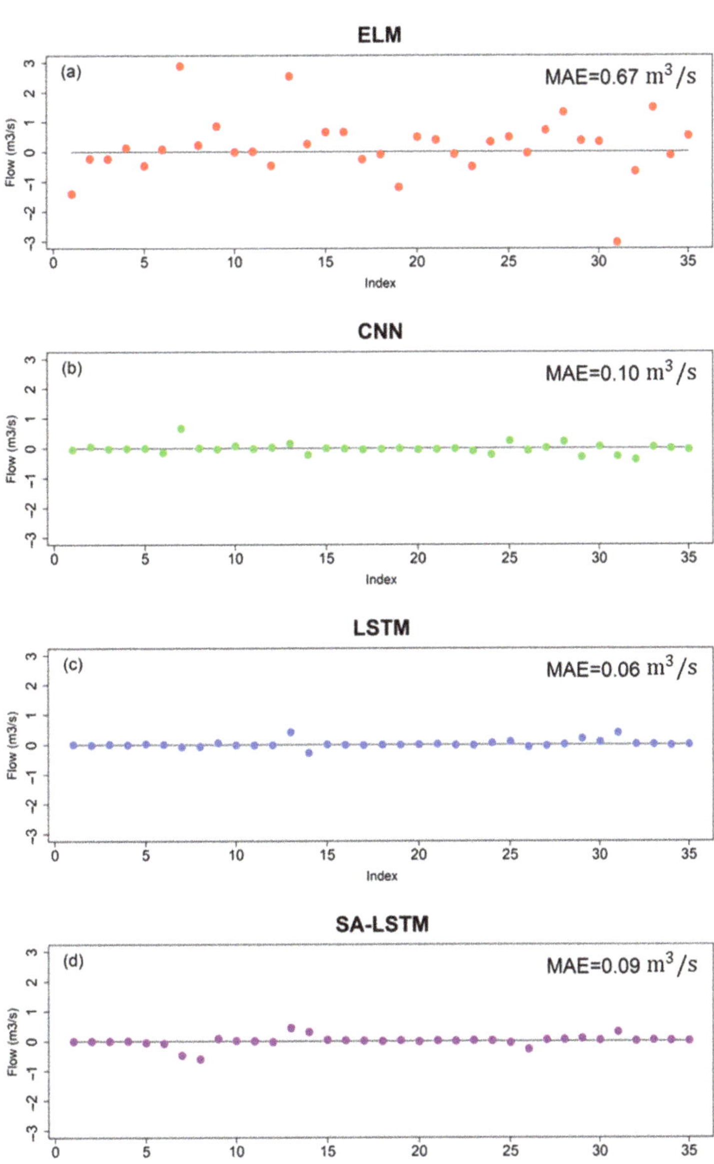

Figure 6. Observed vs. simulated flowrates for the no-flow events by the investigated models during the testing period.

3.3. Predictive Performance for the Extreme High Flow Events

The errors associated with the flowrate estimations of the four models for the three extreme flooding events are summarized in Table 5. For the 2021 flood, which was relatively similar to the 1992 flood included in the training data, and for the 2015 flood, the highest recorded flowrate in the history of the Texas Colorado River headwaters, the LSTM and SA–LSTM models offered the most accurate estimates among the investigated algorithms. The baseline ELM model was outperformed by the deep learning algorithms, with almost 50% underestimation of the largest three extreme flood events.

Table 5. Summary of estimation errors for each model for the three highest flowrates during the testing period.

Error in Flood Estimation%	ELM	CNN	LSTM	SA–LSTM
September 2014	−47%	−37%	−39%	−39%
May 2015	−42%	−43%	−29%	−24%
June 2021	−56%	−25%	+6%	+6%

To further explore the predictive ability of the models of interest for extreme high flows, their performances were compared for the top 30% of high flow events (all the flow events that were greater than or equal to the 70th quantile of all positive flowrates). MAE, RMSE, and KGE metrics were computed (Figure 7) for the four algorithms over extreme high flow estimations.

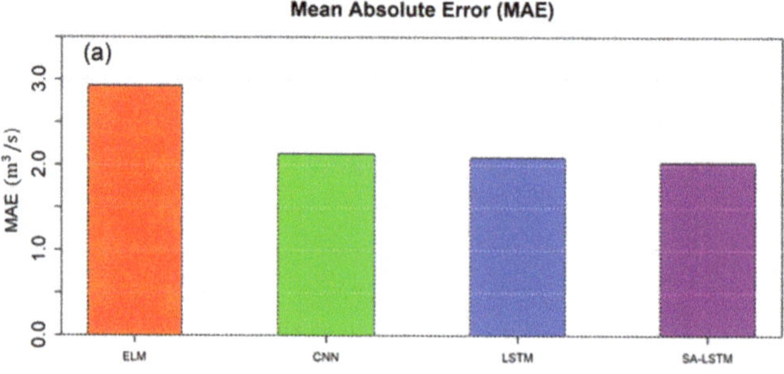

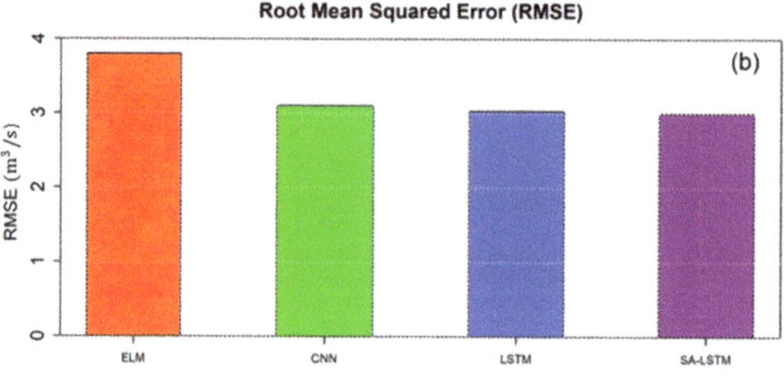

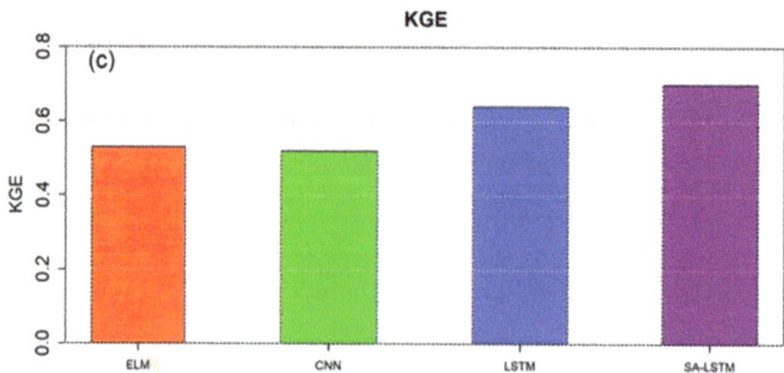

Figure 7. Evaluation of the predictive performance of the models with respect to capturing the extreme high flows during the testing period.

Considering the MAE and RMSE metrics (Figure 7a,b), it is clear that the deep learning algorithms achieved similar estimation errors, which were lower (~0.8 cubic meters per second) than the baseline ELM model. Thus, the deep learning models were identified as appropriate tools for extreme high flow estimation in the headwaters of the Colorado River in Texas and were advantageous compared to the shallow learning ELM counterpart.

A comparison of the KGE scores showed the advantage of the more complex LSTM models (K.G.E. > 0.65) in capturing the extreme high flows compared to the ELM and CNN algorithms (KGE < 0.55). As KGE is a composite metric that accounts for correlation, bias, and variability, it was concluded that the more advanced architecture and complex algorithm of the LSTM units were better alternatives for capturing the extreme high flows in the intermittent headwaters of the Colorado River in Texas.

There are two major limitations to the predictive ability of the data-driven models for estimating extreme high flow events, particularly in an IRES setting. First, as discussed earlier, the investigated data-driven models assume that the entire IRES flow data arise from a single distribution, and fitting a curve in the presence of numerous zero flow entries curtails the predictive performance of the models for the extreme high flowrates, resulting in underestimation of these events.

Second, the flow generation process in IRES is highly climate-driven, and changes in climatic patterns are likely to cause unprecedented flow events (e.g., record-breaking floods, prolonged dry spells) in such streams. This is exemplified by the case of the 2014 and 2015 floods in the headwaters of the Colorado River in Texas, where they broke the previous flood record by twice the magnitude. In such cases, to achieve an accurate streamflow estimation, the model must make a prediction outside its valid domain. The extrapolation problem or severe deviation of model performance when the inputs are dissimilar to the training data is a well-known weakness of the data-driven models, even the more complex deep learning algorithms [147–149]. According to the results, utilizing the more advanced LSTM deep learning models yielded more accurate estimates for the extrapolation cases, making them more reliable alternatives for modeling intermittent headwaters in the face of climate change. However, further research on methods to address the extrapolation problem, removing the burden of the no-flow events on extreme high flow prediction, and reducing the uncertainty associated with extreme flow analysis of IRES is needed.

4. Summary and Conclusions

Reliable streamflow prediction of intermittent rivers and ephemeral streams, such as the headwaters of the Colorado River in Texas, is an essential requirement for a variety of planning and management tasks associated with these streams, from drought analysis to flood warning systems, supply allocation, and riparian ecosystem conservation. In this study, the application of advanced deep learning algorithms, namely CNN, LSTM, and SA–LSTM models, were compared against a baseline ELM model for hydro–meteorological modeling and monthly streamflow prediction at the headwaters of the Texas Colorado River, located in a climate hotspot and exposed to changes in precipitation and temperature variability. The performance of these algorithms was evaluated using a suite of model evaluation metrics and compared over the entire range of flow, as well as for the no-flow events and extreme high flowrates. Here is a list of the major findings of this study for intermittent streamflow prediction at the headwaters of the Texas Colorado River:

- While all the investigated models offered skillful streamflow predictions (as measured by the KGE score above −0.41), overall, the deep learning models clearly outperformed the baseline ELM model with respect to all evaluation metrics. The more advanced models better captured the flow dynamics in the IRES setting and were found to be appropriate tools for streamflow prediction at the site of study.
- None of the investigated data-driven algorithms were able to capture absolute zero flowrates. However, deep learning models, more specifically LSTM and SA–LSTM, estimated closer values to zero and predicted considerably less unrealistic negative flowrates.

- Deep learners also offered more accurate predictions of the extreme high flows, with lower RMSE and MAE errors and higher correlations and KGE scores in comparison to the ELM.
- With respect to the principle of parsimony, the LSTM model is the most appropriate model among the considered alternatives as it outperformed the ELM and CNN models with considerable higher performance metrics and achieved relatively similar results to the SA–LSTM model, despite not having the attention unit and being a slightly simpler methodology.
- LSTM and SA–LSTM models outperformed their counterparts when challenged with the extrapolation problem for the unprecedented record-breaking flood events of 2014 and 2015.
- Despite its simplicity and fast speed, the ELM model was found to provide unreliable streamflow estimations, and its application is not recommended for the studied stream, particularly because it exhibited the most severe underestimation of the extreme high flows.
- The ELM model was found to be prone to overfitting and learning the noise in the training data, which yielded noticeably lower quality of performance during the independent testing period.
- The CNN model, while achieving better evaluation metrics than the baseline ELM model, predicted a large number of negative flowrates and failed to provide accurate estimates of the extreme high flows. Hence, the application of the CNN algorithm is not recommended for the stream of study.
- The SA–LSTM, as the cutting-edge alternative and the most complex tool among the investigated models, offered the best performance in capturing the extreme ends of the IRES streamflow spectrum: no-flow events and extreme floods.
- The pooling mechanism in CNNs and the dropout mechanism for the LSTM-based models were found to be effective in considerably lowering the extent of performance loss from training to testing and controlling overfitting.

According to the results of this study, deep learning algorithms are powerful and effective tools for predicting streamflow in the headwaters of the Colorado River in Texas. The layered architecture and advanced algorithm of these models allow them to model various portions of the IRES flow series, including the extreme hydrologic events, with higher accuracy, enhanced reliability, and a considerably lower extent of overfitting. Deep learning streamflow prediction models offer valuable information about IRES flow dynamics to support various management and planning efforts associated with these growingly important surface water resources. However, modelers and other groups of users should be cautious with the estimations of these data-driven models due to their limitations, such as their inability to capture absolute zero flowrates or their failure to maintain high performance when applied to data dissimilar to the training set (e.g., an unprecedented flood event). Such limitations introduce uncertainties that should be considered when applying data-driven models and interpreting their results, regardless of how advanced their architecture may be. Further research is required to develop methodologies that can capture the complex streamflow generation and cessation processes in IRES, tackle the extrapolation problem with minimal performance loss, and provide reliable intermittent streamflow prediction. Additionally, increasing the quality and quantity of available hydrologic and meteorologic data in IRES sites, such as the headwaters of the Texas Colorado River, can significantly enhance the performance of data-driven models and lead to more effective water planning in these usually water-scarce regions.

Author Contributions: Conceptualization, F.F. and G.M.; methodology, F.F. and G.M.; software, G.M.; validation, F.F.; formal analysis, F.F. and G.M.; data curation, G.M.; writing—original draft preparation, F.F.; writing—review and editing, F.F. and G.M.; visualization, F.F. and G.M.; supervision, F.F. All authors have read and agreed to the published version of the manuscript.

Funding: This research received no external funding.

Data Availability Statement: All data that support the findings of this study are available from the corresponding author upon reasonable request.

Acknowledgments: The authors would like to acknowledge financial support from J.T. and Margaret Talkington Fellowship, TTU Water Resources Center, and the Department of Civil, Environmental, and Construction Engineering at Texas Tech University.

Conflicts of Interest: The authors declare no conflict of interest.

References

1. Datry, T.; Singer, G.; Sauquet, E.; Capdevilla, D.J.; Von Schiller, D.; Subbington, R.; Magrand, C.; Paril, P.; Milisa, M.; Acuña, V. Science and management of intermittent rivers and ephemeral streams (SMIRES). *Res. Ideas Outcomes* **2017**, *3*, 23. [CrossRef]
2. Levick, L.R.; Goodrich, D.C.; Hernandez, M.; Fonseca, J.; Semmens, D.J.; Stromberg, J.C.; Tluczek, M.; Leidy, R.A.; Scianni, M.; Guertin, D.P. *The Ecological and Hydrological Significance of Ephemeral and Intermittent Streams in the Arid and Semi-Arid American Southwest*; US Environmental Protection Agency, Office of Research and Development: Washington, DC, USA, 2008.
3. Eng, K.; Wolock, D.M.; Dettinger, M.D. Sensitivity of intermittent streams to climate variations in the USA. *River Res. Appl.* **2016**, *32*, 885–895. [CrossRef]
4. Gutiérrez-Jurado, K.Y.; Partington, D.; Batelaan, O.; Cook, P.; Shanafield, M. What triggers streamflow for intermittent rivers and ephemeral streams in low-gradient catchments in Mediterranean climates. *Water Resour. Res.* **2019**, *55*, 9926–9946. [CrossRef]
5. Leigh, C.; Boulton, A.J.; Courtwright, J.L.; Fritz, K.; May, C.L.; Walker, R.H.; Datry, T. Ecological research and management of -intermittent rivers: An historical review and future directions. *Freshw. Biol.* **2016**, *61*, 1181–1199. [CrossRef]
6. Jaeger, K.L.; Sutfin, N.A.; Tooth, S.; Michaelides, K.; Singer, M. Chapter 2.1—Geomorphology and Sediment Regimes of Intermittent Rivers and Ephemeral Streams. In *Intermittent Rivers and Ephemeral Streams*; Datry, T., Bonada, N., Boulton, A., Eds.; Academic Press: Cambridge, MA, USA, 2017; pp. 21–49. [CrossRef]
7. Hill, M.J.; Milner, V.S. Ponding in intermittent streams: A refuge for lotic taxa and a habitat for newly colonising taxa? *Sci. Total Environ.* **2018**, *628–629*, 1308–1316. [CrossRef]
8. Tolonen, K.E.; Picazo, F.; Vilmi, A.; Datry, T.; Stubbington, R.; Pařil, P.; Rocha, M.P.; Heino, J. Parallels and contrasts between intermittently freezing and drying streams: From individual adaptations to biodiversity variation. *Freshw. Biol.* **2019**, *64*, 1679–1691. [CrossRef]
9. Catalán, N.; Casas-Ruiz, J.P.; von Schiller, D.; Proia, L.; Obrador, B.; Zwirnmann, E.; Marcé, R. Biodegradation kinetics of dissolved organic matter chromatographic fractions in an intermittent river. *J. Geophys. Res. Biogeosci.* **2017**, *122*, 131–144. [CrossRef]
10. Scordo, F.; Seitz, C.; Melo, W.D.; Piccolo, M.C.; Perillo, G.M.E. Natural and human impacts on the landscape evolution and hydrography of the Chico River basin (Argentinean Patagonia). *Catena* **2020**, *195*, 104783. [CrossRef]
11. Steward, D.R.; Yang, X.; Lauwo, S.Y.; Staggenborg, S.A.; Macpherson, G.L.; Welch, S.M. From precipitation to groundwater baseflow in a native prairie ecosystem: A regional study of the Konza LTER in the Flint Hills of Kansas, USA. *Hydrol. Earth Syst. Sci.* **2011**, *15*, 3181–3194. [CrossRef]
12. Courtwright, J.; May, C.L. Importance of terrestrial subsidies for native brook trout in Appalachian intermittent streams. *Freshw. Biol.* **2013**, *58*, 2423–2438. [CrossRef]
13. Datry, T.; Boulton, A.J.; Bonada, N.; Fritz, K.; Leigh, C.; Sauquet, E.; Tockner, K.; Hugueny, B.; Dahm, C.N. Flow intermittence and ecosystem services in rivers of the Anthropocene. *J. Appl. Ecol.* **2018**, *55*, 353–364. [CrossRef]
14. Karaouzas, I.; Smeti, E.; Vourka, A.; Vardakas, L.; Mentzafou, A.; Tornés, E.; Sabater, S.; Muñoz, I.; Skoulikidis, N.T.; Kalogianni, E. Assessing the ecological effects of water stress and pollution in a temporary river—Implications for water management. *Sci. Total Environ.* **2018**, *618*, 1591–1604. [CrossRef] [PubMed]
15. Vander Vorste, R.; Obedzinski, M.; Pierce, S.N.; Carlson, S.M.; Grantham, T.E. Refuges and ecological traps: Extreme drought threatens persistence of an endangered fish in intermittent streams. *Glob. Chang. Biol.* **2020**, *26*, 3834–3845. [CrossRef] [PubMed]
16. Grey, D.; Sadoff, C.W. Sink or Swim? Water security for growth and development. *Water Policy* **2007**, *9*, 545–571. [CrossRef]
17. Kampf, S.K.; Faulconer, J.; Shaw, J.R.; Lefsky, M.; Wagenbrenner, J.W.; Cooper, D.J. Rainfall thresholds for flow generation in desert ephemeral streams. *Water Resour. Res.* **2018**, *54*, 9935–9950. [CrossRef]
18. Azarnivand, A.; Camporese, M.; Alaghmand, S.; Daly, E. Simulated response of an intermittent stream to rainfall frequency patterns. *Hydrol. Processes* **2020**, *34*, 615–632. [CrossRef]
19. Sauquet, E.; Beaufort, A.; Sarremejane, R.; Thirel, G. Predicting flow intermittence in France under climate change. *Hydrol. Sci. J.* **2021**, *66*, 2046–2059. [CrossRef]
20. Tramblay, Y.; Rutkowska, A.; Sauquet, E.; Sefton, C.; Laaha, G.; Osuch, M.; Albuquerque, T.; Alves, M.H.; Banasik, K.; Beaufort, A.; et al. Trends in flow intermittence for European rivers. *Hydrol. Sci. J.* **2021**, *66*, 37–49. [CrossRef]
21. Zipper, S.C.; Hammond, J.C.; Shanafield, M.; Zimmer, M.; Datry, T.; Jones, C.N.; Kaiser, K.E.; Godsey, S.E.; Burrows, R.M.; Blaszczak, J.R.; et al. Pervasive changes in stream intermittency across the United States. *Environ. Res. Lett.* **2021**, *16*, 084033. [CrossRef]
22. Mix, K.; Groeger, A.W.; Lopes, V.L. Impacts of dam construction on streamflows during drought periods in the Upper Colorado River Basin, Texas. *Lakes Reserv. Sci. Policy Manag. Sustain. Use* **2016**, *21*, 329–337. [CrossRef]
23. Diffenbaugh, N.S.; Giorgi, F.; Pal, J.S. Climate change hotspots in the United States. *Geophys. Res. Lett.* **2008**, *35*, L16709. [CrossRef]

24. Datry, T.; Fritz, K.; Leigh, C. Challenges, developments and perspectives in intermittent river ecology. *Freshw. Biol.* **2016**, *61*, 1171–1180. [CrossRef]

25. Stubbington, R.; Paillex, A.; England, J.; Barthès, A.; Bouchez, A.; Rimet, F.; Sánchez-Montoya, M.M.; Westwood, C.G.; Datry, T. A comparison of biotic groups as dry-phase indicators of ecological quality in intermittent rivers and ephemeral streams. *Ecol. Indic.* **2019**, *97*, 165–174. [CrossRef]

26. Sazib, N.; Bolten, J.; Mladenova, I. Exploring spatiotemporal relations between soil moisture, precipitation, and streamflow for a large set of watersheds using Google Earth Engine. *Water* **2020**, *12*, 1371. [CrossRef]

27. Katz, G.L.; Denslow, M.W.; Stromberg, J.C. The Goldilocks Effect: Intermittent streams sustain more plant species than those with perennial or ephemeral flow. *Freshw. Biol.* **2012**, *57*, 467–480. [CrossRef]

28. Tooth, S.; Nanson, G.C. The role of vegetation in the formation of anabranching channels in an ephemeral river, Northern plains, arid central Australia. *Hydrol. Process.* **2000**, *14*, 3099–3117. [CrossRef]

29. Mehr, A.D. An improved gene expression programming model for streamflow forecasting in intermittent streams. *J. Hydrol.* **2018**, *563*, 669–678. [CrossRef]

30. Chebaane, M.; Salas, J.D.; Boes, D.C. Product periodic autoregressive processes for modeling intermittent monthly streamflows. *Water Resour. Res.* **1995**, *31*, 1513–1518. [CrossRef]

31. Aksoy, H.; Bayazit, M. A model for daily flows of intermittent streams. *Hydrol. Process.* **2000**, *14*, 1725–1744. [CrossRef]

32. Kişi, Ö. Neural networks and wavelet conjunction model for intermittent streamflow forecasting. *J. Hydrol. Eng.* **2009**, *14*, 773–782. [CrossRef]

33. Makwana, J.J.; Tiwari, M.K. Intermittent streamflow forecasting and extreme event modelling using wavelet based artificial neural networks. *Water Resour. Manag.* **2014**, *28*, 4857–4873. [CrossRef]

34. Badrzadeh, H.; Sarukkalige, R.; Jayawardena, A.W. Intermittent stream flow forecasting and modelling with hybrid wavelet neuro-fuzzy model. *Hydrol. Res.* **2017**, *49*, 27–40. [CrossRef]

35. Rahmani-Rezaeieh, A.; Mohammadi, M.; Danandeh Mehr, A. Ensemble gene expression programming: A new approach for evolution of parsimonious streamflow forecasting model. *Theor. Appl. Climatol.* **2020**, *139*, 549–564. [CrossRef]

36. Mehr, A.D.; Gandomi, A.H. MSGP-LASSO: An improved multi-stage genetic programming model for streamflow prediction. *Inf. Sci.* **2021**, *561*, 181–195. [CrossRef]

37. Kisi, O.; Alizamir, M.; Shiri, J. Conjunction Model Design for Intermittent Streamflow Forecasts: Extreme Learning Machine with Discrete Wavelet Transform. In *Intelligent Data Analytics for Decision-Support Systems in Hazard Mitigation*; Springer: Berlin/Heidelberg, Germany, 2021; pp. 171–181.

38. Li, M.; Robertson, D.E.; Wang, Q.J.; Bennett, J.C.; Perraud, J.-M. Reliable hourly streamflow forecasting with emphasis on ephemeral rivers. *J. Hydrol.* **2021**, *598*, 125739. [CrossRef]

39. Alizadeh, B.; Bafti, A.G.; Kamangir, H.; Zhang, Y.; Wright, D.B.; Franz, K.J. A novel attention-based LSTM cell post-processor coupled with bayesian optimization for streamflow prediction. *J. Hydrol.* **2021**, *601*, 126526. [CrossRef]

40. Uhlenbrook, S.; Seibert, J.A.N.; Leibundgut, C.; Rodhe, A. Prediction uncertainty of conceptual rainfall-runoff models caused by problems in identifying model parameters and structure. *Hydrol. Sci. J.* **1999**, *44*, 779–797. [CrossRef]

41. Hapuarachchi, H.; Bari, M.; Kabir, A.; Hasan, M.; Woldemeskel, F.; Gamage, N.; Sunter, P.; Zhang, X.; Robertson, D.; Bennett, J.; et al. Development of a national 7-day ensemble streamflow forecasting service for Australia. *Hydrol. Earth Syst. Sci. Discuss.* **2022**, *2022*, 1–35. [CrossRef]

42. Sneed, E.D.; Folk, R.L. Pebbles in the lower Colorado River, Texas a study in particle morphogenesis. *J. Geol.* **1958**, *66*, 114–150. [CrossRef]

43. Clay, C.; Kleiner, D.J. Colorado River—The Handbook of Texas Online. 2017. Available online: https://www.tshaonline.org/handbook/entries/colorado-river (accessed on 10 July 2022).

44. Samady, M.K. *Continuous Hydrologic Modeling for Analyzing the Effects of Drought on the Lower Colorado River in Texas*; Michigan Technological University: Houghton, MI, USA, 2017.

45. Nielsen-Gammon, J.W. The changing climate of Texas. *Impact Glob. Warm. Tex.* **2011**, *39*, 86.

46. Griffith, G.E.; Bryce, S.; Omernik, J.; Rogers, A. *Ecoregions of Texas*; US Geological Survey: Reston, VA, USA, 2004.

47. US Environmental Protection Agency. Available online: https://www.epa.gov/ (accessed on 10 July 2022).

48. U.S. Geological Survey. Available online: https://www.usgs.gov (accessed on 15 July 2022).

49. Moritz, S.; Sardá, A.; Bartz-Beielstein, T.; Zaefferer, M.; Stork, J. Comparison of different methods for univariate time series imputation in R. *arXiv* **2015**, arXiv:1510.03924.

50. Welch, G.; Bishop, G. *An Introduction to the Kalman Filter*; Department of Computer Science, University of North Carolina at Chapel Hill: Chapel Hill, NC, USA, 1995.

51. Godsey, S.E.; Kirchner, J.W. Dynamic, discontinuous stream networks: Hydrologically driven variations in active drainage density, flowing channels and stream order. *Hydrol. Process.* **2014**, *28*, 5791–5803. [CrossRef]

52. Durighetto, N.; Vingiani, F.; Bertassello, L.E.; Camporese, M.; Botter, G. Intraseasonal drainage network dynamics in a headwater catchment of the Italian alps. *Water Resour. Res.* **2020**, *56*, e2019WR025563. [CrossRef]

53. Botter, G.; Durighetto, N. The Stream Length Duration Curve: A Tool for Characterizing the Time Variability of the Flowing Stream Length. *Water Resour. Res.* **2020**, *56*, e2020WR027282. [CrossRef] [PubMed]

54. Botter, G.; Vingiani, F.; Senatore, A.; Jensen, C.; Weiler, M.; McGuire, K.; Mendicino, G.; Durighetto, N. Hierarchical climate-driven dynamics of the active channel length in temporary streams. *Sci. Rep.* **2021**, *11*, 21503. [CrossRef]
55. PRISM Climate Group, Oregon State University. Available online: https://prism.oregonstate.edu (accessed on 1 June 2022).
56. Thornthwaite, C.W. An approach toward a rational classification of climate. *Geogr. Rev.* **1948**, *38*, 55–94. [CrossRef]
57. Montgomery, D.C.; Jennings, C.L.; Kulahci, M. *Introduction to Time Series Analysis and Forecasting*; John Wiley & Sons: Hoboken, NJ, USA, 2015.
58. Huang, G.-B.; Zhu, Q.-Y.; Siew, C.-K. Extreme learning machine: Theory and applications. *Neurocomputing* **2006**, *70*, 489–501. [CrossRef]
59. Huang, G.-B.; Chen, L.; Siew, C.K. Universal approximation using incremental constructive feedforward networks with random hidden nodes. *IEEE Trans. Neural Netw.* **2006**, *17*, 879–892. [CrossRef]
60. Kisi, O.; Alizamir, M. Modelling reference evapotranspiration using a new wavelet conjunction heuristic method: Wavelet extreme learning machine vs. wavelet neural networks. *Agric. For. Meteorol.* **2018**, *263*, 41–48. [CrossRef]
61. Zhu, S.; Heddam, S.; Wu, S.; Dai, J.; Jia, B. Extreme learning machine-based prediction of daily water temperature for rivers. *Environ. Earth Sci.* **2019**, *78*, 202. [CrossRef]
62. Atiquzzaman, M.; Kandasamy, J. Prediction of hydrological time-series using extreme learning machine. *J. Hydroinform.* **2015**, *18*, 345–353. [CrossRef]
63. Yin, Z.; Feng, Q.; Yang, L.; Deo, R.C.; Wen, X.; Si, J.; Xiao, S. Future projection with an extreme-learning machine and support vector regression of reference evapotranspiration in a mountainous inland watershed in North-West China. *Water* **2017**, *9*, 880. [CrossRef]
64. Niu, W.-J.; Feng, Z.-K.; Zeng, M.; Feng, B.-F.; Min, Y.-W.; Cheng, C.-T.; Zhou, J.-Z. Forecasting reservoir monthly runoff via ensemble empirical mode decomposition and extreme learning machine optimized by an improved gravitational search algorithm. *Appl. Soft Comput.* **2019**, *82*, 105589. [CrossRef]
65. Yaseen, Z.M.; Faris, H.; Al-Ansari, N. Hybridized extreme learning machine model with Salp swarm algorithm: A novel predictive model for hydrological application. *Complexity* **2020**, *2020*, 8206245. [CrossRef]
66. Feng, B.-F.; Xu, Y.-S.; Zhang, T.; Zhang, X. Hydrological time series prediction by extreme learning machine and sparrow search algorithm. *Water Supply* **2021**, *22*, 3143–3157. [CrossRef]
67. Khoi, D.N.; Quan, N.T.; Linh, D.Q.; Nhi, P.T.T.; Thuy, N.T.D. Using machine learning models for predicting the water quality index in the La Buong River, Vietnam. *Water* **2022**, *14*, 1552. [CrossRef]
68. Deo, R.C.; Şahin, M. An extreme learning machine model for the simulation of monthly mean streamflow water level in Eastern Queensland. *Environ. Monit. Assess.* **2016**, *188*, 90. [CrossRef]
69. Mosavi, A.; Ozturk, P.; Chau, K.-W. Flood prediction using machine learning models: Literature review. *Water* **2018**, *10*, 1536. [CrossRef]
70. Yaseen, Z.M.; Sulaiman, S.O.; Deo, R.C.; Chau, K.-W. An enhanced extreme learning machine model for river flow forecasting: State-of-the-art, practical applications in water resource engineering area and future research direction. *J. Hydrol.* **2019**, *569*, 387–408. [CrossRef]
71. Boucher, M.-A.; Quilty, J.; Adamowski, J. Data assimilation for streamflow forecasting using extreme learning machines and multilayer perceptrons. *Water Resour. Res.* **2020**, *56*, e2019WR026226. [CrossRef]
72. Belotti, J.; Mendes, J.J.; Leme, M.; Trojan, F.; Stevan, S.L.; Siqueira, H. Comparative study of forecasting approaches in monthly streamflow series from Brazilian hydroelectric plants using Extreme Learning Machines and Box & Jenkins models. *J. Hydrol. Hydromech.* **2021**, *69*, 180–195. [CrossRef]
73. Abda, Z.; Zerouali, B.; Chettih, M.; Santos, C.A.G.; de Farias, C.A.S.; Elbeltagi, A. Assessing machine learning models for streamflow estimation: A case study in Oued Sebaou watershed (Northern Algeria). *Hydrol. Sci. J.* **2022**, *67*, 1328–1341. [CrossRef]
74. Huang, G.; Huang, G.-B.; Song, S.; You, K. Trends in extreme learning machines: A review. *Neural Netw.* **2015**, *61*, 32–48. [CrossRef] [PubMed]
75. Huang, G.-B.; Chen, L. Enhanced random search based incremental extreme learning machine. *Neurocomputing* **2008**, *71*, 3460–3468. [CrossRef]
76. Zhao, W.; Jiao, L.; Ma, W.; Zhao, J.; Zhao, J.; Liu, H.; Cao, X.; Yang, S. Superpixel-based multiple local CNN for panchromatic and multispectral image classification. *IEEE Trans. Geosci. Remote Sens.* **2017**, *55*, 4141–4156. [CrossRef]
77. Canizo, M.; Triguero, I.; Conde, A.; Onieva, E. Multi-head CNN–RNN for multi-time series anomaly detection: An industrial case study. *Neurocomputing* **2019**, *363*, 246–260. [CrossRef]
78. Shu, X.; Peng, Y.; Ding, W.; Wang, Z.; Wu, J. Multi-step-ahead monthly streamflow forecasting using convolutional neural networks. *Water Resour. Manag.* **2022**, *36*, 3949–3964. [CrossRef]
79. Ghimire, S.; Yaseen, Z.M.; Farooque, A.A.; Deo, R.C.; Zhang, J.; Tao, X. Streamflow prediction using an integrated methodology based on convolutional neural network and long short-term memory networks. *Sci. Rep.* **2021**, *11*, 17497. [CrossRef]
80. Mozo, A.; Ordozgoiti, B.; Gómez-Canaval, S. Forecasting short-term data center network traffic load with convolutional neural networks. *PLoS ONE* **2018**, *13*, e0191939. [CrossRef]
81. Barzegar, R.; Aalami, M.T.; Adamowski, J. Short-term water quality variable prediction using a hybrid CNN–LSTM deep learning model. *Stoch. Environ. Res. Risk Assess.* **2020**, *34*, 415–433. [CrossRef]

82. Baek, S.-S.; Pyo, J.; Chun, J.A. Prediction of water level and water quality using a CNN-LSTM combined deep learning approach. *Water* **2020**, *12*, 3399. [CrossRef]

83. Duan, S.; Ullrich, P.; Shu, L. Using convolutional neural networks for streamflow projection in California. *Front. Water* **2020**, *2*. [CrossRef]

84. Le, X.H.; Nguyen, D.H.; Jung, S.; Yeon, M.; Lee, G. Comparison of deep learning techniques for river streamflow forecasting. *IEEE Access* **2021**, *9*, 71805–71820. [CrossRef]

85. Xu, W.; Chen, J.; Zhang, X.J. Scale effects of the monthly streamflow prediction using a state-of-the-art deep learning model. *Water Resour. Manag.* **2022**, *36*, 3609–3625. [CrossRef]

86. Li, P.; Zhang, J.; Krebs, P. Prediction of flow based on a CNN-LSTM combined deep learning approach. *Water* **2022**, *14*, 993. [CrossRef]

87. Hochreiter, S.; Schmidhuber, J. Long short-term memory. *Neural Comput.* **1997**, *9*, 1735–1780. [CrossRef]

88. Kratzert, F.; Klotz, D.; Brenner, C.; Schulz, K.; Herrnegger, M. Rainfall–runoff modelling using Long Short-Term Memory (LSTM) networks. *Hydrol. Earth Syst. Sci.* **2018**, *22*, 6005–6022. [CrossRef]

89. Sudriani, Y.; Ridwansyah, I.; Rustini, H.A. Long short term memory (LSTM) recurrent neural network (RNN) for discharge level prediction and forecast in Cimandiri River, Indonesia. *IOP Conf. Ser. Earth Environ. Sci.* **2019**, *299*, 012037. [CrossRef]

90. Wu, Q.; Lin, H. Daily urban air quality index forecasting based on variational mode decomposition, sample entropy and LSTM neural network. *Sustain. Cities Soc.* **2019**, *50*, 101657. [CrossRef]

91. Liu, W.; Liu, W.D.; Gu, J. Forecasting oil production using ensemble empirical model decomposition based long short-term memory neural network. *J. Pet. Sci. Eng.* **2020**, *189*, 107013. [CrossRef]

92. Srivastava, N.; Hinton, G.; Krizhevsky, A.; Sutskever, I.; Salakhutdinov, R. Dropout: A simple way to prevent neural networks from overfitting. *J. Mach. Learn. Res.* **2014**, *15*, 1929–1958.

93. Liang, C.; Li, H.; Lei, M.; Du, Q. Dongting lake water level forecast and its relationship with the three gorges dam based on a long short-term memory network. *Water* **2018**, *10*, 1389. [CrossRef]

94. Bowes, B.D.; Sadler, J.M.; Morsy, M.M.; Behl, M.; Goodall, J.L. Forecasting groundwater table in a flood prone coastal city with long short-term memory and recurrent neural networks. *Water* **2019**, *11*, 1098. [CrossRef]

95. Miao, Q.; Pan, B.; Wang, H.; Hsu, K.; Sorooshian, S. Improving monsoon precipitation prediction using combined convolutional and long short term memory neural network. *Water* **2019**, *11*, 977. [CrossRef]

96. Lees, T.; Reece, S.; Kratzert, F.; Klotz, D.; Gauch, M.; De Bruijn, J.; Kumar Sahu, R.; Greve, P.; Slater, L.; Dadson, S.J. Hydrological concept formation inside long short-term memory (LSTM) networks. *Hydrol. Earth Syst. Sci.* **2022**, *26*, 3079–3101. [CrossRef]

97. Hu, C.; Wu, Q.; Li, H.; Jian, S.; Li, N.; Lou, Z. Deep learning with a long short-term memory networks approach for rainfall-runoff simulation. *Water* **2018**, *10*, 1543. [CrossRef]

98. Apaydin, H.; Feizi, H.; Sattari, M.T.; Colak, M.S.; Shamshirband, S.; Chau, K.-W. Comparative analysis of recurrent neural network architectures for reservoir inflow forecasting. *Water* **2020**, *12*, 1500. [CrossRef]

99. Thapa, S.; Zhao, Z.; Li, B.; Lu, L.; Fu, D.; Shi, X.; Tang, B.; Qi, H. Snowmelt-driven streamflow prediction using machine learning techniques (LSTM, NARX, GPR, and SVR). *Water* **2020**, *12*, 1734. [CrossRef]

100. Rahimzad, M.; Nia, A.M.; Zolfonoon, H.; Soltani, J.; Mehr, A.D.; Kwon, H.-H. Performance comparison of an LSTM-based deep learning model versus conventional machine learning algorithms for streamflow forecasting. *Water Resour. Manag.* **2021**, *35*, 4167–4187. [CrossRef]

101. Hunt, K.M.R.; Matthews, G.R.; Pappenberger, F.; Prudhomme, C. Using a long short-term memory (LSTM) neural network to boost river streamflow forecasts over the western United States. *Hydrol. Earth Syst. Sci. Discuss.* **2022**, *2022*, 1–30. [CrossRef]

102. Nogueira Filho, F.J.M.; Souza Filho, F.d.A.; Porto, V.C.; Rocha, R.V.; Sousa Estácio, Á.B.; Martins, E.S.P.R. Deep learning for streamflow regionalization for ungauged basins: Application of long-short-term-memory cells in Semiarid regions. *Water* **2022**, *14*, 1318. [CrossRef]

103. Wang, Y.; Huang, M.; Zhu, X.; Zhao, L. Attention-Based LSTM for Aspect-Level Sentiment Classification. In Proceedings of the 2016 Conference on Empirical Methods in Natural Language Processing, Austin, TX, USA, 1–5 November 2016; pp. 606–615.

104. Katrompas, A.; Metsis, V. Enhancing LSTM Models with Self-attention and Stateful Training. In Proceedings of the SAI Intelligent Systems Conference, Amsterdam, The Netherlands, 1–2 September 2022; pp. 217–235.

105. Jing, R. A self-attention based LSTM network for text classification. *J. Phys. Conf. Ser.* **2019**, *1207*, 012008. [CrossRef]

106. Chen, B.; Li, T.; Ding, W. Detecting deepfake videos based on spatiotemporal attention and convolutional LSTM. *Inf. Sci.* **2022**, *601*, 58–70. [CrossRef]

107. Pei, W.; Baltrusaitis, T.; Tax, D.M.; Morency, L.-P. Temporal attention-gated model for robust sequence classification. *IEEE Conf. Comput. Vis. Pattern Recognit.* **2016**, *1*, 6730–6739.

108. Girihagama, L.; Khaliq, M.N.; Lamontagne, P.; Perdikaris, J.; Roy, R.; Sushama, L.; Elshorbagy, A. Streamflow modelling and forecasting for Canadian watersheds using LSTM networks with attention mechanism. *Neural Comput. Appl.* **2022**. [CrossRef]

109. Yan, L.; Chen, C.; Hang, T.; Hu, Y. A stream prediction model based on attention-LSTM. *Earth Sci. Inform.* **2021**, *14*, 723–733. [CrossRef]

110. R Core Team. *R: A Language and Environment for Statistical Computing*; R Foundation for Statistical Computing: Vienna, Austria, 2022. Available online: https://www.R-project.org/ (accessed on 1 June 2022).

111. Van Rossum, G.; Drake, F.L. *Python 3 Reference Manual*; CreateSpace: Scotts Valley, CA, USA, 2009.

112. Kling, H.; Fuchs, M.; Paulin, M. Runoff conditions in the upper Danube basin under an ensemble of climate change scenarios. *J. Hydrol.* **2012**, *424–425*, 264–277. [CrossRef]
113. Moriasi, D.N.; Arnold, J.G.; Van Liew, M.W.; Bingner, R.L.; Harmel, R.D.; Veith, T.L. Model evaluation guidelines for systematic quantification of accuracy in watershed simulations. *Trans. ASABE* **2007**, *50*, 885–900. [CrossRef]
114. Willmott, C.J. On the validation of models. *Phys. Geogr.* **1981**, *2*, 184–194. [CrossRef]
115. Legates, D.R.; McCabe, G.J., Jr. Evaluating the use of "goodness-of-fit" Measures in hydrologic and hydroclimatic model validation. *Water Resour. Res.* **1999**, *35*, 233–241. [CrossRef]
116. Pearson, K., IV. Contributions to the mathematical theory of evolution. III. Regression, heredity, and panmixia. *Proc. R. Soc. Lond.* **1896**, *59*, 69–71. [CrossRef]
117. Galton, A. *English Prose: From Maundevile to Thackeray*; W. Scott: London, UK; Gage: Toronto, WJ, Canada, 1888; Volume 35.
118. Lee Rodgers, J.; Nicewander, W.A. Thirteen ways to look at the correlation coefficient. *Am. Stat.* **1988**, *42*, 59–66. [CrossRef]
119. Asuero, A.G.; Sayago, A.; González, A.G. The correlation coefficient: An overview. *Crit. Rev. Anal. Chem.* **2006**, *36*, 41–59. [CrossRef]
120. Schober, P.; Boer, C.; Schwarte, L.A. Correlation coefficients: Appropriate use and interpretation. *Anesth. Analg.* **2018**, *126*, 1763–1768. [CrossRef] [PubMed]
121. Draper, N.R.; Smith, H. *Applied Regression Analysis*; John Wiley & Sons: Hoboken, NJ, USA, 1998; Volume 326.
122. Nash, J.E.; Sutcliffe, J.V. River flow forecasting through conceptual models part I—A discussion of principles. *J. Hydrol.* **1970**, *10*, 282–290. [CrossRef]
123. McCuen, R.H.; Knight, Z.; Cutter, A.G. Evaluation of the Nash–Sutcliffe Efficiency Index. *J. Hydrol. Eng.* **2006**, *11*, 597–602. [CrossRef]
124. Lin, F.; Chen, X.; Yao, H. Evaluating the use of Nash-Sutcliffe Efficiency coefficient in goodness-of-fit measures for daily runoff simulation with SWAT. *J. Hydrol. Eng.* **2017**, *22*, 05017023. [CrossRef]
125. Gupta, H.V.; Kling, H.; Yilmaz, K.K.; Martinez, G.F. Decomposition of the mean squared error and NSE performance criteria: Implications for improving hydrological modelling. *J. Hydrol.* **2009**, *377*, 80–91. [CrossRef]
126. Milella, P.; Bisantino, T.; Gentile, F.; Iacobellis, V.; Liuzzi, G.T. Diagnostic analysis of distributed input and parameter datasets in Mediterranean basin streamflow modeling. *J. Hydrol.* **2012**, *472–473*, 262–276. [CrossRef]
127. Zajac, Z.; Revilla-Romero, B.; Salamon, P.; Burek, P.; Hirpa, F.A.; Beck, H. The impact of lake and reservoir parameterization on global streamflow simulation. *J. Hydrol.* **2017**, *548*, 552–568. [CrossRef]
128. Paul, M.; Negahban-Azar, M. Sensitivity and uncertainty analysis for streamflow prediction using multiple optimization algorithms and objective functions: San Joaquin Watershed, California. *Modeling Earth Syst. Environ.* **2018**, *4*, 1509–1525. [CrossRef]
129. Alfieri, L.; Lorini, V.; Hirpa, F.A.; Harrigan, S.; Zsoter, E.; Prudhomme, C.; Salamon, P. A global streamflow reanalysis for 1980–2018. *J. Hydrol. X* **2020**, *6*, 100049. [CrossRef] [PubMed]
130. Hallouin, T.; Bruen, M.; O'Loughlin, F.E. Calibration of hydrological models for ecologically relevant streamflow predictions: A trade-off between fitting well to data and estimating consistent parameter sets? *Hydrol. Earth Syst. Sci.* **2020**, *24*, 1031–1054. [CrossRef]
131. Knoben, W.J.M.; Freer, J.E.; Woods, R.A. Technical note: Inherent benchmark or not? Comparing Nash–Sutcliffe and Kling–Gupta efficiency scores. *Hydrol. Earth Syst. Sci.* **2019**, *23*, 4323–4331. [CrossRef]
132. Zagoruyko, S.; Komodakis, N. Wide residual networks. *arXiv* **2016**, arXiv:1605.07146.
133. Salman, S.; Liu, X. Overfitting mechanism and avoidance in deep neural networks. *arXiv* **2019**, arXiv:1901.06566.
134. Hawkins, D.M. The problem of overfitting. *J. Chem. Inf. Comput. Sci.* **2004**, *44*, 1–12. [CrossRef]
135. Dietterich, T. Overfitting and undercomputing in machine learning. *ACM Comput. Surv. CSUR* **1995**, *27*, 326–327. [CrossRef]
136. Deng, W.-Y.; Zheng, Q.-H.; Chen, L.; Xu, X.-B. Research on extreme learning of neural networks. *Chin. J. Comput.* **2010**, *33*, 279–287.
137. Ding, S.; Zhao, H.; Zhang, Y.; Xu, X.; Nie, R. Extreme learning machine: Algorithm, theory and applications. *Artif. Intell. Rev.* **2015**, *44*, 103–115. [CrossRef]
138. Ashiquzzaman, A.; Tushar, A.K.; Islam, M.; Shon, D.; Im, K.; Park, J.-H.; Lim, D.-S.; Kim, J. Reduction of Overfitting in Diabetes Prediction Using Deep Learning Neural Network. In *IT Convergence and Security 2017*; Springer: Berlin/Heidelberg, Germany, 2018; pp. 35–43.
139. Sheela, K.G.; Deepa, S.N. Review on methods to fix number of hidden neurons in neural networks. *Math. Probl. Eng.* **2013**, *2013*, 425740. [CrossRef]
140. Nair, V.; Hinton, G.E. Rectified Linear Units Improve Restricted Boltzmann Machines. In Proceedings of the 27th International Conference on International Conference on Machine Learning, Haifa, Israel, 21–24 June 2010.
141. Liu, Z.P.; Castagna, J.P. Avoiding Overfitting Caused by Noise Using a Uniform Training Mode. In Proceedings of the IJCNN'99 International Joint Conference on Neural Networks (Cat. No.99CH36339), Washington, DC, USA, 10–16 July 1999; Volume 1783, pp. 1788–1793.
142. Martín-Félez, R.; Xiang, T. Uncooperative gait recognition by learning to rank. *Pattern Recognit.* **2014**, *47*, 3793–3806. [CrossRef]
143. Qian, L.; Hu, L.; Zhao, L.; Wang, T.; Jiang, R. Sequence-dropout block for reducing overfitting problem in image classification. *IEEE Access* **2020**, *8*, 62830–62840. [CrossRef]

144. Bejani, M.M.; Ghatee, M. A systematic review on overfitting control in shallow and deep neural networks. *Artif. Intell. Rev.* **2021**, *54*, 6391–6438. [CrossRef]
145. Uddameri, V.; Singaraju, S.; Karim, A.; Gowda, P.; Bailey, R.; Schipanski, M. Understanding climate-hydrologic-human interactions to guide groundwater model development for southern high plains. *J. Contemp. Water Res. Educ.* **2017**, *162*, 79–99. [CrossRef]
146. De Girolamo, A.M.; Bouraoui, F.; Buffagni, A.; Pappagallo, G.; Lo Porto, A. Hydrology under climate change in a temporary river system: Potential impact on water balance and flow regime. *River Res. Appl.* **2017**, *33*, 1219–1232. [CrossRef]
147. Reichstein, M.; Camps-Valls, G.; Stevens, B.; Jung, M.; Denzler, J.; Carvalhais, N.; Prabhat. Deep learning and process understanding for data-driven Earth system science. *Nature* **2019**, *566*, 195–204. [CrossRef]
148. Meredig, B.; Antono, E.; Church, C.; Hutchinson, M.; Ling, J.; Paradiso, S.; Blaiszik, B.; Foster, I.; Gibbons, B.; Hattrick-Simpers, J. Can machine learning identify the next high-temperature superconductor? Examining extrapolation performance for materials discovery. *Mol. Syst. Des. Eng.* **2018**, *3*, 819–825. [CrossRef]
149. Reyes, K.G.; Maruyama, B. The machine learning revolution in materials? *MRS Bull.* **2019**, *44*, 530–537. [CrossRef]

Article

Artificial Neural Networks and Multiple Linear Regression for Filling in Missing Daily Rainfall Data

Ioannis Papailiou [1], Fotios Spyropoulos [1], Ioannis Trichakis [2,*] and George P. Karatzas [1]

1   School of Chemical and Environmental Engineering, Technical University of Crete, 73100 Chania, Greece
2   European Commission, Joint Research Centre, 21027 Ispra, Italy
*   Correspondence: ioannis.trichakis@ec.europa.eu

Abstract: As demand for more hydrological data has been increasing, there is a need for the development of more accurate and descriptive models. A pending issue regarding the input data of said models is the missing data from observation stations in the field. In this paper, a methodology utilizing ensembles of artificial neural networks is developed with the goal of estimating missing precipitation data in the extended region of Chania, Greece on a daily timestep. In the investigated stations, there have been multiple missing data events, as well as missing data prior to their installation. The methodology presented aims to generate precipitation time series based on observed data from neighboring stations and its results have been compared with a Multiple Linear Regression model as the basis for improvements to standard practice. For each combination of stations missing daily data, an ensemble has been developed. According to the statistical indexes that were calculated, ANN ensembles resulted in increased accuracy compared to the Multiple Linear Regression model. Despite this, the training time of the ensembles was quite long compared to that of the Multiple Linear Regression model, which suggests that increased accuracy comes at the cost of calculation time and processing power. In conclusion, when dealing with missing data in precipitation time series, ANNs yield more accurate results compared to MLR methods but require more time for producing them. The urgency of the required data in essence dictates which method should be used.

Keywords: rainfall time series; artificial neural networks; Multiple Linear Regression; Chania

Citation: Papailiou, I.; Spyropoulos, F.; Trichakis, I.; Karatzas, G.P. Artificial Neural Networks and Multiple Linear Regression for Filling in Missing Daily Rainfall Data. *Water* **2022**, *14*, 2892. https://doi.org/10.3390/w14182892

Academic Editors: Fi-John Chang, Li-Chiu Chang and Jui-Fa Chen

Received: 26 August 2022
Accepted: 13 September 2022
Published: 16 September 2022

Publisher's Note: MDPI stays neutral with regard to jurisdictional claims in published maps and institutional affiliations.

1. Introduction

The successful development of reliable models for predicting the status of water resources of a particular region is inextricably linked to the quantity and quality of the climate and hydrological data used [1]. One of the most critical pieces of data for such a study is the available rainfall data in the area of interest [2]. The possibility of errors or gaps within an available rainfall data time series is real and may be due to errors in the measuring instruments, a possible instrument failure, or an extreme weather event. Therefore, the development of a model capable of accurately simulating, or even complementing, a time series of rainfall data is necessary.

The importance of rainfall data availability is inarguable in hydrological modelling as these data are an essential input parameter in almost any approach. Previous research has supported the notion that the traditional statistical methods for infilling (imputing) missing data may be inefficient for small temporal and spatial scales [3,4]. Thus, an indicator of the success of the model is its outperformance over standard interpolation methods. Such practices have become more nuanced over the years, specifically with the incorporation of weighting factors that compensate for the variation between stations due to the morphological features of each case study [5].

When looking at the recently published scientific literature, Artificial Neural Networks have shown encouraging results in modeling nonlinear problems, such as hydrological processes [6]. They are able to recognize strong seasonal patterns without the need for preprocessing raw data to remove outliers, and there is solid evidence that supports the accuracy of their prediction [7]. A work similar to the current article has been conducted using meteorological data from the internet, with the intent of forecasting future rainfall using multi-layer perceptron (MLP) with back propagation and optimization algorithms [8]. In another work, the MLPs are used for forecasting future precipitation using rainfall data from nearby weather stations as inputs [9]. As an alternative method for monthly rainfall prediction, it has been suggested that the use of ANNs with wavelet regression provides more accurate results compared to models using ANNs, which implies the need for optimization [10]. An alternative to MLPs is Long Short-Term Memory networks, which are a class of recurrent neural networks that have shown promising results in estimating runoff from rainfall. With respect to the problem at hand, the selected neural networks provide a high degree of regression ability. Using recurrent networks, like those used in rainfall runoff modelling [11], would not have a physical meaning, since the relationship between inputs and outputs (daily rainfall values) does not include a temporal delay. Other techniques for filling in missing data in the field of hydrology include K-nearest neighbors (KNN), adaptive neuro-fuzzy inference systems (ANFISs) and random forest regression (RFR) [12–14], but these go beyond the scope of this work and could be considered for future research. Regarding the number of inputs, large numbers of different inputs do not guarantee more accurate results. A genetic algorithm can improve the process of selection when aiming for forecasting, but in this work, in order to reduce computational demands and given the nature of the network, another optimization method was chosen [15]. Apart from genetic algorithms as optimization techniques, others exist, such as particle swarm, cuckoo search, and bat- or kidney-inspired algorithms, depending on the level of strictness demanded [16]. In this paper, optimization is achieved through the use of a competitive algorithm in the creation of each ensemble, corresponding to each combination of missing data from the observation stations. Artificial intelligence tools have been implemented in the past in different scientific fields, from filling in spatially and temporally missing data by using augmented interpolation [17] to using photonic neural networks analysis for the changing morphology of an area [18]. In regions with high unpredictability due to extreme weather conditions, ANNs have been successful in forecasting rainfall [19]; given this fact, ANNs might perform even better in regions with strong seasonal patterns and a temperate climate, such as Crete. In large areas with varied topography, proximity of stations does not always guarantee a correlation between observed rainfall values, especially if the stations belong in two different hydrological catchments [1]. In the current case study, the area is hydrologically homogenous with only a small increase in precipitation at higher elevations [20]. In addition, fluctuations between extreme values can be smoothed out by classifying data either spatially [21] or based on intensity [22], which implies training and using multiple ANNs. Multiple ANNs with targeted training working on their own niche outperform an all-purpose ANN trained with the whole data set, with differences being dependent on the physical problem [23]. As hinted previously, multiple ANNs creating an ensemble might outperform a singular one by minimizing the occurrence of local minimums and individual biases [24]. The most simplified approach to composing an ensemble of neural networks is averaging their results using simple or weighted averages. Previous research has also proposed that the structure of the ANN ensemble can itself become the input of a general regression neural network [25]. This technique can exploit the variability of results produced by biased individuals and increase overall accuracy. In addition, it utilizes a full set of ANNs in which there may be individuals that produce error-increasing results. In order to address this, it is suggested to develop competitive algorithms where ANNs or ensembles are compared to each other and the best-performing method ends up being used for predictions [26]. In the same manner of thinking, elimination of the least significant input variables can be performed in an ensemble by considering the correlation coefficient, which has been mostly applied to climatic variables in forecasting rather than regression-based forecasting [27]. One approach to creating an ensemble with a limited data set is to alternate between training and testing data sets during the training period and eventually average out the ensemble outputs [28]. Another issue arising when working with ANNs, especially ensembles, is the network architecture, since it can greatly impact the performance; in most cases, an optimization algorithm is developed since there is no standard and optimal architecture is defined by trial and error [24]. Finally, one optimization technique which borders on architecture modification is the dropout method which randomly turns off units and their connections during training [29], which shows that random-based optimization might produce adequate results.

This paper aims to develop a methodology to estimate missing daily precipitation values from weather stations. Five weather stations monitoring rainfall in the prefecture of Chania, Greece, were used as a case study. This work focuses on the comparison of ANN ensembles based on multi-layer perceptrons and the more commonly used multiple linear regression (MLR) for completion of time series of daily rainfall data. This way, the results of the ANNs are compared to a technique that is standard practice in the field (MLR) [13]. In this approach, the best ANN from each ensemble imputes the missing data values to end up with a completed dataset for all stations. It is important to state that classification based on different combinations of missing data (henceforth called cases) adds to the accuracy of the model in general, since the ANNs are specialized in each case. This would not be feasible if modeling was done by creating a single ensemble for all stations, or an ensemble for each station. The respective MLR results are calculated as a baseline for comparison.

2. Materials and Methods

2.1. ANN and MLR Creation

The proposed methodology starts from a dataset with missing rainfall data for some stations and results in two completed datasets from the ANN ensembles and the MLR. The first step of the algorithm is to check every date containing recorded data. If a daily dataset has no missing values, then it is included in the dataset which will be used for training and validation of the ANN ensembles and validation of the MLR model. Otherwise, it is added to the dataset meant for imputing. It is important at this point to state that if a daily dataset has no recorded data at any of the stations, then imputation is unfeasible with the proposed methodology, primarily because completion of the time series occurs on a daily timestep by correlating the missing data with the observed data. In addition, a precipitation event is not dependent on a past precipitation event, and since rainfall is the sole input in this model, it was deemed both unnecessary and accuracy-decreasing to impute the time series by correlating data from datasets that correspond to different dates. This is the reason why the completed time series span from the first recorded dataset up to the current day and not further into the past or future.

The outcomes of the separation are two datasets: a complete and an incomplete one. The full daily datasets will be used for the training and validation of the ANN ensemble. Due to the different cases of missing data, it was deemed necessary to create multiple ANNs (multiple layer perceptron) that are specialized to each case, since inputs and outputs for each case differ, which implies a different topology for each case. The inputs and outputs are always daily rainfall values from weather stations, and for each different combination of missing data, the stations with observed values are used as input nodes and the stations with missing values are used as output nodes. In order to increase the accuracy of the model altogether, for each case an ensemble of 10,000 ANNs with one hidden layer was trained, in which the daily datasets for training and validation were randomly selected from the full set. With the use of a competitive algorithm, only one ANN—the best-performing one, according to its test error value—was selected to give outputs, using MATLAB's ANN tool version 2017b. According to the literature [30,31], one hidden layer is sufficient and might also outperform ANNs with multiple hidden layers when used for regression. The competitive algorithm selects the best-performing ANN based on training error and the results are produced solely based on that ANN. The use of ensembles instead of one single ANN addresses any concerns regarding the reliability, performance, and behavior of the proposed approach. The calibration (training and validation) dataset was 80% of the full available dataset with complete records, and the testing dataset consisted of the remaining 20% for all ANNs. After the training and validation are conducted, the ensembles are ready to complete the time series. Similarly, the MLR functions are created by the training and validation dataset for each case. After both processes have completed the time series, all negative values that are generated are turned into zeroes.

The whole process is graphically represented in Figure 1 below.

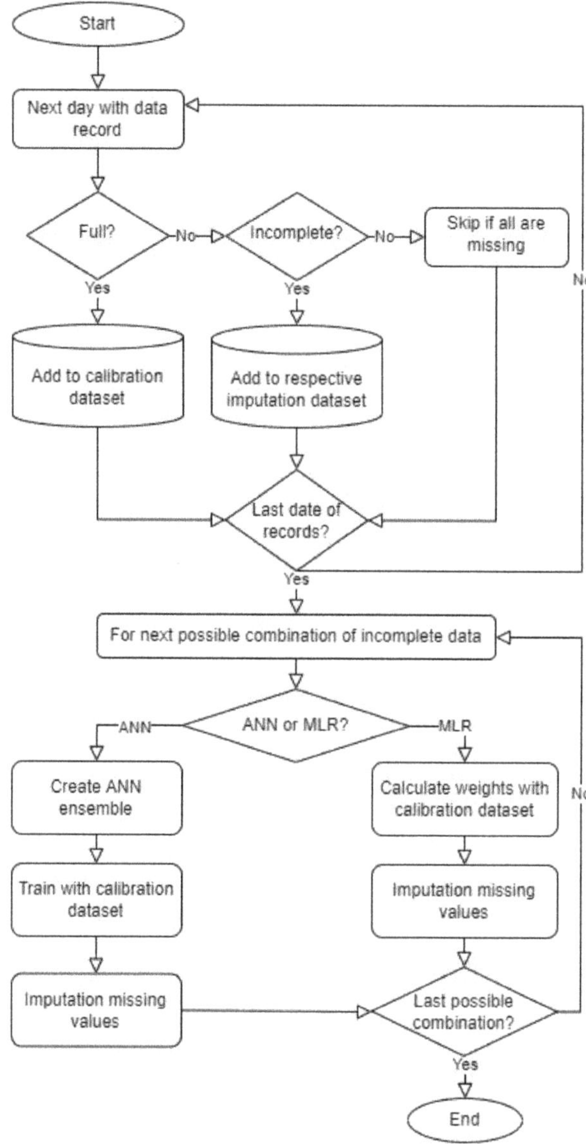

Figure 1. Flowchart of the methodology.

2.2. Model Evaluation

The validity of the results of both models is verified by the calculation of the correlation coefficients between the target and the simulated value. The value of the Nash–Sutcliffe coefficient is calculated, which can take values from minus infinity to one ($-\infty$ to 1), based on which the validity of the model is determined, with a value of one (1) indicating complete agreement between the simulated values given by the model and those observed by the stations. According to the literature, an NSE index value above 0.7 corresponds to a very good estimation [32]. Finally, the Root Mean Square Error is extracted from the model results in each of the cases considered [32].

2.3. Case Study

In the prefecture of Chania, near the northern coast of Crete, there are five automatic weather stations at a relatively close distance (approximately 5 km) to each other, as shown in Figure 2.

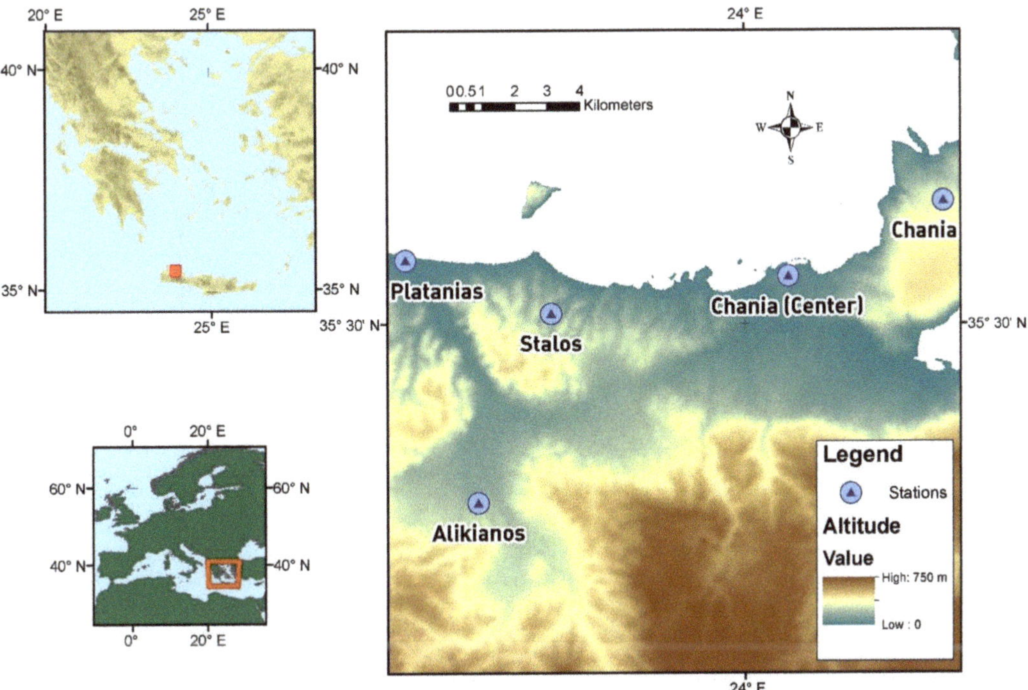

Figure 2. Weather station locations.

Regarding the locations of the stations as shown in Figure 2, the overall highest value of rainfall, historically, has occurred at Alikianos station, while the lowest has occurred at Platanias station. Platanias station has the lowest recorded altitude at 12 m, while Alikianos station is located at 95 m. Chania station (137 m) is located at a higher altitude than Alikianos station (95 m). Although it would be expected that a station at a higher altitude has a greater amount of rainfall, it was observed from the data that Alikianos station has a greater amount of rainfall. The reason for this might be that Alikianos station is furthest from the sea compared to all the other stations considered and is situated at the foot of the Lefka Ori. Platanias, on the other hand, is located a short distance from the sea and at a low altitude.

Table 1 contains a summary of the recorded daily precipitation values available from the automatic weather station NOANN network [33] (in total 15,040 records).

Table 1. Daily data availability and initial operating day of each rainfall station.

Station	Altitude (m)	Number of Data	Start of Data
Alikianos	95	3044	1 September 2012
Chania	137	5448	1 February 2006
Chania (Center)	7	3745	1 October 2010
Platanias	12	2011	1 July 2015
Stalos	93	792	1 November 2018

Based on these records, a timeline showing the availability and gaps in the datasets for the study period is shown in Figure 3. In total, 759 days had a complete dataset and were used for calibration and 4689 days had at least one missing value.

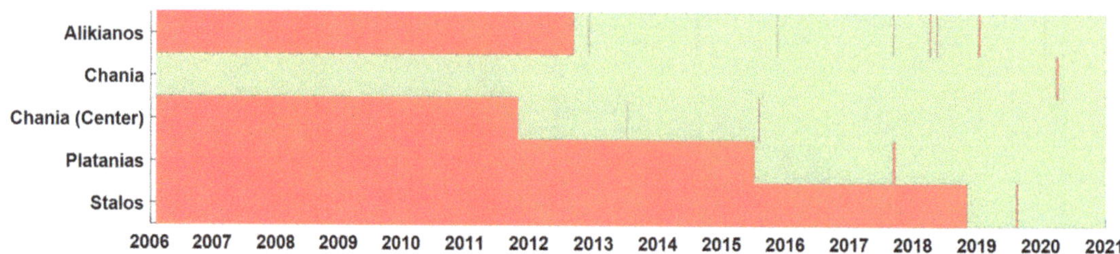

Figure 3. Timeline of daily rainfall data availability and gaps in the datasets (red color indicates gaps).

The recording of the data used in this work starts with the creation of Chania station on 1 February 2006. This means that for the period from 1 February 2006 to 30 September 2010, the available rainfall data originates only from Chania station. As of the next day, on 1 October 2010, when Chania station (Center) was put into operation, the recorded rainfall data come from the two stations previously mentioned. On 1 September 2012 the Alikianos meteorological station was put into function, therefore the recorded rainfall data come from the above three meteorological stations. To continue, on 1 July 2015, the recording of rainfall data from Platanias meteorological station starts, which means that the model input data comes from four stations. Finally, on 1 November 2018, the last station, Stalos, was put into operation. Therefore, for the next period, we have logging data from all five stations until 31 December 2020. It is worth noting that the period of time that a station is in operation is not always the same as the period of time that it records data, as there may be losses due to errors in the measuring instruments, a possible instrument failure, or an extreme weather phenomenon. This is clearly shown in Figure 3 of the paper.

2.4. Different Combinations of Stations Missing Data (Cases)

There are five rainfall stations in our study and each one of them has a different installation date, from which point on data are available. In addition, there are periods when, for different reasons (maintenance, power cuts, malfunction), one or more daily values are missing from the time series. The values missing for each day, together with the values available, can be categorized into different cases, in order to organize and group the different dates based on different calculation needs.

Figure 4 shows all the possible combinations of stations having or missing a daily record. By having all the possible cases identified, the algorithm is able to create ensembles for cases that have not occurred yet.

In the full, observed dataset, 9 cases occur out of a total of 32 that were theoretically possible. Specifically, the included cases are Case 2, Case 3, Case 6, Case 11, Case 14, Case 15, Case 22, Case 24, and Case 29. In three cases, namely Cases 2, 3 and 6, one station had a missing value; in three other cases, namely Cases 11, 14 and 15, two stations had missing values; in two cases, Cases 22 and 24, three stations had missing values; and in the last case, Case 29, four stations had missing values. The numbering of each case is not derived from the numerical order, but from the corresponding case, as shown in Figure 4. For example, in Case 2 the input precipitation data are the values from Chania, Chania (Center), Platanias and Stalos stations, and the output is the precipitation value for Alikianos station.

Case	Alikianos	Chania	Chania (Center)	Platanias	Stalos
1	1	1	1	1	1
2	0	1	1	1	1
3	1	0	1	1	1
4	1	1	0	1	1
5	1	1	1	0	1
6	1	1	1	1	0
7	0	0	1	1	1
8	0	1	0	1	1
9	1	0	1	1	0
10	0	1	1	0	1
11	0	1	1	1	0
12	1	0	0	1	1
13	1	1	0	0	1
14	1	1	1	0	0
15	1	1	0	1	0
16	1	0	1	0	1
17	0	0	0	1	1
18	0	0	1	0	1
19	0	0	1	1	0
20	0	1	0	0	1
21	0	1	0	1	0
22	0	1	1	0	0
23	1	0	0	0	1
24	1	1	0	0	0
25	1	0	1	0	0
26	1	0	0	1	0
27	0	0	0	0	1
28	1	0	0	0	0
29	0	1	0	0	0
30	0	0	1	0	0
31	0	0	0	1	0
32	0	0	0	0	0

Figure 4. Possible combinations of availability of daily rainfall data. Red indicates that the station in question has no recorded rainfall value for the day of recording. Cases occurring in the dataset are shown in bold.

3. Results

After completing a full run of the algorithm built using the proposed methodology, the incomplete time series of each station receives model-generated data for the full period in which at least one of the five stations has an observed value. In the following charts (Figure 5), the results of the two methodologies are shown for all stations. In the left column, the model-generated values of the ANN have an orange color, and in the right column, the model-generated values of the MLR have a red color, while the observed values in all charts are in a blue color.

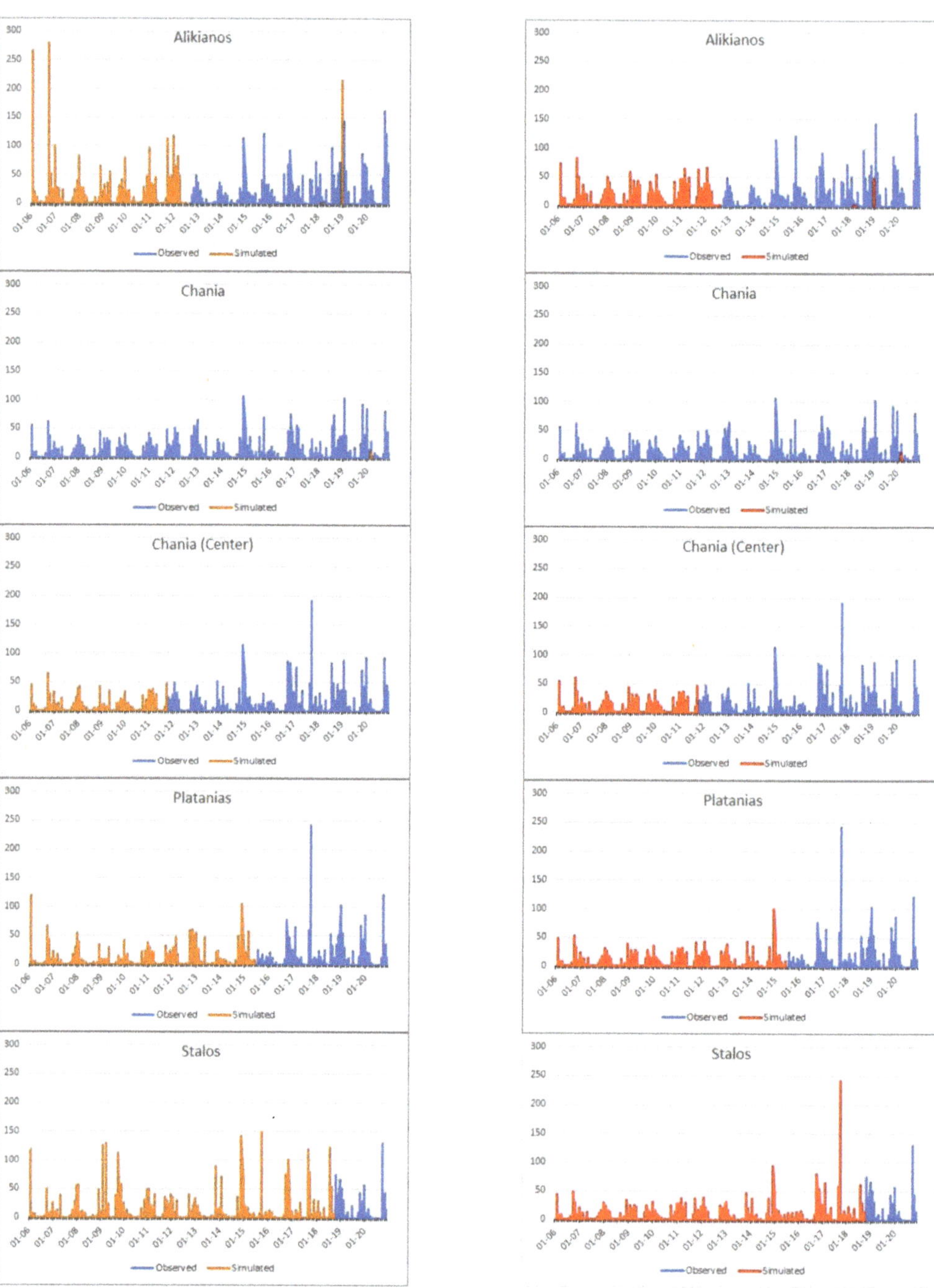

Figure 5. Observed values (blue), and model-generated values from the ANNs (orange) and MLR model (red).

To compare the two methods, three different metrics were used, the root mean square error, the Nash–Sutcliffe efficiency coefficient, and the correlation coefficient. The results are shown on a per-case basis, as the two methods might show different sensitivities to missing data. Comparative tables at the end of each section summarize the results of the testing dataset.

Concerning the computational effort and time needed for the two methods, the ANN did take a considerably large amount of time to optimize its structure (almost 36 h on a PC (Personal Computer) with an Intel i7 8th generation processor). On the other hand, the MLR was significantly faster, and required only a few minutes to run.

3.1. Root Mean Square Error (RMSE)

The RMSE index indicates the deviation between the observed and simulated values and indicates whether the data are clustered around the line of best fit. The models calculate the Root Mean Square Error (RMSE) for each of the cases considered. Regarding the Artificial Neural Network model, the best value is presented in Case 15 and shows an error equal to 1.16 mm, while the worst value is presented in Case 29, with an error value equal to 2.42 mm. The corresponding results of the Multiple Linear Regression model are shown in Case 3 with a value of 2.37 mm and Case 2 with a value of 6.43 mm. Overall, the Artificial Neural Network model shows lower errors, ranging from 42% to 72.6%, compared to the Multiple Linear Regression model.

The following Table 2 contains all the above results aggregated as follows:

Table 2. Root Mean Square Error of testing dataset.

Case	RMSE [mm]	
	ANN	**MLR**
Case 2	1.76	6.43
Case 3	1.22	2.37
Case 6	1.22	2.92
Case 11	2.30	4.99
Case 14	1.24	3.03
Case 15	1.16	2.46
Case 22	2.19	4.47
Case 24	1.83	3.16
Case 29	2.42	4.94

3.2. Nash–Sutcliffe Efficiency

The Nash–Sutcliffe coefficient can take values from minus infinity to one ($-\infty$ to 1), where for these values the following applies:

- If NSE = 1, then there is a complete match between the simulated values given by the model and those observed by the stations;
- If NSE = 0, then the values simulated by the model give the same result as if the average of the observed values of the stations were used as the forecast model for each time point;
- If NSE < 0, then the model is practically unusable, as the values simulated by it give a less accurate result than if the average of the observed values of the stations were used as a predictive model for each time point.

With respect to the calculation of the Nash–Sutcliffe coefficients, the Artificial Neural Network model shows, again, higher overall values ranging from 2.1% to 28.7%. For the Artificial Neural Network model, the best value of the Nash–Sutcliffe coefficient is presented in Case 15, with a value of 0.989, while the worst value is presented in Case 29, with a value of 0.911. The corresponding results for the Multiple Linear Regression model appear in Case 15 with a value of 0.968 and in Case 29 with a value of 0.708.

Similarly, the Nash–Sutcliffe Efficiency values for all cases are presented in the following Table 3 which contains all the results in an aggregated way:

Table 3. Nash–Sutcliffe Efficiencies of testing dataset.

Case	Nash–Sutcliffe Efficiency		Simulated Precipitation Value Station(s)
	ANN	**MLR**	
Case 2	0.967	0.803	Alikianos
Case 3	0.975	0.937	Chania
Case 6	0.981	0.882	Stalos
Case 11	0.954	0.803	Alikianos
	0.957	0.882	Stalos
Case 14	0.976	0.908	Platanias
	0.969	0.845	Stalos
Case 15	0.989	0.968	Chania (Center)
	0.973	0.871	Stalos
Case 22	0.934	0.802	Alikianos
	0.957	0.908	Platanias
	0.927	0.844	Stalos
Case 24	0.975	0.954	Chania (Center)
	0.957	0.869	Platanias
	0.943	0.781	Stalos
Case 29	0.911	0.708	Alikianos
	0.971	0.933	Chania (Center)
	0.968	0.843	Platanias
	0.959	0.748	Stalos

The results show a clear increase in the performance of the Nash–Sutcliffe efficiency when using the ANN instead of MLR. The ANN's performance was also higher when fewer stations were available compared to its MLR counterpart, which had a declining performance especially when one or two stations were available. It is also clear that there is a great correlation between the Chania (Center) and Chania stations, so when one is available, the results for the other are always very good. This is confirmed by the results of Case 3 where only Chania station is missing and from the results of Cases 15, 24 and 29, where station Chania is available, and Chania (Center) is missing.

3.3. Coefficient of Correlation (R)

The Coefficient of Correlation (R) indicates the proportion of variance of the dependent variable derived from the independent variable. A value of one (1) is the maximum value the coefficient can take, which indicates that there is a complete match between the two compared values.

Regarding the calculation of the Correlation Coefficient (R) for each case, the Artificial Neural Network model shows higher overall values ranging from 5.4% to 29.7%. More specifically, the best value of the above coefficient for the Artificial Neural Network model is presented in Case 15, with a value of 0.99274, while the worst value is presented in Case 6, with a value of 0.93957. The corresponding results for the Multiple Linear Regression model appear in Case 3, with a value of 0.93740 and in Case 29, with a value of 0.74782. Similarly, the Coefficients of Correlation for each case are presented in the following Table 4 which contains the aggregated results:

Table 4. Coefficients of Correlation of testing dataset.

Case	Coefficient of Correlation (R)	
	ANN	**MLR**
Case 2	0.98353	0.80337
Case 3	0.98777	0.93740
Case 6	0.99066	0.88198
Case 11	0.97737	0.76844
Case 14	0.98639	0.87493
Case 15	0.99101	0.90800
Case 22	0.96842	0.78287
Case 24	0.97975	0.86749
Case 29	0.96998	0.74782

4. Discussions

This work developed and compared two models for the simulation of precipitation values, which simulated and accurately completed five time series of precipitation data from five meteorological stations in the region of Chania, Crete. The first model was developed using an Artificial Neural Network ensemble approach (similar to other previously published works [6,10,27]), while the second model was developed using the Multiple Linear Regression method, both in a MATLAB environment.

It is observed that the four meteorological stations that are relatively close to the sea, while at the same time are relatively close to each other (Chania, Chania (Centre), Platanias and Stalos), show similar results for their total rainfall values (Figure 5). From a hydrological standpoint, both models present results that are in accordance with the theoretical expectations; the simulated values at the weather stations near the seafront are always lower when compared to those of stations at higher altitude. In addition, there is a small decline in the precipitation values along the west to east axis, which is expected since most of the water load in the clouds is released when they reach the coastal fronts coming from the Western Mediterranean.

Looking at the ANN results, a couple of simulated values might draw the attention of the reader as being exceptionally high and possibly outliers (e.g., October 2006 and January 2019). Nevertheless, the scientific literature and the observed values from already installed stations confirm that these were months with extreme rainfall events, confirming the plausibility of these simulated values. In October 2006, extreme rainfall events occurred throughout the study area, leading to flooding in the city of Chania, serious material damages and one casualty [20]. At that time, the only installed and operating station was the one in Chania, which had a very high observed value of 214.6 mm, one of the highest ever recorded. For the same month, the simulated precipitation value for Alikianos station is 345 mm based on ANNs, while the corresponding value using the MLR method is 194 mm. These values, although they seem quite high for the area concerned, are in accordance with the value observed in Chania. In January and February 2019, other extreme rainfall events occurred with similar results. In 2019, all weather stations were operational, but there was a 10-day gap in the beginning of January in Alikianos station, possibly because of device failure due to the extremity of the rainfall events. Regarding the month of January 2019, the simulated precipitation value for Alikianos station is 692 mm based on ANNs, while the corresponding value using the MLR method is 362 mm. The extremity of those values is confirmed by the literature, while the events continued in February with the Chionis and Oceanida storms [20]. The seemingly high simulated value for January is confirmed by the observed values in February at all weather stations. In Figure 5, the observed values in February are significantly high, with the highest value recorded at Alikianos station (568.8 mm in total) and the next highest value at Chania station (360 mm in total). Based on the above, we conclude that the simulated values for Alikianos are plausible and do not consider them as outliers. Comparing the two models, the results of the ANN model show that it is more capable of simulating extreme weather values compared to the model obtained with the MLR method.

5. Conclusions

Both methods have proven more than adequate for the task of imputation of gaps in the daily rainfall time series. The Nash–Sutcliffe coefficient for both methods is above 0.7 for all cases, a value generally considered as the threshold for very good performance. Nevertheless, throughout this work, the Artificial Neural Network ensembles consistently outperformed the Multiple Linear Regression model. The obvious caveat is the increased time needed for training the ANN model. When comparatively small datasets are available for training (like in this work), the computational effort for training the ANN ensembles is also relatively small (taking just over thirty-six hours). In such cases using ANNs might make more sense, always considering the urgency of the application. In cases where the available dataset is large the training time is expected to increase, but the results will

Water **2022**, *14*, 2892

probably be better than those obtained with Multiple Linear Regression. A decision should be made as to whether accuracy or speed is more important. For increased accuracy, the results of this study suggest using ANNs, for increased speed, the results point to using Multiple Linear Regression. Given the good performance of the ensembles in this work, future work can focus on testing different activation functions like the reLU and tanhLU [34].

Author Contributions: Conceptualization, I.T.; methodology, I.T.; software, I.P. and I.T.; validation, I.P. and I.T.; data curation, I.P.; writing—original draft preparation, I.P. and F.S.; writing—review and editing, F.S. and I.T.; visualization, I.T.; supervision, G.P.K. All authors have read and agreed to the published version of the manuscript.

Funding: This research received no external funding.

Conflicts of Interest: The authors declare no conflict of interest.

References

1. Canchala-Nastar, T.; Carvajal-Escobar, Y.; Alfonso-Morales, W.; Loaiza Cerón, W.; Caicedo, E. Estimation of Missing Data of Monthly Rainfall in Southwestern Colombia Using Artificial Neural Networks. *Data Brief* **2019**, *26*, 104517. [CrossRef] [PubMed]
2. Nkuna, T.R.; Odiyo, J.O. Filling of Missing Rainfall Data in Luvuvhu River Catchment Using Artificial Neural Networks. *Phys. Chem. Earth Parts A/B/C* **2011**, *36*, 830–835. [CrossRef]
3. Tran Anh, D.; Van, S.P.; Dang, T.D.; Hoang, L.P. Downscaling Rainfall Using Deep Learning Long Short-term Memory and Feedforward Neural Network. *Int. J. Climatol.* **2019**, *39*, 4170–4188. [CrossRef]
4. Ben Aissia, M.-A.; Chebana, F.; Ouarda, T.B. Multivariate Missing Data in Hydrology—Review and Applications. *Adv. Water Resour.* **2017**, *110*, 299–309. [CrossRef]
5. Teegavarapu, R.S.V.; Aly, A.; Pathak, C.S.; Ahlquist, J.; Fuelberg, H.; Hood, J. Infilling Missing Precipitation Records Using Variants of Spatial Interpolation and Data-Driven Methods: Use of Optimal Weighting Parameters and Nearest Neighbour-Based Corrections: INFILLING MISSING PRECIPITATION RECORDS. *Int. J. Climatol.* **2018**, *38*, 776–793. [CrossRef]
6. Elshaboury, N.; Elshourbagy, M.; Al-Sakkaf, A.; Abdelkader, E.M. Rainfall Forecasting in Arid Regions Using an Ensemble of Artificial Neural Networks. *J. Phys. Conf. Ser.* **2021**, *1900*, 012015. [CrossRef]
7. Mishra, N.; Soni, H.K.; Sharma, S.; Upadhyay, A.K. A Comprehensive Survey of Data Mining Techniques on Time Series Data for Rainfall Prediction. *J. ICT Res. Appl.* **2017**, *11*, 167–183. [CrossRef]
8. Kashiwao, T.; Nakayama, K.; Ando, S.; Ikeda, K.; Lee, M.; Bahadori, A. A Neural Network-Based Local Rainfall Prediction System Using Meteorological Data on the Internet: A Case Study Using Data from the Japan Meteorological Agency. *Appl. Soft Comput.* **2017**, *56*, 317–330. [CrossRef]
9. Ridwan, W.M.; Sapitang, M.; Aziz, A.; Kushiar, K.F.; Ahmed, A.N.; El-Shafie, A. Rainfall Forecasting Model Using Machine Learning Methods: Case Study Terengganu, Malaysia. *Ain Shams Eng. J.* **2021**, *12*, 1651–1663. [CrossRef]
10. Goyal, M.K. Monthly Rainfall Prediction Using Wavelet Regression and Neural Network: An Analysis of 1901–2002 Data, Assam, India. *Theor. Appl. Climatol.* **2014**, *118*, 25–34. [CrossRef]
11. Hu, C.; Wu, Q.; Li, H.; Jian, S.; Li, N.; Lou, Z. Deep Learning with a Long Short-Term Memory Networks Approach for Rainfall-Runoff Simulation. *Water* **2018**, *10*, 1543. [CrossRef]
12. Qin, Y.; Lou, Y. Hydrological Time Series Anomaly Pattern Detection Based on Isolation Forest. In Proceedings of the 2019 IEEE 3rd Information Technology, Networking, Electronic and Automation Control Conference (ITNEC), Chengdu, China, 15–17 March 2019; pp. 1706–1710.
13. Khazaee Poul, A.; Shourian, M.; Ebrahimi, H. A Comparative Study of MLR, KNN, ANN and ANFIS Models with Wavelet Transform in Monthly Stream Flow Prediction. *Water Resour. Manag.* **2019**, *33*, 2907–2923. [CrossRef]
14. Desai, S.; Ouarda, T.B. Regional Hydrological Frequency Analysis at Ungauged Sites with Random Forest Regression. *J. Hydrol.* **2021**, *594*, 125861. [CrossRef]
15. Haidar, A.; Verma, B. A Novel Approach for Optimizing Climate Features and Network Parameters in Rainfall Forecasting. *Soft Comput.* **2018**, *22*, 8119–8130. [CrossRef]
16. Jaddi, N.S.; Abdullah, S. Optimization of Neural Network Using Kidney-Inspired Algorithm with Control of Filtration Rate and Chaotic Map for Real-World Rainfall Forecasting. *Eng. Appl. Artif. Intell.* **2018**, *67*, 246–259. [CrossRef]
17. Cheng, S.; Lu, F. A Two-Step Method for Missing Spatio-Temporal Data Reconstruction. *ISPRS Int. J. Geo-Inf.* **2017**, *6*, 187. [CrossRef]
18. Yen, M.-H.; Liu, D.-W.; Hsin, Y.-C.; Lin, C.-E.; Chen, C.-C. Application of the Deep Learning for the Prediction of Rainfall in Southern Taiwan. *Sci. Rep.* **2019**, *9*, 12774. [CrossRef]
19. Lee, J.; Kim, C.-G.; Lee, J.; Kim, N.; Kim, H. Application of Artificial Neural Networks to Rainfall Forecasting in the Geum River Basin, Korea. *Water* **2018**, *10*, 1448. [CrossRef]

20. Goumas, C.; Trichakis, I.; Vozinaki, A.-I.; Karatzas, G.P. Flood Risk Assessment and Flow Modeling of the Stalos Stream Area. *J. Hydroinform.* **2022**, *24*, 677–696. [CrossRef]
21. Praveen, B.; Talukdar, S.; Shahfahad, M.S.; Mondal, J.; Sharma, P.; Islam, A.R.M.D.T.; Rahman, A. Analyzing Trend and Forecasting of Rainfall Changes in India Using Non-Parametrical and Machine Learning Approaches. *Sci. Rep.* **2020**, *10*, 10342. [CrossRef]
22. Beritelli, F.; Capizzi, G.; Lo Sciuto, G.; Napoli, C.; Scaglione, F. Rainfall Estimation Based on the Intensity of the Received Signal in a LTE/4G Mobile Terminal by Using a Probabilistic Neural Network. *IEEE Access* **2018**, *6*, 30865–30873. [CrossRef]
23. Jhong, Y.-D.; Chen, C.-S.; Lin, H.-P.; Chen, S.-T. Physical Hybrid Neural Network Model to Forecast Typhoon Floods. *Water* **2018**, *10*, 632. [CrossRef]
24. Alam, K.M.R.; Siddique, N.; Adeli, H. A Dynamic Ensemble Learning Algorithm for Neural Networks. *Neural Comput. Appl.* **2020**, *32*, 8675–8690. [CrossRef]
25. Zhou, J.; Peng, T.; Zhang, C.; Sun, N. Data Pre-Analysis and Ensemble of Various Artificial Neural Networks for Monthly Streamflow Forecasting. *Water* **2018**, *10*, 628. [CrossRef]
26. Haidar, A.; Verma, B.; Sinha, T. A Novel Approach for Optimizing Ensemble Components in Rainfall Prediction. In Proceedings of the 2018 IEEE Congress on Evolutionary Computation (CEC), Rio de Janeiro, Brazil, 8–13 July 2018; pp. 1–8.
27. Kim, T.; Shin, J.; Kim, H.; Heo, J. Ensemble-Based Neural Network Modeling for Hydrologic Forecasts: Addressing Uncertainty in the Model Structure and Input Variable Selection. *Water Resour. Res.* **2020**, *56*, e2019WR026262. [CrossRef]
28. Granata, F.; Di Nunno, F. Forecasting Evapotranspiration in Different Climates Using Ensembles of Recurrent Neural Networks. *Agric. Water Manag.* **2021**, *255*, 107040. [CrossRef]
29. Althoff, D.; Rodrigues, L.N.; Bazame, H.C. Uncertainty Quantification for Hydrological Models Based on Neural Networks: The Dropout Ensemble. *Stoch. Environ. Res. Risk Assess.* **2021**, *35*, 1051–1067. [CrossRef]
30. Bandyopadhyay, G.; Chattopadhyay, S. Single Hidden Layer Artificial Neural Network Models versus Multiple Linear Regression Model in Forecasting the Time Series of Total Ozone. *Int. J. Environ. Sci. Technol.* **2007**, *4*, 141–149. [CrossRef]
31. Zhong, K.; Song, Z.; Jain, P.; Bartlett, P.L.; Dhillon, I.S. Recovery Guarantees for One-Hidden-Layer Neural Networks. *arXiv* **2017**, arXiv:1706.03175.
32. Moriasi, D.N.; Arnold, J.G.; van Liew, M.W.; Bingner, R.L.; Harmel, R.D.; Veith, T.L. Veith Model Evaluation Guidelines for Systematic Quantification of Accuracy in Watershed Simulations. *Trans. ASABE* **2007**, *50*, 885–900. [CrossRef]
33. Lagouvardos, K.; Kotroni, V.; Bezes, A.; Koletsis, I.; Kopania, T.; Lykoudis, S.; Mazarakis, N.; Papagiannaki, K.; Vougioukas, S. The Automatic Weather Stations NOANN Network of the National Observatory of Athens: Operation and Database. *Geosci. Data J.* **2017**, *4*, 4–16. [CrossRef]
34. Shen, S.-L.; Zhang, N.; Zhou, A.; Yin, Z.-Y. Enhancement of Neural Networks with an Alternative Activation Function TanhLU. *Expert Syst. Appl.* **2022**, *199*, 117181. [CrossRef]

Article

A Study on the Optimal Deep Learning Model for Dam Inflow Prediction

Beom-Jin Kim [1], You-Tae Lee [2] and Byung-Hyun Kim [3,*]

[1] Advanced Structures and Seismic Safety Research Division, Korea Atomic Energy Research Institute, Daejeon 34057, Korea
[2] National Drought Information-Analysis Center, Korea Water Resources Corporation, Daejeon 34350, Korea
[3] Department of Civil Engineering, Kyungpook National University, Daegu 41566, Korea
* Correspondence: bhkimc@knu.ac.kr; Tel.: +82-53-950-7819

Abstract: In the midst of climate change, the need for accurate predictions of dam inflow to reduce flood damage along with stable water supply from water resources is increasing. In this study, the process and method of selecting the optimal deep learning model using hydrologic data over the past 20 years to predict dam inflow were shown. The study area is Andong Dam and Imha Dam located upstream of the Nakdong River in South Korea. In order to select the optimal model for predicting the inflow of two dams, sixteen scenarios ($2 \times 2 \times 4$) are generated considering two dams, two climatic conditions, and four deep learning models. During the drought period, the RNN for Andong Dam and the LSTM for Imha Dam were selected as the optimal models for each dam, and the difference between observations was the smallest at 4% and 2%, respectively. In typhoon conditions, the GRU for Andong Dam and the RNN for Imha Dam were selected as optimal models. In the case of Typhoon Maemi, the GRU and the RNN showed a difference of 2% and 6% from the observed maximum inflow, respectively. The optimal recurrent neural network-based models selected in this study showed a closer prediction to the observed inflow than the SFM, which is currently used to predict the inflow of both dams. For the two dams, different optimal models were selected according to watershed characteristics and rainfall under drought and typhoon conditions. In addition, most of the deep learning models were more accurate than the SFM under various typhoon conditions, but the SFM showed better results under certain conditions. Therefore, for efficient dam operation and management, it is necessary to make a rational decision by comparing the inflow predictions of the SFM and deep learning models.

Keywords: deep learning; dam inflow; RNN; LSTM; GRU; hyperparameter

Citation: Kim, B.-J.; Lee, Y.-T.; Kim, B.-H. A Study on the Optimal Deep Learning Model for Dam Inflow Prediction. *Water* **2022**, *14*, 2766. https://doi.org/10.3390/w14172766

Academic Editors: Fi-John Chang, Li-Chiu Chang and Jui-Fa Chen

Received: 18 July 2022
Accepted: 1 September 2022
Published: 5 September 2022

1. Introduction

Due to extreme climatic change, accurate analysis of water resources is increasingly demanded for stable water supply and flood damage mitigation. Among various research subjects, the amount of the dam inflow is an important element in establishing plans for coping with drought, flooding, and operating the dam. The major factors affecting the amount of the inflow are climatic factors, including rainfall, which is the most influential, temperature, and wind speed, as well as topographical factors such as the basin area and the height of the slope [1]. However, recently, local rainfalls, which are difficult to predict, have frequently occurred nationwide. In particular, in Andong and Imha Dams in 2015, the inflow decreased to one-third the level of the average inflow over the past 20 years; and in 2017 and 2018, the discharge rates were adjusted due to entering the drought "attention stage." In addition, in 2020, due to the prolonged rainy season, the inflow increased to more than 40%, and therefore, floodgate discharge was performed at Andong Dam for the first time in 17 years. As such, it is an important issue to predict more accurately and quickly the inflow for two dams, which frequently change in drought and flood conditions every year. The reason for this study is that Andong Dam and Imha Dam are important dams that account for 50% of the water supply in the Nakdong River watershed, but there are few studies that predict the dam inflow using a deep learning model. In addition, although the geographical locations of the two dams are adjacent, dam inflow tends to be different depending on the watershed and precipitation characteristics. In particular, during the typhoon Maisak and Haishen in 2020, an instantaneous inflow greater than the designed flood was observed at Imha Dam. Therefore, it is necessary to accurately predict the inflow using deep learning for the two dams in consideration of climate change. In the past, the amount of inflow was calculated using conceptual and physical models; however, recently, artificial intelligence technology has been used in more and more cases to analyze the amount of inflow. Kim et al. [2] took Chungju Dam and Soyanggang Dam as subjects and used the artificial neural network (ANN) model in predicting the inflow of the dams by applying the meteorological data in their basin areas, and the basin precipitation was calculated using the Thiessen network. This study showed that the model using all rainfall stations in the Thiessen network performed better than using only in-watershed or out-watershed stations. Kim et al. [3] analyzed the average precipitation and the inflow data of Chungju Dam in the Han River basin by applying an ANN model including a back propagation algorithm. This study showed that there was a significant improvement in the model accuracy including the correlation coefficient (CC) when data preprocessing was performed. Mok et al. [4] applied the Long Short-Term Memory (LSTM) and the ANN model to predict the inflow per hour of Yongdam Dam. In this study, the LSTM hyperparameters (sequence, hidden dimension, learning rate, and iteration) were optimized and the model accuracy was improved by applying dam inflow and rainfall as input variables. Lee et al. [5] performed a quantitative evaluation by adjusting and simulating input variables for the Taehwa River basin using recurrent neural network (RNN), time delay neural network (TDNN), and nonlinear autoregressive exogenous (NARX) models. This study improved the Nash–Sutcliffe efficiency (NSE) from 0.530 to 0.988 by adjusting the time delay parameter of the model. Chang et al. [6] introduced recent advances in machine learning in flood prediction and management, and presented an academic approach to flood risk-related modeling. Chang et al. [7] explored the effectiveness of multiple rainfall sources for assimilation-based multi-sensor precipitation estimates and performed multi-step-ahead rainfall forecasts based on the assimilated precipitation. Chakravarti et al. [8] demonstrated that the ANN model could be a promising tool to provide insights from learned relationships as well as accurate modeling of complex processes through a comparison of the runoff generated by rainfall simulator in the laboratory and the predicted runoff of the ANN model. Kao et al. [9] proposed a Long Short-Term Memory based Encoder-Decoder (LSTM-ED) for multi-step-ahead flood prediction for the first time. Shen et al. [10] suggested that hydrology scientists consider research using DL-based data mining to complement traditional approaches. Tokar et al. [11] compared and analyzed the conceptual models and the ANN models, which differed for each basin. After comparing the Watbal model for the Fraser River, the Sacramento Soil Moisture Accounting (SAC-SMA) model for the Raccoon River, and the Simple Conceptual Rainfall–Runoff (SCRR) model for the Little Patuxent River, Colorado, USA, with the ANN model, it was shown as a result that the ANN model together with the existing conceptual model could be utilized for rainfall-discharge prediction. Chen et al. [12] compared and analyzed the hourly precipitation and discharge data for each hour following the hit of 27 typhoons from 2005 and 2009 at the Linbien River Basin, Taiwan, by applying the conventional regression model and the ANN model along with the concept of backpropagation. In statistical evaluation, the ANN model showed better results than the conventional regression analysis model. Coulibaly et al. [13] predicted the inflow of multi-purpose dams by applying rainfall, snowfall, inflow, and temperature as input variables of four models: Multilayer Perceptron (MLP), Input Delayed Neural Network (IDNN), RNN, and Time Delay Current Neural Network (TDRNN).

In this study, a deep learning model was used to predict the inflow of Andong and Imha Dams in the Nakdong River watershed in Korea. To build an optimal prediction model based on inflow and rainfall data over the past 20 years, accuracy and reliability were evaluated by generating various scenarios according to input variables. In addition, the RNN models were applied considering that the dam inflow is time series data and the learning efficiency of the existing ANN model decreases as the number and period of data increase. The prediction model derived from this study is expected to contribute to stable dam operation management and coping with the disaster.

2. Study Methods

2.1. ANN and RNNs

In this study, the ANN model and the RNN model were compared and analyzed to derive an optimal model for predicting dam inflow. The flow chart of this study is shown in Figure 1. Deep learning is one of the algorithms of machine learning and is a more deeply constructed algorithm than conventional neural network structures. Non-linear characteristics between input variables can be estimated and have superior effects over traditional machine learning algorithms. Machine learning is a process in which humans feed the computers a lot of information, and then the computers predict information, while deep learning has the characteristics of the computers learning and predicting it without human's teaching specifically. The typical activation functions used in the hidden layers of deep learning are mainly Sigmoid, tanh (hyperbolic tangent), and Rectified Linear Unit (ReLU). The sigmoid function is a logistic regression function with values between "0" and "1," which is utilized for simple classification problems. The tanh function has a value between "-1" and "1," and as it moves away from the center value, the slope is lost during the backpropagation. For solving this slope loss problem is the ReLU function, and all values below "0" are treated as "0" to stop the learning progress [14].

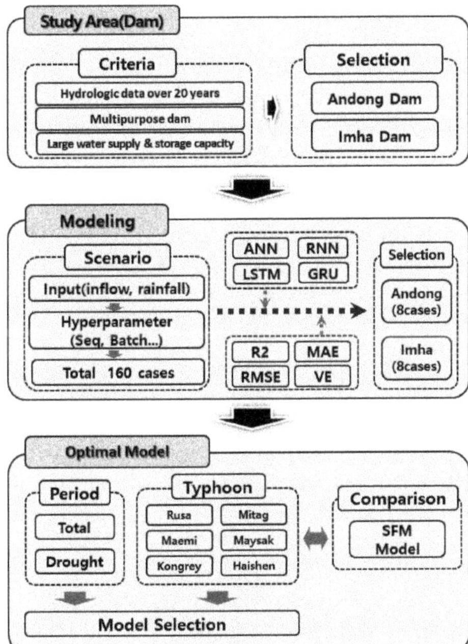

Figure 1. Flow chart of this study.

The RNN is a specialized model in the field of ordered data processing. In particular, time series data are mainly utilized, and the previous output data are cycled back into the input. The following is a comparison of the hidden layer calculation Equation (1) of

Convolution Neural Network (CNN), which processes grid data like an image, and the hidden layer calculation Equation (2) of the RNN.

$$\text{CNN } \; h_t = W_{xh}x_t \tag{1}$$

$$\text{RNN } \; h_t = \tanh\left(W_{hh}h_{t-1} + W_{xh}x_t \right) \tag{2}$$

The RNN has the characteristics of weighing each data individually to determine its importance and memorize it while turning to the next data, but there appears a gradual loss of information of distant past data in the hidden layer; therefore, a method supplemented with a separate memory cell prepared is LSTM [15]. The LSTM is one of the RNN models and is composed of a Forget gate, an Input gate and an Output gate. In order to solve the problem of gradient loss that occurs as the time difference increases in the RNN model, the LSTM model introduces a cell. Information is stored in this cell, and it plays a role in preventing the stored information from being lost in the process of analysis. The gate serves as a filter that allows unnecessary information to be forgotten or necessary information to be stored and passed through the cell. This is represented by Equations (3)–(6). In the forget gate, how much past data will be forgotten is determined, and the input gate plays a role in estimating important values among the incoming data. Output gates are used to keep information from past data and predict them simultaneously.

$$\text{Forget Gate}: \; f_t = \sigma\left(U_f h_{t-1} + W_f x_t + b \right) \tag{3}$$

$$\text{Input Gate}: \; i_t = \sigma(U_i h_{t-1} + W_i x_t + b) \tag{4}$$

$$\text{Output Gate}: \; o_t = \sigma(U_o h_{t-1} + W_o x_t + b) \tag{5}$$

$$h_t = o_t \times \tanh(Cell) \tag{6}$$

where σ is the activation function, U is the input weight, W is the cyclic weight, h_{t-1} is the previous stage output, h_t is the new output value, x_t is the current input vector, and b is the bias.

In addition, the Gated Recurrent Unit (GRU) is a method with the structure improved for processing faster than LSTM [16]. GRUs are configured as Reset gate and Update gate for the advantage of lower learning weights; therefore, faster processing speed with similar performance compared to LSTM is observed. Reset gate determines the ratio of past data to remove, and Update gate determines the discarding past data, such as forget gate of LSTM, and selects only one of $t-1$ and t memory data.

2.2. The Storage Function Model (SFM)

The SFM is one of the rainfall–runoff models, and calculates the runoff from the watershed using the reservoir storage and rainfall as main input variables. In this case, impervious area, infiltration, and groundwater are considered. The model makes the basic assumption that stream channels (I $\sim$ O) have a downward slope and that the watershed receives the same amount of precipitation (R_{ave}) as shown in Figure 2. The runoff from the watershed is calculated by Equation (7) [17].

$$Q_T\left(m^3/s\right) = \frac{1}{3.6} \times f_1 \times A \times q_f + \frac{1}{3.6} \times (f_{sa} - f_1) \times A \times q_s + q_b \tag{7}$$

where f_1 is the primary runoff rate (dimensionless), A is the watershed area (km^2), q_f. is the unit runoff height of runoff area (mm/day), q_s is the unit runoff height of infiltration area (mm/day), f_{sa} is the unit runoff in seepage areas directly infiltrating groundwater, and q_b is the base runoff (m^3/s).

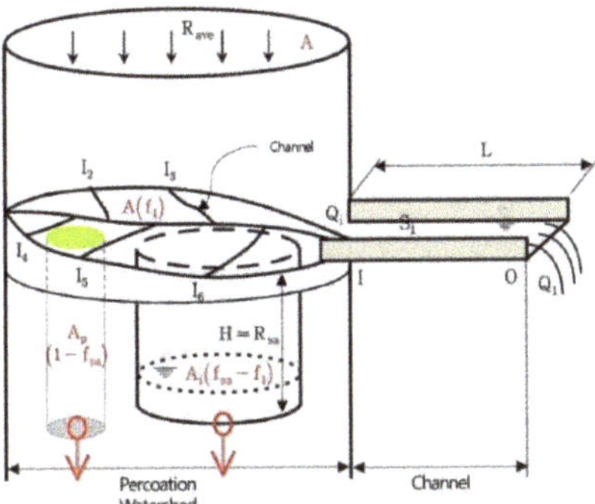

Figure 2. Schematic diagram of the storage function model [17].

Korea Water Resources Corporation (K-water) operates dams through inflow prediction using the SFM, and the parameters of the SFM corresponding to each dam are optimized in consideration of the characteristics of the dam basin [17].

2.3. Study Area

Sufficient learning materials are required to calculate the inflow of dams using deep learning. In this study, Andong Dam and Imha Dam of Nakdong River were selected as the study areas among multi-purpose dams in Korea that have collected hydrological data for more than 20 years and secured the largest amount of water supply and storage capacity in the water system. The locations of Andong Dam and Imha Dam are shown in Figure 3.

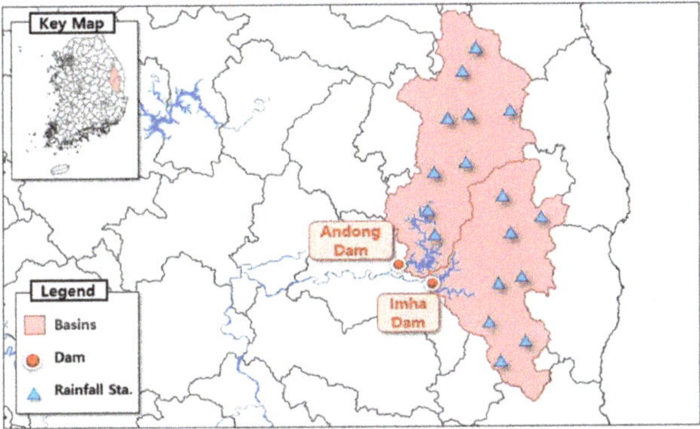

Figure 3. Location and Watershed of Andong and Imha Dams.

Andong Dam was completed in 1976, with a basin area of 1584 km^2 and a total water storage capacity of 1248×10^6 m^3. It was built to reduce flood damage by utilizing 110×10^6 m^3 of flood control capacity and facilities. It is responsible for supplying 926×10^6 m^3 of water annually, including Nakdong River's living water, industrial water, and river maintenance flow. Imha dam was completed in 1993 and has a basin area of 1361 km^2 and a total storage capacity of 595×10^6 m^3. It is 73.0 m-high, with a 515.0 m-long central cutoff-wall type rockfill dam built to prevent flood damage in the mid- and downstream of the Nakdong River and to supply water to the Nakdong River and the southeast coast areas. It supplies 615.3×10^6 m^3 of water annually, including living water, industrial water, and river maintenance flow (Table 1).

Table 1. General status of Andong and Imha Dams.

Category	Andong	Imha
Storage(10^6 m^3)	1248	595
Flood control capacity(10^6 m^3)	110	80
Water supply (10^6 m^3/y)	926.0	615.3
Flood volume (m^3/s)	6480	4500
Discharge (m^3/s)	4600	2500

2.4. Database Buliding

In this study, the time series period required to compare and analyze four models, ANN, RNN, LSTM, and GRU models, was set from 2001 to 2020, and we intend to build an inflow prediction model by utilizing the inflow and precipitation data of Andong and Imha Dams in the subject period. The equations for daily and hourly inflow are as shown in Equations (8) and (9). Rainfall data collected from nine rainfall observatories in Andong Dam basin and eight rainfall observatories in Imha Dam basin were used.

$$\text{Daily inflow}\left(\frac{\text{m}^3}{\text{s}}\right) = \frac{\text{Water Storage}(\text{at } 24:00 \text{ today} - \text{at } 24:00 \text{ the day before}) \times 10^6}{60 \times 60 \times 24} + \text{Daily Average Outflow} \quad (8)$$

$$\text{Hourly inflow}\left(\text{m}^3/\text{s}\right) = \frac{\text{Water Storage}(\text{at fixed time} - \text{at 1hr ago}) \times 10^6}{60 \times 60} + \text{Hourly Average Outflow} \quad (9)$$

Considering the inflow of Andong and Imha Dams from 2001 to 2020, the annual inflow of Andong Dam in 2003 and 2015 was almost six times different. The inflows of Andong and Imha Dams during the flood period accounts for approximately 2/3 of the average annual inflows, and the precipitation and inflow during specific periods, such as the normal season or the drought and flood periods, are different. Therefore, it is necessary to analyze after dividing the seasons into the normal season or the drought and flood periods when selecting the optimal model later. Figure 4 shows the rainfall and inflow of Andong Dam watershed for 20 years.

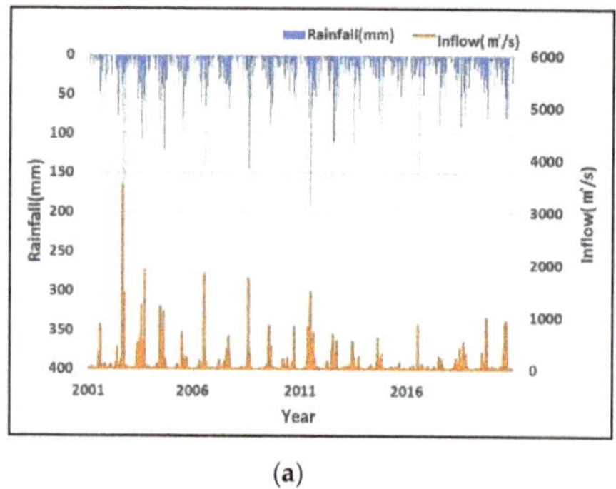

(a)

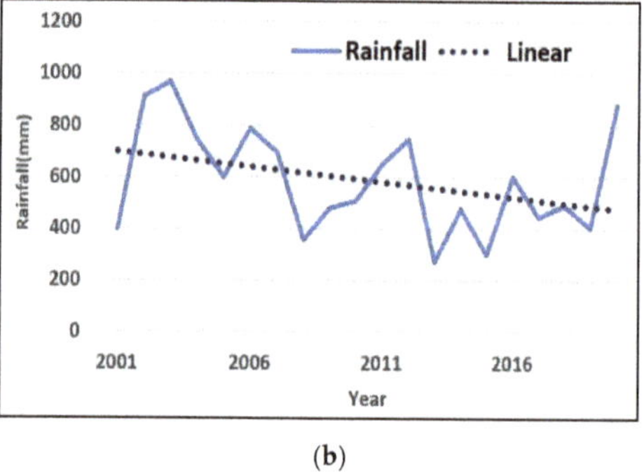

(b)

Figure 4. Inflow and rainfall at Andong Dam. (**a**) Inflow and rainfall for 2001–2020. (**b**) Rainfall for the flood period (21 June–20 September).

There were four releases through Andong–Imha connection tunnel from 2019 to 2020. The corresponding discharge was calculated as the inflow of Andong Dam and, therefore, excluded from data preprocessing. Since the range of inflow and precipitation data is wide, data normalization was used to convert it to a value between 0 and 1 by Min–Max Scaling. In addition, the data for 20 years are divided into a training set, a validation set, and a testing set in a 5:3:2 ratio as shown in Figure 5.

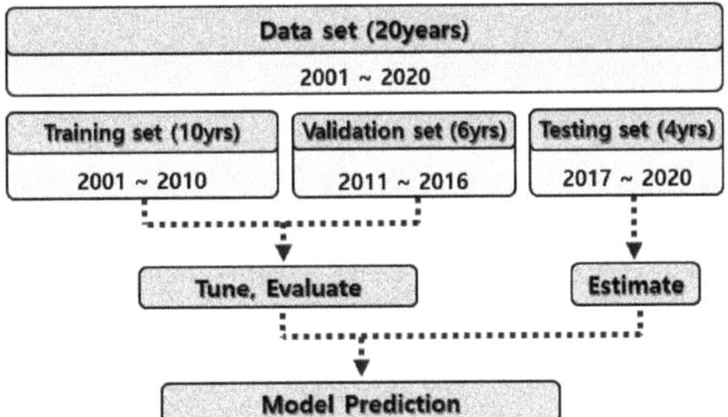

Figure 5. Data analysis flow chart.

2.5. Input and Output Predictors

In this study, precipitation and dam inflow from previous times were used as input data to predict the inflow of the dam. The number of previous times precipitation and inflow are considered for dam inflow prediction is related to the sequence hyperparameter to be described later. For example, if the sequence is 21, 21 precipitations (P_t, P_{t-1}, $\cdots$ P_{t-20}) and 21 dam inflows (Q_t, Q_{t-1}, $\cdots$ Q_{t-20}) are simultaneously considered. P_t and Q_t are precipitation and dam inflow at the current time, respectively, Q_{t+1} is the dam inflow at the next time step to be predicted, and P_{t-1} and Q_{t-1} are the precipitation and dam inflow at the previous time steps to be considered for predicting the dam inflow, respectively. Figure 6 shows a schematic diagram of the input and output data of the model with sequence 21.

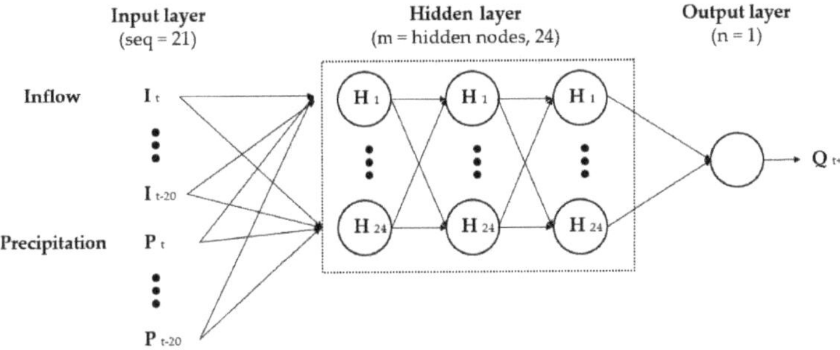

Figure 6. Schematic diagram of the input and output.

2.6. Optional Hyperparameter

In this study, two hyperparameters (Sequence and Batch size) were optimized by applying a grid search at regular intervals as shown in Table 2. The hyperparameters were optimized by applying a grid search at regular intervals shown in Table 2. The trial-and-error method was additionally applied to compensate for the shortcomings of grid search, which can be difficult to find optimal hyperparameters with regular interval application. The trial-and-error method found optimal variables for sequence length and batch size within the range of 1–100 and compared them with the results of grid search. In particular, the reason why the sequence length(hour) was selected as 12 is that for flood control at the multi-purposed dam, outflow discharge is approved by the government one day before the opening of the gate and notified to downstream residents in advance. Among the high-accuracy models, when overfitting occurs compared to the validation data and test data, the dropout method was used to supplement the analysis results. The remaining hyperparameters without grid search were optimized with trial and error. The application ranges of each parameter are shown in Table 2, and Learning rate 0.001, Dropout 0.2, and Hidden layer 3 were applied as optimal values in this study.

Table 2. Application range for hyperparameter optimization.

Optimization			Grid Search	T and E
Sequence		day	7, 14, 21	1~100
		hour	12, 24, 48	
Batch size		day	7, 14, 21, 28, 35	1~100
		hour	12, 24, 36, 48, 60, 72	
Epoch			-	100~500 (Early stop)
Learning rate			-	0.01~0.0001
Dropout			-	0.1~0.25
Hidden layer			-	2~5

The name of the scenario is the first letter of 'dam name–day/time–application model–scenario order or optimization'. As an example, the scenario is named "ADA-S1", which means "Andong–Day–ANN–Scenario No.1", and "ADA-Opt", which means "Andong–Day–ANN–Optimize".

To evaluate the statistical error and accuracy of the model according to the hyperparameter for each model scenario, the coefficient of determination (R^2), mean absolute error (MAE), root mean square error (RMSE), and volume error (VE) presented by Hu et al. [18] were used as performance indicators. Table 3 representatively shows the ANN model results for Andong Dam among 8 cases (2 dams × 4 deep learning models) that analyzed the best performance according to each scenario. Among the various scenarios, ADA-S9 for daily data and AHA-S4 for hourly data were selected.

Table 3. Statistical performance by scenario for the ANN at Andong Dam.

Scenario		Input		Statistical Indices				Selection
		Sequence	Batch	R^2	MAE	RMSE	VE	
Day	ADA-S1	7	7	0.89	12.01	25.17	0.29	
				...				
	ADA-S4	14	14	0.83	9.56	20.10	0.13	
				...				
	ADA-S7	21	21	0.81	9.83	22.37	0.31	
	ADA-S8	21	28	0.86	9.54	28.62	0.33	
	ADA-S9	21	35	0.91	9.40	19.18	0.03	○
	ADA-Opt	20	20	0.82	11.36	24.40	0.28	
Hour	AHA-S1	12	12	0.80	20.70	42.42	0.17	
				...				
	AHA-S4	24	24	0.94	12.26	22.94	0.12	○
	AHA-S5	24	36	0.89	11.34	30.50	0.20	
	AHA-S6	24	48	0.88	11.82	32.52	0.29	
	AHA-S7	48	48	0.91	12.59	27.72	0.27	
				...				
	AHA-Opt	10	10	0.91	11.77	29.45	0.13	

○: Selected optimal scenario.

Table 4 shows the optimal scenario selection and the corresponding R^2 by comparing the observations and simulations for each model. The ANN model of the daily data at

Andong Dam had a correlation R^2 validation indicator of 0.91, which was closest to the observation compared to other models. However, in the peak inflow, the GRU model showed the closest results to the observations. In the peak inflow of the daily data of Imha Dam, LSTM model showed 925.2 m^3/s, least different from the actual inflow. As for the scenario result applying the time data of Andong Dam, the correlation of the ANN model was 0.94, similar to the daily data usage, which was the closest to the observation. Unlike Andong Dam, in Imha Dam, the RNN model showed less difference between actual peak inflow and predicted peak inflow than the ANN model. In particular, it was the smallest in the LSTM model at 34.5 m^3/s.

Table 4. Optimal scenario selection.

Dam/Time		Observed (m^3/s)	Simulated (m^3/s), R^2			
			ANN	**RNN**	**LSTM**	**GRU**
Andong	Day	998.5	ADA-S9	ADR-Opt	ADL-S1	ADG-S1
			696.8	725.7	921.6	956.1
			0.91	0.82	0.81	0.79
	Hour	2629.1	AHA-S4	AHR-S8	AHL-S9	AHG-S6
			1835.3	2327.7	3458.1	3053.5
			0.94	0.86	0.87	0.87
Imha	Day	935.1	IDA-S9	IDR-S4	IDL-Opt	IDG-S5
			653.0	915.17	925.2	988.1
			0.92	0.82	0.79	0.87
	Hour	4890.1	IHA-S4	IHR-S9	IHL-S6	IHG-S7
			3909.0	4226.0	4855.6	4248.5
			0.92	0.95	0.95	0.95

2.7. Performance Evaluation of Optional Scenarios

For the evaluation for the performance evaluation of the scenarios, the RMSE-observed standard deviation ratio (RSR) and the Nash–Sutcliffe efficiency (NSE) were applied among various criteria. The equations for each criterion are shown in the following Equations (10) and (11). With the calculated RSR and NSE, the model performance can be judged based on the general performance rating (Table 5) [19].

$$\text{RSR} = \frac{RMSE}{STDEV_{obs}} = \frac{\sqrt{\left(\sum_{i=1}^{n}(y_i - \hat{y}_i)^2\right)}}{\sqrt{\left(\sum_{i=1}^{n}(y_i - \overline{y_i})^2\right)}} \tag{10}$$

$$\text{NSE} = 1 - \frac{\sum_{i=1}^{n}(y_i - \hat{y}_i)^2}{\sum_{i=1}^{n}(y_i - \overline{y_i})^2} \tag{11}$$

where y_i is the observed value, $\overline{y_i}$ is the mean value, $\hat{y}_i$ is the predicted value, and n is the numbers of data.

Table 5. General performance ratings [19].

Performance Rating	RSR	NSE
Very Good	$0.00 \leq \text{RSR} \leq 0.50$	$0.75 < \text{NSE} \leq 1.00$
Good	$0.50 < \text{RSR} \leq 0.60$	$0.65 < \text{NSE} \leq 0.75$
Satisfactory	$0.60 < \text{RSR} \leq 0.70$	$0.50 < \text{NSE} \leq 0.65$
Unsatisfactory	$\text{RSR} > 0.70$	$\text{NSE} \leq 0.50$

Table 6 shows the RSR and the NSE calculated for the validation and test data of the selected scenarios (Table 4), and the performance ratings evaluated with these values. As a result of having validated the selected scenarios, the RSR value of Andong Dam daily data was low and similar compared to the Imha Dam results, and the evaluation result was "Very Good" in the ANN model and "Good" in the RNN model. In the hourly data, the ANN model showed the lowest result of 0.34, and was evaluated as "Very Good" in all models. Similar to Andong Dam, Imha Dam was evaluated as "Good" in the RNN model except the ANN model. In the hourly data, the evaluation was "Very Good" in all models and the NSE value was above 0.90, deriving reliable results.

Table 6. Performance rating evaluation for selected scenarios.

Case			RSR/NSE			
			ANN	RNN	LSTM	GRU
Andong	Day	Validation	0.31/0.91	0.55/0.70	0.56/0.72	0.54/0.68
		Test	0.31/0.90	0.53/0.68	0.56/0.75	0.56/0.66
		Evaluation	Very Good	Good	Good	Good
	Hour	Validation	0.33/0.99	0.38/0.99	0.38/0.99	0.37/0.99
		Test	0.34/0.89	0.48/0.96	0.48/0.95	0.46/0.96
		Evaluation	Very Good	Very Good	Very Good	Very Good
Imha	Day	Validation	0.36/0.87	0.54/0.68	0.52/0.70	0.59/0.73
		Test	0.36/0.87	0.54/0.66	0.53/0.70	0.58/0.70
		Evaluation	Very Good	Good	Good	Good
	Hour	Validation	0.28/0.99	0.22/0.99	0.20/0.99	0.20/0.99
		Test	0.29/0.91	0.24/0.95	0.25/0.96	0.24/0.96
		Evaluation	Very Good	Very Good	Very Good	Very Good

3. Selection of Optimal Models

3.1. Drought Period

In order to select the optimal model according to the period for Andong Dam and Imha Dam, first, the inflow by quantile for the total test period (2017–2020) was compared.

Then, the analysis results for each quantile of the inflow during the normal and dry season are derived, and the daily inflow from Andong and Imha Dams are used to select the inflow prediction model with the highest reliability during the drought period. In addition, the periods of 28 June–20 August 2017, and 13 February–29 March 2018 in the study area was in the 'caution' stage of drought crisis warning under the "Fundamental Act on Disaster and Safety". Therefore, this period data was used for drought period analysis.

Table 7 shows the inflows of the 1st (25%), 2nd (50%), and 3rd (75%) quartiles and peak inflows of ADA-S9, ADR-Opt, ADL-S1, and ADG-S1, which are the optimal scenarios for Andong Dam (Table 4). Over the total period (2017–2020), the RNN model showed that the 1st, 2nd, and 3rd quartile values were close to the observations, especially within the maximum difference of up to 2 m^3/s. In the drought period (2017–2018), the RNN predicted the 2nd and 3rd quartile inflows and maximum inflows closest to the observations, excluding the 1st quartile values. The difference in the maximum inflow between RNN predictions and observations was 6.25 m^3/s, the smallest difference compared to other RNN models. Figure 7 shows a comparison of the predicted inflow ranges for each model versus the observed ranges for the total and drought periods.

Table 7. Inflow prediction by period at Andong Dam.

Andong		Observed (m³/s)	Simulated (m³/s)			
			ANN	RNN	LSTM	GRU
Total period (2017–2020)	25%	3.70	10.88	5.61	1.56	4.43
	50%	8.12	11.09	8.50	4.43	7.54
	75%	20.41	24.44	21.75	16.09	14.49
Drought period (2017–2018)	25%	3.38	10.88	5.61	1.15	4.65
	50%	6.38	10.88	6.52	2.28	7.26
	75%	14.62	16.65	13.65	10.94	8.67
	Max	299.03	214.77	305.28	241.11	258.09

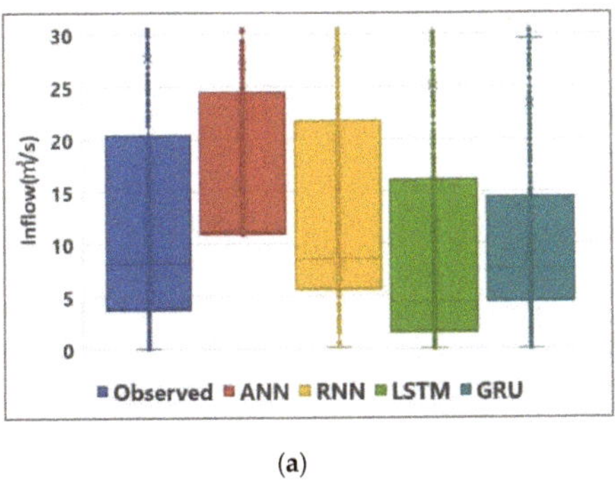
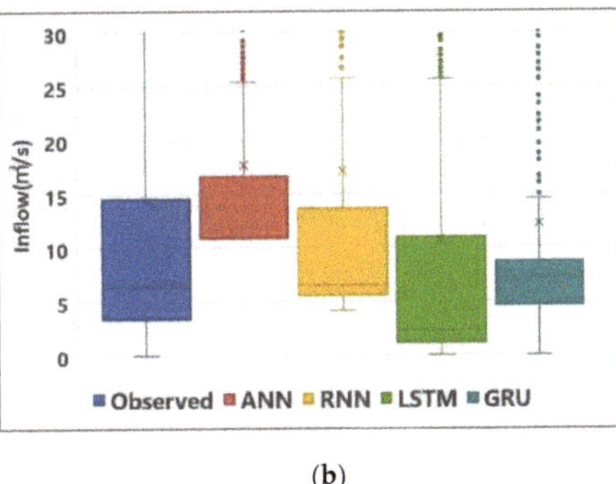

(**a**)   (**b**)

Figure 7. Comparison of predictions and observations for inflow ranges to Andong Dam. (**a**) Total period. (**b**) Drought period.

In the case of Imha Dam, the inflows of the 1st, 2nd and 3rd quartiles and peak inflows were calculated by applying the optimal scenarios (IDA-S9, IDR-S4, IDL-Opt, IDG-S5). Figure 8 shows a comparison of the predicted inflow ranges and the observed ranges of each model for the total and drought periods at Imha Dam. As shown in Table 8 and Figure 8, the prediction of the RNN shows the largest difference from the quartile value of the measured inflow compared to other models. On the other hand, inflow predictions of LSTM have the smallest differences from observations in the 1st and 3rd quartiles during the total period and in the 1st and 2nd quartiles and the maximum during the drought period. In the prediction of the maximum inflow, the difference between observation and prediction was 45.14 m³/s, which showed a difference of approximately 10%. The GRU prediction showed the most accurate result with a difference of 0.27 m³/s from the observation in the 3rd quartile of the drought period. As shown in Table 8, in Imha Dam, LSTM was selected as the optimal model for inflow prediction during the total and drought periods.

As a result of predicting the dam inflow during the drought period, the RNN model for Andong Dam and the LSTM model for Imha Dam were closest to the observed inflow. The reason that the RNN model yielded better results than the LSTM model at Andong Dam lies in the activation function. The existing RNN model uses the tanh function among the activation functions to cause the gradient loss problem. However, in this study, the ReLu function was used to reduce gradient loss during backpropagation learning. The reason that the LSTM model was selected as the optimal model in Imha Dam is that the loss was less than that of the RNN model due to the cells of the LSTM with memory function. In addition, although the watersheds of the two dams are close, the optimal model is different because various factors such as land conditions, river slope, and rainfall characteristics worked. Therefore, it can be seen that the analysis process to find an appropriate model is important by referring to these points.

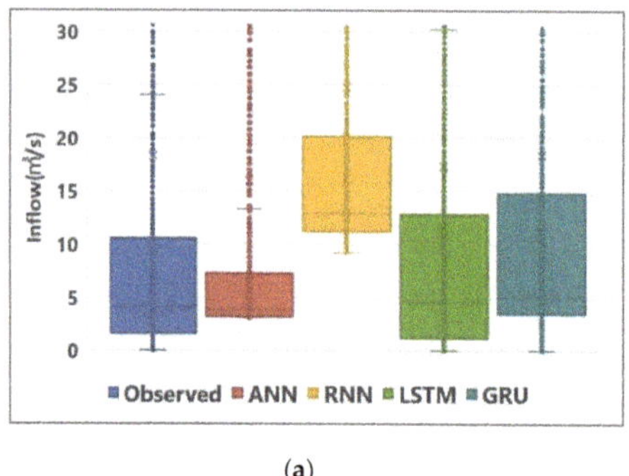
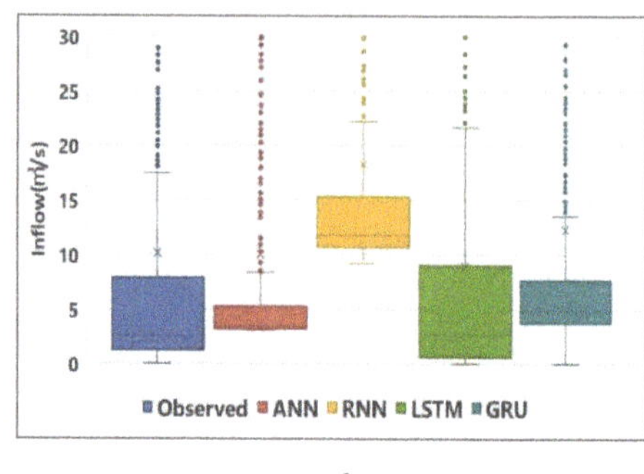

(a)   (b)

Figure 8. Comparison of predictions and observations for inflow ranges to Imha Dam. (**a**) Total period. (**b**) Drought period.

Table 8. Inflow prediction by period at Imha Dam.

Imha		Observed (m^3/s)	Simulated (m^3/s)			
			ANN	RNN	LSTM	GRU
Total period (2017–2020)	25%	1.58	3.20	11.18	1.16	3.49
	50%	4.12	3.87	12.82	4.57	5.08
	75%	10.55	7.27	20.15	12.82	14.79
Drought period (2017–2018)	25%	1.19	3.18	10.72	0.60	3.70
	50%	2.52	3.40	11.80	2.61	4.91
	75%	7.88	5.27	15.26	9.00	7.61
	Max	470.37	652.99	415.17	425.23	388.09

3.2. Typhoons

It is important not only to analyze the normal or drought period using daily data to predict the inflow to the dam, but also to analyze it using hourly data for flood control. In particular, in the case of Imha Dam, the inflow of dams in flood season (21 June–20 September) was 157.9×10^6 m^3 in 2019, while it was 743.6×10^6 m^3 in 2020. In other words, the inflow amount was 4.7 times different even in the same period. Accordingly, by applying the six major typhoon cases to each model, the maximum observed inflow and the prediction of models are compared, and the most accurate model is selected by calculating R^2. Table 9 shows the six major typhoons applied in this study. In particular, after the rainy season in 2020, typhoons occurred consecutively, and approximately 270 mm of rainfall fell in the basins of Andong and Imha Dam, and a maximum of 23.4 mm of rainfall per hour was recorded in the basin of Imha Dam. Among the six typhoon cases, Typhoon Maysak and Haishen in 2020 occurred consecutively and, therefore, are considered to be one case.

Tables 10 and 11 show the peak inflow predicted by each deep learning model using hourly inflow data for Andong Dam and Imha Dam, respectively. In Andong Dam, the GRU predictions had the smallest differences from the peak inflows observed from Typhoons Maemi, Kongrei, and Maysak and Haishen (Table 10). On the other hand, in Imha Dam, the RNN prediction showed the smallest difference from the peak inflow observed in Typhoon Rusa, Kongrei and Mitag (Table 11). Figure 9a,b show the comparison of the observations and predicted inflow by four models for Typhoons Maysak and Haisen in Andong Dam and Imha Dam, respectively. The GRU for Andong Dam and the RNN for Imha Dam were selected as the optimal model based on the maximum inflow prediction and R^2 value under typhoon conditions. However, as the maximum inflow prediction and R^2 values differ greatly depending on the characteristics of each typhoon, such as rainfall strength and preceding rainfall, as shown in Tables 10 and 11, it is considered desirable to compare various models and analyze for future flood simulation.

Table 9. Typhoon cases.

Typhoon	Period	Andong (mm)		Imha (mm)	
		Rainfall	**Hour (Max)**	**Rainfall**	**Hour (Max)**
Rusa	23 August–1 September 2002	165.4	21.9	182.9	29.3
Maemi	6–14 September 2003	251.7	31.5	220.8	26.9
Kongrey	29 September–7 October 2018	94.3	5.1	128.3	10.4
Mitag	28 September–3 October 2019	133.1	12.5	166.6	19.9
Maysak and Haishen	28 August–7 September 2020	268.1	15.0	270.0	23.4

Table 10. Predicted inflow to Andong Dam by typhoon cases.

Typhoon		Observed (m^3/s)	Simulated (m^3/s)			
			ANN	**RNN**	**LSTM**	**GRU**
Rusa	Max	3678	2570	3623	4016	4025
	R^2	-	0.94	0.95	0.94	0.96
Maemi	Max	4522	3161	4267	4339	4597
	R^2	-	0.95	0.94	0.96	0.96
Kongrey	Max	793	549	644	683	699
	R^2	-	0.62	0.77	0.81	0.76
Mitag	Max	1845	1286	1866	2117	1773
	R^2	-	0.91	0.95	0.94	0.95
Maysak andHaishen	Max	2629	1835	2328	3458	3053
	R^2	-	0.80	0.72	0.73	0.90

Table 11. Predicted inflow to Imha Dam by typhoon cases.

Case		Observed (m^3/s)	Simulated (m^3/s)			
			ANN	**RNN**	**LSTM**	**GRU**
Rusa	Max	7113	5677	7102	7014	6709
	R^2	-	0.94	0.96	0.95	0.94
Maemi	Max	6665	5312	6221	6848	6938
	R^2	-	0.95	0.95	0.94	0.92
Kongrey	Max	2584	2086	2458	2174	2222
	R^2	-	0.90	0.97	0.88	0.87
Mitag	Max	3534	2856	3488	3647	2793
	R^2	-	0.96	0.97	0.94	0.95
Maysak and Haishen	Max	4890	3909	4226	4856	4248
	R^2	-	0.91	0.91	0.89	0.90

K-water, which operates Andong Dam and Imha Dam, is currently using the SFM to predict the inflow of the two dams. Therefore, the inflow of the SFM and the predicted inflow of the GRU (Andong Dam) and the RNN (Imha Dam) were compared through analysis according to typhoon conditions. The SFM was calibrated so that the predicted inflow was closest to the observed maximum inflow while adjusting the parameters. In some cases, the R^2 has increased while the maximum predicted inflow has decreases. However, in practical dam operation, the maximum inflow and arrival time are more important factors. Therefore, the calibration was performed to better match the maximum inflow than the R^2 between the prediction and the observation.

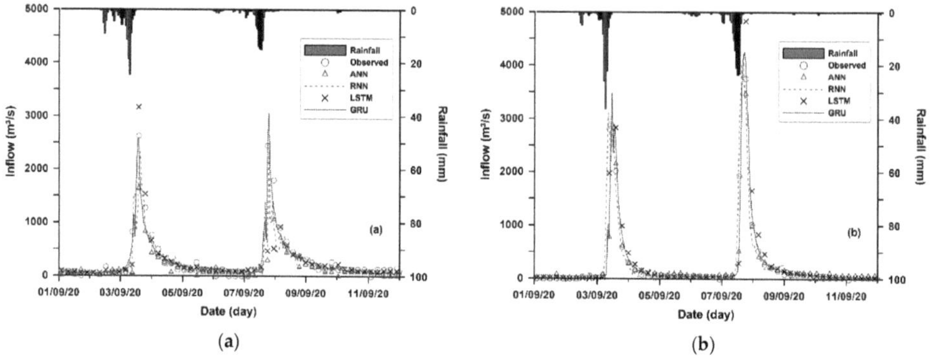

Figure 9. Comparison of observations and predictions for Typhoon Maisak and Haishen. (**a**) Andong Dam. (**b**) Imha Dam.

In Andong Dam, the difference between the predictions and the observations of the maximum inflow for Typhoons Kongrei and Mitag was larger in the SFM than in the GRU. In the case of Imha Dam, the inflow of the SFM was predicted to be lower than the observed value as well as the RNN inflow in all Typhoon conditions (Table 12). These results show that the RNN selected in this study is a reliable model when compared with the results of the SFM currently being used for dam inflow prediction. Overall, the predictions of the deep learning models were closer to the observed maximum inflow than that of the SFM. On the other hand, during Typhoon Maysak and Haishen at Andong Dam, the predictions of the SFM were better in agreement with the observed inflow than those of deep learning models. Therefore, it is necessary to derive more reasonable results through comparison of the predicted values of the SFM and deep learning models when making decisions related to dam operation.

Table 12. Predicted inflow by optimal deep learning model and the SFM in typhoon conditions.

Case		Andong			Imha		
		Observed (m^3/s)	Simulated (m^3/s)		Observed (m^3/s)	Simulated (m^3/s)	
			GRU	SFM		RNN	SFM
Rusa	Max	3628	4025	3799	7113	7102	6098
	R^2	-	0.96	0.96	-	0.96	0.98
Maemi	Max	4522	4597	4267	6665	6221	5767
	R^2	-	0.96	0.92	-	0.95	0.96
Kongrey	Max	793	699	668	2584	2458	2241
	R^2	-	0.76	0.80	-	0.97	0.96
Mitag	Max	1845	1773	1982	3534	3488	3207
	R^2	-	0.95	0.95	-	0.97	0.98
Maysak and Haishen	Max	2629	3053	2486	4890	4226	4011
	R^2	-	0.90	0.89	-	0.91	0.93

4. Discussion

This study showed the process of predicting and analyzing dam inflow using deep learning models. The reason for conducting this study is that it is important to predict the inflow with high accuracy for dam operation in disaster situations such as drought and flood. Most of the prediction results showed that the RNN models had higher accuracy than the ANN model. The reason for these results is that precipitation and inflow are time-series data, and the RNN models circulate the previous results as input variables so that learning is performed continuously without compromising the learning ability relatively. In typhoon and drought conditions, recurrent neural network models (RNN, LSTM, GRU) were selected as optimal models. In comparison with the SFM and the deep learning models, the prediction of most deep learning models was found to be closer to the observed maximum inflow than that of the SFM, but the SFM also showed better results under certain conditions.

These results suggest that even if dam basins are adjacent, different deep learning models may be selected as the optimal model for each dam by various factors including land condition and rainfall characteristics. Therefore, further studies including various factors such as land condition, evaporation, temperature, and wind speed that have not been considered in this study are needed to predict more accurate dam inflow using deep learning model.

5. Conclusions

In this study, for efficient water resource management of Andong Dam and Imha Dam, the optimal model was selected through comparison and validation of deep learning models in predicting the inflow to the two dams. Considering that dam inflow prediction is a time series analysis, RNN models were mainly applied. Four deep learning techniques—ANN, RNN, LSTM, and GRU—were utilized based on dam hydrology data for the past 20 years to predict the inflow of the dams, and optimal input variables were derived through various indicators. In addition,

(1) To evaluate the detailed prediction capability of the deep learning model with each scenario, the data were analyzed according to quartile values after differentiating the entire period and the drought period. To select a deep learning model most suitable to the drought and normal season based on the scenario, predictions and observations for the inflows of the 1st, 2nd and 3rd quartiles and peak inflow were compared using the daily time series data. In Andong Dam, the RNN model produced the closest quartile values to the observed inflow in the total period (2017–2020) and it also derived the closest to the measurements in the normal and drought period. In Imha Dam, the LSTM model showed the closest to the observations in the normal season. During the drought period, the LSTM prediction showed the smallest difference from the observations in the 1st and 2nd quartiles, whereas the GRU prediction showed the smallest difference in the 3rd quartile.

(2) A comparative analysis of six cases of past typhoons showed different predictions depending on the deep learning models. In Andong Dam, the GRU model showed higher accuracy compared to other models in the inflow prediction. In Imha Dam, unlike Andong Dam, the predicted inflow of the RNN showed the highest correlation and the most agreement with the observations. In Typhoon Mitag, R^2 has a high correlation of 0.97 and a difference of 1% between the observations and predictions which is the closest to the measured value compared to other models. As a result of analyzing the selected model, since the dam inflow and precipitation were characterized as time series data, the RNN derived predicted inflow with relatively high reliability.

(3) Compared with the SFM currently used to predict the inflow into the dam, the selected deep learning models derived results that were closer to the observed inflow in the maximum inflow prediction. In predicting future typhoon inflows, using a conceptual or physical model and a deep learning model together will help in efficient decision making.

The appropriate deep learning model varies depending on weather conditions such as drought, typhoon, and torrential rain; therefore, it is important to compare various deep learning models to cope with uncertain future climate change and to manage the operation of reservoirs efficiently and safely. In addition, as the SFM rather than the deep learning model shows better prediction results under certain typhoons, the analytical ability of hands-on workers to utilize deep learning models, as well as existing SFMs is important, as shown in the previous analysis. This study, which analyzed inflow predictions using hydrological data and deep learning models, is expected to contribute to stable dam operation management and disaster response when used as basic data for inflow prediction models of various multi-purpose dams including Andong and Imha Dams.

Author Contributions: Conceptualization and methodology, B.-J.K. and B.-H.K.; validation, Y.-T.L.; writing—original draft preparation, B.-J.K.; writing—review and editing, Y.-T.L. and B.-H.K. All authors have read and agreed to the published version of the manuscript.

Funding: This research was supported by Kyungpook National University Research Fund, 2019.

Institutional Review Board Statement: Not applicable.

Informed Consent Statement: Not applicable.

Data Availability Statement: Data are available on request.

Acknowledgments: The authors thank Kyungpook National University for providing the research fund, 2019.

Conflicts of Interest: The authors declare no conflict of interest.

References

1. Yoon, T.H. *Applied Hydrology*; Cheongmungag: Seoul, Korea, 2011.
2. Kim, S.H.; Kim, K.U.; Hwang, S.H.; Park, J.H.; Lee, J.N.; Kang, M.S. Influence of Rainfall observation Network on Daily Dam Inflow using Artificial Neural Networks. *J. Korean Soc. Agric. Eng.* **2019**, *61*, 63–74.
3. Kim, M.E.; Shon, T.S.; Joo, J.S.; Jang, Y.S.; Shin, H.S. Forecasting of Short-term Runoff with Artificial Neural Network with Pre-processing Techniques. In Preceedings of the Joint Fall Conference & Water Korea, Daegeon, Korea, 2–3 November 2011.
4. Mok, J.Y.; Choi, J.H.; Moon, Y.I. Prediction of Multipose Dam Inflow using Deep Learning. *J. Korea Water Resour. Assoc.* **2020**, *53*, 97–105.
5. Lee, J.Y.; Kim, H.I.; Han, K.Y. Linkage of Hydrological Model and Machine Learning for Real-time Prediction of River Flood. *J. Korean Soc. Civ. Eng.* **2020**, *40*, 303–314.
6. Chang, F.J.; Hsu, K.; Chang, L.C. *Flood Forecasting Using Machine Learning Methods*; MDPI: Basel, Switzerland, 2019; ISBN 978-3-03897-549-6.
7. Chang, F.J.; Chiang, Y.M.; Tsai, M.J.; Shieh, M.C.; Hsu, K.L.; Sorooshian, S. Watershed rainfall forecasting using neuro-fuzzy networks with the assimilation of multi-sensor information. *J. Hydrol.* **2014**, *508*, 374–384. [CrossRef]
8. Chakravarti, A.; Joshi, N.; Panjiar, H. Rainfall Runoff Analysis Using the Artificial Neural Network. *Indian J. Sci. Technol.* **2015**, *8*, 1–7. [CrossRef]
9. Kao, I.F.; Zhou, Y.; Chang, L.C.; Chang, F.J. Exploring a Long Short-Term Memory based Encoder-Decoder framework for multi-step-ahead flood forecasting. *J. Hydrol.* **2020**, *583*, 124631. [CrossRef]
10. Shen, C.; Laloy, E.; Elshorbagy, A.; Albert, A.; Bales, J.; Chang, F.J.; Ganguly, S.; Hsu, K.L.; Kifer, D.; Fang, Z.; et al. HESS Opinions: Incubating deep-learning-powered hydrologic science advances as a community. *Hydrol. Earth Syst. Sci.* **2018**, *22*, 5639–5656.
11. Tokar, A.S.; Markus, M. Precipitation-runoff modeling using artificial neural networks and conceptual models. *J. Hydrol. Eng.* **2000**, *5*, 156–161. [CrossRef]
12. Chen, S.M.; Wang, Y.M.; Tsou, I. Using artificial neural network approach for modelling rainfall–runoff due to typhoon. *J. Earth Syst. Sci.* **2013**, *122*, 399–405. [CrossRef]
13. Coulibaly, P.; Anctil, F.; Bobee, B. Multivariate reservoir inflow forecasting using temporal neural networks. *J. Hydrol. Eng.* **2001**, *6*, 367–376.
14. Chollet, F. *Deep Learning with Python*; Simon and Schuster: New York, NY, USA, 2021; ISBN 1638350094.
15. Hochreiter, S.; Schmidhuber, J. Long short-term memory. *Neural Comput.* **1997**, *9*, 1735–1780. [PubMed]
16. Chung, J.; Gulcehre, C.; Cho, K.; Bengio, Y. Empirical Evaluation of Gated Recurrent Neural Networks on Sequence Modeling. *arXiv* **2014**, arXiv:1412.3555.
17. Korea Water Resources Corporation (K-water). *Hydrometeorology and Watershed Management*; K-water: Daejeon, Korea, 2016.
18. Hu, C.; Wu, Q.; Li, H.; Jian, S.; Li, N.; Lou, Z. Deep Learning with a Long Short-Term Memory Networks Approach for Rainfall-Runoff Simulation. *Water* **2018**, *10*, 1543. [CrossRef]
19. Moriasi, D.N.; Arnold, J.G.; Van Liew, M.W.; Bingner, R.L.; Harmel, R.D.; Veith, T.L. Model evaluation guidelines for systematic quantification of accuracy in watershed simulations. *Trans. ASABE* **2007**, *50*, 885–900.

Article

Improved Monthly and Seasonal Multi-Model Ensemble Precipitation Forecasts in Southwest Asia Using Machine Learning Algorithms

Morteza Pakdaman [1,*], **Iman Babaeian** [1] **and Laurens M. Bouwer** [2,*]

1   Atmospheric Science and Meteorological Research Center (ASMERC), Climate Research Institute (CRI), Mashhad, Iran

2   Climate Service Center Germany (GERICS), Helmholtz-Zentrum Hereon, Fischertwiete 1, 20095 Hamburg, Germany

*   Correspondence: pakdaman.m@gmail.com (M.P.); laurens.bouwer@hereon.de (L.M.B.)

Abstract: Southwest Asia has different climate types including arid, semiarid, Mediterranean, and temperate regions. Due to the complex interactions among components of the Earth system, forecasting precipitation is a difficult task in such large regions. The aim of this paper is to propose a learning approach, based on artificial neural network (ANN) and random forest (RF) algorithms for post-processing the output of forecasting models, in order to provide a multi-model ensemble forecasting of monthly precipitation in southwest Asia. For this purpose, four forecasting models, including GEM-NEMO, NASA-GEOSS2S, CanCM4i, and COLA-RSMAS-CCSM4, included in the North American multi-model ensemble (NMME) project, are considered for the ensemble algorithms. Since each model has nine different lead times, a total of 108 different ANN and RF models are trained for each month of the year. To train the proposed ANN an RF models, the ERA5 reanalysis dataset is employed. To compare the performance of the proposed algorithms, four performance evaluation criteria are calculated for each model. The results indicate that the performance of the ANN and RF post-processing is better than that of the individual NMME models. Moreover, RF outperformed ANN for all lead times and months of the year.

Keywords: multi-model ensemble; artificial neural network; random forest; precipitation; forecasting; persian gulf

Citation: Pakdaman, M.; Babaeian, I.; Bouwer, L.M. Improved Monthly and Seasonal Multi-Model Ensemble Precipitation Forecasts in Southwest Asia Using Machine Learning Algorithms. *Water* **2022**, *14*, 2632. https://doi.org/10.3390/w14172632

Academic Editors: Fi-John Chang, Li-Chiu Chang and Jui-Fa Chen

Received: 7 July 2022
Accepted: 19 August 2022
Published: 26 August 2022

Publisher's Note: MDPI stays neutral with regard to jurisdictional claims in published maps and institutional affiliations.

1. Introduction

The accurate forecasting of precipitation has been an important topic from both theoretical points of view and practical applications. Many researchers around the world are working to improve the accuracy of monthly and seasonal weather forecasts. Such forecasts are essential for various water resource planning practices, as well as related actions for agricultural planning and securing food supplies in many regions of the world. With increasing populations that rely on such water resources and the increasing impact of climatic variations and climate change, requirements for the accuracy of such forecasts are also increasing. Different computational methods are used for improving forecasts that are generally divided into two categories: dynamical and statistical methods for post-processing the outputs of General circulation models (GCMs). The performance of these techniques can vary for different geographical areas with different topographies and climate types.

Among the computational methods for post-processing the output of forecasts, the use of machine learning (ML) techniques is very widespread. ML theory employs several types of tools such as fuzzy systems (e.g., [1]), artificial neural networks, decision trees, and so on. The widespread use of these techniques can be attributed to the ability of these methods to model Big Data and their flexibility in calculations, as well as their increased abilities to use such techniques with available modern computational resources. Examples in hydrology include flood forecasting applications (see e.g., [2]) and climate modelling applications for sub-grid processes ([3]).

In seasonal weather forecasting, [4] employed artificial neural networks (ANNs) for the monthly forecasting of precipitation over Iran, and used the outputs of the North American multi-model ensemble (NMME) project. They also used other machine learning techniques including ANN, support vector regression, decision tree, and random forests for the monthly forecast of precipitation [5]. Ref. [6] used the non-linear autoregressive neural network (NARNN), non-linear input–output (NIO), and NARNN with exogenous input (NARNNX), for annual precipitation forecasting over 27 precipitation stations located in Gilan, Iran. Their results showed that the accuracy of the NARNNX was better than that of the NARNN and NIO. Ref. [7] proposed a model for the monthly forecasting of precipitation for East Azarbaijan in Iran, over a ten-year period using the multilayer perceptron neural network (MLP) and support vector regression (SVR) models. They employed the flow regime optimization algorithm (FRA) for training step in multilayer neural network and support vector machine. Ref. [8] constructed an artificial neural network model for generating probabilistic subseasonal precipitation forecasts over California. They used an artificial neural network (ANN) to establish relationships between the 7-day accumulated precipitation of numerical weather prediction (NWP) ensemble forecast. Ref. [9] tried to use the random-forest-based machine learning algorithm for nowcasting convective rain in Kolkata, India, with a ground-based radiometer. They found that their proposed model is very sensitive to the boundary layer instability, as indicated by the variable importance measure. Their study showed the suitability of a random forest algorithm for nowcasting the application and other forecasting problems. Ref. [10] used the ensemble precipitation forecasts of six numerical models from the THORPEX Interactive Grand Global Ensemble (TIGGE) database, associated with four basins in Iran for 2008–2018, which were extracted and bias-corrected by the quantile mapping (QM) and random forest (RF) methods. Their results demonstrated that most models had better skills in forecasting precipitation depth after bias correction using the RF method, compared to using the QM method and raw forecasts. Ref. [11] used random forests and regression tree for quantitative precipitation estimates with operational dual-polarization radar data. The use of neural networks and machine learning techniques is not limited to precipitation forecasting. For example, Ref. [12] used decision tree and neural networks for lightning prediction. Ref. [13] used a kernel least mean square algorithm for solving fuzzy differential equations and studied its application in earth's energy balance model and climate. Ref. [14] studied the theoretical impact of changing albedo on precipitation and [15] found that fuzzy uncertainty for albedo creates more real results after solving the fuzzy energy balance equation. For some other applications of machine learning in climate modeling, see [16–23].

Southwest Asia contains many geographical features including water bodies of Caspian Sea, Persian Gulf, Oman Sea, Arabian Sea, and elongated mountains of Zagros and Alborz, as well as widespread deserts in Saudi Arabia, Iraq, and Iran (Lut and Dasht-e-Kavir). The region has different types of climates including arid, semiarid, Mediterranean, and temperate regions. Due to complex interactions among components of the earth and weather systems, the forecasting of precipitation is a difficult task in such large regions. Several researches have conducted monthly forecasts of precipitation in this region and countries therein. Ref. [24] studied the attributes of precipitation for the Middle East and southwest Asia during periods of enhanced or reduced tropical eastern Indian Ocean precipitation associated with opposing phases of the Madden–Julian oscillation (MJO). They used multiple estimates of both observed precipitation and MJO state during November–April 1981–2016 to provide a more robust assessment in this data-limited region. Ref. [25] assessed the sensitivity of southwest Asia precipitation during the November–April rainy season to four types of El Niño–Southern Oscillation (ENSO) events, El Niño and La Niña, using an ensemble of climate model simulations forced by 1979–2015 boundary conditions. Ref. [26] studied the potential predictability and skill of boreal winter (December to February: DJF) precipitation over central-southwest Asia by using six models of the North American multi-model ensemble project for the period 1983–2018. Ref. [27] explored the predictability of central-southwest Asia wintertime precipitation based on its time-lagged relationship with the preceding months' (September–October) sea surface temperature (SST), using a canonical correlation analysis (CCA) approach. They showed that the regional potential predictability has a strong dependency on the ENSO phenomenon, and the strengthening (weakening) of this relationship yields forecasts with higher (lower) predictive skill. Based on the literature review, there are various studies in precipitation forecasting for the countries of the Southwest Asian region separately. However, there is no integrated approach based on machine learning methods as well as MME approaches that consider multiple countries simultaneously. Although extending the computations to a larger area increases the computational costs, it can better model large precipitation patterns. On the other hand, using a MME approach can help reduce forecasting error.

The aim of this research is to construct a framework based on ANN and RF methods to improve the performance of monthly forecast of precipitation in Southwest Asia. For this purpose, a multi-model approach is proposed which uses the output of four GCMs, including GEM-NEMO, NASA-GEOSS2S, CanCM4i, and COLA-RSMAS-CCSM4 from the NMME project. We build on previous developments, but include a wider range of ensemble forecasts, and apply these methods to a larger region, to assess its performance. The paper is organized as follows: Section 2 contains information about the datasets and Section 3 illustrates the details of the proposed algorithms. Section 4 contains the results, and finally Section 5 contains concluding remarks.

2. Data

We propose a multi-model ensemble approach for monthly precipitation forecasting in Southwest Asia using the output of four models (GEM-NEMO, NASA-GEOSS2S, CanCM4i, and COLA-RSMAS-CCSM4) from the NMME project. We also need observed precipitation data as a benchmark to compare the forecasts (or, in this case, the hindcasts). In this regard, we use ERA5 data for estimating monthly precipitation. The region of southwest Asia in this paper is contained in the region which is depicted in Figure 1. This region is limited to latitude 22 °N to 42 °N and longitude 39 °E to 70 °E.

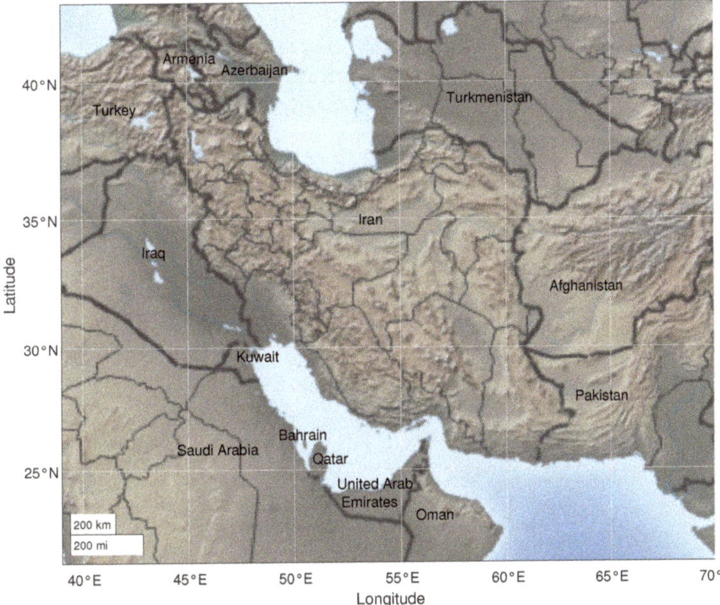

Figure 1. Region of study.

Details of the GCMs and their abbreviation are provided in Table 1. The applied spatial resolution in this study is $1° \times 1°$. The number of members of NEMO, CanCM4i, and CCSM4 is 10, while the number of members of the NASA model is 4. For simplicity, for each of four GCMs, we calculated the average values of all members. Moreover, the number of lead times of NEMO, CanCM4i, and CCSM4 is 12 while the number of lead times of NASA is 9. Thus, we considered the same number of lead times (nine) for all four GCMs. Based on the hindcast period and available data for all four models and also the ERA5 data, the period 1982–2016 was considered for constructing the post-processing model. Data of the four NMME models were downloaded from https://iridl.ldeo.columbia.edu (accessed on 22 January 2022), while the ERA5 data are accessible via https://cds.climate.copernicus.eu (accessed on 22 January 2022). The unit of the amount of precipitation for NMME models was different from the units in the ERA5 data. In preprocessing step, the data were prepared for the post-processing algorithms. No missing data were found in the datasets.

Table 1. Four models which are used in this study for post-processing.

Models	Abbreviation	Members	Lead Times	Hindcast Period
GEM-NEMO	NEMO	10	12 (0.5–11.5 months)	1981–2018
NASA-GEOSS2S	NASA	4	9 (0.5–8.5 months)	1981–2017
CanCM4i	CanCM4i	10	12 (0.5–11.5 months)	1981–2018
COLA-RSMAS-CCSM4	CCSM4	10	12 (0.5–11.5 months)	1982–2021

3. Methods

In this section, we use the ability of ANN and RF for monthly forecast of precipitation in Southwest Asia. Based on [28], MLP neural networks, with hidden layers and sigmoid transfer function

$$s(t) = \frac{1}{1 + \exp(-t)},$$ (1)

are universal approximators, and we can use them for regression tasks. Figure 2 depicts a general architecture of a single (hidden) layer perceptron neural network which was considered in this paper.

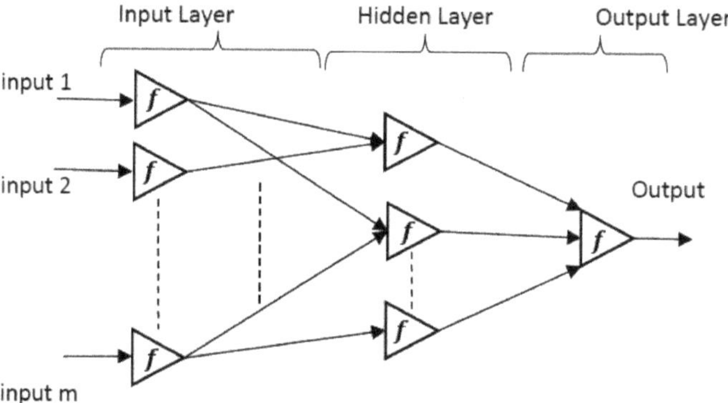

Figure 2. General architecture of a single (hidden) layer perceptron neural network. f is the activation function.

On the other hand, random forest (RF) is an ensemble method in machine learning that can be used for both classification and regression purposes. The base algorithm in RF is a decision tree (DT). Based on the type of the problem, two different types of DT are available: regression DT and classification DT. Thus, based on the type of DT, we can have RF for regression or classification. For more details about the RF, ANN, and DT, see [29]. In this paper, we used the machine learning toolbox of MATLAB 2018b. A general form of the proposed algorithm is depicted in Figure 3, wherein we will simply replace the ensemble learning algorithm by ANN or RF.

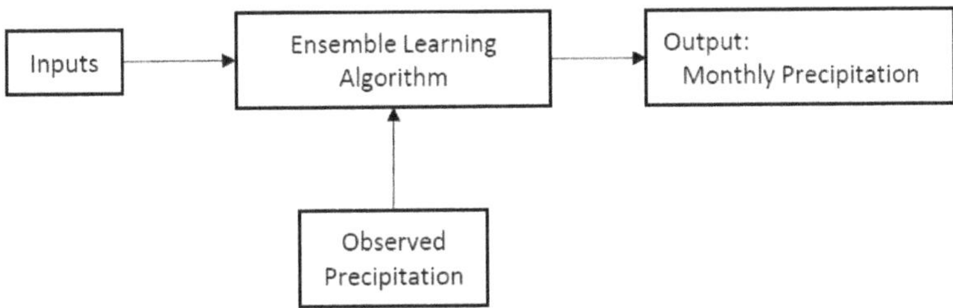

Figure 3. A general architect of the proposed algorithm.

More details about the proposed approach are illustrated in Figure 4. As the figure shows, the proposed algorithm has six inputs and one output. The inputs include the monthly precipitation data of four models from the NMME project, along with the latitude and longitude of the region. In Figure 4, ERA5 data are used to train RF and ANN, and finally the output of the algorithm is an approximation of monthly precipitation.

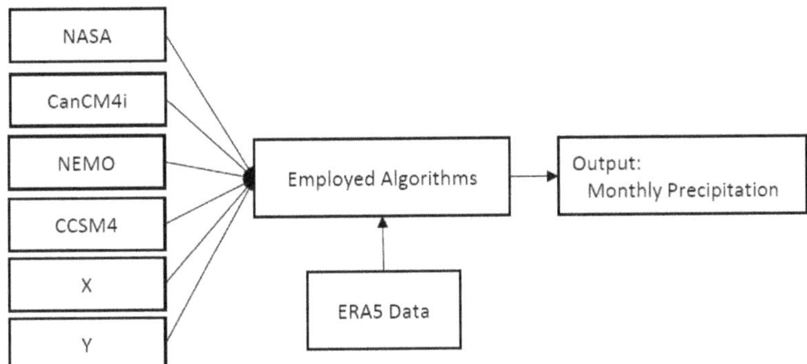

Figure 4. Details of the proposed algorithm.

Suppose that M is the output of the ANN or RF algorithm. Since each climate model generates the forecast data for different lead times, we need to construct a specific post-processing model related to each month and lead time. Suppose that $M_i(m, l)$ is the output of the i-th NMME model M, $i = 1, 2, 3, 4$ for month m and lead time l. We can say that $M_i \in \{\text{CanCM4i}, \text{CCSM4}, \text{NASA}, \text{NEMO}\}$. Also suppose that, $\hat{M}(m, l)$ is the output of the RF or ANN algorithm for month m, and lead time l. Indeed, $\hat{M}(m, l)$ is a function of $M_i(m, l)$, $i = 1, 2, 3, 4$, latitude, and longitude, as depicted in Figure 4. This function can be defined as follows:

$$\hat{M}(m, l) = F(M(m, l), x, y, P), \tag{2}$$

wherein $M(m, l)$ contains all data of $M_i(m, l)$, $i = 1, 2, 3, 4$ (four NMME models), and x and y indicate longitude and latitude, respectively. Furthermore, P indicates the adjustable parameters of RF or ANN. If F indicates the RF, P indicates the adjustable parameters of RF (for example the number of decision trees in the RF algorithm). If F indicates the ANN algorithm, P indicates the weights of input, hidden, and output layers of the ANN algorithm. The optimal values of P (which will be indicated by P^*) will be obtained for the training period. For a seasonal forecast, we do not need to provide separate RF and ANN algorithms. The monthly results of three consecutive months can be aggregated for the seasonal forecast of three months.

For selecting the correct initial months for each month for which the prediction is made, see Table 2. An example is provided in Table 3 for a situation which we have four lead times. This table indicates the correct month number for different initial months.

Table 2. Lead times versus months.

Month \ Lead Time	1	2	...	9
1	$M(1,1)$	$M(1,2)$	...	$M(1,9)$
2	$M(2,1)$	$M(2,2)$	...	$M(2,9)$
⋮	⋮	⋮	...	⋮
12	$M(12,1)$	$M(12,2)$	...	$M(12,9)$

Table 3. An example of determining the target month for a given month and lead time.

Month \ Lead Time	L_1	L_2	L_3	L_4
January	January	February	March	April
February	February	March	April	May
March	March	April	May	June
April	April	May	June	July
May	May	June	July	August
June	June	July	August	September
July	July	August	September	October
August	August	September	October	November
September	September	October	November	December
October	October	November	December	January
November	November	December	January	February
December	December	January	February	March

To compare the performance of the models, the values of the correlation coefficient, root mean squared error (RMSE), Kling–Gupta efficiency (KGE), and Nash–Sutcliffe efficiency (NSE) can be calculated as follows:

$$r = \frac{\sum_{i=1}^{n} (M_i - \overline{M})(O_i - \overline{O})}{\sqrt{\sum_{i=1}^{n} (M_i - \overline{M})^2 \sum_{i=1}^{n} (O_i - \overline{O})^2}}, \tag{3}$$

wherein M and O indicate the values of the model and observation, respectively. Furthermore, $\overline{M}$ and $\overline{O}$ denote the mean values of corresponding values. The $RMSE$ can be calculated as follows:

$$RMSE = \sqrt{\frac{1}{n}\sum_{i=1}^{n}(M_i - O_i)^2}. \tag{4}$$

Kling–Gupta efficiency (KGE) is defined as follows (see [30]):

$$KGE = 1 - \sqrt{(r-1)^2 + \left(\frac{\sigma_M}{\sigma_O} - 1\right)^2 + \left(\frac{\overline{M}}{\overline{O}} - 1\right)^2} \tag{5}$$

wherein σ indicates the standard deviation. The maximum value of KGE is 1 which indicates perfect agreement between the model output and observations. Finally, the NSE (Nash–Sutcliffe efficiency) is calculated as follows ([31]):

$$NSE = 1 - \frac{\sum_{i=1}^{n}(M_i - O_i)^2}{\sum_{i=1}^{n}\left(O_i - \overline{O}\right)^2}. \tag{6}$$

Similar to KGE, the maximum value of NSE is 1 which indicates perfect agreement between model output and observations.

For tuning the hyperparameters of the ANN and RF algorithms, we used an empirical approach. We examined several values for each hyperparameter and selected the best case. For example, to determine the best value for the number of decision trees in the RF, we examined the values $20, 40, 60, \ldots, 200$. The best performance of RF was achieved for the number of trees equal to 100. As another example, for the number of neurons in the hidden layer of the ANN, we examined the values $3, 6, 9, \ldots, 21$. The best performance of ANN (considering the RMSE, KGE, NSE, and the correlation coefficient) was achieved for the number of neurons equal to 15. It must be noted that using more (less) neurons or more (less) DTs may cause overfitting (underfitting).

4. Results and Discussion

4.1. Monthly Forecasts

After applying ANN and RF for post-processing the output of the four NMME models, the values of root mean squared error (RMS), correlation coefficient, Nash–Sutcliffe efficiency (NSE), and Kling–Gupta efficiency (KGE) were calculated to evaluate the performance of the proposed algorithms for each model separately.

The raw forecast performance of the four models (before post-processing), along with the performance of RF and ANN (after post-processing), was calculated for all lead times and months. To train and test the RF and ANN, the dataset was divided into two subsets. Since the hindcast period was 1982–2016 and contains 35 years, the period 1982–2007 (26 years) was considered for training the RF and ANN, while the period 2008–2016 (9 years) was considered to test the RF and ANN. This means that approximately 75% of data were used for training and 25% was used for testing.

Figure 5 depicts the values of KGE for all months and lead times 1–4 while Figure 6 depicts the values of KGE for all months and lead times 5–9. For example, the maximum value of KGE in the first lead time of Figure 5 occurs in May (which predicts May itself) and for both ANN and RF. The maximum value of KGE in the second lead time of Figure 5 occurs in April (which predicts May) and for both ANN and RF. It is also should be noted that all calculated performance indices (KGE, RMSE, and NSE), as well as the correlation coefficient, were calculated for test period 2008–2016. Thus, all generated figures are devoted to period 2008–2016.

As shown in Figure 5, the values of KGE for RF and ANN are higher than the other NMME models. This is true for approximately all months and lead times (except May in lead 4, whereby the KGE of ANN is lower than that of NASA and CanCM4i). However, the values of KGE (for RF) for all months and lead times are higher than those for ANN and also the NMME models.

When considering the RF and ANN methods, Figure 5 shows that for lead time 1, the maximum forecast error occurs in September, which may be due to the fact that September is the transition month from the warm period to the cold period of the year. A similar discussion can be made for Figure 6. As shown in Figure 5, the lowest values of KGE for the NMME models are -0.27 (for CCSM4 and month 8), -0.27 (for NEMO and month 7), -0.31 (for CCSM4 and month 5), and -0.27 (for NEMO and month 5) for lead times 1–4, respectively. These values are improved by applying the RF method to 0.68, 0.66, 0.67, and 0.68 respectively. Similarly, as shown in Figure 6, the lowest values of KGE for the NMME models are -0.27 (for NEMO and month 4), -0.29 (for NEMO and month 3),

−0.21 (for NEMO and month 3), −0.19 (for NEMO and month 2), and −0.13 (for NEMO and month 1) for lead times 5–9, respectively. These values are improved to 0.62 (using the RF and ANN), 0.63 (RF and ANN), 0.4 (RF), 0.41 (ANN), and 0.4 (RF), respectively.

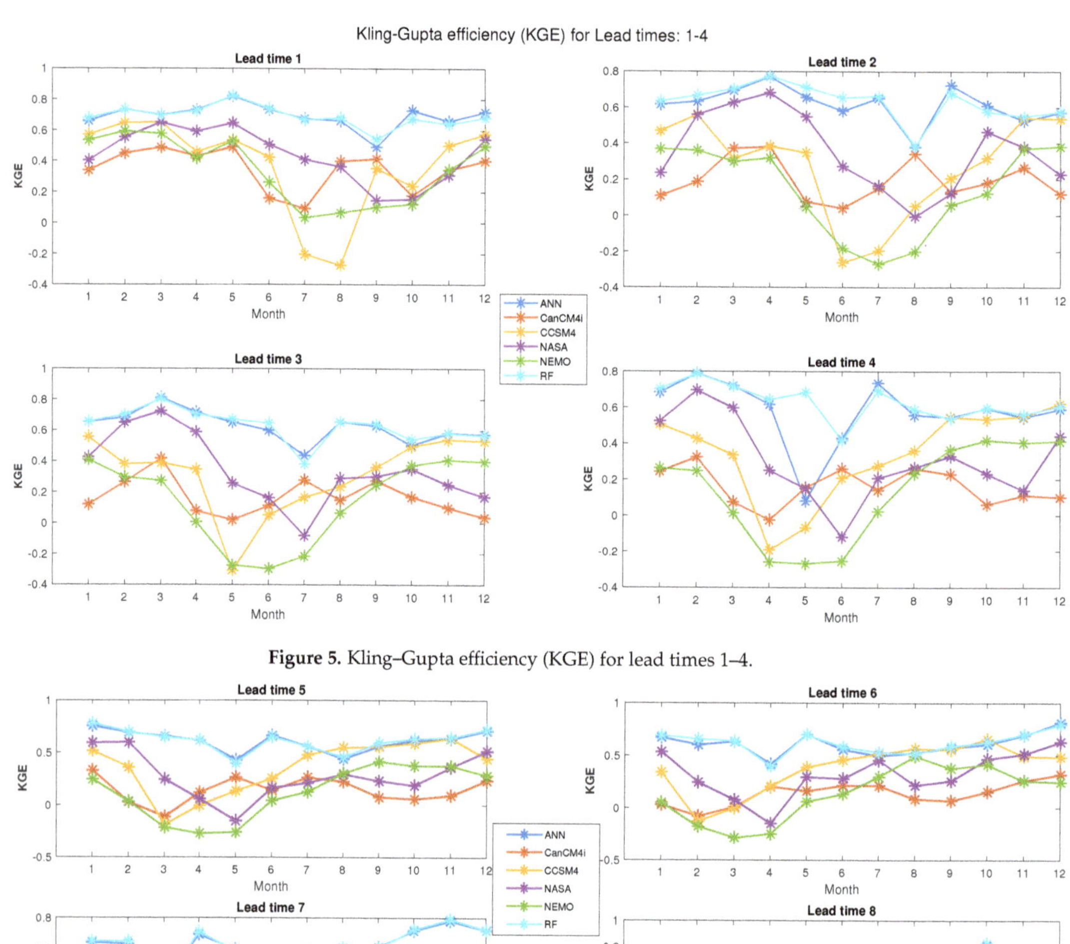

Figure 5. Kling–Gupta efficiency (KGE) for lead times 1–4.

Figure 6. Kling–Gupta efficiency (KGE) for lead times 5–9.

Figure 7 shows the values of NSE for all months and lead times 1–4. As it can be observed in Figure 7, the values of NSE for the RF and ANN are higher than those for the other NMME models. This is true for approximately all months and lead times (except some months of lead 2). However, the values of NSE for the RF algorithm (for all months and lead times) are higher than those for the ANN and also for the NMME models. As shown in Figure 7, the lowest values of NSE for the NMME models are −0.48 (for CCSM4 and month 8), −0.42 (for CCSM4 and month 6), −0.39 (for CCSM4 and month 5), and −0.44 (for CCSM4 and month 4) for lead times 1–4, respectively. These values are improved by the RF method to 0.6, 0.57, 0.61, and 0.57, respectively.

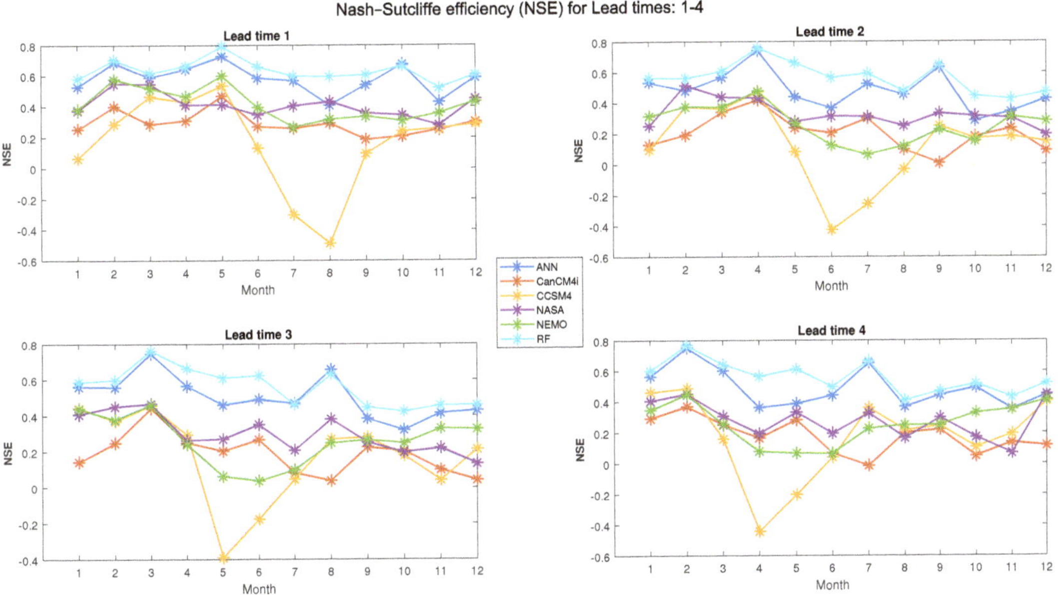

Figure 7. Nash–Sutcliffe efficiency (NSE) for lead times 1–4.

Figure 8 shows the values of RMSE for all months and lead times 1–4. As it can be observed in Figure 8, the values of RMSE for the RF and ANN are lower than those for the other NMME models. This is true for approximately all months and lead times. However, the values of RMSE for the RF algorithm (for all months and lead times) are lower than those for the ANN and also the NMME models. As shown in Figure 8, the maximum values of RMSE for the NMME models are 47.64 (for CCSM4 and month 8), 46.81 (for CCSM4 and month 6), 46.24 (for CCSM4 and month 5), and 47.1 (for CCSM4 and month 4) for lead times 1–4, respectively. These values are improved by the RF method to 24.83, 25.79, 24.5, and 25.8, respectively.

Figure 9 shows the values of correlation coefficient for all months and lead times 1–4. As it can be observed in Figure 9, the correlation coefficient values for the RF and ANN are higher than those for the other NMME models. This is true for approximately all months and lead times. However, the values of correlation coefficient for the RF algorithm (for all months and lead times) are higher than those for the ANN and also the NMME models. As shown in Figure 9, the lowest values of correlation coefficient for the NMME models are 0.41 (for CCSM4 and month 8), 0.27 (for CanCM4i and month 9), 0.3 (for CanCM4i and month 8), and 0.26 (for CanCM4i and month 7) for lead times 1–4, respectively. These values are improved to 0.77 (RF), 0.81 (RF), 0.81 (ANN), and 0.82 (RF), respectively.

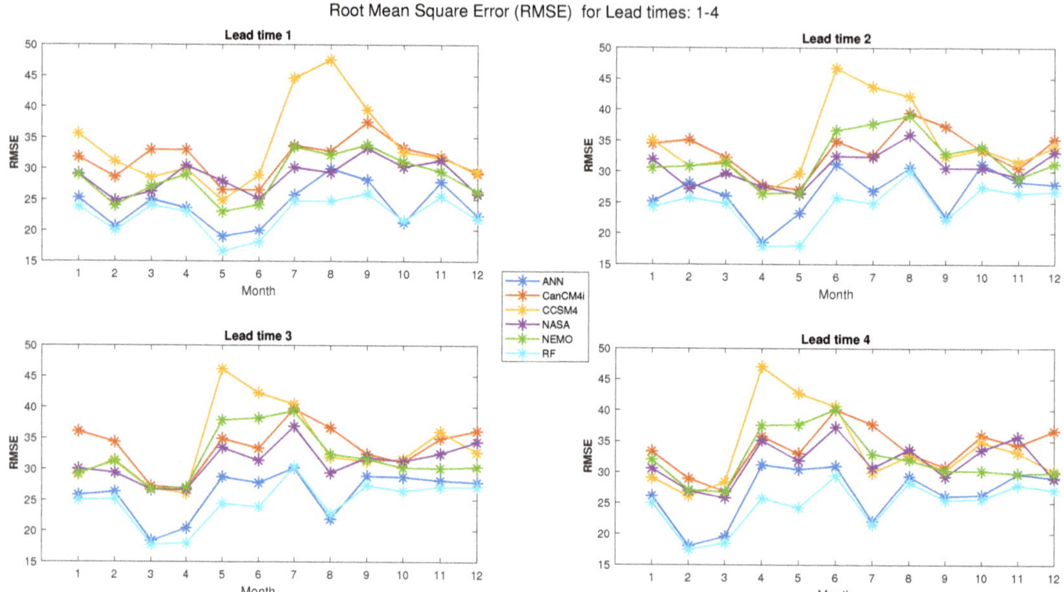

Figure 8. Root mean square error (RMSE) for lead times 1–4.

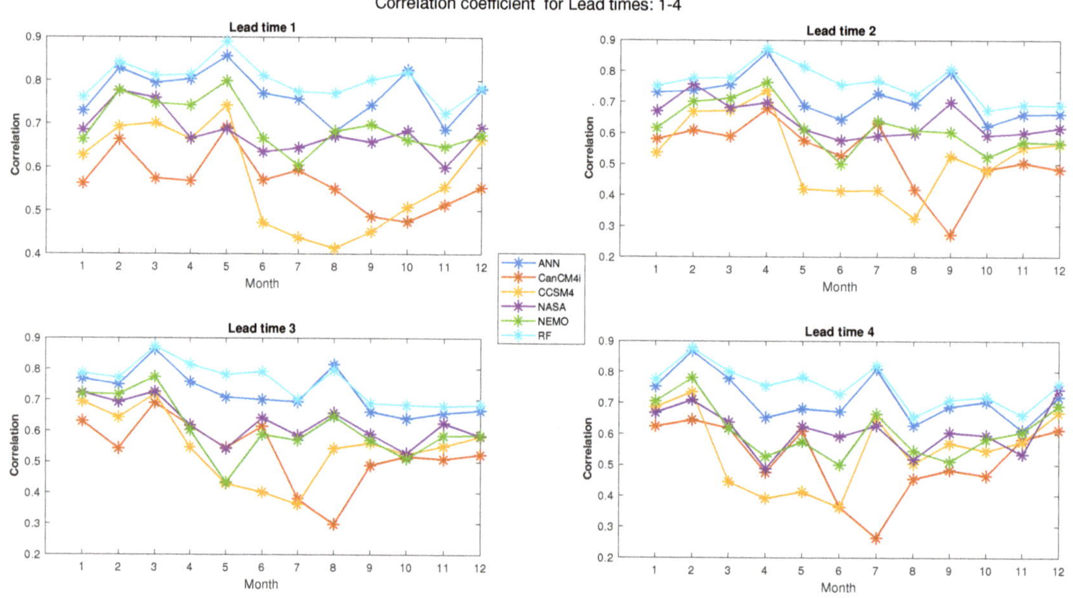

Figure 9. Correlation coefficient for lead times 1–4.

4.2. Seasonal Forecasts

In this section, the skills of the proposed algorithm are evaluated for seasonal forecasts including: DJF (winter), MAM (spring), JJA (summer), and SON (autumn). For this purpose, the lead 1 data of three consecutive months were considered in order to generate the seasonal data. Similar calculations can be performed for the other lead times. Figure 10 depicts the values of RMSE for all seasons and for the test data (2008–2016). Except SON, whereby ANN has lower RMSE than RF, for other seasons, RF outperforms ANN. Furthermore, in comparison to four NMME models, the RMSE of ANN and RF is lower.

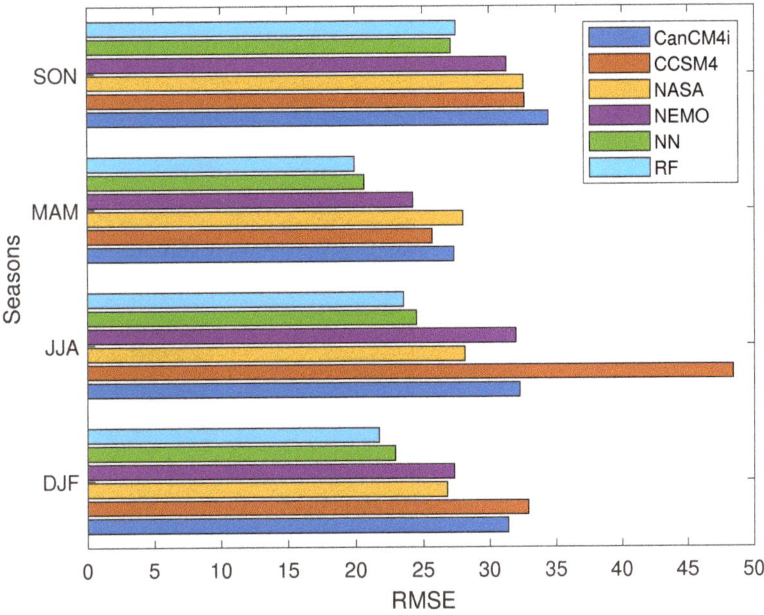

Figure 10. Seasonal RMSE for the models.

Figure 11 depicts the values of NSE for all seasons and for test data. Except SON, for which ANN has a larger NSE than RF, RF outperforms ANN. Moreover, in comparison with four NMME models, the values of NSE for ANN and RF are larger.

Figure 12 depicts the values of correlation coefficient for all seasons and for test data. Except SON, for which ANN has a larger correlation coefficient than RF, for other seasons, RF outperforms ANN. Furthermore, in comparison with four NMME models, the values of correlation coefficient for ANN and RF are larger.

To indicate the performance pattern of the models, during the DJF season, the values of RMSE are plotted for test data in Figure 13. As it can be observed in Figure 13a, the RF algorithm has the best performance among the other models. After the RF algorithm, based on Figure 13b, the ANN algorithm has better performance in comparison with the NMME models. Moreover, it can be inferred that in the raw (NMME) model data, the maximum error occurs in the mountainous areas of Zagros and Alborz and lands in the east and south of the region. However, the ANN and RF methods are able to reduce the error over high-elevation lands, while the ability of the RF method to reduce the error is larger than that for the ANN. However, in both methods, the error increased slightly in some lowland areas.

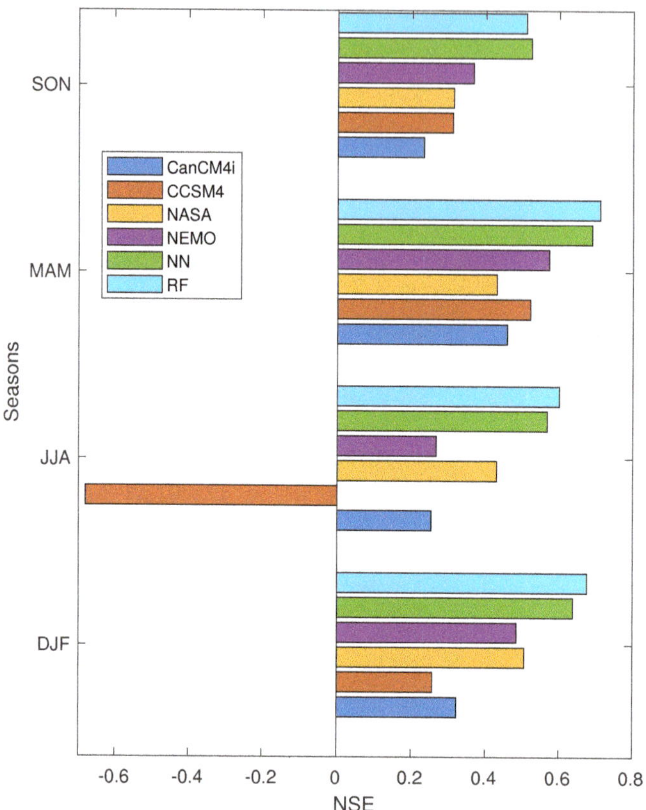

Figure 11. Seasonal NSE for the models.

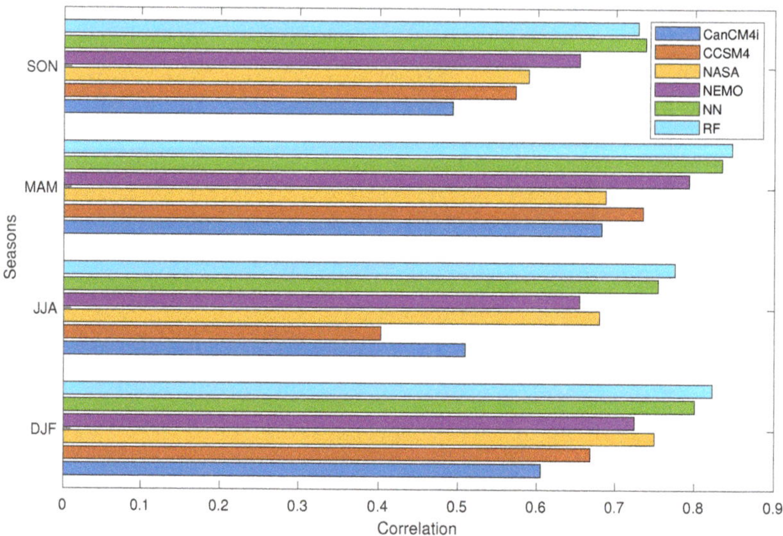

Figure 12. Seasonal correlation coefficient for the models.

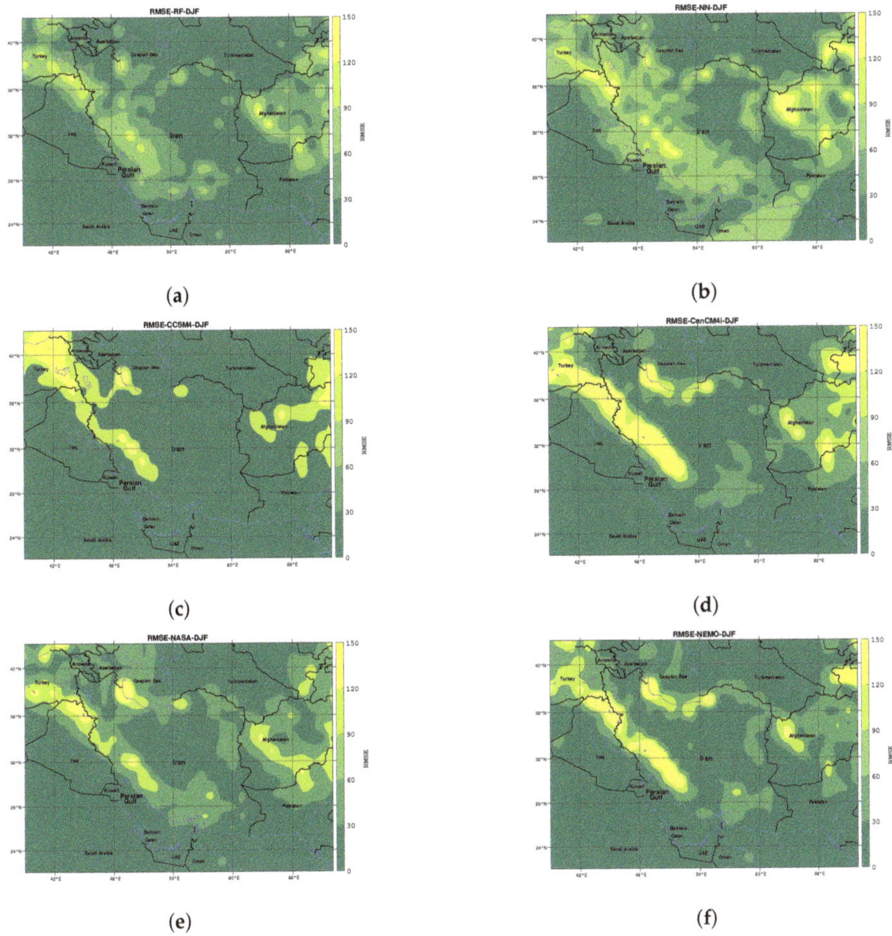

Figure 13. The values of RMSE during the DJF for test data: (d–f) have the same range of RMSE in the color bar, unlike the other models in (a–c).

5. Conclusions

In this paper, a multi-model ensemble approach was proposed for the monthly fore-casting of precipitation using ANN and RF algorithms for southwest Asia. For this purpose, the outputs of four NMME models GEM-NEMO, NASA-GEOSS2S, CanCM4i, and COLA-RSMAS-CCSM4 were post-processed using the ERA5 data as a reference. The period 1982–2007 was used for learning the ANN and RF, while the data from period 2008–2016 were used for testing the algorithms. Since each model contains of nine different lead time, for each month of year, we had nine different datasets. In this paper, we trained 108 ANN and RF models for each lead time and month separately. Four performance evaluation criteria (root mean squared error (RMS), correlation coefficient, Nash–Sutcliffe efficiency (NSE), and Kling–Gupta efficiency (KGE), were calculated for each of 108 models along with NMME models. The results indicate that, approximately, the output of ANN and RF outperforms the NMME models for all months and lead times. Specifically, RF outperforms the ANN and the other four NMME models for all months and all lead times. The proposed algorithms and approach can be used for monthly forecasting precipitation in southwest Asia, but can also be modified for the monthly forecasting of, e.g., temperature and other variables. Moreover, other machine learning techniques could be applied to improve the accuracy of the proposed multi-model approach. The results of this research show that despite the vastness of the studied area and the different climates contained therein, machine learning methods can be used for post-processing and improving forecasts in a multi-model ensemble approach.

Author Contributions: Conceptualization, M.P. and I.B.; formal analysis, M.P. and I.B.; validation, M.P., I.B. and L.M.B.; writing—original draft preparation, M.P., I.B. and L.M.B.; writing—review and editing, M.P. and L.M.B.; funding acquisition, M.P. All authors have read and agreed to the published version of the manuscript.

Funding: This work was supported in part by the Iran National Science Foundation (INSF) under project number 98028362. L.M.B. was supported by the Helmholtz Association through the research programme Changing Earth—Sustaining our Future.

Data Availability Statement: The datasets that are analyzed in this paper are publicly available from the following archives: forecasting data from https://iridl.ldeo.columbia.edu (accessed on 22 January 2022); the ERA5 data from https://cds.climate.copernicus.eu (accessed on 22 January 2022).

Acknowledgments: We thank three anonymous reviewers for reading our paper, and for providing constructive comments that have helped to improve the presentation of our study.

Conflicts of Interest: The authors declare no conflict of interest.

Abbreviations

The following abbreviations are used in this manuscript:

ANN	artificial neural network
DJF	December, January, February (winter)
ERA5	ECMWF re-analysis dataset 5
JJA	June, July, August (summer)
KGE	Kling–Gupta efficiency coefficient
MAM	March, April, May (spring)
NMME	North American multi-model ensemble
NSE	Nash–Sutcliffe efficiency coefficient
RF	Random forest
RMSE	Root mean squared error
SON	September, October, November (autumn)

References

1. Yazdi, H.S.; Pakdaman, M.; Effati, S. Fuzzy circuit analysis. *Int. J. Appl. Eng. Res.* **2008**, *3*, 1061–1072.
2. Yuval, J.; O'Gorman, P.A. Stable machine-learning parameterization of subgrid processes for climate modeling at a range of resolutions. *Nat. Commun.* **2020**, *11*, 3295. [PubMed]
3. Chang, F.J.; Hsu, K.; Chang, L.C. *Flood Forecasting Using Machine Learning Methods*; MDPI: Basel, Switzerland, 2019.
4. Pakdaman, M.; Falamarzi, Y.; Babaeian, I.; Javanshiri, Z. Post-processing of the North American multi-model ensemble for monthly forecast of precipitation based on neural network models. *Theor. Appl. Climatol.* **2020**, *141*, 405–417.
5. Pakdaman, M.; Babaeian, I.; Javanshiri, Z.; Falamarzi, Y. European Multi Model Ensemble (EMME): A New Approach for Monthly Forecast of Precipitation. *Water Resour. Manag.* **2022**, *36*, 611–623.
6. Valipour, M. Optimization of neural networks for precipitation analysis in a humid region to detect drought and wet year alarms. *Meteorol. Appl.* **2016**, *23*, 91–100.
7. Banadkooki, F.B.; Ehteram, M.; Ahmed, A.N.; Fai, C.M.; Afan, H.A.; Ridwam, W.M.; Sefelnasr, A.; El-Shafie, A. Precipitation forecasting using multilayer neural network and support vector machine optimization based on flow regime algorithm taking into account uncertainties of soft computing models. *Sustainability* **2019**, *11*, 6681.
8. Scheuerer, M.; Switanek, M.B.; Worsnop, R.P.; Hamill, T.M. Using artificial neural networks for generating probabilistic subseasonal precipitation forecasts over California. *Mon. Weather Rev.* **2020**, *148*, 3489–3506.
9. Das, S.; Chakraborty, R.; Maitra, A. A random forest algorithm for nowcasting of intense precipitation events. *Adv. Space Res.* **2017**, *60*, 1271–1282.
10. Zarei, M.; Najarchi, M.; Mastouri, R. Bias correction of global ensemble precipitation forecasts by Random Forest method. *Earth Sci. Inform.* **2021**, *14*, 677–689.

11. Shin, K.; Song, J.J.; Bang, W.; Lee, G. Quantitative Precipitation Estimates Using Machine Learning Approaches with Operational Dual-Polarization Radar Data. *Remote Sens.* **2021**, *13*, 694.

12. Pakdaman, M.; Naghab, S.S.; Khazanedari, L.; Malbousi, S.; Falamarzi, Y. Lightning prediction using an ensemble learning approach for northeast of Iran. *J. Atmos. Sol.-Terr. Phys.* **2020**, *209*, 105417. [CrossRef]

13. Pakdaman, M.; Falamarzi, Y.; Yazdi, H.S.; Ahmadian, A.; Salahshour, S.; Ferrara, F. A kernel least mean square algorithm for fuzzy differential equations and its application in earth's energy balance model and climate. *Alex. Eng. J.* **2020**, *59*, 2803–2810. [CrossRef]

14. Doughty, C.E.; Loarie, S.R.; Field, C.B. Theoretical impact of changing albedo on precipitation at the southernmost boundary of the ITCZ in South America. *Earth Interact.* **2012**, *16*, 1–14.

15. Pakdaman, M.; Habibi Nokhandan, M.; Falamarzi, Y. Revisiting albedo from a fuzzy perspective. *Kybernetes* **2021**, *ahead-of-print*. [CrossRef]

16. Han, X.; Wei, Z.; Zhang, B.; Li, Y.; Du, T.; Chen, H. Crop evapotranspiration prediction by considering dynamic change of crop coefficient and the precipitation effect in back-propagation neural network model. *J. Hydrol.* **2021**, *596*, 126104. [CrossRef]

17. Aksoy, H.; Dahamsheh, A. Markov chain-incorporated and synthetic data-supported conditional artificial neural network models for forecasting monthly precipitation in arid regions. *J. Hydrol.* **2018**, *562*, 758–779.

18. Chang, N.B.; Yang, Y.J.; Imen, S.; Mullon, L. Multi-scale quantitative precipitation forecasting using nonlinear and nonstationary teleconnection signals and artificial neural network models. *J. Hydrol.* **2017**, *548*, 305–321. [CrossRef]

19. Tomassetti, B.; Verdecchia, M.; Giorgi, F. NN5: A neural network based approach for the downscaling of precipitation fields–Model description and preliminary results. *J. Hydrol.* **2009**, *367*, 14–26.

20. Chen, H.; Chandrasekar, V.; Tan, H.; Cifelli, R. Rainfall estimation from ground radar and TRMM precipitation radar using hybrid deep neural networks. *Geophys. Res. Lett.* **2019**, *46*, 10669–10678. [CrossRef]

21. Chen, G.; Wang, W.C. Short-term precipitation prediction for contiguous United States using deep learning. *Geophys. Res. Lett.* **2022**, *49*, e2022GL097904. [CrossRef]

22. Shi, X. Enabling smart dynamical downscaling of extreme precipitation events with machine learning. *Geophys. Res. Lett.* **2020**, *47*, e2020GL090309. [CrossRef]

23. Kim, J.W.; Pachepsky, Y.A. Reconstructing missing daily precipitation data using regression trees and artificial neural networks for SWAT streamflow simulation. *J. Hydrol.* **2010**, *394*, 305–314. [CrossRef]

24. Hoell, A.; Cannon, F.; Barlow, M. Middle East and Southwest Asia daily precipitation characteristics associated with the madden–Julian oscillation during boreal winter. *J. Clim.* **2018**, *31*, 8843–8860. [CrossRef]

25. Hoell, A.; Barlow, M.; Xu, T.; Zhang, T. Cold season southwest Asia precipitation sensitivity to El Niño–Southern Oscillation events. *J. Clim.* **2018**, *31*, 4463–4482. [CrossRef]

26. Ehsan, M.A.; Kucharski, F.; Almazroui, M. Potential predictability of boreal winter precipitation over central-southwest Asia in the North American multi-model ensemble. *Clim. Dyn.* **2020**, *54*, 473–490. [CrossRef]

27. Rana, S.; Renwick, J.; McGregor, J.; Singh, A. Seasonal prediction of winter precipitation anomalies over Central Southwest Asia: A canonical correlation analysis approach. *J. Clim.* **2018**, *31*, 727–741. [CrossRef]

28. Cybenko, G. Approximation by superpositions of a sigmoidal function. *Math. Control. Signals Syst.* **1989**, *2*, 303–314. [CrossRef]

29. Murphy, K.P. *Machine Learning: A Probabilistic Perspective*; MIT Press: Cambridge, MA, USA, 2012.

30. Knoben, W.J.; Freer, J.E.; Woods, R.A. Inherent benchmark or not? Comparing Nash–Sutcliffe and Kling–Gupta efficiency scores. *Hydrol. Earth Syst. Sci.* **2019**, *23*, 4323–4331. [CrossRef]

31. Nash, J.E.; Sutcliffe, J.V. River flow forecasting through conceptual models part I—A discussion of principles. *J. Hydrol.* **1970**, *10*, 282–290. [CrossRef]

Article

Deep Reinforcement Learning Ensemble for Detecting Anomaly in Telemetry Water Level Data

Thakolpat Khampuengson [1,2,*] and Wenjia Wang [1]

1 School of Computing Sciences, University of East Anglia, Norwich NR4 7TJ, UK
2 Hydro-Informatics Institutes, Bangkok 10900, Thailand
* Correspondence: t.khampuengson@uea.ac.uk

Abstract: Water levels in rivers are measured by various devices installed mostly in remote locations along the rivers, and the collected data are then transmitted via telemetry systems to a data centre for further analysis and utilisation, including producing early warnings for risk situations. So, the data quality is essential. However, the devices in the telemetry station may malfunction and cause errors in the data, which can result in false alarms or missed true alarms. Finding these errors requires experienced humans with specialised knowledge, which is very time-consuming and also inconsistent. Thus, there is a need to develop an automated approach. In this paper, we firstly investigated the applicability of Deep Reinforcement Learning (DRL). The testing results show that whilst they are more accurate than some other machine learning models, particularly in identifying unknown anomalies, they lacked consistency. Therefore, we proposed an ensemble approach that combines DRL models to improve consistency and also accuracy. Compared with other models, including Multilayer Perceptrons (MLP) and Long Short-Term Memory (LSTM), our ensemble models are not only more accurate in most cases, but more importantly, more reliable.

Keywords: anomaly detection; deep reinforcement learning; telemetry water level; time series; ensemble

Citation: Khampuengson, T.; Wang, W. Deep Reinforcement Learning Ensemble for Detecting Anomaly in Telemetry Water Level Data. *Water* **2022**, *14*, 2492. https://doi.org/10.3390/w14162492

Academic Editors: Fi-John Chang, Li-Chiu Chang and Jui-Fa Chen

Received: 1 July 2022
Accepted: 8 August 2022
Published: 13 August 2022

Publisher's Note: MDPI stays neutral with regard to jurisdictional claims in published maps and institutional affiliations.

1. Introduction

As climate change becomes more apparent, strong storms that bring heavy rainfalls occur with unusual patterns in many parts of the world. They can cause severe floods that result in devastating damages to infrastructure and loss of human life. In Thailand, flooding occurs more frequently and can cause enormous damages and huge economic losses of up to $46.5 billions a year [1]. On the other hand, drought happened in several parts of Thailand in 2015, notably in the Chao Phraya River Basin, the largest river basin in Thailand. This is consistent with a report from the UNDRR (2020) [2] that the ongoing drought crisis from 2015 to 2016 was the most severe drought in Thailand in 20 years. Therefore, it is essential to monitor water levels around the country because they form an important basis for making decisions on early warning.

In order to monitor the water levels in rivers, the Hydro Informatics Institute (HII) has been studying, building, and deploying water level telemetry stations around Thailand since 2013. Every ten minutes, each station transmits the measured data to the HII data centre through cellular or satellite networks. However, the water level data collected from telemetry station sensors might be incorrect due to some factors, such as human or animal activity, malfunctioning equipment, or interference of items surrounding the sensors. Any irregularity in the data might result in an inaccurate decision, such as false alarms or missed true alarms. Although water level data may be manually reviewed before being distributed for further analysis, the procedure necessitates the use of skilled specialists who examine the data from each station and make judgments about any probable abnormalities that may exist. This process is slow, very time-consuming and also unreliable. This motivates us to develop an automated approach that can identify irregularities in a more accurate, efficient, and reliable manner.

In our previous work [3], we studied seven statistics-based models for detecting the anomalies. We found that although an individual model can be used to identify anomalies, it produces too many false alarms for some situations, such as when the water level will dramatically rise before a flood occurs, which is a scenario notably different from the others, and hence led to that the majority of statistical models identify such points as anomalous. We also created two ensembles as the ensemble methods [4], if constructed properly, have been demonstrated to be able to improve accuracy and reliability over individual models. The first ensemble was built with a simple strategy as it just combines some selected models with majority voting as its decision-making function. However, the test results showed that the simple ensemble models did not work well enough, even though they were usually better than most of the basic individual models. We then developed a complex ensemble method. It basically builds an ensemble of some simple ensembles selected from the candidates with some criteria, and these simple ensembles' outputs are combined with a weighted function. The findings indicate that a complex ensemble can improve the accuracy and consistency in recognising both abnormal and normal data.

In recent decades, deep machine learning methods have been demonstrated to be more powerful than conventional machine learning techniques in tackling complex problems such as speech recognition, handwriting recognition, image recognition, and natural language processing. One of these methods is the Long Short-Term Memory (LSTM) [5], outperformed the Multilayer Perceptron (MLP), although trained with only normal data, for detecting anomaly patterns from ECG signals. Moreover, the C-LSTM methods, which integrated a convolutional neural network (CNN), well performed to detect anomaly signals that are difficult to classify in web traffic data as shown in [6]. Another deep neural network based on anomaly detection technique was recently proposed, called DeepAnt, which consists of a time series predictor that uses CNN to predict the values of the next time step and classify the predicted values as normal or abnormal by passing them to the anomaly detector [7].

Reinforcement Learning (RL) is an algorithm that imitates the human learning process. It is based on the self-learning process in which an agent learns by interacting with the environment without any assumptions or rules. With the advantage of being able to learn on their own, it can identify unknown anomalies [8], which gives it an edge over other models. RL has been applied to a variety of applications such as games [9,10], robotics [11,12], natural language processing [13,14], computer vision [15], etc. It has also been used in some studies to detect anomalies in data, such as an experiment [16] that shows the use of the deep Q-function network (DQN) algorithm to detect anomalies in time series. Network intrusion detection systems (NIDS) are developed by [17], based on deep reinforcement learning. They utilised it to identify anomalous traffic on the campus network with a combination of flexible switching, learning, and detection modes. When the detection model performs below the threshold, the model is retrained. In the comparison against three traditional machine learning approaches, their model outperformed on two benchmark datasets, NSL-KDD and UNSW-NB15. A binary imbalanced classification model based on deep reinforcement learning (DRL) was introduced in [18]. They developed the reward function by setting the rewards for the minority class to be greater than the rewards for the majority class, which made DRL paying more attention to the minority class. They compared it to seven imbalanced learning methods and found that it outperformed other models in text datasets and extremely imbalanced data sets.

Although deep learning and RL methods have achieved excellent results in time series, one common issue is that their performance varies and it is hard to predict when they do better and when they perform relatively poor. In order to improve their consistency and accuracy, ensemble methods can be used. One example of such a method is the technique called particle swarm optimization (PSO), which was developed [19] to predict the changing trend of the Mexican Stock Exchange by combining several neural networks.

An ensemble of MLP, Backpropagation network (BPN), and LSTM, as shown in [20], was used to make models for detecting anomalous traffic in a network. The ensemble approach that utilises DRL schemes to maximise investment in stock trading was developed in [21]. They trained a DRL agent and obtained an ensemble trading strategy using three different actor-critics-based algorithms that outperformed the individual algorithm and two baselines in terms of the risk-adjusted return. Another ensemble RL that employed three types of deep neural networks in Q-learning and used ensemble techniques to make the final decision to increase prediction accuracy for wind speed short-term forecasting was suggested [22].

We discovered that none of the DRL methods have been applied to identify anomalies in telemetry water level data. We wonder whether DRL is applicable for identifying abnormalities in telemetry water level data. Even if the final DRL models perform well on training data, there is no guarantee that they will also perform well on testing data. Previous research has shown that combining many models that were trained in different ways may be more accurate than any of the individual models. So, in this paper, we aim to answer the following two research questions.

(Q_1) Is DRL applicable and effective for identifying abnormalities in water level data?

(Q_2) Can we build some ensembles of DRL to improve accuracy and consistency?

To answer them, in this paper, we conducted intensive investigation by evaluated the accuracy of DRL models with real-world data. Then we proposed a strategy to build some ensembles by selecting some suitable DRL models. The testing results show that DRL is applicable for identifying abnormalities in telemetry water level data with the advantage of identifying an unknown anomaly. However, the process of training takes a long time. The constructed ensembles not only improve accuracy and consistency, but also reduce the rate of false alarms.

Thus, the main contributions of this paper are:

(C_1) DRL models have been demonstrated to be able to detect anomalies in telemetry water level data.

(C_2) The ensembles we have constructed in this research with some suitable DRL models and use a weighted decision-making strategy can improve both accuracy and consistency. The proposed approach has a potential to be further developed and implemented for real-world application.

The rest of the paper is organised as follows: Section 2 overviews related work for anomaly detection. Section 3 describes the methodology. Section 4 presents the experiment design-from data preparation, parameters configurations, to evaluation metrics. Results and discussions are provided in Sections 5 and 6; the conclusion and suggestions for further work are summarised in Section 7.

2. Related Work

There are many methods for detecting anomalies in time series data. One basic approach is to use statistics-based methods, as reviewed in [23,24]. For example, simple and exponential smoothing techniques were used to identify anomalies in a continuous data stream of temperature in an industrial steam turbine [25]. But in general whilst they provided a baseline, they have a disadvantage in handling trends and periotics, e.g., the water level will dramatically rise before the flood, which differs considerably from the other data points and may lead to an increased false alarm rate. In addition, they can be affected by the types of anomaly and some work well for a certain type of problem. For example, for missing and outlier values, when the data is normally distributed, the K-means clustering method [26] is usually used, as it is simple and relatively effective. However, there is unfortunately no general guideline for choosing a method for a given problem.

Change Point Detection (CPD) is an important method for time series analysis. It indicates an unexpected and significant change in the analysed time series stream data and has been studied in many fields, as surveyed in [27,28]. However, the CPD has no ability to detect anomalies since not all detected change points are abnormalities. Many studies are being conducted to solve this problem by integrating CPD with other models to increase anomaly detection effectiveness. For example, researchers from [29] presented new techniques, called rule-based decision systems, that combine the results of anomaly detection algorithms with CPD algorithms to produce a confidence score for determining whether or not a data item is indeed anomalous. They tested their suggested method using multivariate water consumption data collected from smart metres, and the findings demonstrated that anomaly detection can be improved. Moreover, it has been proposed to detect anomalies in file transfer by using the CPD to detect the current bandwidth status from the server, then using this to calculate the expected file transfer time. The server administrator has been notified when observed file transfers take longer than expected, which may mean it may have something wrong [30]. The author of [31] investigated the CUSUM algorithm for change point detection to detect SYN flood attacks. The results demonstrated that the proposed algorithm provided robust performance with both high and low intensity attacks. Although change point detection performed well in many domains, the majority of them focused on changes in the behaviour of time series data (sequence anomaly) rather than point anomaly, which is my primary research emphasis. Furthermore, water level data at certain stations is strongly periodic with tidal effects, resulting in numerous data points changing from high tides to low tides each day, which is typical behaviour.

In recent decades, machine learning methods, including deep neural networks (DNNs), have been satisfactorily implemented in various hydrological issues such as outlier detection [32,33], water level prediction [34,35], data imputation [36], flood forecasting [37], streamflow estimation [38], etc. For example, in [39], the authors proposed the R-ANFIS (GL) method for modelling multistep-ahead flood forecasts of the Three Gorges Reservoir (TGR) in China, which was developed by combining the recurrent adaptive-network-based fuzzy inference system (R-ANFIS) with the genetic algorithm and the least square estimator (GL). The authors of [40] presented a flood prediction by comparing the expected typhoon tracking and the historical trajectory of typhoons in Taiwan in order to predict hydrographs from rainfall projections impacted by typhoons. The PCA-SOM-NARX approach was developed by [41] to forecast urban floods, combining the advantages of three models. Principal component analysis was used to derive the geographical distributions of urban floods (PCA). To construct a topological feature map, high-dimensional inundation recordings were grouped using a self-organizing map (SOM). To build 10-minute-ahead, multistep flood prediction models, nonlinear autoregressive with exogenous inputs (NARX) was utilised. The results showed that not only did the PCA-SOM-NARX approach produce more stable and accurate multistep-ahead flood inundation depth forecasts, but it was also more indicative of the geographical distribution of inundation caused by heavy rain events. Even though we can use forecasting methods to find anomalies by using prediction error as a threshold to classify data points as normal or not, it may take time to find the suitable threshold for each station.

An autoencoder is an unsupervised learning neural network. It is comprised of two parts: an *encoder* and a *decoder*. The encoder uses the concepts of dimension reduction algorithms to convert the original data into the different representations with the underlying structure of the data remaining and ignoring the noise. Meanwhile, the decoder reconstructs the data from the output of the encoder with as close of a resemblance as possible to the original data. An autoencoder is effectively used to solve many applied problems, from face recognition [42,43] and anomaly detection [44–47] to noise reduction [48–50]. In the time series domain, the authors of [51] proposed two autoencoder ensemble frameworks for unsupervised outlier identification in time series data based on sparsely connected recurrent neural networks, which addressed the issues from [52] given the poor results when using an autoencoder with time series data. In one of the frameworks called the Independent Framework, multiple autoencoders are trained independently of one another, whereas in the other framework, the Shared Framework, multiple autoencoders are trained jointly in a manner that is multitask learning. They experimented by using univariate and multivariate real-world datasets. Experimental results revealed that the suggested autoencoder ensembles with a shared framework outperform baselines and state-of-the-art approaches. However, a disadvantage of this method is its high memory consumption when training many autoencoders together. In the hydrological domain, the authors of [53] presented the SAE-RNN model, which combined the stacked autoencoder (SAE) with a recurrent neural network (RNN) for multistep-ahead flood inundation forecasting. They started with SAE to encode the high dimensionality of input datasets (flood inundation depths), then utilised an LSTM-based RNN model to predict multistep-ahead flood characteristics based on regional rainfall patterns, and then decoded the output by SAE into regional flood inundation depths. They conducted experiments on datasets of flood inundation depths gathered in Yilan County, Taiwan, and the findings demonstrated that SAE-RNN can reliably estimate regional inundation depths in practical applications.

Time series based on ensemble methods have recently attracted attention. In a study by [54], they introduced the method EN-RTON2, which is an ensemble model with real-time updating using online learning and a submodel for real-time water level forecasts. However, they experimented with fewer datasets, a smaller number of records, and lower data frequency than our datasets. Furthermore, the authors offered no indication of the time necessary for training models and forecasting, which may be inadequate in our case given the number of stations and frequency of data transmission. The ensemble models were proposed by [55], which applied the sliding window based ensemble method to find the anomaly pattern in sensor data for preventing machine failure. They used a combination of classical clustering algorithms and the principle of biclustering to construct clusters representing different types of structure. Then they used these structures in a one-class classifier to detect outliers. The accuracy of these methods was tested on a time series of real-world datasets from the production of industry. The results have verified the accuracy and the validity of the proposed methods.

Despite the fact that numerous studies have used different anomaly detection techniques to tackle problems in many domains, only a few have focused on finding anomalies in water level data. Furthermore, the various employed sensors, installation area, frequency of data transmission, and measurement purposes lead to a variety of types of anomalies. As a result, techniques that perform well with one set of data may not work well with another.

3. Materials and Methods

This section describes firstly how deep reinforcement learning is constructed for detecting anomalies in water level telemetry data; and then how an ensemble can be built effectively by selecting suitable individual models to improve the accuracy of anomaly detection. The frameworks of these investigations were implemented with Python and their code can be accessed via GitHub (https://github.com/khaitao/RL-Anomaly-Detection-Water-Level, The last check on 5 August 2022).

3.1. Reinforcement Learning (RL)

Reinforcement learning (RL) is a branch of machine learning and it is one of the most active areas of research in artificial intelligence (AI), which is growing rapidly with a wide variety of algorithms. It is goal-oriented learning. The learner, or agent, learns from the result, or rewards, of its actions without being taught what actions to take. The way in which the agent decides which action to perform depends on the policy, which can be in the form of a lookup table or a complex search process. So, a policy function defines the agent's behaviour in an environment.

Most techniques that are used to find the optimal policy for resolving the RL problem are based on the Markov decision process (MDP), whereby the probability of next state s' depends only on the current state s and action a. It is represented by five important variables [56]:

- A finite set of states (S), which may be discrete or continuous.
- A finite set of actions (A). The agent takes an action a from the action set A, $a \in A$.
- A transition probability ($T(s, a, s')$), which is the probability to get from state s to another state s' with action a.
- A reward probability ($R(s, a, s') \in \mathbb{R}$), which is the reward after going from state s to another state s' with action a.
- A discount factor (γ), which focuses on controls the important immediate and future rewards and lies within 0 to 1, $\gamma \in [0, 1]$.

The goal of learning is to maximise the expected cumulative reward in each episode. The agent should try to maximise the reward from any state s. The total reward R at state s as the sum of current rewards and the total discounted reward at the next state s', which can be represented as follow:

$$R(s) = R(s, a, s') + \gamma R(s')$$

The algorithm that has been widely used in RL is Q-learning. It tries to maximize the values from Q-function, as shown in Equation (1), which can be approximated using the Bellman equation, which represents how good it is for an agent to perform a particular action in a state s.

$$NewQ(s, a) = Q(s, a) + \alpha(r + \gamma \max Q'(s', a') - Q(s, a)) \tag{1}$$

where α is the learning rate, and $\max Q'(s', a')$ is the highest Q value between possible actions from the new state s'.

3.1.1. Deep Q-Learning Network

Q-learning has a limitation: it does not perform well with many states and actions. Furthermore, going through all the actions in each state would be time-consuming. Therefore, the deep Q-learning network [57] (DQN) has been developed to solve those issues by using a neural network (NN). The Q-value is approximated by an NN with weights w, instead of finding the optimal Q-value through all possible state-action pairs, and errors are minimized through gradient descent. The overall process of DRL is depicted in Figure 1.

An agent usually does not know what action is best at the beginning of training. It may select the greatest action that is the best based on history (exploitation) or may explore new possibilities that may be better or worse (exploration). However, when should an agent "exploit" rather than "explore"? This remains a challenge since if the chosen action results in a faulty selection, an agent may get stuck in incorrect learning for a time. The epsilon-greedy algorithm is a simple way to balance exploration and exploitation. It does this by randomly choosing between exploration and exploitation and using the hyperparameter ϵ to switch between random action and Q-values, as shown in Equation (2). The normal procedure is to begin with $\epsilon = 1.0$ and gradually lower it to a small value, such as 0.01.

$$a = \begin{cases} \text{select a random action } a & \text{with probability } \epsilon \\ argmax_a Q(s, a) & \text{otherwise} \end{cases} \tag{2}$$

Moreover, we make a transition from one state s to the next state s' by performing some action a and receive a reward r as $T(s, a, s')$. So, neural networks may overfit with correlated experience from those transitions. So, we saved the transition information in a buffer called *replay memory* and trained the DQN with a random transition in replay memory instead of training with last transitions. It will reduce the correlated experience of learning each time, and then it will reduce the overfitting of the model.

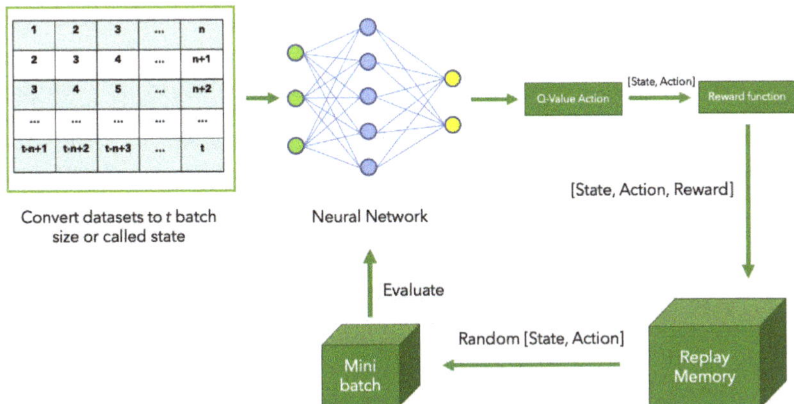

Figure 1. Overall process of DRL (Deep Reinforcement Learning).

3.1.2. Deep Reinforcement Learning Model (DRL)

The action of the DRL agent is to determine whether or not a data point is an abnormality. We assigned a value of 1 to the anomaly class and a value of 0 to the normal class. DQN was chosen as our reinforcement learning strategy. When state s is received, an MLP is used as the RL agent's brain to generate Q-value, which is then followed by the Q function. The epsilon decay approach is used for exploration and exploitation. In order to explore the entire environment space, we use the greedy factor ϵ to determine whether our DRL agent should follow the Q function or randomly select an action.

For each iteration, DQN receives the set of states S and predicts the label for training the DRL model. The transition is stored in replay memory. In each epoch, a mini batch of replay memory is sampled and used to train the model for loss minimization. Moreover, whether the model will learn well or not depends on the rewards function. The good reward function has an effect on the model's performance. If we offer a high reward for correctly identifying normal data in datasets, DRL may identify all data as normal in order to get the highest score. If, on the other hand, we give a high reward for finding outliers, DRL might label all data as outliers to get the best score.

Since our datasets are imbalanced, we will give the reward of the minority class higher than the majority class and give the penalty when our model misclassifies [18]. This will impact on the results in Q-values, then the model will select the best action to maximize the rewards. The reward function is defined below

$$rewards = \begin{cases} A & \text{predicted anomaly correct} \\ B & \text{predicted wrong} \\ C & \text{predicted normal correct} \end{cases} \tag{3}$$

A general issue in training neural networks is to determine how long they should be trained. Too few epochs may result in the model learning insufficiently, whereas too many epochs may result in the model overfitting. So, the performance of the model must be monitored during training by evaluating it on a validation data set at the end of each epoch and updating the model if the performance of the model on a validation is better than at the previous epoch. In our experiments, we selected 5 criteria as the conditions for generating the models: four performance metrics and the maximum number of epochs. The four measures are F1-score, the reward of each epoch, accuracy, and validation loss values. In the end, we will have five models: the finished training model (DRL), the models with the highest F1-score (DRL_{F1}), the models with the highest rewards (DRL_{Rwd}), the model with the highest accuracy (DRL_{Acc}), and the model with the lowest validation loss values (DRL_{Valid}).

3.1.3. Ensemble Methods

In general, the capacity of an individual model is limited and may have only learned some parts of the problem, and hence may make mistakes in the areas where it has not learned sufficiently. Therefore, it can be useful to combine some individual models to form an ensemble to allow them to work collectively to compensate for each other's weaknesses. Many studies [4,58–60] have shown that if an ensemble is built with diverse models and appropriate decision-making functions, it can improve the accuracy of classification and also reliability. In our research, we created multiple ensembles by selecting suitable DRL models that had been generated from the previous experiments. We investigated two combining methods to aggregate the outputs from the member models of an ensemble: simple majority voting and weighted voting algorithms.

- *Majority Voting*: The predictions of each model in an ensemble have to be aggregated, and the final prediction is the class that gets the most votes. Each of our ensembles will be built with an odd number of classifiers in order to avoid a tie situation in voting.
- *Weighted Voting*: As the performance of individual models is usually different, treating them all the same way in decision marking appears unlogical, so we devised a weighted voting mechanism to take this difference into consideration when making a final decision in an ensemble. With the weighted voting method, the contribution from a model is weighed by its performance. For a model m_i, after it has been trained with the training data, its weight score w_i is derived by using its F_1 score that is calculated on the given validation dataset; we then have a set of F1-scores of each model, $F_1 m = \{F_1 m_1, F_1 m_2, ..., F_1 m_M\}$. Then, these F1-scores are ranked to find the maximum and minimum scores. Finally, we calculate the normalised weighting score w_i for module m_i using the equation below:

$$w_i = \frac{F_1 m_i - min(F_1 m)}{max(F_1 m) - min(F_1 m)}, \ \forall \, i = 1, ..., M \tag{4}$$

The output of an ensemble, $\Phi(x)$, is calculated by multiplying the weight with the output of an individual module and taking the argument of maxima as follows:

$$\Phi(x) = argmax \sum_{i=1}^{M} w_i m_i(x) \tag{5}$$

where M is the number of models in an ensemble, and $m_i(x)$ is the predicted class of model i.

3.2. Data Labelling

Water level data from telemetry stations were unlabelled for anomalies. It is then necessary to assign ground truth labels to all anomalies and normal data points in each time series of water level data in order to train the models with supervised algorithms. This was manually done by a group of the domain experts at the HII in a manner similar to the ensemble approach. Each specialist looked at the data and identified all the anomalies based on their experience. Then their judgements were aggregated by taking a consensus to decide if a data point is an anomaly or not.

3.3. Datasets

Since the DRL algorithm takes a lot of time for training on the computing facilities that we had, we were limited to consider some relatively small datasets. After data preprocessing, the 8 stations from the HII telemetry water level station were chosen for use in this experiment, including CPY011, CPY012, CPY013, CPY014, CPY015, CPY016, CPY017, and YOM009. We chose the datasets from May and June for CPY011, CPY012, CPY013, CPY015, CPY016, and CPY017 in 2016 and similar months in 2015 for CPY014 and YOM009 because they have a low percentage of missing data. Figure 2 shows the water levels of these eight stations. It is visually clear that station YOM009 has very different behaviour from the others because it is located in a different region.

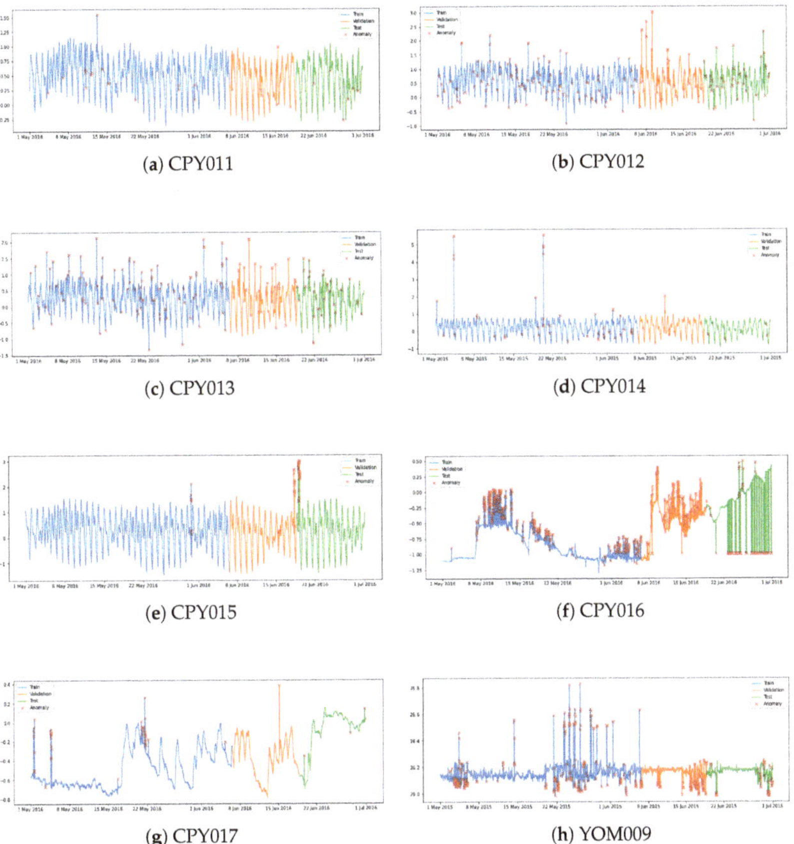

Figure 2. Water level data from eight stations: CPY011, CPY012, CPY013, CPY014, CPY015, CPY016, CPY017, and YOM009 (**a–h**). The different colours show the partitions of the data for training (blue), validation (orange) and testing (green). The anomalies are indicated by red crosses, x.

All the data are normalised and divided into 3 subsets, with the first 60% of a time series for training, the next 20% for validating, and the last 20% for testing, respectively. Table 1 shows the demographics of one partition of the data from each station. As can be seen, in general, the rates of anomalies are quite low for most stations, but the variances are considerably large. For example, they varied from 0.14% to 7.22% in the training data.

Table 1. Demographic summary of the water level data of 8 stations used in this research.

Code	Training		Validating		Testing		Total	
	Rec.	Anomaly	Rec.	Anomaly(%)	Rec.	Anomaly(%)	Rec.	Anomaly(%)
CPY011	5142	16(0.31%)	1713	7(0.41%)	1714	7(0.41%)	8569	30(0.37%)
CPY012	5142	101(1.96%)	1713	41(2.39%)	1714	34(1.98%)	8569	176(2.05%)
CPY013	5142	97(1.89%)	1713	33(1.93%)	1714	28(1.63%)	8569	158(1.84%)
CPY014	5142	49(0.95%)	1713	7(0.41%)	1714	4(0.23%)	8569	60(0.70%)
CPY015	5142	7(0.14%)	1713	15(0.88%)	1714	34(1.98%)	8569	56(0.65%)
CPY016	5142	367(7.14%)	1713	220(12.84%)	1714	107(6.24%)	8569	694(8.10%)
CPY017	5142	42(0.82%)	1713	2(0.12%)	1714	3(0.18%)	8569	47(0.55%)
YOM009	5142	417(7.22%)	1713	173(10.97%)	1714	81(3.79%)	8569	624(7.28%)
Avg.	5142	137(2.66%)	1713	62(3.63%)	1714	37(2.17%)	8569	231(2.69%)

3.4. Evaluation Metrics and Comparison Methods

As our task is basically a classification problem. We therefore chose some commonly used measures: *Recall, Precision,* and *F1,* to evaluate the accuracy of models. They are defined by the following equations, based on the confusion matrix shown in Table 2.

Table 2. Confusion matrix of classification results.

Actual/Predicted	**Anomaly**	**Normal**
Anomaly	TP	FN
Normal	FP	TN

$$Recall = \frac{TP}{TP + FN}$$

$$Precision = \frac{TP}{TP + FP}$$

$$F1 = 2\frac{Precision * Recall}{Precision + Recall}$$

where TP, FP, FN, and TN denote the number of true positive—correct predictions for anomaly data, false positive—the number of incorrect predictions for anomaly data, false negative—the number of incorrect predictions for normal data, and true negative—the number of correct predictions for normal data, respectively.

To make statistical comparisons, we implemented a statistically rigorous test for multiple classifiers across many datasets. This approach was initially described in [61] and is intended to examine the statistical significance of classifiers. This technique takes the strategy of testing the null hypothesis against the alternative hypothesis. The null hypothesis states that no difference exists between the average rankings of k algorithms on N datasets. The alternative hypothesis is that at least one algorithm's average rank differs.

In the first place, the k methods are ranked according to their performance over the N datasets; then, the average ranking of each algorithm is calculated. To test the null hypothesis, the Friedman test is calculated using Equation (6).

$$\chi_F^2 = \frac{12N}{k(k+1)} \left[\sum_j R_j^2 - \frac{k(k+1)^2}{4} \right] \tag{6}$$

where R_j is the rank of the jth of k algorithms on N datasets and the statistic is estimate using a chi-squared distribution with $k - 1$ degrees of freedom.

If the null hypothesis is rejected at the selected significance level α, the post-hoc Nemenyi test is used to compare all classifiers to each other. The Nemenyi test is similar to the Tukey test for ANOVA and uses a critical difference (CD), which is presented in Equation (7)

$$CD = q_\alpha \sqrt{\frac{k(k=1)}{6N}} \tag{7}$$

where q_α is calculated by the difference in the range of standard deviations between the smallest valued sample and the largest valued sample. The results of these tests are often visualised using a critical difference (CD) diagram. Classifiers are shown on a number line based on their average rank across all datasets, and bold CD lines are used to connect classifiers that are not significantly different.

In comparison, the performance of our approach, MLP, and LSTM have been used with the same number of hidden layers and the number of neurons in each hidden layer.

4. Experiment Design and Setting

4.1. Four Sets of Experiments

We designed four sets of experiments to test DRL models and ensemble models. (1) to train various DRL models and test them with the different data sampled from the same water level monitoring stations; (2) to train various DRL models with the data from a station and then test them with the data from other stations; (3) to build several ensembles by selecting different numbers of the DRL models and test them with the testing data from the same stations; and (4) to test the ensembles with the data from different stations. The purpose of doing these cross-station testing is to check and evaluate the generalisation ability of the DLR models and the ensembles.

4.2. Parameter Setting

For the DRL model, a multilayer perceptron network was used in the Q-network with the following parameters: the number of input nodes in the input layer was 36, one hidden layer with 18 nodes, and 2 nodes in the output layer. Moreover, epsilon-greedy policy (ϵ) was used for exploration from 0.1 to 0.0001. The size of replay memory is 50,000, discount factor of intermediate rewards γ was 0.99. The Adam algorithm was used to optimise the parameters of Q-Network and the learning rate was 0.001. The batch size was 256, training with 100, 500, 1000, 5000, and 10,000 episodes. The episode was over when the number of incorrectly identified anomalies was greater than the number of certain anomalies in the training set or had been trained on all the samples in the training set. We set the reward function parameters for A, B, and C to be 0.9, -0.1, and 0.1, respectively. Furthermore, the window size of 6 was chosen to save time during the training process.

For comparison, MLP and LSTM were used with the identical structures as we used in DRL. They were trained using 100 epochs with early stopping to avoid overfitting. For each setting, the experiments were repeated 10 times with variations, and then the means and standard deviations of the results are reported in the next section.

4.3. Computing Facilities

All the experiments were coded with Python Programming Language (V3.6) (Python Software Foundation, https://www.python.org/, accessed on 30 June 2022) and Tensor-Flow 2.8, and run on a personal computer with an Intel Core i5-7500 CPU @ 3.4 GHz, 32 GB RAM, 64-Bit Operating System.

5. Results

5.1. Accuracies of DRL Models

For each station, various DRL models were generated over a range of epochs from 100 to 10,000, with the intention of investigating how well our proposed DRL method learns at the different points of training. The results are shown in Table 3.

Using the CPY011 dataset, we observed that DRL and DRL_{Rwd} with 1000 training iterations not only earned the highest F1-score of 0.8333, 0.7143 recall, and 1.0000 precision but also provided the highest average F1-score of 0.7433. However, after 1000 epochs of training, the performance of all models, with the exception of DRL_{Valid} decreased and then rose when 10,000 epochs were used.

Table 3. The performance of DRL when increasing the learning epochs (the best F1-score of each row shown in bold).

Station	Epochs	DRL			DRL_{F1}			DRL_{Rwd}			DRL_{Acc}			DRL_{Valid}		
		Recall	Prec	F1	Recall	Prec	F1	Recall	Prec	F1	Recall	Prec	F1	Recall	Prec	F1
CPY011	100	0.8571	0.5455	0.6667	0.8571	0.7500	**0.8000**	0.8571	0.5455	0.6667	0.8571	0.7500	**0.8000**	0.7143	0.6250	0.6667
	500	0.8571	0.7500	**0.8000**	0.8572	0.5455	0.6667	0.8571	0.7500	**0.8000**	0.8571	0.5455	0.6667	0.8571	0.6667	0.7500
	1000	0.7143	1.0000	**0.8333**	0.8571	0.6667	0.7500	0.7143	1.0000	**0.8333**	0.8571	0.6667	0.7500	0.4286	0.2727	0.3333
	5000	0.7143	0.6250	0.6667	0.8571	0.5000	0.6316	0.7143	0.6250	0.6667	0.8571	0.5000	0.6316	0.7143	0.7143	**0.7143**
	10,000	0.8571	0.6667	**0.7500**	1.0000	0.5833	0.7368	0.8571	0.6667	**0.7500**	1.0000	0.5833	0.7368	0.8571	0.4000	0.5455
	Avg	0.8000	0.7174	**0.7433**	0.8857	0.6091	0.7170	0.8000	0.7174	**0.7433**	0.8857	0.6091	0.7170	0.7143	0.5357	0.6020
	Std	0.0782	0.1743	0.0760	0.0639	0.0997	0.0674	0.0782	0.1743	0.0760	0.0639	0.0997	0.0674	0.1749	0.1901	0.1689
CPY012	100	0.7059	0.4000	0.5106	0.7647	0.7027	**0.7324**	0.6764	0.3898	0.4946	0.7059	0.6857	0.6957	0.7647	0.7027	**0.7324**
	500	0.7647	0.6341	0.6933	0.7941	0.7297	**0.7606**	0.7353	0.6250	0.6757	0.7941	0.7297	**0.7606**	0.7941	0.7297	**0.7606**
	1000	0.6765	0.6571	0.6667	0.7647	0.6667	**0.7123**	0.6176	0.6000	0.6087	0.7647	0.6667	**0.7123**	0.7647	0.4062	0.5306
	5000	0.7059	0.7059	0.7059	0.6176	0.6176	0.6176	0.7059	0.7273	**0.7164**	0.6765	0.6970	0.6866	0.7059	0.7273	**0.7164**
	10,000	0.6471	0.7586	0.6984	0.7059	0.8000	0.7500	0.7059	0.7742	0.7385	0.7059	0.8276	0.7619	0.7941	0.7714	**0.7826**
	Avg	0.7000	0.6311	0.6550	0.7294	0.7033	0.7146	0.6882	0.6233	0.6468	0.7294	0.7213	**0.7234**	0.7647	0.6675	0.7045
	Std	0.0436	0.1378	0.0821	0.0702	0.0684	0.0572	0.0446	0.1489	0.0984	0.0483	0.0637	0.0357	0.0360	0.1481	0.1005
CPY013	100	0.8710	0.3506	0.5000	0.8710	0.6136	**0.7200**	0.9032	0.3836	0.5385	0.8710	0.6136	**0.7200**	0.6774	0.1214	0.2059
	500	0.6774	0.3684	0.4773	0.8065	0.5000	0.6173	0.7419	0.3966	0.5169	0.8065	0.5000	0.6173	0.8387	0.4906	**0.6190**
	1000	0.8065	0.5952	0.6849	0.8065	0.5682	0.6667	0.7097	0.6111	0.6567	0.8065	0.5682	0.6667	0.9677	0.5556	**0.7059**
	5000	0.7742	0.5714	0.6575	0.6774	0.6774	**0.6774**	0.8065	0.5682	0.6667	0.6774	0.6364	0.6562	0.6774	0.5250	0.5915
	10,000	0.7097	0.6667	0.6875	0.8387	0.7647	**0.8000**	0.7742	0.6857	0.7273	0.8387	0.7647	**0.8000**	0.7419	0.6571	0.6970
	Avg	0.7678	0.5105	0.6014	0.8000	0.6248	**0.6963**	0.7871	0.5290	0.6212	0.8000	0.6166	0.6920	0.7806	0.4699	0.5639
	Std	0.0770	0.1423	0.1039	0.0736	0.1015	0.0685	0.0743	0.1337	0.0899	0.0736	0.0978	0.0706	0.1237	0.2045	0.2061
CPY014	100	0.7500	0.7500	**0.7500**	0.7500	0.7500	**0.7500**	0.7500	0.7500	**0.7500**	0.7500	0.7500	**0.7500**	0.7500	0.7500	**0.7500**
	500	0.7500	0.3750	0.5000	0.7500	0.7500	0.7500	0.7500	1.0000	**0.8571**	0.7500	0.7500	0.7500	0.7500	1.0000	**0.8571**
	1000	0.7500	0.3750	0.5000	0.7500	0.5000	**0.6000**	0.7500	0.3750	0.5000	0.7500	0.5000	**0.6000**	0.7500	0.5000	**0.6000**
	5000	0.7500	0.3333	0.4615	0.7500	0.5000	**0.6000**	0.7500	0.3333	0.4615	0.7500	0.5000	**0.6000**	0.7500	0.4286	0.5455
	10,000	0.2500	0.1667	0.2000	0.7500	0.6000	**0.6667**	0.2500	0.1667	0.2000	0.7500	0.6000	**0.6667**	0.7500	0.4286	0.5455
	Avg	0.6500	0.4000	0.4823	0.7500	0.6200	**0.6733**	0.6500	0.5250	0.5537	0.7500	0.6200	**0.6733**	0.7500	0.6214	0.6596
	Std	0.2236	0.2137	0.1952	0.0000	0.1255	0.0751	0.2236	0.3405	0.2584	0.0000	0.1255	0.0751	0.0000	0.2495	0.1385
CPY015	100	0.2353	0.4706	0.3137	0.3235	0.5238	0.4000	0.2353	0.4706	0.3137	0.3235	0.5238	**0.4000**	0.1765	0.3529	0.2353
	500	0.2059	0.4667	0.2857	0.3235	0.5000	**0.3929**	0.2059	0.4667	0.2857	0.3235	0.5000	**0.3929**	0.1471	0.3846	0.2128
	1000	0.3824	0.4483	0.4127	0.3824	0.5000	**0.4333**	0.3824	0.4483	0.4127	0.3824	0.5000	**0.4333**	0.4118	0.4516	0.4308
	5000	0.2353	0.4706	0.3137	0.3235	0.5238	**0.4000**	0.2353	0.4706	0.3137	0.3235	0.5238	**0.4000**	0.2353	0.4706	0.3137
	10,000	0.3824	0.4483	0.4127	0.3824	0.5200	**0.4407**	0.3824	0.4483	0.4127	0.3824	0.5200	**0.4407**	0.3824	0.4483	0.4127
	Avg	0.2883	0.4609	0.3477	0.3471	0.5135	**0.4134**	0.2883	0.4609	0.3477	0.3471	0.5135	**0.4134**	0.2706	0.4216	0.3211
	Std	0.0868	0.0116	0.0604	0.0323	0.0124	0.0219	0.0868	0.0116	0.0604	0.0323	0.0124	0.0219	0.1202	0.0503	0.0995
CPY016	100	0.6636	0.3100	0.4226	0.5981	0.5203	**0.5565**	0.6916	0.2960	0.4146	0.5981	0.5203	**0.5565**	0.6168	0.4889	0.5455
	500	0.6636	0.2763	0.3901	0.6355	0.4048	0.4945	0.6449	0.2727	0.3833	0.5981	0.5161	**0.5541**	0.5047	0.5094	0.5070
	1000	0.6355	0.2547	0.3636	0.5981	0.5333	0.5639	0.6449	0.2644	0.3750	0.6168	0.5641	**0.5893**	0.6168	0.4342	0.5097
	5000	0.5888	0.2727	0.3728	0.5888	0.6238	**0.6058**	0.3084	0.2089	0.2491	0.5421	0.6105	0.5743	0.5234	0.2902	0.3733
	10,000	0.5794	0.2366	0.3360	0.6168	0.4177	0.4981	0.6355	0.2208	0.3277	0.6262	0.5447	**0.5826**	0.6168	0.4177	0.4981
	Avg	0.6262	0.2701	0.3770	0.6075	0.5000	0.5438	0.5851	0.2526	0.3499	0.5963	0.5511	**0.5714**	0.5757	0.4281	0.4867
	Std	0.0402	0.0274	0.0321	0.0187	0.0904	0.0472	0.1562	0.0366	0.0644	0.0326	0.0384	0.0156	0.0567	0.0858	0.0659
CPY017	100	1.0000	0.7500	**0.8571**	1.0000	0.5000	0.6667	1.0000	0.7500	**0.8571**	1.0000	0.7500	**0.8571**	1.0000	0.7500	**0.8571**
	500	1.0000	0.7500	**0.8571**	1.0000	0.5000	0.6667	1.0000	0.7500	**0.8571**	1.0000	0.3750	0.5455	0.6667	0.5000	0.5714
	1000	1.0000	0.2143	0.3529	1.0000	0.7500	**0.8571**	1.0000	0.2000	0.3333	1.0000	0.7500	**0.8571**	0.0000	0.0000	-
	5000	1.0000	0.5000	**0.6667**	1.0000	0.4286	0.6000	1.0000	0.5000	**0.6667**	1.0000	0.4286	0.6000	1.0000	0.3333	0.5000
	10,000	0.6667	0.6667	0.6667	0.6667	0.2857	0.4000	0.6667	0.6667	0.6667	1.0000	0.6000	**0.7500**	0.6667	0.6667	0.6667
	Avg	0.9333	0.5762	0.6801	0.9333	0.4929	0.6381	0.9333	0.5733	0.6762	1.0000	0.5807	**0.7219**	0.6667	0.4500	0.6488
	Std	0.1491	0.2266	0.2062	0.1491	0.1683	0.1641	0.1491	0.2323	0.2140	0.0000	0.1755	0.1443	0.4082	0.2982	0.1547
YOM009	100	0.6308	0.3178	0.4227	0.5692	0.3033	0.3957	0.6462	0.3182	**0.4264**	0.5692	0.3033	0.3957	0.5692	0.3394	0.4253
	500	0.5846	0.2734	0.3725	0.6769	0.3121	0.4272	0.6923	0.3020	0.4206	0.4769	0.4769	**0.4769**	0.4769	0.4769	**0.4769**
	1000	0.5385	0.2966	0.3825	0.6769	0.2973	0.4131	0.5538	0.3103	0.3978	0.4615	0.3947	**0.4255**	0.5846	0.3016	0.3979
	5000	0.5538	0.1818	0.2738	0.6615	0.3644	**0.4699**	0.6154	0.2581	0.3636	0.4923	0.4103	0.4476	0.5077	0.4177	0.4583
	10,000	0.4769	0.2627	0.3388	0.5692	0.2741	0.3700	0.4769	0.2605	0.3370	0.5385	0.2917	0.3784	0.4308	0.4308	**0.4308**
	Avg	0.5569	0.2665	0.3581	0.6307	0.3102	0.4152	0.5969	0.2898	0.3891	0.5077	0.3754	0.4248	0.5138	0.3933	**0.4378**
	Std	0.0570	0.0519	0.0558	0.0565	0.0334	0.0373	0.0839	0.0285	0.0382	0.0449	0.0776	0.0395	s0.0640	0.0712	0.0306

The top models to identify anomalies on the CPY012 dataset are DRL_{Valid}, with a maximum F1-score of 0.7826 after 10,000 training epochs. However, DRL_{Acc} obtained the greatest average F1-score with 0.7234. Meanwhile, 10,000 training epochs with DRL_{F1} and DRL_{Acc} delivered the highest F1-score for identifying anomalies in CPY013 data, at 0.8000 F1-score. Furthermore, DRL_{F1} provided the highest average F1-score of 0.6963.

With just 500 epochs of training on CPY014 data, DRL_{Rwd} and DRL_{Valid} delivered the best F1-score of 0.8571. However, the maximum average F1-score achieved by DRL_{F1} and DRL_{Acc} was just 0.6733. When looking at the results on CPY015 data, the best models are DRL_{F1} and DRL_{Acc}. This is shown by the fact that their F1-scores were the highest in many training epochs.

DRL_{Acc} was the best model for detecting anomalies in CPY016 data since it not only had the greatest F1-score in almost every training epoch but also had the highest average F1-score of 0.5714. Meanwhile, every model scored the best F1-score of 0.8571, 100 percent recall, and 0.7500 accuracy when trained with 100 epochs on CPY017, with the exception of the DRL_{F1} model, which achieved just 0.6667 F1-score. While the best models for detecting anomalies on YOM009 are DRL_{Acc} and DRL_{Valid}, which both have the same F1-score of 0.4769, the worst models are DRL while training with 5000 iterations at a 0.2728 F1-score, 0.5538 recall, and 0.1818 precision.

Figure 3 shows the comparison of the critical differences between the different DRL models. The number associated with each algorithm is the average rank of the DRL models on each type of dataset, and solid bars represent groups of classifiers with no significant difference. There is no statistically significant difference across the models, with DRL_{Acc} ranking first, followed by DRL_{F1}, DRL, DRL_{Rwd}, and DRL_{Valid} ranking last.

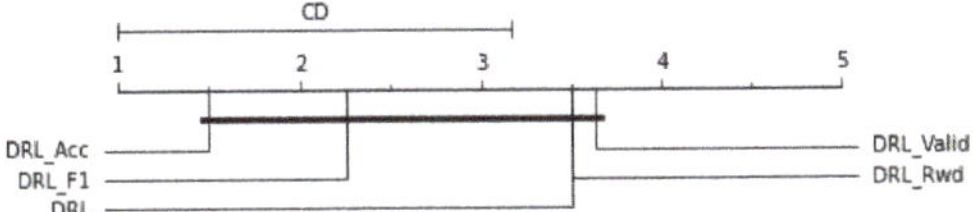

Figure 3. A critical difference diagram for 5 different DRL models on different datasets of telemetry water level data.

Figure 4 also shows a line graph of the F1-score as the number of epochs of training from each model increases. We can observe that as the number of epochs is increased, the performance of all deep reinforcement learning models using data from CPY012, CPY013, and CPY015 tends to improve. When training with CPY014 data, on the other hand, the F1-score of each model tends to stay the same or go down as the number of epochs goes up. In the case of trained models with CPY016 data, the F1-score of each model tends to stabilise and slightly decrease, with the exception of DRL_{Valid}, which tends to grow after 5000 epochs of training. When we looked at the models that were trained with the CPY017 dataset, the F1-score of DRL_{F1} went up after training with 1000 epochs and then went down. Other models, however, went up when training with more epochs, even though the performance of some models went down after 1000 epochs, while the F1-score of models that have been trained with CPY011 and YOM009 remained stable when training with more epochs.

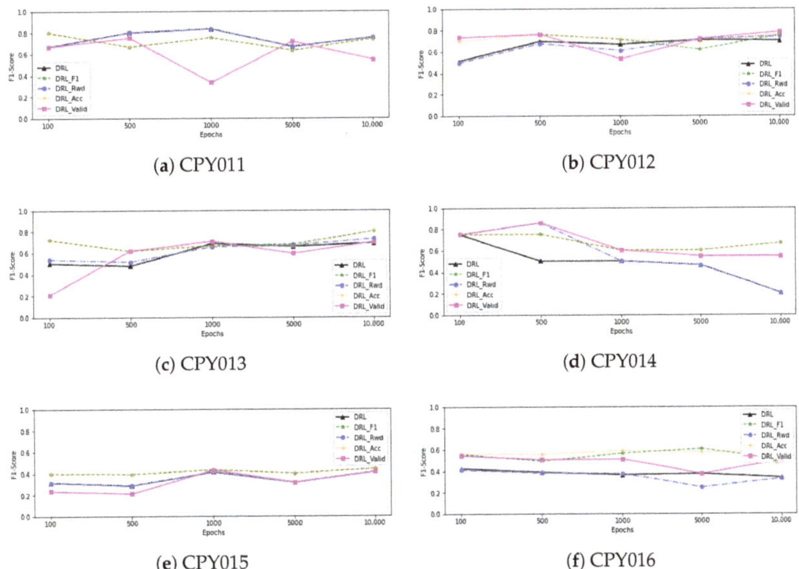

Figure 4. *Cont.*

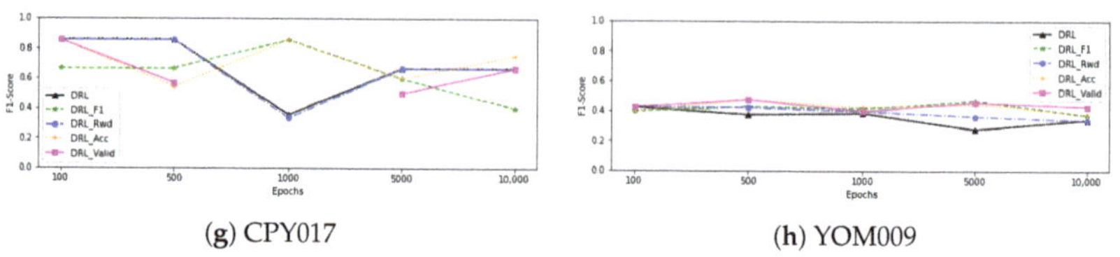

(g) CPY017   **(h)** YOM009

Figure 4. F1-score when increasing the learning epochs at each station.

Figure 5 shows the findings of the best DRL model for each station. We can observe that the DRL model performs well, capturing the majority of abnormalities in testing datasets. However, it still did not work well when there were anomalies in data that changed frequently, like when there were anomalies in YOM009 data between 29 June and 1 July 2015, and in CPY015 data on 19 June 2016.

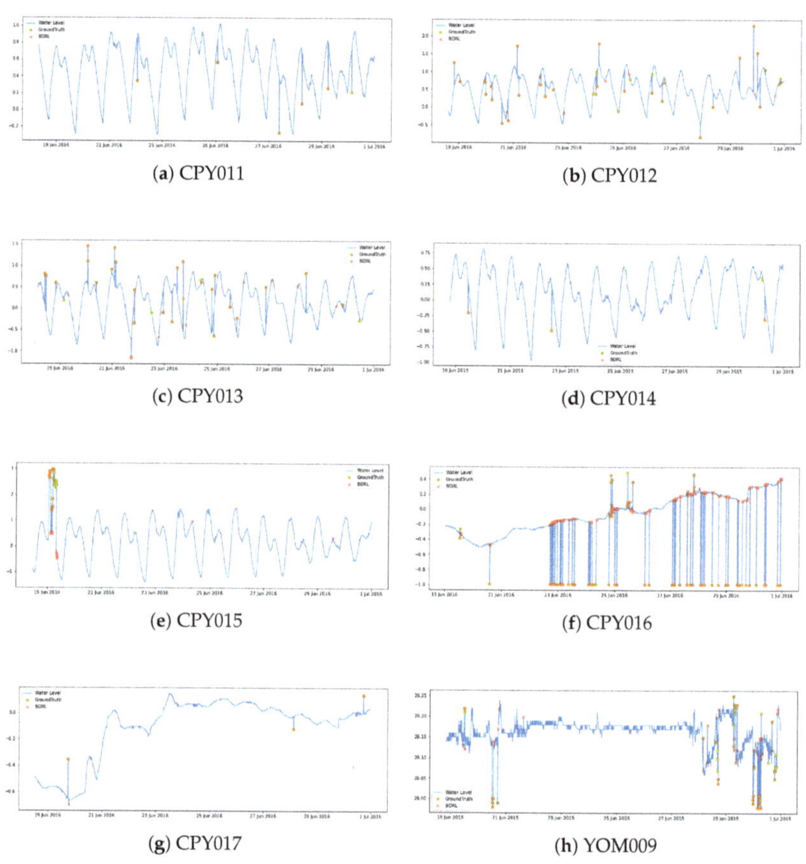

(a) CPY011   **(b)** CPY012

(c) CPY013   **(d)** CPY014

(e) CPY015   **(f)** CPY016

(g) CPY017   **(h)** YOM009

Figure 5. Anomaly detection from the best DRL model of each station. **(a)** CPY011 with DRL; **(b)** CPY012 with DRL_{Valid}; **(c)** CPY013 with DRL_{F1}; **(d)** CPY014 with DRL_{Rwd}; **(e)** CPY015 with DRL_{F1}; **(f)** CPY016 with DRL_{F1}; **(g)** CPY017 with DRL; **(h)** YOM009 with DRL_{Acc}.

5.2. Performance on the Same Station

We evaluated the performance of our techniques with MLP and LSTM models on eight telemetry water level datasets. The data in each station is first divided into training, validating, and testing parts in a 6:2:2 ratio. The results were averaged after being run ten times and then were compared to the averaged DRL models of each station as shown in Table 4. It demonstrated that DRL_{F1} and DRL_{Acc} had the highest average F1-scores for detecting anomalies on CPY015, with F1-scores of 0.4133. MLP had the greatest average F1-score when it came to detecting anomalies on CPY011, CPY012, and CPY014 with scores of 0.8505, 0.7822, and 0.8571, respectively. On the other stations, LSTM was the top performing model. According to the CD diagram in Figure 6, the best LSTM model had the greatest ranking of performance, followed by DRL_{Acc} and MLP.

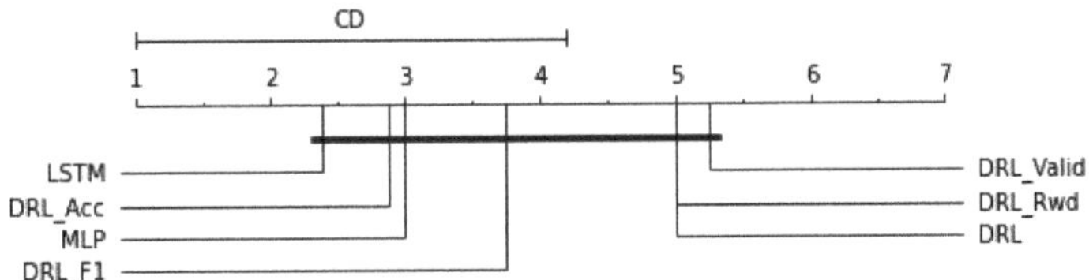

Figure 6. A critical difference diagram of each model.

Table 4. The mean F1-scores and standard deviation of all DRL, MLP, and LSTM models when testing with the dataset from different stations (the best F1-score of each row is shown in bold).

Station	DRL	DRL_{F1}	DRL_{Rwd}	DRL_{Acc}	DRL_{Valid}	MLP	LSTM
CPY011	0.7433 (±0.08)	0.7170 (±0.07)	0.7433 (±0.08)	0.7170 (±0.07)	0.6020 (±0.17)	**0.8505 (±0.06)**	0.8167 (±0.04)
CPY012	0.6550 (±0.08)	0.7146 (±0.06)	0.6468 (±0.10)	0.7234 (±0.04)	0.7045 (±0.10)	**0.7822 (±0.03)**	0.7753 (±0.02)
CPY013	0.6014 (±0.10)	0.6963 (±0.07)	0.6212 (±0.09)	0.6920 (±0.07)	0.5639 (±0.21)	0.6998 (±0.03)	**0.7265 (±0.02)**
CPY014	0.4823 (±0.20)	0.6733 (±0.08)	0.5537 (±0.26)	0.6733 (±0.08)	0.6596 (±0.14)	**0.8571 (±0.00)**	**0.8571 (±0.00)**
CPY015	0.3477 (±0.06)	**0.4134 (±0.02)**	0.3477 (±0.06)	**0.4134 (±0.02)**	0.3211 (±0.10)	0.2220 (±0.10)	0.3276 (±0.09)
CPY016	0.3770 (±0.03)	0.5438 (±0.05)	0.3499 (±0.06)	0.5714 (±0.02)	0.4867 (±0.07)	0.5651 (±0.14)	**0.6252 (±0.06)**
CPY017	0.6801 (±0.21)	0.6381 (±0.16)	0.6762 (±0.21)	0.7219 (±0.14)	0.6488 (±0.15)	0.9778 (±0.07)	**0.9857 (±0.05)**
YOM009	0.3581 (±0.06)	0.4152 (±0.04)	0.3891 (±0.04)	0.4248 (±0.04)	**0.4378 (±0.03)**	0.2358 (±0.05)	0.2596 (±0.06)

We discovered that DRL_{F1} and DRL_{Acc} had the highest average F1-scores for detecting anomalies on CPY015, with F1-scores of 0.4133. MLP had the greatest average F1-score when it came to detecting anomalies on CPY011, CPY012, and CPY014 with scores of 0.8505, 0.7822, and 0.8571, respectively. On the other stations, LSTM was the top performing model. The LSTM model has the highest ranking of performance, according to the CD diagram in Figure 6, followed by DRL_{Acc} and MLP.

Since RL models need time to learn until they have enough knowledge to do their task, time costing is the one important thing that we need to be interested in. We calculate the time spent by the best deep learning models (BDRL) and comparative models, as shown in Table 5. The MLP model requires the least training time per epoch, with an average of 0.30 s, followed by the LSTM model at 0.64 s, and the DRL model at 17.56 s. For MLP and LSTM training with early stopping, they needed an average of 12 and 15 training epochs, respectively, while our method requires around 4638 epochs to get optimal results. It means that the MLP model took an average of 2.97 s to train, while LSTM took 9.20 s and DRL took an average of 78,756 s, which is about 22 h.

Table 5. The number of training epochs and the time spent on each epoch for each model.

Station	Training Epochs			Time (s./epochs)			Total Time (s.)		
	MLP	**LSTM**	**BDRL**	**MLP**	**LSTM**	**BDRL**	**MLP**	**LSTM**	**BDRL**
CPY011	12	17	1000	0.17	0.64	18.66	2.04	10.88	18,666
CPY012	15	19	10,000	0.16	0.62	18.20	2.40	11.78	182,000
CPY013	11	17	10,000	0.18	0.64	17.88	1.98	10.88	178,000
CPY014	11	6	500	0.55	0.83	18.32	6.05	4.98	2490
CPY015	7	11	10,000	0.24	0.62	16.03	1.68	6.82	160,300
CPY016	17	21	5000	0.17	0.51	16.58	2.89	10.71	82,900
CPY017	13	17	100	0.20	0.56	16.74	2.6	9.52	1674
YOM009	6	11	500	0.69	0.73	18.82	4.14	8.03	4015
avg	12	15	4638	0.30	0.64	17.65	2.97	9.20	78,756

5.3. Performance on the Different Station

After generating various models on some stations' data and testing them with the same stations, we tested these models with the data collected from different stations with the intention of examining their generalisation ability. The F1-scores of each model are provided in Table 6.

Table 6. The F1-scores of the best DRL models when testing with the dataset from same station (show in the bracket) and different stations, while the average F1-scores and standard deviations of each station were calculated without their own scores.

Tested Dataset	Trained Dataset							
	CPY011	**CPY012**	**CPY013**	**CPY014**	**CPY015**	**CPY016**	**CPY017**	**YOM009**
CPY011	(0.8333)	0.1667	0.1967	0.0255	0.4138	0.0596	0.2000	0.1556
CPY012	0.6000	(0.7826)	0.7164	0.1136	0.0533	0.5088	0.6667	0.6667
CPY013	0.6531	0.6667	(0.8000)	0.1875	0.1034	0.5421	0.5970	0.5952
CPY014	0.4000	0.8571	0.8571	(0.8571)	0.0000	0.2727	0.5871	0.6000
CPY015	0.0000	0.1538	0.1474	0.0672	(0.4407)	0.0900	0.2078	0.0839
CPY016	0.5369	0.6129	0.6550	0.1276	0.4103	(0.6058)	0.1203	0.6703
CPY017	0.5000	0.3529	0.5455	0.2308	0.0000	0.1765	(0.8571)	0.4000
YOM009	0.0000	0.3297	0.1905	0.0771	0.0000	0.4189	0.3158	(0.4769)
avg	0.3843	0.4485	0.4727	0.1185	0.1401	0.2955	0.3850	**0.4531**
std	0.2742	0.2680	0.2908	0.0713	0.1896	0.1975	0.2257	0.2457

Using DRL_{Rwd}-the best model for detecting anomalies by training with CPY011 data and then identifying anomalies from other stations, we can see that, though it works rather well, with F1-scores ranging from 0.4 on CPY014 to 0.65 on CPY013 data, it is unable to detect anomalies on CPY015 and YOM009. Using the BDRL model of the CPY012 training dataset, DRL_{Valid}, although it provided good performance when identifying anomalies in the CPY013, CPY014, and CPY016 datasets with F1-scores greater than 0.61, especially CPY014 with a 0.8571 f1-score, which more than detected anomalies on its own dataset, it provided poor performance, with an F1-score lower than 0.4000, when detecting anomalies in other stations. Similar to DRL_{F1}, which was trained using CPY013 data, it not only performs well when recognising anomalies on its own dataset but also when detecting anomalies on the CPY014 dataset, with an F1-score of 0.8571. The BDRL model, DRL_{Rwd}, that was trained with CPY014 did the worst when it was used to find anomalies in other stations' data, with an F1-score of less than 0.23 for every dataset and the lowest F1-score of only 0.0255 for CPY011. Similar to the best model on CPY015 datasets, which performed poorly, with the highest F1-score on CPY011 data being 0.4138 and being unable to identify anomalies on CPY014, CPY017, and YOM009. Meanwhile, the best model for detecting anomalies on CPY016 data performed the best for detecting anomalies on CPY013 with a 0.5421 F1-score. The model that was trained on CPY017 did the best of finding anomalies in data from CPY012, CPY013, and CPY014 with an F1-score greater than 0.58. While the best model from the YOM009 training dataset achieved a low F1-score on CPY011, CPY015, and CPY017, 0.0839 is the lowest F1-score. However, when it was used to find outliers on COY012, CPY013, CPY014, and CPY016 with F1-scores higher than 0.59, it did better than its own training data.

It is worth noting that models trained using CPY014 and CPY015 data perform poorly when used to identify anomalies from other stations. This may be due to the fact that the actual number of anomalies in those stations are relatively low and most of them are kind of extreme outliers, as shown in Figure 2, so the models were trained with only those kinds of anomalies, which may not be enough for the model to learn. In contrast to YOM009, which has a many number and types of anomalies for model to learn, as a result, it can identify abnormalities on CPY012, CPY013, CPY014, and CPY016 better than other models that were trained with another station.

Then, we tested MLP and LSTM using data from different stations to compare our method to the candidate models. Table 7 represents the results of the MLP models when tested with the datasets from the same and different stations. Using the CPY011 dataset, the MLP models achieved the highest F1-score of 0.5430 on CPY016, despite their being unable to identify anomalies on CPY014 and YOM009. Similar to finding anomalies on CPY012, it offered good results with F1-scores of more than 0.63, with the exception of CPY011, CPY015, and YOOM009, which produced F1-scores of less than 0.4. The best MLP of the CPY013 training dataset provided the highest F1-score on the CPY014 dataset (0.8571 F1-score) and the lowest on CPY015 (0.2093 F1-score). Anomalies on the YOM009 dataset were the most difficult for the MLP models trained on CPY014 to detect, with an F1-score of just 0.1818. However, it performed excellent results in identifying anomalies on CPY017 with a 1.0000 F1-score. Meanwhile, the MLP model on the CPY015 dataset performed poorly when detecting abnormalities from other stations. On the other hand, the MLP models that were trained on CPY016 and CPY017 generated good results when used to identify anomalies from other stations, despite still performing poorly in some stations. In contrast, the MLP model trained on YOM009 worked well when used to detect abnormalities on other stations but performed badly when detecting anomalies on its own data. Furthermore, it performed well on CPY017 data, with a 1.000 F1-score.

Table 7. The F1-scores of the MLP models when testing with the dataset from the same station (shown in the bracket) and different stations.

Tested Dataset	Trained Dataset							
	CPY011	CPY012	CPY013	CPY014	CPY015	CPY016	CPY017	YOM009
CPY011	(0.9231)	0.4000	0.2917	0.7778	0.0000	0.1373	0.7059	0.6087
CPY012	0.4889	(0.8387)	0.8060	0.7500	0.0541	0.6923	0.7273	0.7857
CPY013	0.4390	0.6786	(0.7500)	0.7000	0.0000	0.7123	0.6909	0.7719
CPY014	0.0000	0.8571	0.8571	(0.8571)	0.0000	0.6000	0.8571	0.8571
CPY015	0.1951	0.1509	0.2093	0.2593	(0.3529)	0.1420	0.3077	0.2414
CPY016	0.5430	0.6380	0.6587	0.6322	0.0915	(0.6629)	0.5665	0.6550
CPY017	0.5000	0.6667	0.6000	1.0000	0.0000	0.2857	(1.0000)	1.0000
YOM009	0.0000	0.2308	0.2955	0.1818	0.0000	0.4404	0.1772	(0.2857)
avg	0.3094	0.5174	0.5312	0.6144	0.0208	0.4300	0.5761	**0.7028**
std	0.2396	0.2609	0.2643	0.2929	0.0371	0.2473	0.2460	0.2408

In the case of the LSTM model, as depicted in Table 8. They performed well, with an average F1-score of more than 0.42 for each station except CPY015, which had an average F1-score of 0.1099. However, they generated poor performances in some stations, such as the LSTM of CPY016 that achieved an F1-score of only 0.1754 when used to detect anomalies on the CPY011 dataset, and it was unable to detect anomalies on CPY014, CPY017, and YOM009 datasets with the LSTM that had been trained on the CPY015 dataset. However, it provided excellent performance when detecting anomalies on CPY017 with the LSTM that has been trained on the CPY014 dataset. When the LSTM was trained on YOM009, it did well at finding anomalies from other stations, especially CPY014 and CPY017, with an F1-score of 0.8571.

Table 8. The F1-scores of the LSTM models when testing with the dataset from the same station (shown in the bracket) and different stations.

Tested Dataset	Trained Dataset							
	CPY011	**CPY012**	**CPY013**	**CPY014**	**CPY015**	**CPY016**	**CPY017**	**YOM009**
CPY011	(0.8571)	0.4828	0.2333	0.5000	0.3077	0.1754	0.7059	0.5185
CPY012	0.6400	(0.8387)	0.8308	0.7368	0.0444	0.7536	0.7500	0.7458
CPY013	0.5652	0.7333	(0.7463)	0.7213	0.1081	0.6857	0.7333	0.7188
CPY014	0.4000	0.8571	0.8571	(0.8571)	0.0000	0.8571	0.8571	0.8571
CPY015	0.3043	0.3636	0.2340	0.3939	(0.4364)	0.2045	0.3704	0.2564
CPY016	0.5679	0.6897	0.6824	0.5634	0.3089	(0.6630)	0.6296	0.6359
CPY017	0.5000	1.0000	0.5455	1.0000	0.0000	0.3333	(1.0000)	0.8571
YOM009	0.0000	0.2619	0.3146	0.2727	0.0000	0.4190	0.2857	(0.3542)
avg	0.4253	0.6269	0.5282	0.5983	0.1099	0.4898	0.6189	**0.6557**
std	0.2190	0.2679	0.2717	0.2430	0.1410	0.2747	0.2111	0.2129

Furthermore, we generated a bar chart to compare the average F1-score from each model when tested with the data collected from different stations, as shown in Figure 7. When evaluated with data from other stations, the models trained with CPY012 and CPY013 produced an average F1-score greater than 0.4. The models trained on CPY015 earned poor performance when used to identify anomalies from other stations, with an average F1-score lower than 0.2. DRL models that were trained with CPY015 outperform other models in detecting anomalies in data from other stations. LSTM models trained on CPY011, CPY012, CPY016, and CPY017, on the other hand, outperform other models in detecting abnormalities on other datasets. When trained with data from CPY013, CPY014, and YOM009, MLP had the best F1-score for finding outliers in other datasets.

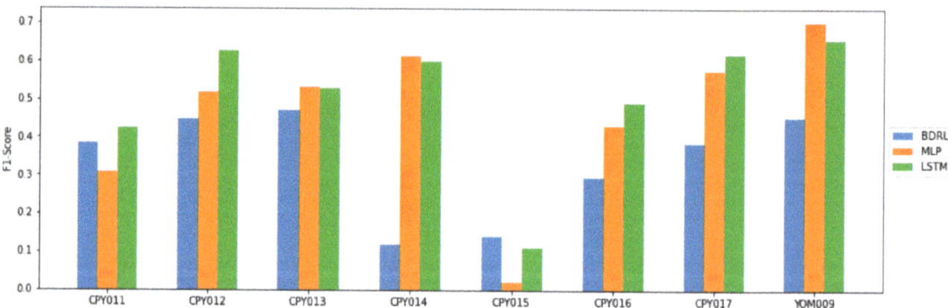

Figure 7. Bar charts of average F1-scores of the DRL, MLP, and LSTM when tested with the data collected from different stations.

5.4. Ensemble Results

Since we have multiple RL models after each epoch of training, and since each model performs the best in each of the criteria, we then built an ensemble that combined the decisions of all RL models, with the aim of generating a better final decision. In model selection, we select all five models and select the three models with the highest ranking in F1-score to build our ensemble model. For decision making, we used majority voting and weighted voting strategies to make a final decision. So, we have 4 ensemble models for each epoch of training, including a majority voting ensemble model with 3 ($EDRL_3$) and 5 ($EDRL_5$) models, and a weighted ensemble model with 3 ($WEDRL_3$) and 5 ($WEDRL_5$) models.

5.4.1. Performance on the Same Station

The results of our ensemble models are shown in Table 9 demonstrated that ensemble with majority voting and weighted voting that were generated from the top three DRL models of CPY011 provided the best with 0.8333 F1-score, while $WDRL_3$ that was generated from the DRL model after trained with 10,000 epochs is the best model to detect anomalies in CPY012 datasets with an F1-score of 0.7941. The ensemble model of CPY013 that performs the best is $EDRL_3$ and $WEDRL_3$ at 0.8000. The best ensemble model for identifying anomalies in CPY014 datasets is the ensemble model that provided the F1-score of 0.8571. With CPY015 data, the models with the highest F1-score are $EDRL_3$, $WEDRL_3$, and $WEDRL_5$. These models were built based on the individual DRL model, which was trained for 10,000 iterations. Meanwhile, $WEDRL_3$ got the highest F1-score of 0.5922 for CPY016 by combining the best three DRL models that were trained over 5000 iterations. With CPY017, $EDRL_5$ outperforms other ensemble models with a 100 percent in every metric. The ensemble results of YOM009, $WEDRL_5$, offered the highest performance with an F1-score of 0.5032 that was generated from the DRL model after 500 epochs of training.

Table 9. The performance of ensemble models (the best F1-score of each row is shown in bold).

Station	Epochs	$EDRL_3$			$EDRL_5$			$WEDRL_3$			$WEDRL_5$		
		Recall	Prec	F1	Recall	Prec	F1	Recall	Prec	F1	Recall	Prec	F1
CPY011	100	0.8571	0.7500	**0.8000**	0.7143	0.6250	0.6667	0.8571	0.7500	**0.8000**	0.8571	0.7500	**0.8000**
	500	0.8571	0.7500	**0.8000**	0.8571	0.6667	0.7500	0.8571	0.7500	**0.8000**	0.8571	0.6667	0.7500
	1000	0.7143	1.0000	**0.8333**	0.7143	0.8333	0.7692	0.7143	1.0000	**0.8333**	0.8571	0.7500	0.8000
	5000	0.7143	0.6250	0.6667	0.7143	0.6250	0.6667	0.7143	0.7143	**0.7143**	0.7143	0.7143	**0.7143**
	10,000	0.8571	0.6667	0.7500	1.0000	0.7778	**0.8750**	0.8571	0.6667	0.7500	0.8571	0.6000	0.7059
	Avg	0.8000	0.7583	0.7700	0.8000	0.7056	0.7455	0.8000	0.7762	**0.7795**	0.8285	0.6962	0.7540
	std	0.0782	0.1455	0.0650	0.1278	0.0949	0.0863	0.0782	0.1297	0.0471	0.0639	0.0637	0.0451
CPY012	100	0.7647	0.7027	0.7324	0.7353	0.7353	**0.7353**	0.7647	0.7027	0.7324	0.7647	0.6842	0.7222
	500	0.7941	0.7297	**0.7606**	0.7941	0.7297	**0.7606**	0.7941	0.7297	**0.7606**	0.7941	0.7105	0.7500
	1000	0.7647	0.6667	0.7123	0.7353	0.8621	**0.7937**	0.7647	0.6667	0.7123	0.7353	0.6410	0.6849
	5000	0.7059	0.7273	0.7164	0.7059	0.7500	0.7273	0.7059	0.7273	0.7164	0.7353	0.7353	**0.7353**
	10,000	0.7059	0.8276	0.7619	0.7059	0.8276	0.7619	0.7941	0.7941	**0.7941**	0.7353	0.8333	0.7812
	Avg	0.7471	0.7308	0.7367	0.7353	0.7809	**0.7558**	0.7647	0.7241	0.7432	0.7529	0.7209	0.7347
	std	0.0394	0.0598	0.0236	0.0360	0.0601	0.0261	0.0360	0.0466	0.0342	0.0263	0.0719	0.0355
CPY013	100	0.8710	0.6136	**0.7200**	0.8387	0.6190	0.7123	0.8710	0.6136	**0.7200**	0.9032	0.5957	0.7179
	500	0.8065	0.5000	0.6173	0.7742	0.5714	0.6575	0.8387	0.4906	0.6190	0.8065	0.5556	**0.6579**
	1000	0.8065	0.6098	0.6944	0.8065	0.6098	0.6944	0.9355	0.6042	**0.7342**	0.9032	0.6087	0.7273
	5000	0.8065	0.5952	0.6849	0.7097	0.5789	0.6377	0.7097	0.6471	0.6769	0.7742	0.6486	**0.7059**
	10,000	0.8387	0.7647	**0.8000**	0.7419	0.6765	0.7077	0.8387	0.7647	**0.8000**	0.8387	0.7429	0.7879
	Avg	0.8258	0.6167	0.7033	0.7742	0.6111	0.6819	0.8387	0.6240	0.7100	0.8452	0.6303	**0.7194**
	std	0.0288	0.0949	0.0660	0.0510	0.0417	0.0328	0.0822	0.0983	0.0674	0.0577	0.0712	0.0467
CPY014	100	0.7500	0.7500	0.7500	0.7500	1.0000	**0.8571**	0.7500	1.0000	**0.8571**	0.7500	0.7500	0.7500
	500	0.7500	1.0000	**0.8571**	0.7500	1.0000	**0.8571**	0.7500	1.0000	**0.8571**	0.7500	1.0000	**0.8571**
	1000	0.7500	0.5000	0.6000	0.7500	0.6000	**0.6667**	0.7500	0.3750	0.5000	0.7500	0.3750	0.5000
	5000	0.7500	0.5000	**0.6000**	0.7500	0.5000	**0.6000**	0.7500	0.5000	**0.6000**	0.7500	0.3750	0.5000
	10,000	0.7500	0.6000	0.6667	0.7500	0.6000	0.6667	0.7500	0.6000	0.6667	0.7500	0.7500	**0.7500**
	Avg	0.7500	0.6700	0.6948	0.7500	0.7400	**0.7295**	0.7500	0.6950	0.6962	0.7500	0.6500	0.6714
	std	0.0000	0.2110	0.1097	0.0000	0.2408	0.1196	0.0000	0.2896	0.1584	0.0000	0.2710	0.1625
CPY015	100	0.3235	0.5238	**0.4000**	0.2647	0.5000	0.3462	0.3235	0.5238	**0.4000**	0.2647	0.4737	0.3396
	500	0.3235	0.5000	**0.3929**	0.2059	0.4667	0.2857	0.3235	0.5000	**0.3929**	0.2941	0.5263	0.3774
	1000	0.3824	0.5000	0.4333	0.4118	0.4828	**0.4444**	0.3824	0.5000	0.4333	0.3824	0.4643	0.4194
	5000	0.3235	0.5238	**0.4000**	0.2353	0.4706	0.3137	0.3235	0.5238	**0.4000**	0.3235	0.5238	**0.4000**
	10,000	0.3824	0.5200	**0.4407**	0.3824	0.4483	0.4127	0.3824	0.5200	**0.4407**	0.3824	0.5200	**0.4407**
	Avg	0.3471	0.5135	**0.4134**	0.3000	0.4737	0.3605	0.3471	0.5135	**0.4134**	0.3294	0.5016	0.3954
	std	0.0323	0.0124	0.0219	0.0916	0.0192	0.0666	0.0323	0.0124	0.0219	0.0526	0.0300	0.0390
CPY016	100	0.5981	0.5203	0.5565	0.6075	0.4962	0.5462	0.5981	0.5203	0.5565	0.6168	0.5366	**0.5739**
	500	0.5981	0.5203	0.5565	0.6168	0.3952	0.4818	0.6075	0.5603	**0.5830**	0.5981	0.5333	0.5639
	1000	0.6168	0.5323	0.5714	0.6542	0.4636	0.5426	0.6168	0.5546	**0.5841**	0.6168	0.5455	0.5789
	5000	0.5701	0.6100	0.5894	0.5514	0.4275	0.4816	0.5701	0.6162	**0.5922**	0.5888	0.3987	0.4755
	10,000	0.6168	0.4177	0.4981	0.6262	0.4295	0.5095	0.6262	0.5447	**0.5826**	0.6449	0.5111	0.5702
	Avg	0.6000	0.5201	0.5544	0.6112	0.4424	0.5123	0.6037	0.5592	**0.5797**	0.6131	0.5050	0.5525
	std	0.0191	0.0684	0.0343	0.0377	0.0386	0.0314	0.0215	0.0353	0.0135	0.0215	0.0608	0.0434
CPY017	100	1.0000	0.7500	**0.8571**	1.0000	0.7500	**0.8571**	1.0000	0.7500	**0.8571**	1.0000	0.7500	**0.8571**
	500	1.0000	0.7500	0.8571	1.0000	1.0000	**1.0000**	1.0000	0.7500	0.8571	1.0000	0.7500	0.8571
	1000	1.0000	0.7500	**0.8571**	1.0000	0.2308	0.3750	1.0000	0.7500	**0.8571**	1.0000	0.7500	**0.8571**
	5000	1.0000	0.5000	**0.6667**	1.0000	0.5000	**0.6667**	1.0000	0.5000	**0.6667**	1.0000	0.5000	**0.6667**
	10,000	0.6667	0.6667	0.6667	0.6667	0.6667	0.6667	1.0000	0.6000	**0.7500**	0.6667	0.6667	0.6667
	Avg	0.9333	0.6833	0.7809	0.9333	0.6295	0.7131	1.0000	0.6700	**0.7976**	0.9333	0.6833	0.7809
	std	0.1491	0.1087	0.1043	0.1491	0.2868	0.2354	0.0000	0.1151	0.0866	0.1491	0.1087	0.1043

Table 9. *Cont.*

Station	Epochs	EDRL$_3$			EDRL$_5$			WEDRL$_3$			WEDRL$_5$		
		Recall	Prec	F1	Recall	Prec	F1	Recall	Prec	F1	Recall	Prec	F1
YOM009	100	0.6308	0.3254	**0.4293**	0.5846	0.3115	0.4064	0.6154	0.3226	0.4233	0.6462	0.3281	0.4352
	500	0.4769	0.4769	0.4769	0.6154	0.3960	0.4819	0.4769	0.4769	0.4769	0.6000	0.4333	**0.5032**
	1000	0.5538	0.3600	0.4364	0.5538	0.3186	0.4045	0.5231	0.3778	**0.4387**	0.4923	0.3299	0.3951
	5000	0.5538	0.4091	0.4706	0.5846	0.4086	0.4810	0.6308	0.3981	**0.4881**	0.5538	0.3830	0.4528
	10,000	0.5385	0.2991	0.3846	0.4923	0.3048	0.3765	0.4154	0.3971	**0.4060**	0.4615	0.3158	0.3750
	Avg	0.5508	0.3741	0.4396	0.5661	0.3479	0.4301	0.5323	0.3945	**0.4466**	0.5508	0.3580	0.4323
	std	0.0548	0.0707	0.0371	0.0467	0.0501	0.0484	0.0914	0.0554	0.0350	0.0757	0.0494	0.0503

Figure 8 depicts line charts that indicate the F1-score of each ensemble model that was trained using data from each station. It was clear from the results that the ensemble models not only delivered good performances and had a tendency to either improve or keep their F1-scores steady but also reduced the false alarms by increasing the precision scores. When we compared the results of each training epoch of the individual DRL model and the ensemble model, as shown in Tables 3 and 9, we discovered that ensemble models performed better than every single DRL model in many training epochs. In particular, $EDRL_5$ on the CPY017 with 500 training epochs generated an excellent score of 1.0000 in every metrics index, resulting from a 25% increase in accuracy and a 15% increase in F1-score. Meanwhile, $EDRL_5$ on the CPY011 with 10,000 training epochs improved the performance of the best individual model with an F1-score from 0.75 to 0.8750, reached 1.00 in terms of recall, and increased precision by 20%. By combining the DRL models trained on only 500 epochs, the ensemble model on YOM009 got the highest F1-score of 0.5032.

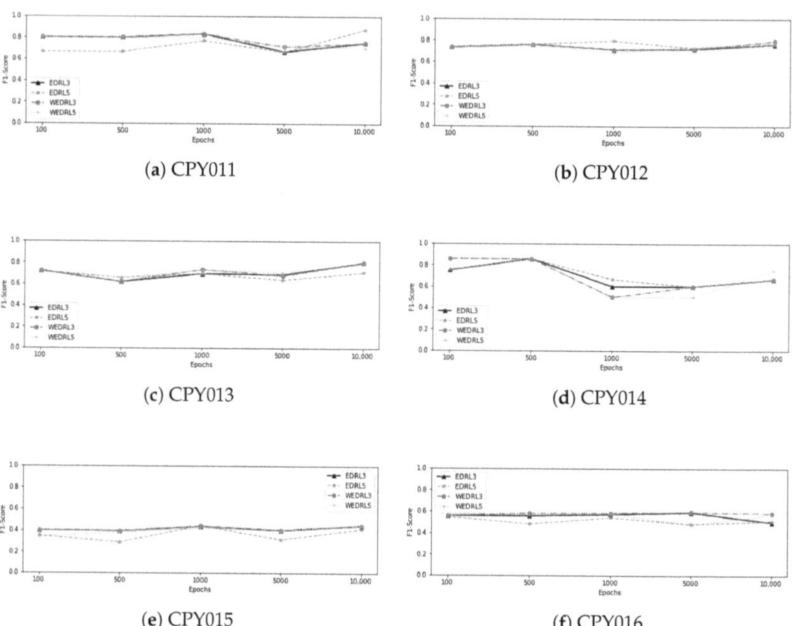

Figure 8. *Cont.*

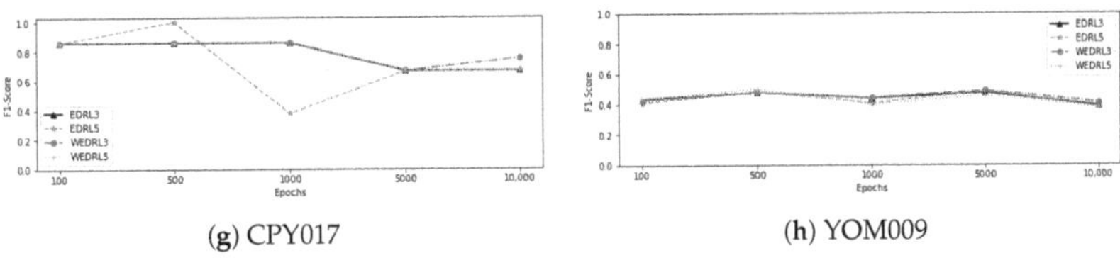

(**g**) CPY017   (**h**) YOM009

Figure 8. F1-score of ensemble model when increasing the learning epochs at CPY011, CPY012, CPY013, CPY014, CPY015, CPY016, CPY017, and YOM009 (**a–h**).

As shown in Table 10, we evaluated the average F1-score of each individual DRL model and ensemble of DRL models against the other neural network models. We can see that the LSTM model was the best model when detecting anomalies on CPY013, CPY014, CPY016, and CPY017, while $WEDRL_3$ provided the highest average F1-score on CPY015 and YOM009. The highest F1-score was 0.4134 on CPY015, which was provided by DRL_{F1}, DRL_{Acc}, $EDRL_3$, and $WEDRL_3$. Although MLP and LSTM beat other models in many datasets, $WEDRL_3$ has the greatest average ranking, as shown in Figure 9. In other words, the ensemble model not only has the potential to improve the performance of a single model, but it also has a higher reliability to deliver excellent performance than a single model.

Table 10. The mean F1-scores and standard deviations of all of the DRL, MLP, LSTM, and ensemble of DRL-based models when testing with the dataset from different stations (the best F1-score of each station is shown in bold).

Models	CPY011	CPY012	CPY013	CPY014	CPY015	CPY016	CPY017	YOM009
DRL	0.7433 (±0.08)	0.6550 (±0.08)	0.6014 (±0.10)	0.4823 (±0.20)	0.3477 (±0.06)	0.3770 (±0.03)	0.6801 (±0.21)	0.3581 (±0.06)
DRL_{F1}	0.7170 (±0.07)	0.7146 (±0.06)	0.6963 (±0.07)	0.6733 (±0.08)	**0.4134 (±0.02)**	0.5438 (±0.05)	0.6381 (±0.16)	0.4152 (±0.04)
$DRLRwd$	0.7433 (±0.08)	0.6468 (±0.10)	0.6212 (±0.09)	0.5537 (±0.26)	0.3477 (±0.06)	0.3499 (±0.06)	0.6762 (±0.21)	0.3891 (±0.04)
DRL_{Acc}	0.7170 (±0.07)	0.7234 (±0.04)	0.6920 (±0.07)	0.6733 (±0.08)	**0.4134 (±0.02)**	0.5714 (±0.02)	0.7219 (±0.14)	0.4248 (±0.04)
DRL_{Valid}	0.6020 (±0.17)	0.7045 (±0.10)	0.5639 (±0.21)	0.6596 (±0.14)	0.3211 (±0.10)	0.4867 (±0.07)	0.6488 (±0.15)	0.4378 (±0.03)
$EDRL_3$	0.7700 (±0.06)	0.7367 (±0.02)	0.7033 (±0.07)	0.6948 (±0.11)	**0.4134 (±0.02)**	0.5544 (±0.03)	0.7809 (±0.10)	0.4396 (±0.04)
$EDRL_5$	0.7455 (±0.09)	0.7558 (±0.03)	0.6819 (±0.03)	0.7295 (±0.12)	0.3605 (±0.07)	0.5123 (±0.03)	0.7131 (±0.24)	0.4301 (±0.05)
$WEDRL_3$	0.7795 (±0.05)	0.7432 (±0.03)	0.7100 (±0.07)	0.6962 (±0.16)	**0.4134 (±0.02)**	0.5797 (±0.01)	0.7976 (±0.09)	**0.4466 (±0.03)**
$WEDRL_5$	0.7540 (±0.05)	0.7347 (±0.04)	0.7194 (±0.05)	0.6714 (±0.16)	0.3954 (±0.04)	0.5525 (±0.04)	0.7619 (±0.10)	0.4323 (±0.05)
MLP	**0.8505 (±0.06)**	**0.7822 (±0.03)**	0.6998 (±0.03)	**0.8571 (±0.00)**	0.2220 (±0.10)	0.5651 (±0.14)	0.9778 (±0.07)	0.2358 (±0.05)
LSTM	0.8167 (±0.04)	0.7753 (±0.02)	**0.7265 (±0.02)**	**0.8571 (±0.00)**	0.3276 (±0.09)	**0.6252 (±0.06)**	**0.9857 (±0.05)**	0.2596 (±0.06)

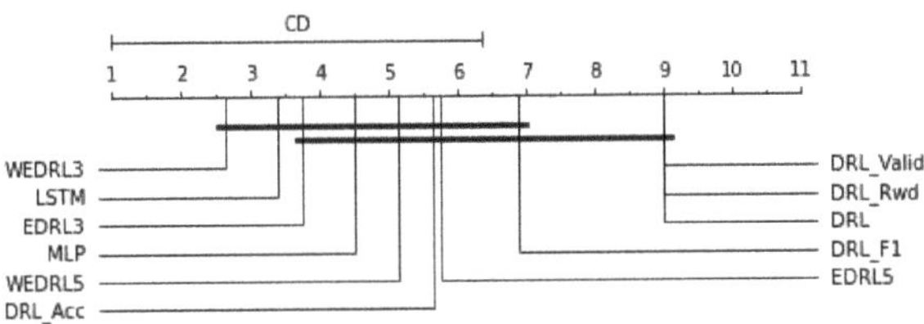

Figure 9. Critical difference diagram.

5.4.2. Performance on the Different Station

We then tested the generalisation ability of the best ensemble ($WEDRL_3$) with the data collected from different stations. The F1-score of each model is depicted in Table 11. We can observe that the ensemble model that was created from the model trained on CPY011 data performed well not only on their own dataset but also on CPY017, with an F1-score of 0.8200, similarly to $WEDRL_3$ on CPY012 and CPY013, which recognised anomalies on CPY014 better than their own dataset with F1-scores of 0.8421 and 0.8143, respectively. Inversely, the ensemble model on CPY014, CPY015, and CPY016 trained datasets provided poor performance when used to detect anomalies on other stations. Even though the ensemble model trained on the CPY017 dataset got an F1-score of more than 0.5 on CPY012, CPY013, and CPY014, it did not do well on many stations, with an F1-score of less than

0.3. $WEDRL_3$ scored badly not just on their own dataset but also on others, with F1-scores ranging from 0.0739 on CPY015 to 0.5748 on CPY016.

Table 11. The mean F1-scores of the $WEDRL_3$ models when testing with the dataset from the same station (shown in the bracket) and different stations.

Tested Dataset	Trained Dataset							
	CPY011	CPY012	CPY013	CPY014	CPY015	CPY016	CPY017	YOM009
CPY011	(0.7795)	0.2772	0.2278	0.1718	0.4345	0.0788	0.2830	0.0900
CPY012	0.7183	(0.7432)	0.6817	0.4624	0.2182	0.5355	0.6419	0.5096
CPY013	0.6920	0.7281	(0.7101)	0.5512	0.2713	0.5625	0.5959	0.4816
CPY014	0.7276	0.8421	0.8143	(0.6962)	0.5714	0.5471	0.7948	0.3450
CPY015	0.3748	0.2683	0.1664	0.1482	(0.4134)	0.1290	0.1994	0.0739
CPY016	0.5813	0.6283	0.6450	0.3327	0.3647	(0.5797)	0.4711	0.5748
CPY017	0.8200	0.4423	0.4514	0.3803	0.3545	0.2075	(0.7976)	0.2931
YOM009	0.1084	0.3261	0.3109	0.2619	0.0568	0.4613	0.3375	(0.4466)
avg	**0.6002**	0.5320	0.5010	0.3756	0.3356	0.3877	0.5152	0.3518
std	0.2426	0.2310	0.2451	0.1887	0.1551	0.2121	0.2292	0.1889

5.4.3. Ensemble with All Seven Models

Then, to learn more about how well the ensemble worked, we combined our developed DRL models with MLP and LSTM models. In model selection, we selected all seven models and selected the five and three models with the highest ranking in F1-score to build our ensemble model. We used the same strategy to make a final decision. So, we have 6 ensemble model for each epochs of training include majority voting ensemble model with 3 ($E3$), 5 ($E5$), and 7 ($E7$) model, and weighted ensemble model with 3 ($WE3$), 5 ($WE5$), and 7 ($WE7$) models, and the results are displayed in Table 12.

We can see that, on the CPY011 dataset, the ensemble of the top three models (E3) earned the greatest F1-score of 0.9231 with every epoch of training. On CPY012, the greatest F1-score of 0.8438 was obtained by E5 and WE7 with models trained with 10,000 epochs, and E7 with models trained with 500 epochs, while E3 and WE3 models trained with 10,000 epochs performed the best in identifying anomalies on the CPY013 dataset. With the CPY014 dataset, all ensemble models gave an F1-score of 0.8571, with the exception of the ensemble with majority voting of all seven models trained with 10,000 epochs, which performed badly with an F1-score of 0.6667. WE7 surpassed other ensemble models on the CPY015 and CPY016 datasets, with the greatest F1-score of 0.4615 and 0.6704, respectively. Every ensemble model on CPY017 produced outstanding results with a 1.0000 F1-score, particularly E3, WE3, WE5, and WE7, which produced excellent results with all training epochs. The weighted ensemble with 5 models (WE5) trained with 500 epochs performed the best on the YOM009 dataset, with a 0.5032 F1-score.

As indicated in Table 13, we averaged the F1-score of each individual model and ensemble model to compare their performance. We can observe that E3 not only performed the best model with the greatest average F1-score on all datasets but also excellently performed with a 1.0000 F1-score on CPY017 and YOM009. Among the models tested on the CPY014 dataset, the best F1-score of 0.8571 was achieved by MLP, LSTM, E3, E5, WE3, WE5, and WE7. In contrast, on the CPY015 dataset, the model with DRL-based (DRL_{F1}, DRI_{Acc}, $EDRL_3$, and $WEDRL_3$) generated the highest F1-score of 0.4134. Furthermore, as shown in Figure 10, the CD diagram was chosen to make a statistical comparison of our results, which revealed that E3 had the highest ranking, and the ensemble model that combined all seven individual models outperformed both the individual model and the ensemble model created using DRL models. It also demonstrated the ability of ensemble methods to improve the performance of individual DRL models because it represented a significant difference from individual models (DRL, DRL_{Rwd}, and DRL_{Valid}).

Table 12. The performance of the ensemble models built by combining DRL and candidate models (the best F1-score of each row is shown in bold).

Station	Epochs	E3			E5			E7			WE3			WE5			WE7		
		Recall	Prec	F1	Recall	Prec	F1	Recall	Prec	F1	Recall	Prec	F1	Recall	Prec	F1	Recall	Prec	F1
CPY011	100	0.8571	1.0000	**0.9231**	0.8571	0.7500	0.8000	0.8571	0.7500	0.8000	0.8571	0.8571	0.8571	0.8571	0.8571	0.8571	0.8571	0.8571	0.8571
	500	0.8571	1.0000	**0.9231**	0.8571	0.7500	0.8000	0.8571	0.7500	0.8000	0.8571	0.8571	0.8571	0.8571	0.8571	0.8571	0.8571	0.8571	0.8571
	1000	0.8571	1.0000	**0.9231**	0.8571	1.0000	**0.9231**	0.8571	1.0000	**0.9231**	0.8571	0.8571	0.8571	0.8571	0.8571	0.8571	0.8571	0.8571	0.8571
	5000	0.8571	1.0000	**0.9231**	0.7143	0.7143	0.7143	0.8571	0.7500	0.8000	0.8571	0.8571	0.8571	0.8571	0.8571	0.8571	0.8571	0.8571	0.8571
	10,000	0.8571	1.0000	**0.9231**	0.8571	0.8571	0.8571	0.8571	0.8571	0.8571	0.8571	0.8571	0.8571	0.8571	0.8571	0.8571	0.8571	0.8571	0.8571
	avg	0.8571	1.0000	**0.9231**	0.8285	0.8143	0.8189	0.8571	0.8214	0.8360	0.8571	0.8571	0.8571	0.8571	0.8571	0.8571	0.8571	0.8571	0.8571
	std	0.0000	0.0000	0.0000	0.0639	0.1168	0.0774	0.0000	0.1101	0.0546	0.0000	0.0000	0.0000	0.0000	0.0000	0.0000	0.0000	0.0000	0.0000
CPY012	100	0.7647	0.9286	**0.8387**	0.7647	0.8387	0.8000	0.7941	0.7714	0.7826	0.7647	0.8966	0.8254	0.7647	0.8966	0.8254	0.7647	0.8966	0.8254
	500	0.7647	0.9286	0.8387	0.7941	0.7297	0.7606	0.7941	0.9000	**0.8438**	0.7647	0.8966	0.8254	0.7647	0.8966	0.8254	0.7647	0.8966	0.8254
	1000	0.7647	0.9286	**0.8387**	0.7647	0.8966	0.8254	0.7647	0.8966	0.8254	0.7647	0.8966	0.8254	0.7647	0.8966	0.8254	0.7647	0.8966	0.8254
	5000	0.7647	0.9286	**0.8387**	0.7059	0.7500	0.7273	0.7353	0.8333	0.7812	0.7647	0.8966	0.8254	0.7647	0.8966	0.8254	0.7647	0.8966	0.8254
	10,000	0.7647	0.9286	0.8387	0.7941	0.9000	**0.8438**	0.7059	0.8276	0.7619	0.7647	0.8966	0.8254	0.7647	0.8966	0.8254	0.7941	0.9000	**0.8438**
	avg	0.7647	0.9286	**0.8387**	0.7647	0.8230	0.7914	0.7588	0.8458	0.7990	0.7647	0.8966	0.8254	0.7647	0.8966	0.8254	0.7706	0.8973	0.8291
	std	0.0000	0.0000	0.0000	0.0360	0.0800	0.0475	0.0383	0.0537	0.0342	0.0000	0.0000	0.0000	0.0000	0.0000	0.0000	0.0131	0.0015	0.0082
CPY013	100	0.7742	0.7273	0.7500	0.8387	0.6842	**0.7536**	0.8065	0.6757	0.7353	0.7419	0.6765	0.7077	0.7742	0.6857	0.7273	0.7742	0.6857	0.7273
	500	0.7742	0.7273	**0.7500**	0.8387	0.6341	0.7222	0.7419	0.6970	0.7188	0.7419	0.6765	0.7077	0.7419	0.6765	0.7077	0.7419	0.6765	0.7077
	1000	0.7742	0.7273	0.7500	0.8710	0.6750	**0.7606**	0.8065	0.6410	0.7143	0.7419	0.6765	0.7077	0.7742	0.6857	0.7273	0.7742	0.6857	0.7273
	5000	0.7742	0.7273	**0.7500**	0.7419	0.6765	0.7077	0.7419	0.6765	0.7077	0.7419	0.6765	0.7077	0.7419	0.6765	0.7077	0.7742	0.6857	0.7273
	10,000	0.8387	0.7647	**0.8000**	0.7742	0.7742	0.7742	0.7742	0.7742	0.7742	0.8387	0.7647	**0.8000**	0.7742	0.7273	0.7500	0.7419	0.6765	0.7077
	avg	0.7871	0.7348	**0.7600**	0.8129	0.6888	0.7437	0.7742	0.6929	0.7301	0.7613	0.6941	0.7262	0.7613	0.6903	0.7240	0.7613	0.6820	0.7195
	std	0.0288	0.0167	0.0224	0.0530	0.0516	0.0277	0.0323	0.0497	0.0267	0.0433	0.0394	0.0413	0.0177	0.0212	0.0175	0.0177	0.0050	0.0107
CPY014	100	0.7500	1.0000	**0.8571**	0.7500	1.0000	**0.8571**	0.7500	1.0000	**0.8571**	0.7500	1.0000	**0.8571**	0.7500	1.0000	**0.8571**	0.7500	1.0000	**0.8571**
	500	0.7500	1.0000	**0.8571**	0.7500	1.0000	**0.8571**	0.7500	1.0000	**0.8571**	0.7500	1.0000	**0.8571**	0.7500	1.0000	**0.8571**	0.7500	1.0000	**0.8571**
	1000	0.7500	1.0000	**0.8571**	0.7500	1.0000	**0.8571**	0.7500	1.0000	**0.8571**	0.7500	1.0000	**0.8571**	0.7500	1.0000	**0.8571**	0.7500	1.0000	**0.8571**
	5000	0.7500	1.0000	**0.8571**	0.7500	1.0000	**0.8571**	0.7500	1.0000	**0.8571**	0.7500	1.0000	**0.8571**	0.7500	1.0000	**0.8571**	0.7500	1.0000	**0.8571**
	10,000	0.7500	1.0000	**0.8571**	0.7500	1.0000	**0.8571**	0.7500	0.6000	0.6667	0.7500	1.0000	**0.8571**	0.7500	1.0000	**0.8571**	0.7500	1.0000	**0.8571**
	avg	0.7500	1.0000	**0.8571**	0.7500	1.0000	**0.8571**	0.7500	0.9200	0.8190	0.7500	1.0000	**0.8571**	0.7500	1.0000	**0.8571**	0.7500	1.0000	**0.8571**
	std	0.0000	0.0000	0.0000	0.0000	0.0000	0.0000	0.0000	0.1789	0.0851	0.0000	0.0000	0.0000	0.0000	0.0000	0.0000	0.0000	0.0000	0.0000
CPY015	100	0.3235	0.5238	**0.4000**	0.2647	0.5000	0.3462	0.2647	0.5000	0.3462	0.3235	0.5238	**0.4000**	0.2647	0.4737	0.3396	0.2353	0.4444	0.3077
	500	0.3235	0.5000	**0.3929**	0.2353	0.5000	0.3200	0.2353	0.5000	0.3200	0.3235	0.5000	**0.3929**	0.2941	0.5263	0.3774	0.2647	0.5000	0.3462
	1000	0.3824	0.5000	0.4333	0.4118	0.4828	**0.4444**	0.2647	0.4286	0.3273	0.3824	0.5000	0.4333	0.3824	0.4643	0.4194	0.3824	0.4333	0.4062
	5000	0.3235	0.5238	**0.4000**	0.2647	0.5000	0.3462	0.2647	0.5000	0.3462	0.3235	0.5238	**0.4000**	0.2941	0.5000	0.3704	0.2941	0.5000	0.3704
	10,000	0.3824	0.5200	0.4407	0.3824	0.4483	0.4127	0.3529	0.4800	0.4068	0.3824	0.5200	0.4407	0.3824	0.5200	0.4407	0.4412	0.4839	**0.4615**
	avg	0.3471	0.5135	**0.4134**	0.3118	0.4862	0.3739	0.2765	0.4817	0.3493	0.3471	0.5135	**0.4134**	0.3235	0.4969	0.3895	0.3235	0.4723	0.3784
	std	0.0323	0.0124	0.0219	0.0795	0.0225	0.0522	0.0446	0.0309	0.0342	0.0323	0.0124	0.0219	0.0551	0.0274	0.0404	0.0858	0.0315	0.0587
CPY016	100	0.5421	0.8529	**0.6629**	0.5981	0.5470	0.5714	0.6075	0.5462	0.5752	0.5421	0.8406	0.6591	0.5421	0.8286	0.6554	0.5421	0.8286	0.6554
	500	0.5421	0.8529	**0.6629**	0.5981	0.6154	0.6066	0.5888	0.6632	0.6238	0.5421	0.8406	0.6591	0.5421	0.8406	0.6591	0.5421	0.8286	0.6554
	1000	0.5421	0.8529	**0.6629**	0.5794	0.5794	0.5794	0.6075	0.5372	0.5702	0.5421	0.8406	0.6591	0.5421	0.8286	0.6554	0.5421	0.8286	0.6554
	5000	0.5421	0.8529	0.6629	0.5607	0.7059	0.6250	0.5607	0.6122	0.5854	0.5421	0.8406	0.6591	0.5421	0.8286	0.6554	0.5607	0.8333	**0.6704**
	10,000	0.5421	0.8529	**0.6629**	0.6075	0.5752	0.5909	0.5981	0.4638	0.5224	0.5421	0.8406	0.6591	0.5421	0.8286	0.6554	0.5421	0.8529	**0.6629**
	avg	0.5421	0.8529	**0.6629**	0.5888	0.6046	0.5947	0.5925	0.5645	0.5754	0.5421	0.8406	0.6591	0.5421	0.8310	0.6561	0.5458	0.8368	0.6606
	std	0.0000	0.0000	0.0000	0.0187	0.0616	0.0215	0.0194	0.0762	0.0363	0.0000	0.0000	0.0000	0.0000	0.0054	0.0017	0.0083	0.0103	0.0063
CPY017	100	1.0000	1.0000	**1.0000**	1.0000	1.0000	**1.0000**	1.0000	0.7500	0.8571	1.0000	1.0000	**1.0000**	1.0000	1.0000	**1.0000**	1.0000	1.0000	**1.0000**
	500	1.0000	1.0000	**1.0000**	1.0000	1.0000	**1.0000**	1.0000	1.0000	**1.0000**	1.0000	1.0000	**1.0000**	1.0000	1.0000	**1.0000**	1.0000	1.0000	**1.0000**
	1000	1.0000	1.0000	**1.0000**	1.0000	0.7500	0.8571	1.0000	0.7500	0.8571	1.0000	1.0000	**1.0000**	1.0000	1.0000	**1.0000**	1.0000	1.0000	**1.0000**
	5000	1.0000	1.0000	**1.0000**	1.0000	0.7500	0.8571	1.0000	0.7500	0.8571	1.0000	1.0000	**1.0000**	1.0000	1.0000	**1.0000**	1.0000	1.0000	**1.0000**
	10,000	1.0000	1.0000	**1.0000**	1.0000	1.0000	**1.0000**	0.6667	1.0000	0.8000	1.0000	1.0000	**1.0000**	1.0000	1.0000	**1.0000**	1.0000	1.0000	**1.0000**
	avg	1.0000	1.0000	**1.0000**	1.0000	0.9000	0.9428	0.9333	0.8500	0.8743	1.0000	1.0000	**1.0000**	1.0000	1.0000	**1.0000**	1.0000	1.0000	**1.0000**
	std	0.0000	0.0000	0.0000	0.0000	0.1369	0.0783	0.1491	0.1369	0.0745	0.0000	0.0000	0.0000	0.0000	0.0000	0.0000	0.0000	0.0000	0.0000
YOM009	100	0.6308	0.3254	0.4293	0.5846	0.3115	0.4064	0.3538	0.4694	0.4035	0.6154	0.3226	0.4233	0.6462	0.3281	**0.4352**	0.5846	0.3065	0.4021
	500	0.4769	0.4769	0.4769	0.6154	0.3960	0.4819	0.4769	0.3875	0.4035	0.4769	0.4769	0.4769	0.6000	0.4333	**0.5032**	0.6154	0.4082	0.4908
	1000	0.5538	0.3600	0.4364	0.5538	0.3186	0.4045	0.4769	0.3778	**0.4387**	0.5231	0.3778	0.4387	0.4923	0.3299	0.3951	0.5538	0.3396	0.4211
	5000	0.5538	0.4091	0.4706	0.4923	0.4384	0.4638	0.4462	0.4531	0.4496	0.6308	0.3981	**0.4881**	0.5538	0.3789	0.4500	0.5385	0.4430	0.4861
	10,000	0.5385	0.2991	0.3846	0.4923	0.3048	0.3765	0.4154	0.3293	0.3673	0.4154	0.3971	**0.4060**	0.4615	0.3158	0.3750	0.5538	0.3186	0.4045
	avg	0.5508	0.3741	0.4396	0.5477	0.3539	0.4266	0.4492	0.3939	0.4124	0.5323	0.3945	**0.4466**	0.5508	0.3572	0.4317	0.5692	0.3632	0.4409
	std	0.0548	0.0707	0.0371	0.0550	0.0599	0.0443	0.0741	0.0661	0.0305	0.0914	0.0554	0.0350	0.0757	0.0489	0.0500	0.0308	0.0595	0.0440

Table 13. The mean F1-scores and standard deviation of all models when testing with the dataset from different stations (the best F1-score of each station is shown in bold).

Models	CPY011	CPY012	CPY013	CPY014	CPY015	CPY016	CPY017	YOM009
DRL	0.7433 (±0.08)	0.6550 (±0.08)	0.6014 (±0.10)	0.4823 (±0.20)	0.3477 (±0.06)	0.3770 (±0.03)	0.6801 (±0.21)	0.3581 (±0.06)
DRL_{F1}	0.7170 (±0.07)	0.7146 (±0.06)	0.6963 (±0.07)	0.6733 (±0.08)	**0.4134 (±0.02)**	0.5438 (±0.05)	0.6381 (±0.16)	0.4152 (±0.04)
DRL_{Rwd}	0.7433 (±0.08)	0.6468 (±0.10)	0.6212 (±0.09)	0.5537 (±0.26)	0.3477 (±0.06)	0.3499 (±0.06)	0.6762 (±0.21)	0.3891 (±0.04)
DRL_{Acc}	0.7170 (±0.07)	0.7234 (±0.04)	0.6920 (±0.07)	0.6733 (±0.08)	**0.4134 (±0.02)**	0.5714 (±0.02)	0.7219 (±0.14)	0.4248 (±0.04)
DRL_{Valid}	0.6020 (±0.17)	0.7045 (±0.10)	0.5639 (±0.21)	0.6596 (±0.14)	0.3211 (±0.10)	0.4867 (±0.07)	0.6488 (±0.15)	0.4378 (±0.03)
MLP	0.8505 (±0.06)	0.7822 (±0.03)	0.6998 (±0.02)	**0.8571 (±0.00)**	0.2220 (±0.10)	0.5651 (±0.14)	0.9778 (±0.07)	0.2358 (±0.05)
LSTM	0.8167 (±0.04)	0.7753 (±0.02)	0.7265 (±0.02)	**0.8571 (±0.00)**	0.3276 (±0.09)	0.6252 (±0.06)	0.9857 (±0.05)	0.2596 (±0.06)
$EDRL_3$	0.7700 (±0.06)	0.7367 (±0.02)	0.7033 (±0.07)	0.6948 (±0.11)	**0.4134 (±0.02)**	0.5544 (±0.03)	0.7809 (±0.10)	0.4396 (±0.04)
$EDRL_5$	0.7455 (±0.09)	0.7558 (±0.03)	0.6819 (±0.03)	0.7295 (±0.12)	0.3605 (±0.07)	0.5123 (±0.03)	0.7131 (±0.24)	0.4301 (±0.05)
E3	**0.9231 (±0.00)**	**0.8387 (±0.00)**	**0.7600 (±0.02)**	**0.8571 (±0.00)**	**0.4134 (±0.02)**	**0.6629 (±0.00)**	**1.0000 (±0.00)**	1.0000 (±0.04)
E5	0.8189 (±0.08)	0.7914 (±0.05)	0.7437 (±0.03)	**0.8571 (±0.00)**	0.3739 (±0.05)	0.5947 (±0.02)	0.9428 (±0.08)	0.9428 (±0.04)
E7	0.8360 (±0.05)	0.7990 (±0.03)	0.7301 (±0.03)	0.8190 (±0.09)	0.3493 (±0.03)	0.5754 (±0.04)	0.8743 (±0.07)	0.8743 (±0.03)
$WEDRL_3$	0.7795 (±0.05)	0.7432 (±0.03)	0.7100 (±0.07)	0.6962 (±0.16)	**0.4134 (±0.02)**	0.5797 (±0.05)	0.7976 (±0.09)	0.4466 (±0.03)
$WEDRL_5$	0.7540 (±0.05)	0.7347 (±0.02)	0.7194 (±0.05)	0.6714 (±0.16)	0.3954 (±0.04)	0.5525 (±0.04)	0.7619 (±0.10)	0.4323 (±0.05)
WE3	0.8571 (±0.00)	0.8254 (±0.00)	0.7262 (±0.04)	**0.8571 (±0.00)**	**0.4134 (±0.02)**	0.6591 (±0.00)	**1.0000 (±0.00)**	1.0000 (±0.03)
WE5	0.8571 (±0.00)	0.8254 (±0.00)	0.7240 (±0.02)	**0.8571 (±0.00)**	0.3895 (±0.04)	0.6561 (±0.00)	**1.0000 (±0.00)**	1.0000 (±0.05)
WE7	0.8571 (±0.00)	0.8291 (±0.01)	0.7195 (±0.01)	**0.8571 (±0.00)**	0.3784 (±0.06)	0.6606 (±0.01)	**1.0000 (±0.00)**	1.0000 (±0.04)

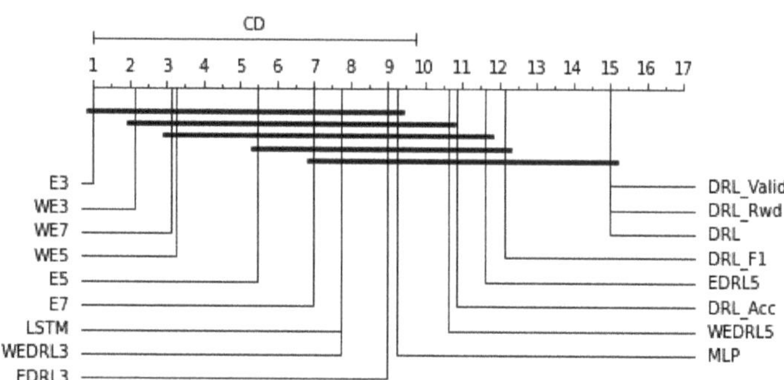

Figure 10. A critical difference diagram.

We then tested the generalisation ability of ensemble models with the data collected from different stations. The F1-score of each station is depicted in Table 14. Using ensemble E3 with CPY011 data, to identify anomalies from other stations, we can see that it works well with F1-scores of more than 0.5800, but it performed poorly at detecting anomalies on CPY015 and YOM009 with F1-scores of 0.3444 and 0.1017, respectively. E3 on CPY012 performed well when detecting anomalies on CPY014 with a 0.8635 F1-score. Similarly, E3 on CPY013 provided a higher F1-score on their own dataset when detecting anomalies on CPY012 and CPY014 with an F1-score of 0.8060 and 0.8571, respectively. The best ensemble on CPY014 generated excellent performance when identifying anomalies on CPY017 data. In contrast, E3 on CPY015 performed poorly on YOM009 with an F1-score of only 0.0437. While considered E3 on CPY016, although it provided good performance with an F1-score higher than 0.6 on CPY012, CPY013, and CPY014, it performed poorly on CPY011, CPY015, CPY017, and YOM009 with an F1-score lower than 0.45. E3 on CPY017 provided good results with an F1-score of more than 0.69, except on CPY015, CPY016, and YOM009 with an F1-score lower than 0.56. Meanwhile, E3 on YOM009 generated an F1-score on its own of only 0.43, but it performed excellently when detecting anomalies on CPY017 and other datasets with an F1-score higher than 0.65, except on CPY015 with an F1-score of 0.2414.

Table 14. The F1-scores of the E3 models when testing with the dataset from the same station (shown in the bracket) and different stations.

Tested Dataset	Trained Dataset							
	CPY011	CPY012	CPY013	CPY014	CPY015	CPY016	CPY017	YOM009
CPY011	(0.9231)	0.4000	0.3222	0.7778	0.4028	0.1373	0.7059	0.6087
CPY012	0.7183	(0.8387)	0.8060	0.7500	0.1508	0.6923	0.7273	0.7857
CPY013	0.6920	0.7350	(0.7600)	0.7000	0.2117	0.7123	0.6909	0.7719
CPY014	0.6895	0.8635	0.8571	(0.8571)	0.5714	0.6000	0.8571	0.8571
CPY015	0.3444	0.2192	0.2093	0.2593	(0.4132)	0.1420	0.3077	0.2414
CPY016	0.5840	0.6398	0.6603	0.6322	0.3495	(0.6629)	0.5665	0.6580
CPY017	0.7600	0.6667	0.6000	1.0000	0.3651	0.2857	(1.0000)	1.0000
YOM009	0.1017	0.2948	0.3221	0.2908	0.0437	0.4500	0.3230	(0.4396)
avg	0.6016	0.5822	0.5671	0.6584	0.3135	0.4603	0.6473	**0.6703**
std	0.2603	0.2468	0.2496	0.2607	0.1682	0.2432	0.2412	0.2415

6. Discussion

We can observe that when the number of training epochs increases, the performance of each model grows or decreases in each epoch, then drops and bounces back. This might indicate that our model is still learning or is learning too much—that is, it is difficult to decide when it is time to stop training.

Even though DRL can do better than other models, it is time-consuming—at least 50 times slower than MLP models on average—because we have to train it until it performs well enough and we cannot predict how long that will take. The size of the windows must also be taken into account. A larger window size takes more time than a smaller window size. The window size has an effect on the comparison of data in windows to identify the anomaly. Additionally, we may add additional neural networks to improve the accuracy of our technique, but training will take longer.

DRL does better than other models when it is trained on datasets with a low number of outliers. This proves the ability to detect unknown anomalies. However, its performance is insufficient, which may be due to an imbalance in our dataset. As a result, models may lack sufficient information to explore and leverage knowledge for adaptive detection of unknown abnormalities.

Moreover, the neural structure that works well with one station may not function well with another. Hence, the problems of this topic include determining the suitable neural structure for each station. Furthermore, the primary parameter that requires further attention is the reward function, since a suitable reward will impact the model's learning process.

In the case of ensemble models, when all of the individual models in an ensemble perform similarly, majority voting is the best method for determining the final decision. However, when the accuracies of individual models are different, the weighted voting is the best way to utilise the strengths of the good models in making a decision. Furthermore, the ensemble model can also reduce the false alarm rate, as seen by an increased precision score. It should be noted that, although single models performed well on certain stations, they did poorly on others, such as the LSTM model. As a result, we cannot rely on a single model since we do not know if it is the best or not. The ensemble models, on the other hand, are more reliable, even though they may not produce the best accuracy for every station. On the whole, nevertheless, most ensembles, such as $WEDRL_3$ performed consistently very well and their accuracies are always ranked highly at every station, whilst the individual models: DRL, MLP and LSTM, are not consistent through out all the stations.

7. Conclusions

In this research, we firstly investigated how deep reinforcement learning (DRL) can be applied to detect anomalies in water level data and then devised two strategies to construct more effective and reliable ensembles. For DRL, we defined a reward function as it plays a key role in determining the success of an RL. We developed ensemble models with five deep reinforcement learning models, generated by the same DRL algorithm but with different criteria of performance measurement. We tested our ensemble approach on telemetry water level data from eight different stations. We compared our approach to two different neural network models. Moreover, we demonstrate the ability to detect unknown anomalies by using the trained model to detect anomalies from other stations' data.

The results indicate that DRL_{Acc} models are the best individual DRL models, but they performed slightly poor than LSTM. When tested on different stations, LSTM still does better than others, but its accuracy is not satisfactory. When compared to an ensemble approach, LSTM was more accurate in some stations than other ensembles with DRL models, but less accurate in some others. On the whole, the statistical results from the CD diagram showed that our ensemble approach with only 3 members of DRL models, $WEDRL_3$, was superior. Furthermore, all ensemble models that were combined by selecting models from 5 DRL models, MLP, and LSTM outperformed both the best individual model, LSTM, and the best ensemble using DRL models, $WEDRL_3$. This is supported by the highest F1-score and rankings with the CD diagram. It is clear that ensemble methods not only increased the accuracy of a single model but also provided a higher reliability of performance.

In conclusion, DRL is applicable for detecting anomalies in telemetry water level data with added benefit of detecting unknown anomalies. Our ensemble construction methods can be used to build ensemble models from selected single DRL models in order to increase the accuracy and reliability. In general, the ensembles are consistent in producing more accurate classification, although they may not always achieve the best results. Moreover, they are superior in reducing the number of false alarms in identifying abnormalities in water level data, which is very important in real application. The next stage in our study will be to develop more effective and efficient techniques for correcting the identified anomalies in the data.

Author Contributions: Conceptualization, T.K. and W.W.; methodology, T.K. and W.W.; formal analysis, T.K. and W.W.; investigation, T.K. and W.W.; resources, T.K.; writing and revision: T.K. and final revision: W.W.; project administration, T.K. All authors have read and agreed to the published version of the manuscript.

Funding: This research received no external funding.

Institutional Review Board Statement: Not applicable.

Informed Consent Statement: Not applicable.

Data Availability Statement: Data available on request due to restriction, e.g., privacy. The data presented in this study are available on request from the Hydro-Informatics Institute (HII).

Acknowledgments: The authors would like to thank the Hydro-Informatics Institute of Ministry of Higher Education, Science, Research and Innovation, Thailand, for providing the scholarship for Thakolpat Khampuengson to do his Ph.D. at the university of East Anglia.

Conflicts of Interest: The authors declare no conflict of interest.

Abbreviations

The following abbreviations are used in this manuscript:

CNN	Convolutional Neural Network
BDRL	Best Deep Reinforcement Learning model
DQN	Deep Q-Learning Network
DRL	Deep Reinforcement Learning
HII	Hydro Informatics Institute
LSTM	Long-Short Term Memory
MLP	Multilayer Perceptron
NN	Neural Network
RL	Reinforcement Learning

References

1. World Bank. *Thai Flood 2011: Rapid Assessment for Resilient Recovery and Reconstruction Planning*; World Bank: Washington, DC, USA, 2012.
2. UNDRR. *Disaster Risk Reduction in Thailand: Status Report 2020*; United Nations Office for Disaster Risk Reduction (UNDRR): Geneva, Switzerland, 2020.
3. Khampuengson, T.; Bagnall, A.; Wang, W. Developing Ensemble Methods for Detecting Anomalies in Water Level Data. In Proceedings of the International Conference on Innovative Techniques and Applications of Artificial Intelligence, Cambridge, UK, 15–17 December 2020; Springer: Berlin/Heidelberg, Germany, 2020; pp. 145–151.
4. Wang, W. Some Fundamental Issues in Ensemble Methods. In Proceedings of the IEEE World Congress on Computational Intelligence, Hong Kong, China, 1–8 June 2008; pp. 2243–2250. [CrossRef]
5. Chauhan, S.; Vig, L. Anomaly detection in ECG time signals via deep long short-term memory networks. In Proceedings of the 2015 IEEE International Conference on Data Science and Advanced Analytics (DSAA), Porto, Portugal, 6–9 October 2021; IEEE: Piscataway, NJ, USA, 2015; pp. 1–7.
6. Kim, T.Y.; Cho, S.B. Web traffic anomaly detection using C-LSTM neural networks. *Expert Syst. Appl.* **2018**, *106*, 66–76. [CrossRef]
7. Munir, M.; Siddiqui, S.A.; Dengel, A.; Ahmed, S. DeepAnT: A deep learning approach for unsupervised anomaly detection in time series. *IEEE Access* **2018**, *7*, 1991–2005. [CrossRef]
8. Pang, G.; van den Hengel, A.; Shen, C.; Cao, L. Deep reinforcement learning for unknown anomaly detection. *arXiv* **2020**, arXiv:2009.06847.
9. Mnih, V.; Kavukcuoglu, K.; Silver, D.; Graves, A.; Antonoglou, I.; Wierstra, D.; Riedmiller, M. Playing atari with deep reinforcement learning. *arXiv* **2013**, arXiv:1312.5602.

10. Silver, D.; Schrittwieser, J.; Simonyan, K.; Antonoglou, I.; Huang, A.; Guez, A.; Hubert, T.; Baker, L.; Lai, M.; Bolton, A.; et al. Mastering the game of go without human knowledge. *Nature* **2017**, *550*, 354–359. [CrossRef]

11. Kormushev, P.; Calinon, S.; Caldwell, D.G. Reinforcement learning in robotics: Applications and real-world challenges. *Robotics* **2013**, *2*, 122–148. [CrossRef]

12. Polydoros, A.S.; Nalpantidis, L. Survey of model-based reinforcement learning: Applications on robotics. *J. Intell. Robot. Syst.* **2017**, *86*, 153–173. [CrossRef]

13. Sharma, A.R.; Kaushik, P. Literature survey of statistical, deep and reinforcement learning in natural language processing. In Proceedings of the 2017 International Conference on Computing, Communication and Automation (ICCCA), Greater Noida, India, 5–6 May 2017; IEEE: Piscataway, NJ, USA, 2017; pp. 350–354.

14. Luketina, J.; Nardelli, N.; Farquhar, G.; Foerster, J.; Andreas, J.; Grefenstette, E.; Whiteson, S.; Rocktäschel, T. A survey of reinforcement learning informed by natural language. *arXiv* **2019**, arXiv:1906.03926.

15. Le, N.; Rathour, V.S.; Yamazaki, K.; Luu, K.; Savvides, M. Deep reinforcement learning in computer vision: A comprehensive survey. *Artif. Intell. Rev.* **2021**, *55*, 2733–2819. [CrossRef]

16. Huang, C.; Wu, Y.; Zuo, Y.; Pei, K.; Min, G. Towards experienced anomaly detector through reinforcement learning. In Proceedings of the Thirty-Second AAAI Conference on Artificial Intelligence, New Orleans, LA, USA, 2–7 February 2018.

17. Hsu, Y.F.; Matsuoka, M. A deep reinforcement learning approach for anomaly network intrusion detection system. In Proceedings of the 2020 IEEE 9th International Conference on Cloud Networking (CloudNet), Piscataway, NJ, USA, 9–11 November 2020; IEEE: Piscataway, NJ, USA, 2020; pp. 1–6.

18. Lin, E.; Chen, Q.; Qi, X. Deep reinforcement learning for imbalanced classification. *Appl. Intell.* **2020**, *50*, 2488–2502. [CrossRef]

19. Pulido, M.; Melin, P.; Castillo, O. Particle swarm optimization of ensemble neural networks with fuzzy aggregation for time series prediction of the Mexican Stock Exchange. *Inf. Sci.* **2014**, *280*, 188–204. [CrossRef]

20. Ikram, S.T.; Cherukuri, A.K.; Poorva, B.; Ushasree, P.S.; Zhang, Y.; Liu, X.; Li, G. Anomaly detection using XGBoost ensemble of deep neural network models. *Cybern. Inf. Technol.* **2021**, *21*, 175–188. [CrossRef]

21. Yang, H.; Liu, X.Y.; Zhong, S.; Walid, A. Deep reinforcement learning for automated stock trading: An ensemble strategy. In Proceedings of the First ACM International Conference on AI in Finance, New York, NY, USA, 15–16 October 2020; pp. 1–8.

22. Liu, H.; Yu, C.; Wu, H.; Duan, Z.; Yan, G. A new hybrid ensemble deep reinforcement learning model for wind speed short term forecasting. *Energy* **2020**, *202*, 117794. [CrossRef]

23. Rousseeuw, P.J.; Hubert, M. Robust statistics for outlier detection. *Wiley Interdiscip. Rev. Data Min. Knowl. Discov.* **2011**, *1*, 73–79. [CrossRef]

24. Zimek, A.; Filzmoser, P. There and back again: Outlier detection between statistical reasoning and data mining algorithms. *Wiley Interdiscip. Rev. Data Min. Knowl. Discov.* **2018**, *8*, e1280. [CrossRef]

25. Kumar, A.; Srivastava, A.; Bansal, N.; Goel, A. Real time data anomaly detection in operating engines by statistical smoothing technique. In Proceedings of the 2012 25th IEEE Canadian Conference on Electrical & Computer Engineering (CCECE), Montreal, QC, Canada, 29 April–2 May 2012; IEEE: Piscataway, NJ, USA, 2012; pp. 1–5.

26. Lin, J.; Sheng, G.; Yan, Y.; Zhang, Q.; Jiang, X. Online Monitoring Data Cleaning of Transformer Considering Time Series Correlation. In Proceedings of the 2018 IEEE/PES Transmission and Distribution Conference and Exposition (T&D), Denver, CO, USA, 16–19 April 2018; IEEE: Piscataway, NJ, USA, 2018; pp. 1–9.

27. Aminikhanghahi, S.; Cook, D.J. A survey of methods for time series change point detection. *Knowl. Inf. Syst.* **2017**, *51*, 339–367. [CrossRef]

28. Truong, C.; Oudre, L.; Vayatis, N. Selective review of offline change point detection methods. *Signal Process.* **2020**, *167*, 107299. [CrossRef]

29. Apostol, E.S.; Truică, C.O.; Pop, F.; Esposito, C. Change point enhanced anomaly detection for IoT time series data. *Water* **2021**, *13*, 1633. [CrossRef]

30. Dao, C.; Liu, X.; Sim, A.; Tull, C.; Wu, K. Modeling data transfers: Change point and anomaly detection. In Proceedings of the 2018 IEEE 38th International Conference on Distributed Computing Systems (ICDCS), Vienna, Austria, 2–6 July 2018; IEEE: Piscataway, NJ, USA, 2018; pp. 1589–1594.

31. Siris, V.A.; Papagalou, F. Application of anomaly detection algorithms for detecting SYN flooding attacks. In Proceedings of the IEEE Global Telecommunications Conference, 2004. GLOBECOM'04, Dallas, TX, USA, 29 November–3 December 2004; IEEE: Piscataway, NJ, USA, 2004; Volume 4, pp. 2050–2054.

32. Yu, Y.; Zhu, Y.; Li, S.; Wan, D. Time series outlier detection based on sliding window prediction. *Math. Probl. Eng.* **2014**, *2014*, 879736. [CrossRef]

33. van de Wiel, L.; van Es, D.M.; Feelders, A.J. Real-Time Outlier Detection in Time Series Data of Water Sensors. In *Advanced Analytics and Learning on Temporal Data*; Lemaire, V., Malinowski, S., Bagnall, A., Guyet, T., Tavenard, R., Ifrim, G., Eds.; Springer International Publishing: Cham, Switzerland, 2020; pp. 155–170.

34. Yang, J.H.; Cheng, C.H.; Chan, C.P. A time-series water level forecasting model based on imputation and variable selection method. *Comput. Intell. Neurosci.* **2017**, *2017*, 8734214. [CrossRef]

35. Park, K.; Jung, Y.; Seong, Y.; Lee, S. Development of Deep Learning Models to Improve the Accuracy of Water Levels Time Series Prediction through Multivariate Hydrological Data. *Water* **2022**, *14*, 469. [CrossRef]

36. Vu, M.; Jardani, A.; Massei, N.; Fournier, M. Reconstruction of missing groundwater level data by using Long Short-Term Memory (LSTM) deep neural network. *J. Hydrol.* **2021**, *597*, 125776. [CrossRef]
37. Chang, L.C.; Chang, F.J.; Yang, S.N.; Kao, I.F.; Ku, Y.Y.; Kuo, C.L.; Amin, I.M.Z.b.M. Building an intelligent hydroinformatics integration platform for regional flood inundation warning systems. *Water* **2019**, *11*, 9. [CrossRef]
38. Liu, Y.; Hou, G.; Huang, F.; Qin, H.; Wang, B.; Yi, L. Directed graph deep neural network for multi-step daily streamflow forecasting. *J. Hydrol.* **2022**, *607*, 127515. [CrossRef]
39. Zhou, Y.; Guo, S.; Chang, F.J. Explore an evolutionary recurrent ANFIS for modelling multi-step-ahead flood forecasts. *J. Hydrol.* **2019**, *570*, 343–355. [CrossRef]
40. Chang, L.C.; Chang, F.J.; Yang, S.N.; Tsai, F.H.; Chang, T.H.; Herricks, E.E. Self-organizing maps of typhoon tracks allow for flood forecasts up to two days in advance. *Nat. Commun.* **2020**, *11*, 1–13. [CrossRef]
41. Chang, L.C.; Liou, J.Y.; Chang, F.J. Spatial-temporal flood inundation nowcasts by fusing machine learning methods and principal component analysis. *J. Hydrol.* **2022**, *612*, 128086. [CrossRef]
42. Gao, S.; Zhang, Y.; Jia, K.; Lu, J.; Zhang, Y. Single sample face recognition via learning deep supervised autoencoders. *IEEE Trans. Inf. Forensics Secur.* **2015**, *10*, 2108–2118. [CrossRef]
43. Xu, C.; Liu, Q.; Ye, M. Age invariant face recognition and retrieval by coupled auto-encoder networks. *Neurocomputing* **2017**, *222*, 62–71. [CrossRef]
44. Pereira, J.; Silveira, M. Unsupervised anomaly detection in energy time series data using variational recurrent autoencoders with attention. In Proceedings of the 2018 17th IEEE International Conference on Machine Learning and Applications (ICMLA), Orlando, FL, USA, 17–20 December 2018; IEEE: Piscataway, NJ, USA, 2018; pp. 1275–1282.
45. Zhou, C.; Paffenroth, R.C. Anomaly detection with robust deep autoencoders. In Proceedings of the 23rd ACM SIGKDD International Conference on Knowledge Discovery and Data Mining, Halifax, NS, Canada, 13–17 August 2017; pp. 665–674.
46. Zong, B.; Song, Q.; Min, M.R.; Cheng, W.; Lumezanu, C.; Cho, D.; Chen, H. Deep autoencoding gaussian mixture model for unsupervised anomaly detection. In Proceedings of the International Conference on Learning Representations, Vancouver, BC, Canada, 30 April–3 May 3, 2018.
47. Gong, D.; Liu, L.; Le, V.; Saha, B.; Mansour, M.R.; Venkatesh, S.; Hengel, A.v.d. Memorizing normality to detect anomaly: Memory-augmented deep autoencoder for unsupervised anomaly detection. In Proceedings of the IEEE International Conference on Computer Vision, Seoul, Korea, 27 October–2 November 2019; pp. 1705–1714.
48. Jiang, G.; Xie, P.; He, H.; Yan, J. Wind turbine fault detection using a denoising autoencoder with temporal information. *IEEE/Asme Trans. Mechatronics* **2017**, *23*, 89–100. [CrossRef]
49. Maas, A.; Le, Q.V.; O'neil, T.M.; Vinyals, O.; Nguyen, P.; Ng, A.Y. Recurrent Neural Networks for Noise Reduction in Robust ASR. 2012. Available online: https://www.google.com.hk/url?sa=t&rct=j&q=&esrc=s&source=web&cd=&ved=2ahUKEwj249fInrf5AhUvEqYKHRHxBiQQFnoECAkQAQ&url=http%3A%2F%2Fai.stanford.edu%2F~amaas%2Fpapers%2Fdrnn_intrspch2012_final.pdf&usg=AOvVaw2_oWylziqsFnVhhUBT_o8v (accessed on 30 June 2022).
50. Chiang, H.T.; Hsieh, Y.Y.; Fu, S.W.; Hung, K.H.; Tsao, Y.; Chien, S.Y. Noise reduction in ECG signals using fully convolutional denoising autoencoders. *IEEE Access* **2019**, *7*, 60806–60813. [CrossRef]
51. Kieu, T.; Yang, B.; Guo, C.; Jensen, C.S. Outlier Detection for Time Series with Recurrent Autoencoder Ensembles. In Proceedings of the Twenty-Eighth International Joint Conference on Artificial Intelligence, Macao, China, 10–16 August 2019; pp. 2725–2732.
52. Chen, J.; Sathe, S.; Aggarwal, C.; Turaga, D. Outlier detection with autoencoder ensembles. In Proceedings of the 2017 SIAM International Conference on Data Mining, Houston, TX, USA, 27–29 April 2017; SIAM: Philadelphia, PA, USA, 2017; pp. 90–98.
53. Kao, I.F.; Liou, J.Y.; Lee, M.H.; Chang, F.J. Fusing stacked autoencoder and long short-term memory for regional multistep-ahead flood inundation forecasts. *J. Hydrol.* **2021**, *598*, 126371. [CrossRef]
54. Yu, L.; Tan, S.K.; Chua, L.H. Online ensemble modeling for real time water level forecasts. *Water Resour. Manag.* **2017**, *31*, 1105–1119. [CrossRef]
55. Iftikhar, N.; Baattrup-Andersen, T.; Nordbjerg, F.E.; Jeppesen, K. Outlier detection in sensor data using ensemble learning. *Procedia Comput. Sci.* **2020**, *176*, 1160–1169. [CrossRef]
56. Atienza, R. *Advanced Deep Learning with Keras: Apply Deep Learning Techniques, Autoencoders, GANs, Variational Autoencoders, Deep Reinforcement Learning, Policy Gradients, and More*; Packt Publishing Ltd.: Birmingham, UK, 2018; p. 272.
57. Mnih, V.; Kavukcuoglu, K.; Silver, D.; Rusu, A.A.; Veness, J.; Bellemare, M.G.; Graves, A.; Riedmiller, M.; Fidjeland, A.K.; Ostrovski, G.; et al. Human-level control through deep reinforcement learning. *Nature* **2015**, *518*, 529–533. [CrossRef] [PubMed]
58. Wan, Y.; Chen, J.; Xu, C.Y.; Xie, P.; Qi, W.; Li, D.; Zhang, S. Performance dependence of multi-model combination methods on hydrological model calibration strategy and ensemble size. *J. Hydrol.* **2021**, *603*, 127065. [CrossRef]
59. Casciaro, G.; Ferrari, F.; Mazzino, A. Comparing novel strategies of Ensemble Model Output Statistics (EMOS) for calibrating wind speed/power forecasts. *arXiv* **2021**, arXiv:2108.12174.
60. Marathe, A.; Walambe, R.; Kotecha, K. Evaluating the performance of ensemble methods and voting strategies for dense 2D pedestrian detection in the wild. In Proceedings of the IEEE/CVF International Conference on Computer Vision, Montreal, BC, Canada, 11–17 October 2021; pp. 3575–3584.
61. Demšar, J. Statistical comparisons of classifiers over multiple data sets. *J. Mach. Learn. Res.* **2006**, *7*, 1–30.

Review

Geospatial Artificial Intelligence (GeoAI) in the Integrated Hydrological and Fluvial Systems Modeling: Review of Current Applications and Trends

Carlos Gonzales-Inca [1,*], Mikel Calle [1,2], Danny Croghan [3], Ali Torabi Haghighi [3], Hannu Marttila [3], Jari Silander [4] and Petteri Alho [1]

1   Department of Geography and Geology, University of Turku, FI-20014 Turun Yliopisto, Finland; mcanav@utu.fi (M.C.); mipeal@utu.fi (P.A.)
2   Turku Collegium of Sciences, Medicine and Technology, University of Turku, FI-20014 Turun Yliopisto, Finland
3   Water, Energy and Environmental Engineering Research Unit, University of Oulu, FI-90014 Oulu, Finland; danny.croghan@oulu.fi (D.C.); ali.torabihaghighi@oulu.fi (A.T.H.); hannu.marttila@oulu.fi (H.M.)
4   Finnish Environmental Institute (SYKE), FI-00790 Helsinki, Finland; jari.silander@syke.fi
*   Correspondence: cagoin@utu.fi

Abstract: This paper reviews the current GeoAI and machine learning applications in hydrological and hydraulic modeling, hydrological optimization problems, water quality modeling, and fluvial geomorphic and morphodynamic mapping. GeoAI effectively harnesses the vast amount of spatial and non-spatial data collected with the new automatic technologies. The fast development of GeoAI provides multiple methods and techniques, although it also makes comparisons between different methods challenging. Overall, selecting a particular GeoAI method depends on the application's objective, data availability, and user expertise. GeoAI has shown advantages in non-linear modeling, computational efficiency, integration of multiple data sources, high accurate prediction capability, and the unraveling of new hydrological patterns and processes. A major drawback in most GeoAI models is the adequate model setting and low physical interpretability, explainability, and model generalization. The most recent research on hydrological GeoAI has focused on integrating the physical-based models' principles with the GeoAI methods and on the progress towards autonomous prediction and forecasting systems.

Keywords: GeoAI; artificial intelligence; machine learning; hydrological; hydraulic; fluvial; water quality; geomorphic; modeling

Citation: Gonzales-Inca, C.; Calle, M.; Croghan, D.; Torabi Haghighi, A.; Marttila, H.; Silander, J.; Alho, P. Geospatial Artificial Intelligence (GeoAI) in the Integrated Hydrological and Fluvial Systems Modeling: Review of Current Applications and Trends. *Water* **2022**, *14*, 2211. https://doi.org/10.3390/w14142211

Academic Editors: Fi-John Chang, Li-Chiu Chang and Jui-Fa Chen

Received: 20 April 2022
Accepted: 24 June 2022
Published: 13 July 2022

Publisher's Note: MDPI stays neutral with regard to jurisdictional claims in published maps and institutional affiliations.

1. Introduction

Hydrology and fluvial research are inexact fields of science, with a large extent of epistemic uncertainty and limited knowledge about the system's complexity, structure, and functioning [1]. Both disciplines have been hindered by the limited quality and availability of data [2]. Nowadays, access to temporally and spatially high-resolution hydrological and fluvial data has substantially increased, mainly due to advances in the use of automatic sensors in monitoring, environmental 3D scanners, and high-resolution remote sensing from different sources, producing 'big data'. The use of big hydrological data requires the development and the use of the applications of new geospatial tools for computational analytics and hydrological models. Technologies under the geospatial artificial intelligence (GeoAI) concept, such as machine learning (ML) and parallel computing, provide the means to utilize this spatial and non-spatial dataset effectively and also to enhance integrated hydrological and fluvial systems modeling [3].

Hydrological and fluvial modeling took a giant leap forward when the computer revolution started in the 1960s [4–6]. Since then, engineers and scientists have developed a wide range of hydrological models with different levels of complexity, including empirical and physical-based models [7–10]. Similarly, several 1D, 2D, and 3D numerical hydraulic models are available to model fluvial processes, such as flow characteristics, sediment transport, flood extent, and water depth [5,11]. Currently, despite a wide range of physical-based model availability, those models still have challenges in adequately and accurately modeling the complex and non-linear hydrological and hydraulic processes occurring in nature [5]. In addition, the application of these models has been restricted to small areas due to limited data availability, challenges in representing spatially varying parameters, and computational intensity [12]. Alternatively, GeoAI and ML data-driven models, such as artificial neural networks (ANNs) and long short-term memory (LSTM) deep learning show promising results for hydrological and hydraulic prediction and forecasting in natural environments and at a large geographical scale [13]. They can represent the non-linear processes and provide high-accuracy predictions [14,15]. ML application in hydrological predictions dates to the 1990s [16], but the development of the new GeoAI and ML algorithms, particularly the deep learning techniques, alongside new data collection technologies, has substantially increased in recent years [17,18]. Moreover, there are new studies on developing hybrid models (ML and physical-based models) [14,19,20] and physical process-guided ML methods [21–24]. Therefore, a review of the potential of the new GeoAI and ML methods for integrated hydrological and fluvial systems modeling is needed to guide scientists and practitioners to select the proper tools and to be aware of current and potential future methodologies.

Existing reviews of GeoAI and ML applications in hydrological modeling and fluvial studies have covered specific topics, such as the prediction of runoff, floods, and water quality [25–29]. Other reviews have focused on applying a particular GeoAI and ML method [30–33]. However, an overarching review of GeoAI and ML in hydrology is lacking. We aim to review the most recent GeoAI and ML method applications in hydrological, hydraulic, water quality, and fluvial process modeling.

This broad review on using GeoAI in hydrological and fluvial processes modeling provides a critical assessment of the technical development, the potential, and the limitations of the models and the current research trends and gaps from the standpoint of hydrology and fluvial system researchers. The review identified more than 1300 publications over the last two decades, published mainly in water resources, civil and environmental engineering, geosciences, and environmental sciences journals.

2. Review Methodology and Outline

The application of GeoAI in hydrological and fluvial systems research has substantially increased and diversified in recent years, comprising a wide range of topics. Therefore, a systematic review is challenging. In this review, we adopted a scoping review methodology [34,35]. The scoping review supports consistent and structured literature searches to capture relevant information and provides a comprehensive overview of the current applications and research. We explored four categories of GeoAI applications: (1) hydrological and hydraulic modeling; (2) hydrological model calibration and modeling optimization problems; (3) water quality modeling; and (4) fluvial geomorphology and morphodynamic mapping. We searched the literature from the Web of Science, including Scopus, Springer Link, Wiley Online Library, and MDPI. We used the Boolean operators (AND and OR), the proximity operators (NEAR and PRE), and nested logic (use of parentheses) to constrain the literature search to those works containing topic keywords combined with GeoAI methods in the article's title, abstract, or keywords. Because the GeoAI terminologies are diverse and commonly phrase-named, e.g., deep neural networks, deep convolutional neural networks, and deep learning, we used proximity operators to find records containing all the terms within a defined number (n) of word neighbors. In the case of a more common GeoAI phrase name, such as machine learning or artificial intelligence, we used double quotation marks to indicate that the words should not be searched separately. In addition, we used left and right truncation or shortening to account for the GeoAI or topic names that vary in prefixes and suffixes, using the asterisk symbol, e.g., *morpho* to select papers containing the terms geomorphologic, geomorphometry, morphodynamic, or hydromorphological. An example of the searching query construction in the Web of Science database for a GeoAI application in fluvial geomorphology and morphodynamic studies is: (TS = (Deep NEAR/3 neural NEAR/3 learn*) OR TS = (Artificial NEAR/3 Neural NEAR/3 Network*) OR TS = "Artificial Intelligence" OR TS = AI OR TS = geoAI OR TS = "Machine learning") AND TS = fluvial AND TS = *morpho*. We used different keywords combined with GeoAI terminologies according to the hydrological subfields of interest, e.g., hydrological, optimization, calibration, water quality, nutrient, pollutant, sediment, etc. In some cases, we also used the NOT operator to further refine the searching by excluding papers containing certain words from other topics or subfields. For example, we excluded water quality terminologies to search for papers focusing purely on hydrological modeling. Figure 1 shows the yearly publication statistics of the GeoAI and ML applications in the different hydrological subfields.

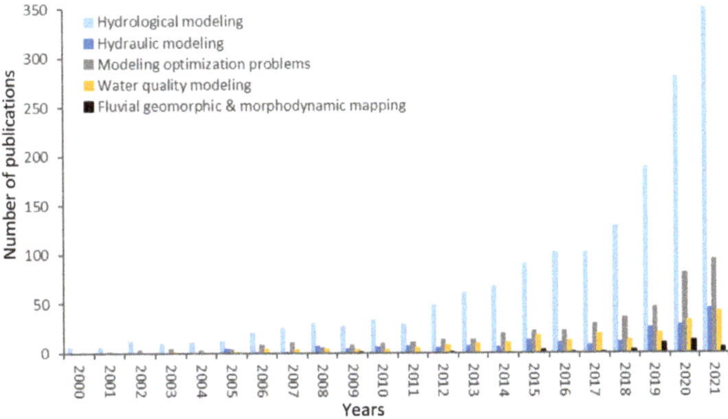

Figure 1. The yearly number of publications found in Web of Science (2000–2021) on GeoAI and machine learning applications in the different hydrological subfields.

We applied additional selection criteria in each database, including only peer-reviewed journal publications. After gathering the initial list, we briefly reviewed them to select only papers within our review's scope. The publication list was further filtered to ensure that the selected publications provided relevant information about GeoAI applications in hydrological and fluvial studies. We thoroughly studied the selected papers to extract information about the GeoAI model performance, the software used, and the advantages and limitations. Based on this information, we further discussed the comparison of GeoAI methods with the conventional and physical-based hydrological models (Section 5) and identified further opportunities and future trends in applying GeoAI methods in hydrological and fluvial studies (Section 6).

3. Brief Introduction to Geospatial Artificial Intelligence

GeoAI is an emerging discipline that combines innovations in spatial data science, AI, ML, and big geospatial data [36]. GeoAI is the study, development, and application of intelligent computer programs to automatic geospatial and non-spatial data processing; it models geospatial association and interaction, predicts spatial dynamics phenomena, provides spatial reasoning, and discovers spatio-temporal patterns and trends [37,38]. GeoAI includes the methods, techniques, and tools of AI and ML to carry out geospatial modeling, such as spatial hydrological prediction and fluvial landform classifications. The GeoAI and ML methods (henceforth, GeoAI) can be grouped into unsupervised learning (clustering and dimension reduction), supervised learning (regression and classification), and modeling optimization problems (see Table 1). A detailed theoretical and mathematical description of the GeoAI and ML methods is given by Hastie et al. [39], Goodfellow et al. [40], and Lee et al. [41].

Table 1. General classification of geospatial artificial intelligence (GeoAI) and machine learning methods.

Unsupervised Learning	Supervised Learning	Modeling Optimization Problems
Clustering:	**Regression and Classification:**	**Evolutionary Computing:**
- K-mean - K-medoids - Fuzzy C-means - Density-based spatial clustering of applications with noise (DBSCAN)	- Linear and logistic regression - Least absolute shrinkage and selection operator (LASSO) - Classification and regression trees (CART) - Generalized boosted regression - k-nearest neighbors algorithm (k-NN) - Support vector machine (SVM) - Random forest (RF) - Naive Bayesian	Genetic algorithm (GA) - Non-dominated sorted genetic algorithm II (NSGA-II) - The genetically adaptive multi-objective method (AMALGAM) Genetic programing (GP)
Dimension reduction unsupervised/ semi-supervised depth learning:	- Bayesian network - Multilayer perceptron (MLP) - Artificial neural networks (ANNs) - Adaptive neuro-fuzzy inference system (ANFIS)	**Metaheuristic methods:**
- Self-organized map (SOM) - t-distributed stochastic neighbor embedding (t-SNE) - Uniform manifold approximation and projection (UMAP) - Restricted Boltzmann machine - AutoEncoder - Generative adversarial networks (GANs)	- Restricted Boltzmann machine - Convolutional neural network (CNN, depth learning) - Recurrent neural network (RNN) - Gated recurrent units networks (GRUs) - Long short-term memory networks (LSTMs)	Particle swan optimization (PSO) algorithm Artificial bee colony (ABC) Ant colony optimization (ACO) Gray wolf optimization algorithm Whale optimization algorithm (WOA)
		Self-supervised learning/Reinforced learning (RL)
		Single agent/Multi-agent RL Model-based RL Model-free RL

Unsupervised clustering techniques are oriented towards automatically grouping or clustering the input data [42]. Several GeoAI clustering algorithms are used for geospatial and time-series data clustering (Table 1). A clustering algorithm does not require prior knowledge about the types and number of classes. More advanced dimension reduction clustering algorithms, such as autoencoders, can be used for data compression, reconstruction, and anomalies detection [43].

GeoAI regression techniques are oriented towards evaluating the relationship between response variables (dependent) and with one or more causative/independent variables (predictors). There is a wide range of methods and techniques in this category, ranging from traditional regression methods to ensemble and boosting regression trees, e.g., random forest, boosted regression, SVM, the traditional ANN, and deep learning methods [39,40,44].

The GeoAI supervised learning techniques are oriented towards identifying classes or categories. They learn from the given set of observations, called training data, and based on that, classify new observations into predefined classes. Unlike regression-based ML algorithms, the output variable of the classification is a category. The values represent class names or labels [39]. Several GeoAI classification methods are available (Table 1), e.g., SVM, random forest, ANN, and deep learning. GeoAI classification is widely used in remote sensing image classification, landform pattern recognition, and change detection.

An ML optimization algorithm is applied to find the best solution out of the solution space [45]. The ML optimization algorithm plays an essential role in optimizing the objective function, e.g., identifying the optimal parameter values of a complex model. ML optimization algorithms can be broadly categorized into evolutionary computing and metaheuristic methods (Table 1). ML optimization shows a wide range of applications, e.g., catchment models, parameter calibration by identifying the optimal set of parameter values and scale of analysis, identification of the best management scenarios for a multi-objective operation, etc. Additionally, the reinforced learning method is another approach for problem optimization.

It enables an agent(s) to learn in a dynamic environment by defining states, actions, and maximum rewards, using feedback from an agent's actions and experiences [46].

4. Current GeoAI Applications in Integrated Hydrological and Fluvial Systems Modeling

GeoAI applications in hydrology and fluvial studies are rapidly increasing and replacing the traditional methods. A reason for rapid GeoAI adoption in hydrological sciences might be linked to the progress in collecting big hydrological datasets, using automatic sensors with internet transmission, or the internet of things (IoT). Similarly, the evolution and increase in earth observation satellites (conventional and nanosatellites), unmanned aerial vehicles (UAV), light detection and ranging (LiDAR), and other surveying technologies produce high-resolution geospatial data, allowing better landscape characterization. GeoAI allows the harnessing of big and high-dimensional data to better understand the hydrological processes in a particular system. Specifically, GeoAI provides new data analytic tools to the entire data processing cycle, such as sensor data fusion, hydrological modeling, data assimilation, multi-objective scenario optimization, smart decision support, evaluation of climate change impact, construction of early warning systems, and geo-visualization. Therefore, GeoAI greatly enhances and supports decision making in integrated water resources management (IWRM) and nexus approaches [47]. Figure 2 depicts a GeoAI application model for a smart IWRM support system.

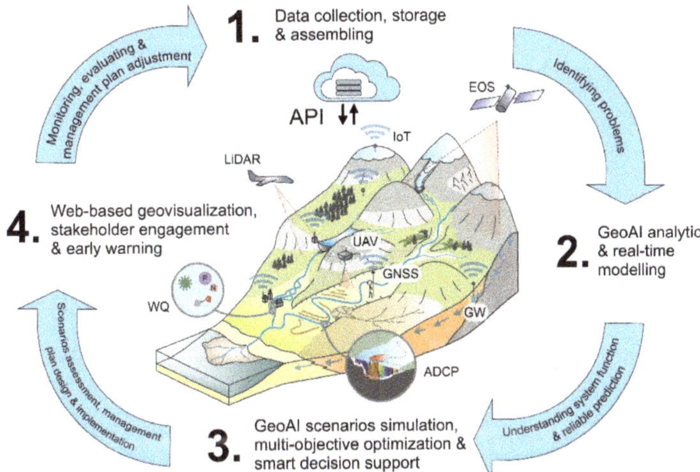

Figure 2. A GeoAI application model for a smart decision support system for integrated water resources management (IWRM). (1) Internet of things (IoT) supports real-time, high-frequency, hydrological monitoring. The data are stored in a cloud platform and accessed by an application programming interface (API). These data can be used for the real-time identification of problems in the system, e.g., a river basin. (2) GeoAI provides data analytic and online real-time modeling tools for hydrological system analysis and prediction. (3) GeoAI also supports multi-objective, multi-scenario optimization modeling, which in turn is the basis of smart decision support systems for IWRM. (4) Geovisualization in web mapping and mobile apps can be used for data dissemination and stakeholder engagements and implementing early warning systems. The smart IWRM system can be closed with the evaluation and adjustment of the IWRM plan and the improvement of the hydrological monitoring system. WQ (automatic water quality monitoring), ADCP (acoustic doppler current profiler for water current velocity measurement and river bathymetry), GW (automatic groundwater monitoring in wells), UAV (unmanned aerial vehicle for very high-resolution land cover mapping and surface elevation models), EOS (earth observation system for environmental condition monitoring), LiDAR (LiDAR survey of high-resolution topography data), and GNSS (use of global navigation satellite systems for ground truth data collection).

This section provides an overview of the current applications of the GeoAI methods and techniques and their advantages and limitations in many hydrological subfields such as hydrological and hydraulic modeling, optimization problems for hydrological model calibration and decision-making support, surface water quality, and fluvial geomorphological studies.

4.1. Hydrological and Hydraulic Modeling

Water flow in the catchments and river networks is a complex and stochastic process, operating in different spatio-temporal scales and characterized by non-stationarity, dynamism, and non-linearity [48,49]. These properties have limited the development of a reliable hydrological and hydraulic prediction model that can be generalized to a large geographical area. The increasing sensor-based, high-frequency (sub-hourly) hydrological data collection and the high spatial and temporal resolution mapping of land cover and topography have enhanced the understanding of hydrological processes. This fact has led to the development of more sophisticated physical-based hydrological models. However, these models are computationally expensive and limited to small-scale applications. Alternatively, several data-driven GeoAI methods have emerged for hydrological and hydraulic classification and prediction at multiple spatial-temporal scales. Table 2 shows examples of the current applications of the GeoAI in hydrological and hydraulic modeling.

Table 2. Selected GeoAI applications in hydrological and hydraulic modeling.

Method and Software	Objectives, Advantages, and Limitations	Reference
ANN and RF and permuted feature importance *Software:* Coded in r, available online: https://gitlab.com/lennartschmidt/floodmagnitude (accessed on 21 June 2022).	*Objective:* To compare ANN and RF flood prediction at the national level in Germany *Advantages:* ANN and RF achieved higher prediction accuracy for a large area with a large dataset than linear models. They reflect basic hydrological principles. *Limitations:* Heterogeneity of results across algorithms. The non-uniqueness/equifinality problem was identified due to the ML model setting.	[50]
ANN, ANFIS, wavelet neural networks, and hybrid ANFIS with wavelets transformation *Software:* MATLAB toolboxes	*Objective:* Hourly rainfall-runoff forecasting in Richmond River, Australia *Advantages:* Several ML methods compared, showing that hybrid ANFIS wavelet-based models significantly outperform ANFIS and ANN. *Limitations:* Only rainfall and runoff data were used in the modeling. Catchment physical features were not included.	[51]
RF and hybrid RF and the hydromad hydrological model. *Software:* RF package in R and hydromad package (available online: https://hydromad.catchment.org) (accessed on 21 June 2022).	*Objective:* To study the RF model performance vs. hydromad conceptual models in USA and Canada. *Advantages:* The RF model is simple and outperformed existing conceptual flood-forecasting models, predicting low and medium flood magnitudes. *Limitations:* The RF models exhibit inaccuracies for higher flood events, and their performance depends on the catchment characteristics.	[52]
ANN *Software:* Matlab toolboxes (e.g., wavelet)	*Objective:* Comparison of ANN and ARMA, combining them with wavelet analysis, empirical model decomposition, and singular spectrum analysis, for hindcasting and forecasting of monthly streamflow in two Chinese basins. *Advantages:* In hindcasting, the hybrid ANN and ARAM models performed better than the non-hybrid ones. *Limitations:* Hybrid models were not suitable for monthly streamflow forecasting, needing further refinement.	[53]

Table 2. *Cont.*

Method and Software	Objectives, Advantages, and Limitations	Reference
SOM and modified SOM (MSOM) *Software:* Not stated	*Objective:* To compare SOM and the MSOM clustering approaches, to deal with missing values, identifying groundwater exchange areas between the Serra Geral and Guarani aquifers (Brazil). *Advantages:* The MSOM showed higher accuracy in mapping hydrochemistry and groundwater physical properties, quantifying relationships in a large set of variables, which would not be revealed with conventional multivariate statistics. *Limitations:* SOM requires extensive data to obtain accurate and informative clusters.	[54]
CNN with transfer learning in time-series flood prediction *Software:* Not stated	*Objective:* To introduce a new CNN transfer-learning model as a conversion tool between time-series and image data to predict water levels in flood events. *Advantages:* CNN showed acceptable agreement with the observed water level data. Quantitative improvement in the CNN transfer learning appeared in the reduction in computational costs. *Limitations:* CNN captured higher peaks poorly. CNN was not as good as a fully connected deep neural network or RNN, especially when predicting the highest peaks.	[55]
Deep neural network (DNN) and data augmentation. *Software:* Not stated	*Objective:* To assess the feasibility of DNN in urban flood mapping, integrated with the stormwater management model (SWMM). Study area was two small urban catchments in Seoul, Korea. *Advantages:* The DNN was about 300 times faster than SWMM. Data augmentation could improve the poor predictive power of DNN. *Limitations:* Limited amount of input data needed to be improved by applying data augmentation.	[56]
CNN-based segmentation, VGG16, U-net, and Segnet *Software:* Not stated	*Objective:* To learn the spatiotemporal patterns of the mismatch between total water storage anomalies derived from the GRACE satellite mission and those simulated by the NOAH land surface model. Study area was India. *Advantages:* CNN models significantly improve the match with modeled and satellite-based observations of terrestrial total water storage. *Limitations:* Current grid resolution is relatively coarse.	[57]
LSTM and sequence-to-sequence (seq2seq) model *Software:* uses the Keras and TensorFlow packages in Python 3	*Objective:* To present a prediction model based on LSTM and the seq2seq structure to estimate hourly rainfall-runoff. Study area was two small watersheds in Iowa, US. *Advantages:* The LSTM-based seq2seq model was demonstrated to be an effective method for rainfall-runoff predictions and applicable to different watersheds. *Limitations:* Needs sufficient predictive power.	[58]
Transfer entropy (TE), ANN, LSTM, random forest regression (RFR), and support vector regression (SVR) *Software:* Not stated	*Objective:* To predict discharge with ML models and identify dominant drivers of discharge and their timescales using sensor data and TE. Study area was the Dry Creek Experimental Watershed, ID, USA. *Advantages:* The LSTM model is effective in identifying the key lag and aggregation scales for predicting discharge. TE was able to identify dominant streamflow controls and the relative importance of different mechanisms of streamflow generation. *Limitations:* Restricting ML models based on dominant timescales undercuts their skill at learning these timescales internally.	[59]
ANN and K-nearest neighbor *Software:* Not stated	*Objective:* To introduce the coupled ANN with the K-nearest neighbor hybrid machine learning (HML) for flood forecast. The study area was the Tunxi watershed, Anhui, China. *Advantages:* HML model showed satisfactory performance and reliable stability and predicted discharge continuously without accuracy loss. *Limitations:* Peak flow was not well captured.	[60]

Table 2. *Cont.*

Method and Software	Objectives, Advantages, and Limitations	Reference
Extreme learning machine (ELM) and the multilayer perceptron (MLP) for data assimilation with GR4J lumped hydrological model. *Software:* Not stated	*Objective:* To test two new ELM and MLP data assimilation methods in the rainfall-runoff models. The study catchments were Mistassibi (Canada), Schwuerbitz (Germany), Ourthe (Belgium), and Los Idolos (Mexico). *Advantages:* It shows that ELM and MLP can be successfully used for data assimilation, with a noticeable improvement over the GR4J and OpenLoop hydrological models for all studied catchments. *Limitations:* The ELM, MLP, and hydrological models are loosely coupled and simulated on an open-loop approach, and no feedback from the model output is considered. Furthermore, few observed variables in the ANN training are used (discharge and temperature).	[61]
Bayesian and variational data assimilation hybrid algorithm called OPTIMISTS (Optimized PareTo Inverse Modeling through Integrated STochastic Search) *Software:* code available online: https://github.com/felherc/OPTIMISTS (accessed on 21 June 2022).	*Objective:* To introduce the new data assimilation algorithm OPTIMISTS and to test it using the DHSVM hydrological models in the Blue River and Indiantown watershed, USA. *Advantages:* OPTIMISTS combines the features from Bayesian and variational approaches. OPTIMISTS produced probabilistic forecasts efficiently, with the combined advantages of allowing for fast, non-Gaussian, non-linear, and high-resolution prediction and for the balancing of the imperfect observation. *Limitations:* The model seems to be under development.	[62]

4.1.1. Hydrological System Classification

The classification of the different types of hydrological systems is one of the most widely applied modeling tasks in hydrology and ecohydrology. It aims to find similarities between different hydrological systems, e.g., those based on the hydrological response, the hydromorphological and climatic characteristics, and other variables. Unsupervised GeoAI algorithms, such as K-mean clustering [63] and SOM [64,65], have been applied to catchment classification. Both algorithms organize multidimensional input data through linear and non-linear techniques, depending on the intrinsic similarity of the data themselves. Several studies highlight the SOM nonlinear techniques for producing robust and consistent hydrological classification [66–68], even though the classification consistency is highly influenced by the quality of the input variables [69]. Additionally, where training data is available, supervised GeoAI methods have produced highly accurate and biophysically meaningful catchment classification [70,71].

4.1.2. Hydrological Data Fusion and Geospatial Downscaling

Integrated hydrological modeling requires the extensive data collocation of different components of the hydrological system in various spatial and temporal scales. Therefore, it is necessary to complete data and/or create new data by integrating several datasets from different sources, resolutions, and measurement noisiness [72]. This approach is called data fusion. Data fusion can increase the measurement quality and reliability, estimate unmeasured states, and increase spatial and temporal coverage. Several probabilistic and GeoAI data fusion techniques are available [73,74]. The commonly used GeoAI techniques in data fusion are non-linear Bayesian regression, ANN, RF, and deep learning [73–76]. These methods provide several advantages in representing non-linear, complex, and lagged relationships in different hydrological datasets. GeoAI data fusion is also applied to automatic data denoising and anomaly detection and remote sensing data fusion [77]. GeoAI data fusion is also used in rain, soil moisture, and discharge data generation. Sist et al. [78] introduce the ANN-based data fusion of multispectral (visible and infrared) satellite data with radar (microwave) satellite data to improve rainy area mapping and the estimation of the precipitation amount. Zhuo and Han [79] used data fusion to generate soil moisture products from satellite data, land surface temperature, and multi-angle surface brightness reflectance and were able to significantly increase the availability of daily soil moisture products. Fehri et al. [80] used the best linear unbiased predictor data fusion technique to generate discharge data from crowdsourced data and existing monitoring systems. There are more examples of using data fusion in data integration in areas other than in improving knowledge, which could be the next step to be further explored.

Environmental geospatial data, particularly remote sensing data, are usually measured at different spatial and temporal scales; high-temporal resolution data are usually measured at coarse (low) spatial resolution, and fine (high) spatial resolution data are obtained with low temporal frequency [77,81]. Therefore, combining the different datasets by downscaling methods is necessary to generate spatio-temporal high-resolution data. GeoAI-based downscaling has shown several advantages. For example, CNN is frequently used for downscaling coarse-resolution to fine-resolution precipitation products, using different static and dynamic variables as predictors [82,83]. These studies have shown that CNN achieves different degrees of accuracy, depending on the precipitation rate and the condition complexity; it has, e.g., lower accuracy in extreme wet conditions [83]. Other studies have shown a higher downscaling accuracy of GeoAI methods by having a spatial component in the model, e.g., spatial RF vs. RF in downscaling daily fractional snow cover [84] and land surface temperature from MODIS data [85–87].

4.1.3. Spatial Prediction of Hydrological Variables

The application of GeoAI in hydrological spatial prediction is diverse; it can be used, for example, in the risk mapping of hydrological extremes such as flood and drought [88–90]. In particular, GeoAI is widely applied in flood mapping, using satellite imagery, UAVs, high resolution LiDAR topographic data, and automatic water level sensors [91–93]. The common GeoAI algorithms used, e.g., in flood prediction are SVM, RF, ANN, and deep learning [92–94]. The selection of the methods is variable and depends on the mapping objective, the system complexity, and the data availability [91]. In areas with limited data and/or complex systems, where nonlinear methods are not easily interpretable, ANFIS soft computing has been applied with good prediction accuracy and strong generalization ability [95]. ANFIS combines data and expert knowledge through a set of fuzzy semantic conditional rules [96–98].

Another GeoAI application is the spatial prediction of hydrological model variables, e.g., saturated hydraulic conductivity [99,100] and weather data [101]. This is particularly useful as spatial hydrological variables are not available. Thus, they can only be predicted using points observation and surrogate spatial data such as remote sensing data. GeoAI spatial prediction has shown advantages in modeling nonlinear processes. However, the prediction quality depends on the quality and quantity of the observed data points and the applied GeoAI method [102].

4.1.4. Hydrological Process Modeling

GeoAI has shown the potential for accurate hydrological modeling, such as for rainfall-runoff, river discharge, soil moisture dynamics, and groundwater table fluctuation [95,103,104]. The non-linear nature of these processes is challenging to model with simple empirical and physical-based models. Therefore, GeoAI methods such as ANNs have proved to be better for modeling complex hydrological processes and forecasting them in the short and long term and in different management scenarios [26]. However, traditional ANNs do not model sequential order data such as time-series data. Therefore, a further development for the temporal dynamics of hydrological sequential events is the RNN and LSTM neural networks. RNN and LSTM use the previous information in the sequence to produce the current output, although RNN is better designed to model short sequences only. In the case of long temporal sequences of the antecedent conditions, LSTM is preferred. LSTM uses an additional 'memory cell' compared to RNN to maintain information for long sequences or periods of time [105,106]. This memory cell lets the model learn longer-term dependencies, e.g., the effects of antecedent soil moisture conditions on runoff generation [105,107]. LSTM is advantageous for modeling hydrological processes in regions with strong seasonality, such as a northern climate with varying winter conditions [91,105]. The LTSM model also allows the use of multiple time-series predictors, such as precipitation, temperature, discharge, and time [58,108]. A further extension of LSTM is created by combining it with CNN. In CNN, learning is achieved through convolving an input with filter layers to speed up parameter optimization [107,109]. Combining CNNs and LSTM encodes both the spatial and the temporal information [87,110]. LSTM techniques can also be coupled with other signal-processing algorithms such as wavelet transformation (WT). WT is applied to time-series data decomposition, e.g., the decomposition of high- and low-frequency flow signals, the identification of seasonality and trend, the decomposition of non-stationary signals, and data denoising [30]. Denoised data are used as inputs for the LSTM model [111].

Another approach is to use a physical-based model coupled with GeoAI, e.g., for runoff and flood prediction [19,112,113]. Overall, the output of the physical-based model is used as the input for GeoAI model training. For example, Noori and Kalin [14] used the SWAT model to simulate daily streamflow and estimate baseflow and stormflow, which were used as inputs for ANNs. The benefit of this approach is that once the model is trained, it can perform orders of magnitude faster than the original physical-based models without impairing prediction accuracy [17]. Another benefit of the hybrid modeling is that a trained model, e.g., in catchment hydrological modeling, can achieve better performance for other catchments than the uncalibrated process-based models [105,112].

Overall, most of the GeoAI models achieved higher prediction accuracy than the physical-based hydrological models. However, there are several types of GeoAI algorithms, with different architectures and mathematical formulations (e.g., ANN, CNN, and LSTM) to perform similar tasks. In addition, different types of predictor variables and data sampling sizes are used, making the GeoAI model performance comparison challenging. GeoAI models are less physically interpretable, as they do not explicitly represent the physical laws governing the hydrological processes. Therefore, their causal inference is still limited. GeoAI applications are currently oriented towards hydrological prediction. GeoAI has the potential to provide accurate and timely information which is applicable to large areas, and using data from IoT sensors and cloud computing, it can deliver real-time prediction [114].

4.1.5. Hydraulic Modeling

The new generation of very-high-resolution river bathymetry has improved the 1D, 2D, and 3D hydraulic modeling of rivers [115,116]. River hydraulic models have been widely used in the estimation of flood extent, water depth and velocity, sediment transport, and the assessment of fluvial morphodynamics [5,11,117]. However, very complex hydraulic models (3D) are data and computationally demanding and restricted to small-scale applications. Hydraulic modeling is sometimes inconsistent and does not represent all the bio-physical processes occurring in the natural fluvial environment [118,119]. In addition, the numerical solving approach of the hydraulic model results in high numerical instability due to sensitivity to the initial and boundary conditions, model structure, and spatial and temporal discretization [120]. Thus, the GeoAI method has emerged as a promising tool for hydraulic modeling in large-scale and natural systems [19,119,121,122]. Emerging deep learning applications in computer fluid dynamics have also shown potential for the modeling of turbulent and complex flow structures [123–125]. Additionally, coupling the hydraulic model with the Bayesian GeoAI methods improves hydraulic modeling over a broad range of spatiotemporal scales and physical processes [126].

4.1.6. Hydrological Data Assimilation

Hydrological data assimilation (DA) is a state estimation theory that assumes that models are an imperfect representation of the system and that hydrological data might contain noise. Both can also contain different types of information and be complementary [127]. DA aims to harness the information in the hydrological model and in the observations to approximate the true state of the system, considering its uncertainty statistically [127–129]. DA methods include linear dynamics (e.g., Kalman filter, the most popular state estimation method) and nonlinear dynamics [127]. The DA methods can be related to ML. Data fusion and DA use similar techniques, but the problem formulation differs [130].

In hydrological modeling, the ML-based DA is the most common type of coupling of ML and the physical-based model, the so-called loosely hybrid hydrological model [131]. DA updates the state system predicted by a physical-based model at a given time or place with observational data, using Bayesian approximation such as the ensemble Kalman filter (EnKF) [127] or ML methods, e.g., ANN, RNN, and LSTM [132]. Both the DA and the ML methods solve an inverse problem, expressed as the model y = h(x,w), where h is the model function, x represents the state/feature variable, w is the parameters/weights of the model, and y is the observations/labels in DA/ML, respectively. DA is oriented to find the true state of the system (x) from the observation and ML is commonly oriented to find model parameters or weight (w) from the observation. DA holds w constant to estimate x; ML holds x constant to estimate w; see [133] for a detailed revision.

Many studies have shown that ANN data assimilation outperforms conventional DA, particularly for complex and non-linear response systems [61]. An additional development of ML-based DA methods is the so-called deep DA [132], which trains deep learning neural networks such as LSTM for high dynamic systems. Deep DA has shown potential for accurate prediction for periods or sites where observations are unavailable and conventional DA cannot be applied to reduce the model error [132].

4.2. Modeling Optimization Problems for Hydrological Model Calibration and Decision Support System

4.2.1. Hydrological Model Calibration

In hydrological modeling, the inverse modeling approach is widely applied. In inverse modeling, the model features and parameter values are unknown, and those are identified by minimizing the error between the model output and the observed data [134,135]. The model feature identification includes the definition of the main hydrological processes, the mathematical equations representing it, the boundary conditions, and the time regime [136]. The parameter identifications encompass the identifying of the model optimal parameter set values that reproduce the observed data acceptably [136]. In highly parameterized models, identifying the optimal values of the parameters is challenging and represents a substantial part of the modeling work. Usually, there is not a single set of optimal values of parameters that can simulate the observed data well but a set of optimal parameters values that can achieve similar model performance. This modeling phenomenon is called the non-uniqueness or equifinality problem [137]. The hydrological model calibration often requires specialized optimization algorithms, and several ML-based calibration algorithms have been developed to support model calibration (see Table 3 for examples).

Table 3. Selected GeoAI applications for model calibration and decision support systems.

ML Method	Objectives, Advantages, and Limitations	Reference
Non-dominated sorting genetic algorithm II (NSGA-II), particle swarm optimization (MPSO), the Pareto envelope-based selection algorithm II (PESA-II), the strength Pareto evolutionary algorithm II (SPEA-II) with the combined objective function and genetic algorithm. *Software:* Not stated	*Objective:* To compare optimization techniques to calibrate conceptual hydrological models. *Advantages:* All techniques perform well, better results gained than when using a single-objective algorithm. The NSGA-II with two indicators performed better than the MPSO with one indicator. Limitations: Depending on the selected performance indicators, the best model varied.	[138]
The multi-objective evolution algorithm MODE-ACM and the enhanced Pareto multi-objective differential evolution algorithm (EPMODE). *Software:* Not stated	*Objective:* To introduce the MODE-ACM and the enhanced EPMODE model. To test model efficacy by comparing the NSGA-II and SPEA2 model. *Advantages:* The EPMODE and the MODE-ACM were both reliable and showed better performance than the NSGA-II and SPEA2 models. *Limitation:* Very complex model.	[139]

Table 3. *Cont.*

ML Method	Objectives, Advantages, and Limitations	Reference
The multi-objective particle swarm optimization (MOPSO), NSGA-II and the multi-objective shuffled complex evolution metropolis (MOSCEM-UA). *Software:* Not stated	*Objective:* Comparison of three multi-objective algorithms for hydrological model calibration. *Advantages:* All three algorithms are able to find Pareto sets of solutions. The most uniform distribution of the solutions was derived with MOSCEM-U as the NSGA-II has the shortest Pareto optimal front and the MOPSO has the maximal extent of the obtained non-dominated front. *Limitations:* The rate of convergence with the optimal solutions varies across the three algorithms.	[140]
MOSCEM-UA and shuffled complex evolution metropolis (SCEM-UA). *Software:* Not stated	*Objective:* To calibrate hydrological models using more effective and efficient multi-objective algorithms called MOSCEM-UA. Advantages: MOSCEM-UA allows multi-objective calibration, preventing the collapse of the algorithm to a single region of highest attraction. It combines the complex shuffling and the probabilistic covariance-annealing search procedure of the SCEM-UA algorithm. *Limitations:* Challenging to make sure that a diverse and large initial population size is provided, which supports multiple objective approaches.	[141]
ANN coupled with SWAT model. *Software:* SWAT-ANN, available online: https://zenodo.org/record/3699658#.YewXid_RZaQ (accessed on 21 June 2022).	*Objective:* Evaluation of SWAT-ANN rainfall-runoff simulation in a catchment, Italy. *Advantages:* The SWAT-ANN prediction accuracy was acceptable and useful in the absence of observational data. *Limitations:* The overall model calibration is only evaluated by model residual error.	[142]
Genetic algorithm *Software:* Multi-objective evolutionary sensitivity handling algorithm (MOESHA), programed in Python	*Objective:* River discharge modeling with a hydrological model called EXP-HYDRO, applied in a small catchment in Wales, UK. *Advantages:* The MOESHA algorithm determines well the optimal distribution of the model parameters that maximize model robustness and minimize error; it also estimates model parameter uncertainty. *Disadvantages:* Computationally expensive.	[143]
Multi-algorithm, genetically adaptive multi-objective method (AMALGAM), single evolutionary multi-objective optimization (SPEA-II and NSGA-II). *Software:* AMALGAM code in Visual Basic	*Objective:* Comparative study of SWAT model calibration using single- and multi-evolutionary optimization algorithms. The study areas are the Yellow River Headwaters Watershed (Tibet Plateau, China), the Reynolds Creek Experimental Watershed (Idaho, USA), the Little River Experimental Watershed (Georgia, USA), and the Mahantango Creek Experimental Watershed (Pennsylvania, USA). *Advantage:* The advantage of the multi-evolutionary algorithm AMALGAM calibrating the SWAT model has been demonstrated. AMALGAM provides fast, reliable, and computationally efficient solutions to multi-objective optimization problems. *Limitations:* With a small number of run schemes, multiple trials might be needed for implementing AMALGAM. The solution contribution of the different algorithms used in AMALGAM varies in the different watersheds.	[144]
Single-objective and multi-objective particle swarm optimization (PSO) algorithms. *Software:* Not stated	*Objective:* Automatic calibration of hydrologic engineering center-hydrologic modeling systems (HEC-HMS) rainfall-runoff model. The study area is the Tamar sub-basin of the Gorganroud River Basin, Iran. *Advantages:* Multi-objective PSO calibration outperforms the single-objective one. However, an appropriate combination of objective functions is essential in multi-objective calibration. *Limitations:* Increasing the number of objective functions did not necessarily lead to a better performance than the bi-objective calibration. The increasing number of objective functions also introduces computational challenges. Insufficient data and flood events seriously affect the calibration performance.	[145]

Table 3. *Cont.*

ML Method	Objectives, Advantages, and Limitations	Reference
The deep learning gradient-based optimizer. *Software:* Python code, available online: https://github.com/ckrapu/gr4j_theano (accessed on 21 June 2022).	*Objective:* Estimating unknown parameters for GR4J conceptual hydrological model. *Advantages:* The deep learning gradient-based optimizer is effective for high-dimensional inverse estimation problems for hydrological models. *Limitations:* Need for hydrological model to be established at the site and high computational effort.	[146]
Partial least squares regression (PLSR) and ANFIS. *Software:* Not stated	*Objective:* Development of a smart irrigation decision support system, using PLSR and ANFIS as reasoning engines. The study area is in the southeast of Spain. *Advantages:* ANFIS prediction was better than PLSR for water requirement estimation and identifying timely crop irrigation needs. *Limitations:* The model has been tested with a few soil and weather variables.	[147]
Reinforcement learning (RL) *Software:* Python code, available online: https://github.com/kLabUM/rl-storm-control (accessed on 21 June 2022).	*Objective:* To implement a smart stormwater real-time system control based on RL. *Advantages:* The RL-based model learned the control valve strategy in a distributed stormwater system by interacting with the system it controls under thousands of simulated storm scenarios. It effectively tries various control strategies until it achieves target water level and flow set points. *Limitations:* RL performance is highly sensitive to the RL agent reward formulation and requires a significant amount of computational resources to achieve a good performance.	[148]

Hydrological models are often calibrated with a single objective function, although adequate and fast multi-objective optimization techniques exist, which better support the several output variables [141]. There are many optimization algorithms, meta-heuristic and ML-based, for model parameter calibration, such as particle swarm optimization (PSO), grey wolf optimization (GWO), genetic algorithms (GAs), genetic programming (GP), strength Pareto evolutionary algorithms (SPEA), micro-genetic algorithms (micro-GA), and Pareto-archived evolution strategies (PAES). Depending on the selected performance indicators of the model, the best model for hydrologists varied. According to the free lunch theorem [149], this is not expected to change for a while; it proposes that no one model fits all. In any case, all the models performed well. See Yusoff et al. [150] and Ibrahim et al. [45] for a specific review of optimization algorithms.

Meta-heuristic optimization algorithms, which are mostly inspired by the biological/behavioral strategies of animals, provide a good solution to optimization problems, particularly with incomplete or imperfect information or limited computational capacity [151]. An advantage of these algorithms is that they make relatively few assumptions about the optimization problems and reduce the computational demand by randomly sampling a subset of solutions, which otherwise would be too large to be iterated entirely [151]. However, some meta-heuristic algorithms such as PSO may not guarantee that a globally optimal solution will be found, particularly when the number of decision variables or dimensions being optimized is large [45]. The GA is inspired by genetic evolutionary concepts, such as the non-dominated sorted genetic algorithm II (NSGA-II). The genetically adaptive multi-objective method (AMALGAM) [152] has been applied for multi-objective, multi-site calibration and to solve highly non-linear optimization problems [144,153]. AMALGAM is a multi-algorithm that blends the attributes of several optimization algorithms (NSGA-II, PSO, the adaptive metropolis search, and differential evolution) [144]. The GA has been shown to be well-suited for hydrological models, such as the SWAT semi-distributed hydrological models, which cannot be adequately calibrated by gradient-based calibration algorithms [144,153,154]. The objective function for each solution in a GA can be assessed in parallel computation, providing computational efficiency [144]. Additional calibration methods based on deep learning have also been developed, outperforming many of the existing evolutionary and regionalization methods [20,146].

4.2.2. Decision Support System for Integrated Water Resources Management

Integrated water resources management (IWRM) deals with multiple actors to consensually and communicatively integrate decisions in a hydrological unit to ensure equitable economic development and social welfare while assuring hydrological system sustainability [155]. IWRM demands quality and timely information. Hence, increasing automation with GeoAI-based decision support systems is thought to enhance IWRM [17,156]. Multi-objective and scenario analysis are typical applications of GeoAI techniques in IWRM to find solutions for conflicting objectives, forecast the impact of management strategies, and optimize hydrological system operation [157,158]. We found widespread applications of GeoAI in reservoir and water distribution optimization using ANN [159,160], assembled and deep learning algorithms, and genetic programming [161,162]. Another application is found in building a smart irrigation decision support system [147]. Here, partial least square regression and the adaptive network-based fuzzy inference system (ANFIS) are proposed as reasoning engines for automated decisions. An additional example of artificial intelligence application is the adaptive intelligent dynamic urban water resource planning [158]. It uses Markov's decision process to tackle complex water management problems, predicting water demand, scheduling management, financial planning, tariff adjustment, and the optimization of water supply operations [158] (See Table 3). Overall, the GeoAI-based IWRM integrates various types of algorithms to perform different tasks, such as prediction and forecasting using various types of geospatial data, and optimization algorithms for management scenarios with multiple objectives. Algorithms such as ANFIS are used for system reasoning to automate the decision support [157,158,163]. ANFIS allows the mimicking of human reasoning and decision making based on a set of fuzzy IF-THEN rules. ANFIS has the learning capability to approximate nonlinear functions and can self-improve in order to adjust the membership function parameters directly from the data [164].

4.3. Automatic Water Quality Monitoring and Spatio-Temporal Prediction
4.3.1. Automatic Water Quality Monitoring

The data collection of water quality with wireless sensor networks and internet of things (IoT) technologies is rapidly increasing and providing very-high-frequency WQ data (sub-hourly) [165,166]. There is evidence that the high-frequency data better represent the dynamics variation of river discharge and sediment and solute fluxes [167]. It enables the early mitigation of floods and drinking water problems [168,169]. High-frequency data can also lead to a more precise and accurate classification of the biochemical status of rivers and lakes [170]. However, such sensors and devices are subject to failures, poor calibration, and inaccurate data recording in certain conditions [171,172]. Therefore, automatic data quality control, error and anomaly detection, sensor drift compensation, and uncertainty assessment are important [171–173]. GeoAI showed advantages in managing WQ sensor networks and sensor data fusion, such as fault detection, data correction, and upgrades from different monitoring sensors by data fusion [174]. See Table 4 for selected examples of GeoAI applications on WQ monitoring. Additional applications of GeoAI are in the detection, localization, and quantification of pollutant critical sources and critical periods of loading in monitoring networks [175,176]. The most common GeoAI algorithms for WQ sensor fusion are based on Bayesian algorithms, fuzzy set theory, genetic programming, ANN, and LSTM [177–180].

Table 4. Selected GeoAI applications in monitoring and spatio-temporal prediction of water quality.

Method and Software	Objective, Advantages, and Limitations	Reference
ANN *Software:* ANNs computed and fitted with the Keras package in R. Available online: https://github.com/benoit-liquet/AD_ANN (accessed on 21 June 2022).	*Objective:* To use ANN to detect anomalies caused by the technical error of in situ sensor monitoring of conductivity and turbidity. *Advantage:* Semi-supervised ANN very well suited to identifying short-term anomalous events. Supervised classification able to identify long-term anomalies. *Limitations:* High rate of false positive detection. Large dataset required due to relative scarcity of anomalous events to train data.	[174]
MLP, SVM-SMO, lazy-instance-based learning K nearest-neighbor (IBK), KStar, RF, random tree, and REPTree. *Software:* Python	*Objective:* To compare ML techniques for soft sensor monitoring of biological oxygen demand (BOD). *Advantage:* IBK algorithm was best for estimating BOD based on turbidity, dissolved oxygen, pH, and water temperature sensors. IBK algorithm can also be used within a low-cost system to allow incorporation into IoT-based WQ systems. *Limitations:* Overloading of servers may occur in IoT-based systems if prediction algorithms run on the cloud for a large number of sensing nodes.	[181]
Feed-forward ANN (FF-ANN). *Software:* Python using the scikit-learn library.	*Objective:* To compare FF-ANN and traditional methods for drift correction. *Advantages:* The FF-ANN model outperformed traditional methods for drift calibration and may increase calibration lifetime and reduce calibration frequency for water quality sensors. *Limitations:* High error rate apparent when using logistic function.	[172]
RF, SVM, and logistic regression *Software:* Python	*Objective:* Evaluation of 3D wide-area WQ monitoring and analysis systems, using unmanned surface vehicles (USV) and ML algorithms. *Advantage:* RF demonstrated high precision for estimating WQ status, using WQ measured data (turbidity, total dissolved solid, and pH) by USV at multiple points and different water depth levels. *Limitations:* USV drifting control, working performance, and system accuracy were evaluated just in one environmental condition (a small lake) and for a few WQ parameters.	[177]
Gradient boosting (XGBoost), RF, coupled with denoising model called CEEMDAN. *Software:* Not stated	*Objective:* Hourly prediction of several WQ parameters in Tualatin River, Oregon, USA. *Advantages:* The best model performance depended on the predicted WQ variable. CEEMDAN-RF and CEEMDAN-XGBoost show better performance, less errors, and higher stability than simple RF and XGBoost. New error metric is introduced for model performance evaluation and compared with conventional methods of model evaluation. *Limitations:* The prediction model only depends on time-series data and no other explanatory variables were included.	[182]
Multi-layer perceptron, radial basis function ANN, ANFIS, and ANFIS with wavelet denoising technique (WDT-ANFIS). *Software:* wavelet and fuzzy logic toolboxes of MATLAB	*Objective:* Prediction of WQ parameters such as ammoniacal nitrogen, suspended solid, and pH in Johor river, Malaysia. *Advantages:* Several ML algorithms were compared, where the WDT-ANFIS model advantage was well illustrated. *Limitations:* Complex models and not possible to identify a single network structure that can best predict WQ parameters.	[28]
Decision tree (DT), RF, and deep cascade forest (DCF), trained by big data. *Software:* Python. RF and DT built using the package Scikit-learn v.019. DCF was built with the package gcForest	*Objective:* Comparison of 7 traditional learning models vs. 3 ensemble models for prediction of 6 levels of WQ parameters for major rivers and lakes in China. *Advantages:* DT, RF, DCF trained by big data all had significantly better prediction of WQ compared with traditional learning models. DCF had the best performance overall for prediction of all 6 levels of WQ. *Limitations:* DCF unable to learn directly from big raw data.	[183]

Table 4. *Cont.*

Method and Software	Objective, Advantages, and Limitations	Reference
RF *Software:* Python using the libraries Scikit-Learn (0.20.1) and MLxtend (0.13.0)	*Objective:* Comparison of random forest to multiple linear regression for estimation of high-frequency nutrient concentration. *Advantages:* RF outperformed linear models when more than one predictor was included. *Limitations:* At least 3 predictors required to identify clear benefit of using RF compared to multiple linear regressions.	[184]
Integrated LSTM, using cross-correlation and association rules (apriori). *Software:* Not stated	*Objective:* GeoAI system to identify and trace point sources of pollution in Shandong Province, China. *Advantages:* LSTM algorithm had high prediction accuracy for tracing the main point sources of pollution. *Limitations:* GeoAI model was not aware of the change in aquatic environment conditions. It requires multi-dimensional and multi-spatial perspectives to identify, analyze, and respond to data.	[185]
ANN *Software:* Mathematica	*Objective:* Modeling daily total organic carbon (TOC), total nitrogen (TN), total phosphorous (TP), and predicting future fluxes under climate change scenarios for two streams in Finland. *Advantages:* ANN model managed to recreate most dynamics in TOC, TN, and TP. *Limitations:* ANN model struggled to capture extreme values for TOC, TN, and TP.	[186]
Least square-SVM (LS-SVM) and multivariate adaptive regression spline (MARS). *Software:* Not stated	*Objective:* Prediction of 5-day biochemical oxygen demand (BOD) and chemical oxygen demand (COD) in natural streams in Karoun River, southwest Iran. *Advantages:* LS-SVM and MARS models performed better in terms of external validation criteria and F test compared with multiple-regression-based models and ANN and ANFIS equations. *Limitations:* Intensive amount of data collection required for a wide variety of parameters.	[187]
ANN using feed-forward network with Levenberg–Marquardt back-propagation learning. *Software:* Not stated	*Objective:* Hybrid approach using a SWAT model as an input to an ANN to simulate monthly nitrate, ammonium, and phosphate loads in Atlanta, GA, USA. *Advantages:* Hybrid model outperformed standalone SWAT and ANN models for prediction of monthly loads. Hybrid models are useful for predictions in unmonitored catchments. *Limitations:* Hybrid model for nitrate had substantially better predictions than the ammonium and phosphate models. Large peaks of ammonium and phosphate were underestimated.	[188]
ANN compared with SVM and group method of data handling (GMDH). *Software:* Not stated	*Objective:* ML techniques compared for predicting various WQ components in Tireh River, southwest Iran. *Advantages:* SVM models were most accurate, with less data dispersion. *Limitations:* All models overestimated some properties.	[189]
Twelve hybrid data mining algorithms compared, spanning two main groups. Group one featured decision tree algorithms, and group two featured meta-classifier or hybrid algorithms. *Software:* Not stated	*Objective:* Comparison of 12 hybrid GeoAI models predicting WQ indices in Talar Catchment, Iran. *Advantages:* Hybrid models showed improved predictive power compared to standalone algorithms. Hybrid bagging random tree (BA-RT) model showed greatest predictive power and was able to produce reliable results despite a dataset spanning a short time period. *Limitations:* BA-RT model struggled to accurately predict extreme WQ index values, while most models also overestimated WQI values.	[190]

Table 4. *Cont.*

Method and Software	Objective, Advantages, and Limitations	Reference
SVM and RF *Software:* e1071 R package used to build SVM model and random forest R package used to build RF model.	*Objective:* SVM and RF algorithms compared to predict high-frequency variation in stream solutes in Hubbard Brook, New Hampshire, USA. *Advantages:* Both ML algorithms were capable of effectively predicting concentrations of major ions. *Limitations:* Solutes with atmospheric, episodic, or strong biotic and abiotic controls were much more poorly predicted than solutes least affected by ecosystem dynamics.	[191]

Many WQ parameters cannot easily be measured in situ and in real time for various reasons, such as high-cost sensors, low sampling rate, multiple processing stages, and the requirement of frequent cleaning and calibration. Therefore, a common practice is the estimation of a particular WQ parameter value based on other surrogate parameters, called soft sensors [181,183,184]. ML techniques showed higher accuracy in implementing soft sensors than conventional regression-based models [181,183,184,192].

The ML method has also shown an advantage in automatic hysteresis pattern analysis using high-frequent water quality data with, e.g., restricted Boltzmann ANN [193]. A more detailed hysteresis pattern classification allows the gaining of new insights into WQ pollutants sources and drivers, the influence of catchment and riverine features, the effect of antecedent conditions, and the influence of changes in rainfall and snowmelt patterns [193].

4.3.2. Spatio-Temporal Water Quality Prediction

We found diverse applications of the GeoAI methods in WQ spatio-temporal pattern analysis, the classification of WQ, and the prediction of WQ variables and the pollutant loading estimation. A detailed review of the ML application in WQ prediction is found in Rajaee et al. [27], Naloufi et al. [29], and Chen et al. [194]. Table 4 shows examplesof GeoAI applications for this purpose. Commonly used GeoAI for WQ prediction and classification are unsupervised clustering such as k-means, density-based spatial clustering of applications with noise (DBSCAN), and SOM, but also time-series segmentation such as dynamic time warping [195]. Supervised ML classification and prediction algorithms for WQ are RF, SVM, the Bayesian network, and ANN, and deep learning such as LSTM is also frequently used [190,196,197].

High-frequency WQ monitoring data contains noise signals due to random and systematic errors, impairing the WQ prediction accuracy. Hence, combining data denoising techniques such as Fourier and wavelet transform with GeoAI improves WQ prediction. For example, Song et al. [198] found that combining synchro-squeezed wavelet transform and an LSTM network substantially improved the WQ parameter prediction. Similarly, Najah Ahmed et al. [28] integrated wavelet discrete transform with the artificial neuro-fuzzy inference system (WDT-ANFIS) to obtain high-accuracy prediction of river WQ parameters.

Additionally, the WQ data usually have temporal autocorrelation and multi-collinearity between the WQ parameters. To consider these characteristics in the prediction models, Zhou et al. (2020) [199] proposed an ML model based on t-distributed stochastic neighbor embedding (t-SNE) and self-attention bidirectional LSTM (SA-Bi-LSTM), demonstrating substantial WQ prediction improvement. Another promising approach is uniform manifold approximation and projection (UMAP) for multidimensional WQ data ordination and classification. Unlike other dimension reduction methods, UMAP retains a global and local information structure, and the data ordination is bio-physically meaningful [200].

Inland water has naturally high spatial variation. It requires complex spatial prediction models and large datasets. The GeoAI have shown breakthroughs in spatial WQ prediction by combining field observations, remote sensing data, or UAV imagery. For example, using deep learning, RF, genetic algorithm—RF, adaptive boosting (AdaBoost), genetic algorithm—AdaBoost and the genetic algorithm—extreme gradient boosting (GA-

XGBoost) [183,194]. However, these models usually demand extensive training data, which are restricted to a few pilot areas or intensely monitored areas.

Another approach in WQ prediction is the application of hybrid models and the integration of physical-based models with GeoAI methods, such as SVM, RF, ANN, and LSTM. Hybrid models usually outperformed physical-based models. For example, Noori et al. [188] found substantial improvement in monthly nitrate, ammonium, and phosphate load prediction when using hybrid SWAT-ANN models. Hybrid models are also helpful for unmonitored catchment predictions [188]. The hybrid model also improves GeoAI explanatory and generalization capability, although some disadvantages observed in the physical-based model, such as extreme values not being well predicted, persisted in the hybrid models. Similarly, the process-guided recurrent neural network (RNN), which combines the biophysical principles of the process-based model and RNN, modeled the seasonal variation of lake phosphorus loading with lower bias and better reproduced the long-term changes of phosphorus loading compared to using the physical-based model and RNN independently [21].

Overall, the GeoAI water quality prediction depends not only on the selected algorithms and settings but also on the WQ parameters, data size, and training data quality for the learning models [183,188,191].

4.4. Machine Learning in Fluvial Geomorphic and Morphodynamic Mapping

Fluvial geomorphology triggered the quantitative dynamic paradigm [201] as an approach to quantifying and understanding the processes of the fluvial environment [5]. The simultaneous development of techniques such as multispectral satellite images, synthetic aperture radar (SAR), LiDAR, UAV imagery, structure from motion photogrammetry (SfM), multibeam sonar (sound navigation and ranging), among others, has resulted in an unprecedented, seamless characterization and quantification of the fluvial environment and its dynamics [202–204]. This geospatial dataset explosion, as in many other disciplines, has resulted in the perfect foundation for applying GeoAI methods in fluvial geomorphology. Here, we reviewed the recent GeoAI applications in fluvial geomorphological studies. Table 5 shows selected examples of GeoAIapplications in fluvial geomorphic studies.

Table 5. Selected GeoAI applications in fluvial geomorphic and morphodynamic mapping.

Method and Software	Objective, Advantages and Limitations	Reference
NASNet CNN *Software:* Python based. CNN supervised classification available for PyQGIS	*Objective:* To classify fluvial scenes in 11 rivers in Canada, Italy, Japan, the United Kingdom, and Costa Rica. *Advantages:* The NasNet-CNN model outperformed other supervised classifiers, e.g., maximum likelihood, MLP, and RF. The NasNet-CNN model can be transferable to other rivers with no training data, obtaining good classification accuracy of fluvial land cover. *Limitations:* CNN is sensitive to the hyperparameters definition. CNN was also affected by imbalanced training data size in land cover classes. Some features were misclassified.	[205]
Fuzzy-CNN *Software:* Python based. Dependency packages: TensorFlow	*Objective:* To predict vegetation, bare sediment, and water bodies at a sub-pixel scale with Sentinel-2 images, trained with high resolution UAV images. Study areas were Sesia, Po, and Paglia Rivers in Italy. *Advantages:* Fuzzy-CNN models were successfully used to provide continuous and crisp subpixel classification of Sentinel-2 imagery. The model was transferable to satellite images with different acquisition time. It can be used for annual change detection. *Limitations:* UAV reference data obtained with manual OBIA was highly time-consuming. The process was computationally expensive due to the "super-resolution" process used to feed the CNN. The model was tested only in Mediterranean drainage basins.	[206]

Table 5. *Cont.*

Method and Software	Objective, Advantages and Limitations	Reference
ANN *Software:* MATLAB-based software. Leaf area index calculation (LAIC)	*Objective:* Hydromorphological features classification using very high resolution UAV images (2.5 cm) in a reach of the River Dee, Wales, UK. *Advantages:* ANN LAIC model showed satisfactory classification accuracy and potential to identify multiple hydromorphological classes that can be attributed to site features based, e.g., on their hydraulic, habitat, or vegetation types. The model settings seem to be transferable to other rivers without training data. *Limitations:* The algorithm showed misclassification of small fluvial entities, e.g., shallow water areas with rippled surface, water areas affected by tree shadow, and vegetated banks and/or areas obscured by brown submerged vegetation. The authors provided very limited information about the model parameter setting in the paper.	[207]
CNN *Software:* Matlab- based software. Visualization and image processing for environmental research (VIPER)	*Objective:* To measure river wetted width (RWW) with a novel approach at the subpixel scale by using MODIS and Landsat OLI images in Bay of Bengal, India, and Landsat TM in Columbia River, USA. *Advantages:* CNN-based sub-pixel scale classification resulted in a more accurate estimation of RWW than the conventional hard image classification. *Limitations:* No full spectral unmixing was possible due to the spectral variations of land cover classes and the nonlinear mixture phenomenon. Misclassification issues were reported when shadows, bridges, or trees were located along riverbanks. In such situations, RWW was unmeasured.	[208]
GEOBIA, EL, RF, extra tree (ET), gradient tree boosting (GTB), extreme gradient boosting (XGB). Then, it was combined with a voting classifier. *Software:* Python-based with scikit-learn package.	*Objective:* To map the main hydromorphological types that characterize fluvial landscapes in Europe by using Copernicus image mosaic and EU DEM. Target classes: water, sediment bars, riparian vegetation, other floodplain units. *Advantages:* RF outperformed any other tested classifier, e.g., ET, GTB, and XGB. Hierarchical object-based segmentation is robust for combining spectral and topographical data at different spatial resolutions and enhancing low spectral resolutions. Area-based validations were the preferred method to validate the quality of the object-based maps. *Limitations:* Vegetation units and sediment bars were not very well classified. Main source of error was related to the high mixture of riparian vegetation, sediment bars, and other floodplain features.	[209]
Object-based RF and pixel-based RF, combined with recursive feature elimination and PCAs *Software:* rpart and caret R packages, and EnMAP-Box (environmental mapping and analysis program) 2.1.1 software.	*Objective:* To reveal uncertainty, overfitting, and efficiency of terrain attribute identification in fluvial landforms using morphometric variables derived from a LiDAR DTM from Tisza River, Hungary. *Advantages:* Object-based RF method had a better classification accuracy (95%) than the pixel-based RF method (78%) when identifying 4 different river landforms (crevasse channels, swales, point bars, levees). Overfitting was controlled in the study by limiting the number of input variables. *Limitations:* Object-based RF classifications needed visual interpretation, field observations, and high-resolution data. PCAs did not help to select more efficient and important variables.	[210]

Table 5. *Cont.*

Method and Software	Objective, Advantages and Limitations	Reference
RF and SVM *Software:* eCognition developer 9 software.	*Objective:* Semi-automatic map of riverscape units and in-stream microhabitats, providing continuous, objective, and multi-scale classification using very-high-resolution near-infrared aerial imagery and LiDAR DTM from the Orco River, Italy. *Advantages:* RF better identified riverscape elements, e.g., channel and bars, while SVM did better when classifying in-stream meso-habitats. Topographical data, in particular detrended DTM (DDTM), was a relevant data source for an accurate classification of the riverscape units. Near-infrared imagery combined with DDTM was the best predictor. *Limitations:* Extensive expert-based training was necessary for detailed post-classification. Several subjective rules added to the process. Most confusion in the classification was detected between the floodplain and sparse vegetation classes, where the DDTM was not helpful.	[211]
K-means clustering *Software:* ArcGIS based. Geoprocessing tool (multivariate clustering)	*Objective:* To delineate valley bottom extent across large catchments and automatically classify valley bottom segments of variable length by using DEM-based derivatives from Richmond River, Australia. *Advantages:* The k-means successfully clustered the entire river network into 6 valley bottom segments of varying length. *Limitations:* The resulting cluster is unlabeled and needs expert recognition. Only used a low number of the variables selected (slope and valley bottom width). The model was validated with a basin-scale expert mapping of valley types. This is time-consuming and not available for other areas. The model was only proposed as a preliminary assessment for further studies.	[212]
Modified Hebbian algorithms and k-mean clustering. *Software:* On-line batch Hebbian algorithm and CoSA (clustering of sparse approximations) packages.	*Objective:* To investigate the applicability of ML classifier in Arctic regions using DigitalGlobe Worldview-2 visible/near-infrared, high-resolution imagery from Mackenzie River, Canada, and Selawik and Barrow Rivers in Alaska (USA). *Advantages:* Allows automatic discretization of landscape units in large areas. Useful as a preliminary method to learn which scale of clustering is suitable to study different processes or focuses of the study (e.g., hydrology versus vegetation). Capable of detecting vegetation changes as it recognizes vegetation levels in different classes. *Limitations:* No error assessment was performed, nor was there ground truth validation. The selection of an appropriate number of clusters depends on the expert's decision.	[213]
SVM, RF, ANN, partial least squares, multivariate adaptive regression splines, flexible discriminant analysis, k-NN, regression tree, bagged trees, linear discriminant analysis, regularized linear discriminant analysis, and naive Bayes. *Software:* Caret and h2o R packages.	*Objective:* To extrapolate a geomorphic classification of channel types to a regional stream network using DEMs and thematic maps (e.g., lithology, soil, stream network, etc.) from Sacramento River, USA. *Advantages:* Multiple algorithms compared. RF outperformed other models with more accuracy and stability and lower entropy in reach-scale river type classification. Rigorous approach in model design and evaluation of performance with 20×10-fold cross validation used for clarification of some black box aspects of ML. *Limitations:* Needs large expert-based field survey data. It is unclear if ML is able to integrate predictors at different scales and to show different uncertainty across the watershed.	[214]

Table 5. *Cont.*

Method and Software	Objective, Advantages and Limitations	Reference
RF and RF combined with recursive feature elimination (RF-RFE). *Software:* R program	*Objective:* To detect structural and/or neotectonic controls influencing the knickpoints of the drainage network using DEM and thematic maps (geology and geomorphology) from Abaeté Watershed, Brazil. *Advantages:* Simple and reproducible methodology that provides causal relationship of knickpoint formation to lithological contacts and neotectonic configuration and activity. RF succeeded in partially predicting geomorphic indices (e.g., stream length gradient index or normalized channel steepness index) and can be used to predict them in unsampled areas without overfitting. *Limitations:* Low performance of the methods, obtaining $R^2 = 0.38$ as the highest correlation between predicted and direct estimation values of geomorphic indices. It may have been affected by the selection of the covariables.	[215]
Template-matching (object-based) algorithm (TMA) and pixel-based SVM. *Software:* Feature Analyst (Overwatch Systems Ltd.). Not specified for SVM.	*Objective:* To delineate water surface boundaries and assess the influence of river and bank characteristics in the efficacy of a template-matching compared to a pixel-based algorithm, using high-resolution images with false-color infrared, from the Brazos River, USA. *Advantages:* Both algorithms adequately delineate the water surface. SVM performed better and handled complex and noisy class relationships. TMA performed better than SVM in spatially complex channel morphologies (e.g., partially submerged sediment deposits, sediment bar structures) due to its capability to incorporate both spectral and spatial information. *Limitations:* Validation relies on expert knowledge and previous maps. Selection of ancillary data types depends on expert decision and the delineation accuracy of TMA. In addition, the low spectral dimension of the images limited the pixel-based classification. Both algorithms encounter problems when classifying multiple complex features (e.g., overhanging trees) and illumination conditions (e.g., shading). The TMA performance was less spatially consistent than that of SVM.	[216]
SOM *Software:* R based (R v3.5.1). Package "kohonen" v3.0.7	*Objective:* To produce waterbody typology from 22 GIS-derived continuous catchment characteristics to capture the dominant controls that influence river reaches across England and Wales. *Advantages:* SOM-based water body topology reflects catchment functional feature controls on river reaches. The method is extendable to other areas where reach-level monitoring is relevant. The SOM combined with hierarchical clustering can be applied over a wide range of catchment, e.g., a national-level waterbody typology map. *Limitations:* The method could not isolate individual effects from catchment controls as they are dependent on each other. It does not detect temporal change and local controls such as dams, channelization, and others not taken into account.	[217]
U-Net convolutional neural networks (CNNs). *Software:* Not stated.	*Objective:* To introduce the BathyNet framework, a photogrammetry and radiometric-based combined retrieval of water depth using U-Net CNNs. Study area was Lech river, Augsburg, Bavaria, Germany. *Advantages:* U-Net CNNs approximate arbitrary functions and include spatial context. The U-Net CNN-based depth retrieval outperformed traditional regression-based optical inversion methods. *Limitations:* U-Net CNNs require large amounts of training data and their application might be unfeasible in areas with scarce water-depth field samples.	[218]

The current state-of-the-art of GeoAI in fluvial geomorphology consists of an automatic extraction of fluvial features at a fine scale by integrating larger and multidimensional datasets, using unsupervised classifiers (e.g., K-means, SOM), supervised classifiers (e.g., RF, SVM, ANN, deep learning, CNN), or by combining both methods, e.g., K-means with ANN. Most of the reviewed articles were focused on the development of the methods and workflow, the testing of new applications, or the comparison of algorithm performances [205,207,209], rather than the study of fluvial processes and underlying dynamics. These applications of GeoAI provide the basis to the discovery of new fluvial patterns and trends and increase knowledge about fluvial environments (e.g., Ling et al. 2019; Guillon et al. 2020, Heasley et al. 2020) [208,214,217].

Overall, GeoAI outperforms conventional methods of fluvial landform classification, reaching a classification accuracy of over 80%. Most common applications are found in river channels and water body mapping [208,216], the classification of riverine landforms and vegetation successions [213,214,219,220], the estimation of catchment hydrogeomorphic characteristics (e.g., valley bottom, floodplain, and terrace) [212,221], and benthic and fish habitat mapping [207,211,222,223].

Another application of GeoAI is the integration of multiple techniques to provide more accurate and very-high-resolution data for fluvial studies. For example, the fluvial environment is highly dynamic and demands frequent bathymetry surveys to understand the change and morphodynamic drivers in lakes and rivers. Emerging technologies, such as acoustic Doppler current profiler (ADCP), green LiDAR, high-resolution image radiometric model, and 3D cloud points generation with SfM, allow more frequent and accurate bathymetry mapping [203,204]. However, each approach has limitations, e.g., ADCP collects data only from areas where the sensor has passed, and it does not provide continuous spatial scanning. It does not measure near-bank areas, and it is subject to the acoustic side-lobe effect [224]. Photogrammetry and the green LiDAR method are sensitive to water turbidity and light penetration in the water column [225,226]. Therefore, multisource bathymetry modeling using the GeoAI method increases the bathymetric data accuracy and reduces uncertainties due to data quality in change detection. For example, ADCP data, image radiometric-based water depth, and SfM depth data can be integrated using U-Net convolutional neural networks [218,227].

The GeoAI approach, when using multi-temporal remote sensing data, allows the mapping of a broader fluvial landscape and its change, thereby revealing spatiotemporal scales of fluvial morphodynamics, as in e.g., Van Iersel et al. [228], Hemmelder et al. [229], and Boothroyd et al. [230]. There are different GeoAI approaches for automatic change detection using multi-temporal images such as generative adversarial networks (GAN), autoencoder, CNN, and others, as presented by Shi et al. [231].

Although GeoAI has been rapidly adopted in fluvial geomorphological studies, a wide spectrum of workflows and software is found; many GeoAI approaches seem to be under development and in the testing stage. Therefore, without a general, consistent, and robust workflow among them, it is difficult to generalize and compare the GeoAI methods performance and overall accuracies, as well as the study results.

The current limitations of GeoAI methods in fluvial studies are that the classification quality is highly dependent on expert knowledge. The unsupervised classification output is often inconsistent, and the cluster classes do not have direct geomorphic or fluvial process meaning and need a post-classification labeling. Supervised GeoAI classifiers require a large training sampling, and the training data quality is highly dependent on expert knowledge. In addition, many of the studies using GeoAI to classify fluvial landform or river typologies have been conducted in areas where an extensive quantity of previous studies and data collection exists [212,214]. Therefore, its application in poorly sampled areas is somewhat limited.

In many cases, GeoAI is enhanced with the use of fine-scale fluvial geomorphic mapping, e.g., LiDAR or UAV-based images, which are still restricted to pilot areas, mostly in Western countries. In addition, several different landform class names are used to rename fine-scale fluvial landforms, and therefore, a standardized fluvial landform taxonomy is lacking [232].

Another limitation of supervised GeoAI applications is the misclassification of elements out of the GeoAI training range, as presented, e.g., in Carbonneau et al. [205]. Moreover, the use of very different methods for assessing the GeoAI algorithm's performance and accuracy may lead to inconsistencies in the validity of results, e.g., map cross-tabulation often uses limited validation points rather than areal-based reference data, due to the lack of geomorphological reference maps at a very fine scale. Another issue with regard to performance and accuracy assessments is the use of scalar error statistics, such as root mean square error, which may not be reliable in fluvial mapping. Here the resulting error is a complex combination of random and systematic components, and the isotropy and stationary assumptions do not apply to the fluvial process [233]. It is also heavily influenced by a small percentage of classification errors, which lead to incorrect rankings of overall model performances or to prediction error [206]. Therefore, a more consistent and comparable GeoAI-based fluvial mapping accuracy assessment is needed.

5. GeoAI Causal and Predictive Inference Capability

5.1. Renewed Data-Driven Research

Observational and experimental studies have been the basis of understanding the empirical relationships of physical processes occurring in the earth and the development of the mechanical or physical-based models to predict them [234]. With the substantial increase in observational data and the development of GeoAI methods, empirical studies have been renewed with data-driven models [17,235]. Unlike traditional statistical models, GeoAI methods do not rely on a formal assumption about the data structure and types of data distribution such as normality. They are more flexible and adaptable for nonlinear and high-dimensional data. GeoAI methods automatically identify and exploit correlations and patterns (classification) in the data to make predictions. For example, an ANN, with many hidden layers and free parameters estimated by training and arbitrary fitting curves, converts inputs to outputs by simply minimizing error variances [39].

To date, in most of the GeoAI applications for hydrological studies the cause–effect relationship inference has been limited, because the multiple driven factors and interactions between the used variables and scales are not explicitly represented in the models [50,123]. In addition, the GeoAI and ML internal hyper-parameter optimization is not explicitly stated in most of the modeling studies. For this reason, GeoAI methods are often called "black-box" models [236]. See Table 6 for the characteristics of physical-based and GeoAI hydrological models. Therefore, causal inferences might be questionable without robust assumptions and the veracity of the assumed data structure [237]. Thus, most GeoAI models are mostly inductive approaches, mainly oriented for operational prediction and forecasting work, such as early warning systems. Nevertheless, GeoAI models have the potential to reveal unknown associations and complex patterns of hydrology processes by integrating more high-dimensional and multi-source data than traditional methods. By implementing proper model interpretation and explainability methods, they can also extend GeoAI applications for causal inference [236,237].

Table 6. Characteristics of physical-based and GeoAI hydrological models.

Process-Based Model	GeoAI Model
Based on general physical laws.	A data-driven approach, inductive model building, may not fully respect physical laws.
All input variables and parameter ranges are well-structured and known.	Unstructured data, not all input variables' role in the model is known, making the output less interpretable.
Limited to the current state-of-the-art.	Able to reveal unknown associations and patterns.
It is a deductive, hypothesis testing approach. It can be used for causal inference.	It is an inductive, exploratory approach. Their use in causal inference depends on the GeoAI model building and selected variables.

Table 6. *Cont.*

Process-Based Model	GeoAI Model
No uniqueness problem due to inverse modeling in model parameterization.	No uniqueness problem due to GeoAI hyper-parameter optimizations.
Mostly deterministic, the system is represented by the average values of variables.	Deterministic and probabilistic, depending on the GeoAI method, variables can be treated as probabilistic.
Reductionist, considerable simplification of complex processes which can result in prediction bias.	Integrative, GeoAI can be integrated into several types of observation and may be able to reveal patterns not represented by a physical-based model. Therefore, GeoAI prediction can be less biased.
It is assumed to be of general application.	It is assumed to be applied only within the range of the training data.
Fixed to the model basis data requirement, and unable to deal with multisource data.	Flexible to data input, from minimal input to big data. GeoAI can maximize the use of all types of available data, from different sources, types, and quality.
High computing demand for high-frequency and large-scale modeling.	High computing efficiency and suitable for high-frequency and large-scale modeling.
One-time calibration, once the model is calibrated, the parameters are usually fixed.	Continuous learning model, the model calibration is constantly updated with past and new data.
Well-defined framework for model performance, uncertainty, and error propagation evaluation.	Diverse and developing approaches for model performance, uncertainty, and error propagation evaluation.

5.2. Generalization of GeoAI Prediction

GeoAI models may only be applicable within their specific training data or calibrated ranges [238], unless the modeling scheme and variables used can be argued to be generally valid, e.g., representing general laws such as conservation and momentum laws that govern natural processes [234]. GeoAI modeling generalization is also a challenging problem from the perspective of model performance assessment, depending on the model complexity, variables, and training dataset size. A very simple model cannot learn the problem being modeled (underfitting problem), whereas a highly complex model with a large dataset might overfit the training dataset (overfitting problem). Both cases are not generalizable or applicable to new datasets. Current GeoAI generalization approaches are based on finding an optimal tradeoff between training and validation accuracy, using regularization, weight decay, ensembles, and other approaches in the model training stage [40]. However, the decision boundary in complex models becomes sensitive to data size and outliers, model architecture, and hyper-parameter optimization. It has also been observed that different sets of the model architecture and hyper-parameters can produce a similar model performance, leading to the non-uniqueness modeling problem [50,239].

5.3. GeoAI Data Requirement for Reliably Prediction

GeoAI models depend on the quality and quantity of the data. The amount of data required for them depends on many factors, such as the complexity of the hydrological system and the applied GeoAI algorithm [47]. A complex system with more sophisticated GeoAI methods will demand a large and multidimensional dataset [42]. For example, deep/extreme learning algorithms usually demand large sample sizes to compute acceptable results [240]. Current hydrological and geospatial data are increasing rapidly, fostered by the development of automatic monitoring systems and land surveying technologies. However, the data quantity (volume) and quality (veracity and value) vary; the data types are diverse (unstructured, structures, spatial, non-spatial, etc.), and the datasets usually come from different sources. Datasets with these characteristics are called big data [241,242] and require advanced and new methodologies to integrate them with GeoAI models properly.

5.4. GeoAI Capacity to Provide Novel Physical Insights

GeoAI data-driven research and data mining are increasingly used to gain information from data, elucidate systems behavior, reveal new insights about the system functioning, and detect change in the system responses [17]. There are several examples of GeoAI applications in hydrological modeling [91,105,107,243]. Recent studies applying deep learning to rainfall-runoff simulation indicate that there is significantly more information in large-scale hydrological data sets than hydrologists have been able to translate into theory or models [129]. GeoAI has also revealed new hydrological patterns and trends, using heterogeneous data from different sources and quality [244,245]. Therefore, novel data-driven modeling provides the potential to gain new information and knowledge and a better understanding of the hydrological system and its changes [129,235].

6. GeoAI Research Trends in Integrated Hydrological and Fluvial Systems Modeling

6.1. Toward Transdisciplinary GeoAI Research in Hydrological Modeling

Nowadays, earth science has mostly adopted GeoAI approaches developed in other fields, particularly computer science. GeoAI is also an active field of research in advanced hydrological modeling, providing new insights into hydrological system functioning, advantages in computational efficiency, and prediction accuracy. Nevertheless, it depends on how the hydrological GeoAI model has been set up by the user, the quantity and quality of the data, and the types and number of variables used.

GeoAI methods can be integrated with other data analysis techniques, e.g., Fourier and wavelet transformation, to remove noise and provide better hydrological feature extraction [30,198]. Hence a transdisciplinary approach is demanded to ensure insightful research on GeoAI applications in hydrological and fluvial studies [235]. This is particularly relevant as the complexity of the GeoAI models is increasing continuously, and model parametrization and parallel computing solutions require expert knowledge for proper GeoAI technology adoption [18,240]. Similarly, these issues also arise when hydrological science principles are not explicitly integrated with the GeoAI data-driven models, resulting in a limited explainability of the underlying physical laws that govern hydrological and hydraulic processes [50,129].

6.2. Augmenting GeoAI Prediction Capability with Open Data and Crowdsourced Data

GeoAI models demand a large amount of training data. Although data collection technology has progressed substantially, only a few geographical areas or pilot hydrological systems are well equipped. For example, very few catchments have implemented IoT hydrological monitoring technology. GeoAI models will demand a rapid and massive increase in data collection. The current open-access policy of many governmental environmental agencies, related to climatological, hydrological, and environmental data, enhances the data-driven research and GeoAI applications, particularly in Western countries. Similarly, open access to high-resolution topographical and earth observation data (e.g., NASA and the ESA-EU Copernicus Programme) also accelerates the development of GeoAI-based hydrological models [241,246]. Additionally, the current trend of implementing open-access training libraries, e.g., training data for land cover classification, is valuable, but more specialized hydrogeomorphic labeled data are still under development.

Citizen science also plays a key role in complementing and increasing data collection worldwide. There are several examples of how hydrological crowdsourcing enhances hydrological data availability for scientific research, using images and social media data [247] and low-cost data loggers [248,249], but the success and quality of hydrological crowdsourcing are variable, depending on the regions, the instrument used, and the variables reported [250]. GeoAI-based hydrological model development will benefit from crowdsourcing data collection.

6.3. From Physical-Based and GeoAI Hybrid Models to Fully Integrated GeoAI–Physical-Based Models

Physical-based and GeoAI hydrological models have had different paths of development. As discussed previously, a physical-based model is derived from empirical and experimental research; meanwhile, a GeoAI model is derived from data sciences techniques.

Physical-based and GeoAI models are not complementary per se, but in many cases, the integration of both approaches has shown a great potential to improve hydrological modeling [18,129]. Currently, there is a different level of integration; most of them are still so-called loose integrated models, where the GeoAI and the physical-based models work independently. The GeoAI method is used for data preparation and the refining of physical-based models, e.g., data fusion, ad hoc parameter optimizations, and data assimilation. In some cases, the outputs of physical-based models are used to train GeoAI models [19,188,251]. Currently, full GeoAI-physical integration is under development, embedding machine learning solutions into physical-based models or developing physically guided GeoAI models; see, e.g., Hanson et al. [21]. Both approaches tend to overcome current GeoAI model limitations by providing more physical explanatory power, physically consistent and robust prediction, and a high level of generalization.

6.4. From Small-Scale to Global-Scale Hydrological Modeling

In recent years, substantial attention has been paid to large-scale and global-scale hydrological modeling [252–254]. Although only experimental catchments have sufficient data to perform a reliable hydrological prediction, the global availability of climatological, hydrological, and remote sensing data allows for the parametrizing of the global-level hydrological model. This planet-wide dataset can only be handled thanks to a combined advancement in GeoAI application and cloud computing development, e.g., Google Earth Engine, CoLab, SEPAL [255], and many other national high-performance computer clusters. However, global-scale hydrological modeling still involves a high level of prediction uncertainty [256,257], but current progress in the development of physical-based GeoAI models and remote sensing data assimilation can improve global modeling accuracy.

6.5. Automation of Hydrological and Fluvial System Modeling

GeoAI applications are increasing the automation of hydrological prediction and forecasting [258]. Some hydrological modeling has already applied internal self-calibration [259–261]. Similarly, there is also substantial progress in developing automated machine learning (autoML) by self-tuning the models' hyperparameters, such as, e.g., autotune and AUTO-SKLEARN [262,263]. The hyperparameters drive both the efficiency of the model training process and the resulting model quality [262]. Therefore, a self-tuning module will enhance a more rapid adoption of GeoAI models in hydrological modeling, and the integration of physical-based and GeoAI models can improve autonomous hydrological prediction.

Similarly, self-supervised image classification, particularly that developed in the robotic field [264], is rapidly being adopted in hydrological studies in, e.g., satellite image classification, fluvial landform classification, and landform change detection. Self-supervised models use automatically generated pseudo-labels, significantly reducing manual labeling, one of the most time-consuming tasks in supervised classification [265]. Self-supervised image classification is enhanced by machine learning methods such as autoencoder and the generative adversarial network (GAN). Autoencoder enhances image quality and reduces noise by dimension reduction and retaining latent features [266]. GAN is a promising technique to further automate high-dimensional image classification with limited data training. GAN generates new data instances that resemble the existing training data by the competition between a generator and a discriminator [267]. Several examples show the advantages of incorporating GAN models in hydrological classification [267–269] or combining it with autoencoder [270]. Integrating GAN with an LSTM network model [271–273]; combining GAN with an ANN fuzzy model [274] was also found to improve the automated hydrological and weather prediction using satellite data.

6.6. GeoAI-Based Multi-Dimensional Geo-Visualization and Digital Twin

Hydrological systems are complex by nature and have been challenged to comprehensively and effectively convey spatial and non-spatial hydrological information. The explosion of high-dimensional, multi-source, spatio-temporal hydrological data demands new ways of multi-dimensional geo-visualization [275]. The GeoAI model optimizes the transformation of multi-dimensional data into conventional 3D geo-visualization (x, y, and z features), but also into 4D (including temporal dimension) and 5D geo-visualization (including geographical scale). The 4D and 5D visualization is crucial for dynamic and interactive web-based geo-visualization [276]. The GeoAI also supports building hydrological digital twins, integrating IoT sensors, and multi-scale satellite and close-range remote sensing data, with web-based hydrological GeoAI models for real-time prediction and geo-visualization. A 'digital twin' is a comprehensive digital emulator of the real-world system that aims to optimize the design and operations of complex processes through a highly interconnected workflow [277]. Hydrological digital twins support the correct implementation of the IWRM actions, including natural disaster response, nexus approaches, and adaptation to climate change. Those actions require approaches underpinned by a deeper analysis of river basin systems functioning, scaling-up field-based knowledge, and new digital solutions to provide real-time, high-resolution information [278]. Additionally, the advance in web-mapping services (WMS) and mobile app development with interactive geo-visualization [279] enhances hydrological information dissemination for decision-makers, stakeholders, and the general public engagement.

7. Conclusions

GeoAI applications in integrated hydrological and fluvial system modeling have steadily increased in recent years. We found plenty of GeoAI applications in hydrological and fluvial studies. The main applications were for assessing GeoAI hydrological prediction and classification performance, comparing GeoAI methods with hydrological physical-based models and integrating physical-based models with GeoAI. A wide range of GeoAI methods are currently applied in this field, e.g., RF, SVM, ANN, LSTM, GAN, GA, and meta-heuristic algorithms. The selection of a particular algorithm depends on the application objective, data availability, and user expertise.

Overall, GeoAI applications showed advantages in non-linear modeling, computational efficiency, integration of heterogeneous data sources, high-accuracy prediction, and the unraveling of new hydrological patterns or in detecting changes using high-dimensional, multi-source geospatial data. GeoAI methods seem particularly relevant for complex systems and large geographical-scale modeling. A significant disadvantage of GeoAI models is the low level of physical interpretability, explainability, and model generalization. Therefore, current research trends focus on integrating the physical-based model with GeoAI methods to bridge data-driven and theory-driven knowledge generation. Several levels of model integrations exist, but a full physical-based GeoAI model is still under development. The GeoAI models have shown high potential for autonomous hydrological prediction and forecasting and early warning systems.

Author Contributions: Conceptualization, C.G.-I., J.S., M.C. and H.M.; methodology, C.G.-I., M.C. and P.A.; analysis, C.G.-I., J.S., M.C., D.C. and H.M.; investigation, C.G.-I., J.S., D.C., M.C. and H.M.; resources, P.A.; writing—original draft preparation, C.G.-I., J.S., D.C., M.C. and H.M.; writing—review and editing, C.G.-I., J.S., H.M., D.C., M.C., A.T.H. and P.A.; visualization, C.G.-I. and M.C.; funding acquisition, P.A. All authors have read and agreed to the published version of the manuscript.

Funding: This work was funded by Academy of Finland, grant number 337279, 346161, 347701 and 346165 (NextGenerationEU). MC was funded by the Turku Collegium of Science, Medicine and Technology (TCSMT), University of Turku, Finland.

Institutional Review Board Statement: Not applicable.

Informed Consent Statement: Not applicable.

Data Availability Statement: Not applicable.

Acknowledgments: We kindly thank the Freshwater Competence Centre (Available online: www.freshwatercompetencecentre.com (accessed on 21 June 2022)).

Conflicts of Interest: The authors declare no conflict of interest.

Abbreviations

AI	Artificial Intelligence
ANN	Artificial Neural Network
CNN	Convolution Neural Network
DA	Data Assimilation
DEM	Digital Elevation Model
GA	Genetic Algorithm
GAN	Generative Adversarial Networks
GeoAI	Geospatial Artificial Intelligence
GP	Genetic programing
IoT	Internet of Things
IWRM	Integrated Water Resources Management
LiDAR	Light Detection and Ranging
LSTM	Long Short-term Memory Networks
ML	Machine Learning
RF	Random Forest (RF)
RL	Reinforced Learning
RNN	Recurrent Neural Network (RNN)
SVM	Support Vector Machine (SVM)
UAV	Unmanned Aerial Vehicles
WQ	Water Quality

References

1. Beven, K. Towards a Methodology for Testing Models as Hypotheses in the Inexact Sciences. *Proc. R. Soc. Math. Phys. Eng. Sci.* **2019**, *475*, 20180862. [CrossRef] [PubMed]
2. McMillan, H.K.; Westerberg, I.K.; Krueger, T. Hydrological Data Uncertainty and Its Implications. *WIREs Water* **2018**, *5*, e1319. [CrossRef]
3. Sun, A.Y.; Scanlon, B.R. How Can Big Data and Machine Learning Benefit Environment and Water Management: A Survey of Methods, Applications, and Future Directions. *Environ. Res. Lett.* **2019**, *14*, 073001. [CrossRef]
4. Todini, E. History and Perspectives of Hydrological Catchment Modelling. *Hydrol. Res.* **2011**, *42*, 73–85. [CrossRef]
5. Coulthard, T.J.; Van De Wiel, M.J. Modelling River History and Evolution. *Philos. Trans. R. Soc. Math. Phys. Eng. Sci.* **2012**, *370*, 2123–2142. [CrossRef]
6. Singh, V.P. Hydrologic Modeling: Progress and Future Directions. *Geosci. Lett.* **2018**, *5*, 15. [CrossRef]
7. Koren, V.; Smith, M.; Cui, Z. Physically-Based Modifications to the Sacramento Soil Moisture Accounting Model. Part A: Modeling the Effects of Frozen Ground on the Runoff Generation Process. *J. Hydrol.* **2014**, *519*, 3475–3491. [CrossRef]
8. Arnold, J.; Moriasi, D.; Gassman, P.; Abbaspour, K.; White, M.; Srinivasan, R.; Santhi, C.; Harmel, R.; van Griensven, A.; Van Liew, M.; et al. SWAT: Model Use, Calibration, and Validation. *Trans. ASABE* **2012**, *55*, 1491–1508. [CrossRef]
9. Gochis, D.J.; Yu, W.; Yates, D. The WRF-Hydro Model Technical Description and User-s Guide. *NCAR Tech. Doc.* **2013**. [CrossRef]
10. Brunner, P.; Simmons, C.T. HydroGeoSphere: A Fully Integrated, Physically Based Hydrological Model. *Groundwater* **2012**, *50*, 170–176. [CrossRef]
11. Lane, S.N.; Ferguson, R.I. Modelling Reach-Scale Fluvial Flows. In *Computational Fluid Dynamics*; John Wiley & Sons, Ltd.: Hoboken, NJ, USA, 2005; pp. 215–269, ISBN 978-0-470-01519-3.
12. Clark, M.P.; Bierkens, M.F.P.; Samaniego, L.; Woods, R.A.; Uijlenhoet, R.; Bennett, K.E.; Pauwels, V.R.N.; Cai, X.; Wood, A.W.; Peters-Lidard, C.D. The Evolution of Process-Based Hydrologic Models: Historical Challenges and the Collective Quest for Physical Realism. *Hydrol. Earth Syst. Sci.* **2017**, *21*, 3427–3440. [CrossRef]
13. Hauswirth, S.M.; Bierkens, M.F.P.; Beijk, V.; Wanders, N. The Potential of Data Driven Approaches for Quantifying Hydrological Extremes. *Adv. Water Resour.* **2021**, *155*, 104017. [CrossRef]
14. Noori, N.; Kalin, L. Coupling SWAT and ANN Models for Enhanced Daily Streamflow Prediction. *J. Hydrol.* **2016**, *533*, 141–151. [CrossRef]
15. Li, X.; Yan, D.; Wang, K.; Weng, B.; Qin, T.; Liu, S. Flood Risk Assessment of Global Watersheds Based on Multiple Machine Learning Models. *Water* **2019**, *11*, 1654. [CrossRef]

16. Hsu, K.; Gupta, H.V.; Sorooshian, S. Artificial Neural Network Modeling of the Rainfall-Runoff Process. *Water Resour. Res.* **1995**, *31*, 2517–2530. [CrossRef]
17. Reichstein, M.; Camps-Valls, G.; Stevens, B.; Jung, M.; Denzler, J.; Carvalhais, N. Prabhat Deep Learning and Process Understanding for Data-Driven Earth System Science. *Nature* **2019**, *566*, 195–204. [CrossRef]
18. Razavi, S. Deep Learning, Explained: Fundamentals, Explainability, and Bridgeability to Process-Based Modelling. *Environ. Model. Softw.* **2021**, *144*, 105159. [CrossRef]
19. Hosseiny, H.; Nazari, F.; Smith, V.; Nataraj, C. A Framework for Modeling Flood Depth Using a Hybrid of Hydraulics and Machine Learning. *Sci. Rep.* **2020**, *10*, 8222. [CrossRef]
20. Tsai, W.-P.; Feng, D.; Pan, M.; Beck, H.; Lawson, K.; Yang, Y.; Liu, J.; Shen, C. From Calibration to Parameter Learning: Harnessing the Scaling Effects of Big Data in Geoscientific Modeling. *Nat. Commun.* **2021**, *12*, 5988. [CrossRef]
21. Hanson, P.C.; Stillman, A.B.; Jia, X.; Karpatne, A.; Dugan, H.A.; Carey, C.C.; Stachelek, J.; Ward, N.K.; Zhang, Y.; Read, J.S.; et al. Predicting Lake Surface Water Phosphorus Dynamics Using Process-Guided Machine Learning. *Ecol. Model.* **2020**, *430*, 109136. [CrossRef]
22. Lu, D.; Konapala, G.; Painter, S.L.; Kao, S.-C.; Gangrade, S. Streamflow Simulation in Data-Scarce Basins Using Bayesian and Physics-Informed Machine Learning Models. *J. Hydrometeorol.* **2021**, *22*, 1421–1438. [CrossRef]
23. Maxwell, R.M.; Condon, L.E.; Melchior, P. A Physics-Informed, Machine Learning Emulator of a 2D Surface Water Model: What Temporal Networks and Simulation-Based Inference Can Help Us Learn about Hydrologic Processes. *Water* **2021**, *13*, 3633. [CrossRef]
24. Tran, H.; Leonarduzzi, E.; De la Fuente, L.; Hull, R.B.; Bansal, V.; Chennault, C.; Gentine, P.; Melchior, P.; Condon, L.E.; Maxwell, R.M. Development of a Deep Learning Emulator for a Distributed Groundwater–Surface Water Model: ParFlow-ML. *Water* **2021**, *13*, 3393. [CrossRef]
25. Madaeni, F.; Lhissou, R.; Chokmani, K.; Raymond, S.; Gauthier, Y. Ice Jam Formation, Breakup and Prediction Methods Based on Hydroclimatic Data Using Artificial Intelligence: A Review. *Cold Reg. Sci. Technol.* **2020**, *174*, 103032. [CrossRef]
26. Zounemat-Kermani, M.; Matta, E.; Cominola, A.; Xia, X.; Zhang, Q.; Liang, Q.; Hinkelmann, R. Neurocomputing in Surface Water Hydrology and Hydraulics: A Review of Two Decades Retrospective, Current Status and Future Prospects. *J. Hydrol.* **2020**, *588*, 125085. [CrossRef]
27. Rajaee, T.; Khani, S.; Ravansalar, M. Artificial Intelligence-Based Single and Hybrid Models for Prediction of Water Quality in Rivers: A Review. *Chemom. Intell. Lab. Syst.* **2020**, *200*, 103978. [CrossRef]
28. Najah Ahmed, A.; Binti Othman, F.; Abdulmohsin Afan, H.; Khaleel Ibrahim, R.; Ming Fai, C.; Shabbir Hossain, M.; Ehteram, M.; Elshafie, A. Machine Learning Methods for Better Water Quality Prediction. *J. Hydrol.* **2019**, *578*, 124084. [CrossRef]
29. Naloufi, M.; Lucas, F.S.; Souihi, S.; Servais, P.; Janne, A.; Wanderley Matos De Abreu, T. Evaluating the Performance of Machine Learning Approaches to Predict the Microbial Quality of Surface Waters and to Optimize the Sampling Effort. *Water* **2021**, *13*, 2457. [CrossRef]
30. Nourani, V.; Hosseini Baghanam, A.; Adamowski, J.; Kisi, O. Applications of Hybrid Wavelet–Artificial Intelligence Models in Hydrology: A Review. *J. Hydrol.* **2014**, *514*, 358–377. [CrossRef]
31. Sit, M.; Demiray, B.Z.; Xiang, Z.; Ewing, G.J.; Sermet, Y.; Demir, I. A Comprehensive Review of Deep Learning Applications in Hydrology and Water Resources. *Water Sci. Technol.* **2020**, *82*, 2635–2670. [CrossRef]
32. Kambalimath, S.; Deka, P.C. A Basic Review of Fuzzy Logic Applications in Hydrology and Water Resources. *Appl. Water Sci.* **2020**, *10*, 191. [CrossRef]
33. Zounemat-Kermani, M.; Batelaan, O.; Fadaee, M.; Hinkelmann, R. Ensemble Machine Learning Paradigms in Hydrology: A Review. *J. Hydrol.* **2021**, *598*, 126266. [CrossRef]
34. Pham, M.T.; Rajić, A.; Greig, J.D.; Sargeant, J.M.; Papadopoulos, A.; McEwen, S.A. A Scoping Review of Scoping Reviews: Advancing the Approach and Enhancing the Consistency. *Res. Synth. Methods* **2014**, *5*, 371–385. [CrossRef] [PubMed]
35. Sucharew, H.; Macaluso, M. Methods for Research Evidence Synthesis: The Scoping Review Approach. *J. Hosp. Med.* **2019**, *14*, 416–418. [CrossRef] [PubMed]
36. VoPham, T.; Hart, J.E.; Laden, F.; Chiang, Y.-Y. Emerging Trends in Geospatial Artificial Intelligence (GeoAI): Potential Applications for Environmental Epidemiology. *Environ. Health* **2018**, *17*, 40. [CrossRef]
37. Janowicz, K.; Gao, S.; McKenzie, G.; Hu, Y.; Bhaduri, B. GeoAI: Spatially Explicit Artificial Intelligence Techniques for Geographic Knowledge Discovery and Beyond. *Int. J. Geogr. Inf. Sci.* **2020**, *34*, 625–636. [CrossRef]
38. Hu, Y.; Gao, S.; Lunga, D.; Li, W.; Newsam, S.; Bhaduri, B. GeoAI at ACM SIGSPATIAL: Progress, Challenges, and Future Directions. *Sigspatial Spec.* **2019**, *11*, 5–15. [CrossRef]
39. Hastie, T.; Tibshirani, R.; Friedman, J. Introduction. In *The Elements of Statistical Learning: Data Mining, Inference, and Prediction*; Hastie, T., Tibshirani, R., Friedman, J., Eds.; Springer Series in Statistics; Springer: New York, NY, USA, 2009; pp. 1–8, ISBN 978-0-387-84858-7.
40. Goodfellow, I.; Bengio, Y.; Courville, A. *Deep Learning*; Adaptive Computation and Machine Learning Series; MIT Press: Cambridge, MA, USA, 2016; ISBN 978-0-262-03561-3.
41. Lee, T.; Singh, V.P.; Cho, K.H. *Deep Learning for Hydrometeorology and Environmental Science*; Water Science and Technology Library; Springer International Publishing: Cham, Switzerland, 2021; ISBN 978-3-030-64776-6.

42. Sarker, I.H. Machine Learning: Algorithms, Real-World Applications and Research Directions. *SN Comput. Sci.* **2021**, *2*, 160. [CrossRef]

43. Nicholaus, I.T.; Park, J.R.; Jung, K.; Lee, J.S.; Kang, D.-K. Anomaly Detection of Water Level Using Deep Autoencoder. *Sensors* **2021**, *21*, 6679. [CrossRef]

44. Elith, J.; Leathwick, J.R.; Hastie, T. A Working Guide to Boosted Regression Trees. *J. Anim. Ecol.* **2008**, *77*, 802–813. [CrossRef]

45. Ibrahim, K.S.M.H.; Huang, Y.F.; Ahmed, A.N.; Koo, C.H.; El-Shafie, A. A Review of the Hybrid Artificial Intelligence and Optimization Modelling of Hydrological Streamflow Forecasting. *Alex. Eng. J.* **2022**, *61*, 279–303. [CrossRef]

46. Jia, J.; Wang, W. Review of Reinforcement Learning Research. In Proceedings of the 2020 35th Youth Academic Annual Conference of Chinese Association of Automation (YAC), Zhanjiang, China, 16–18 October 2020; pp. 186–191.

47. Lange, H.; Sippel, S. Machine Learning Applications in Hydrology. In *Forest-Water Interactions*; Levia, D.F., Carlyle-Moses, D.E., Iida, S., Michalzik, B., Nanko, K., Tischer, A., Eds.; Ecological Studies; Springer International Publishing: Cham, Switzerland, 2020; pp. 233–257, ISBN 978-3-030-26086-6.

48. Troch, P.A.; Carrillo, G.A.; Heidbüchel, I.; Rajagopal, S.; Switanek, M.; Volkmann, T.H.M.; Yaeger, M. Dealing with Landscape Heterogeneity in Watershed Hydrology: A Review of Recent Progress toward New Hydrological Theory. *Geogr. Compass* **2009**, *3*, 375–392. [CrossRef]

49. Sivakumar, B. Nonlinear Dynamics and Chaos in Hydrologic Systems: Latest Developments and a Look Forward. *Stoch. Environ. Res. Risk Assess.* **2009**, *23*, 1027–1036. [CrossRef]

50. Schmidt, L.; Heße, F.; Attinger, S.; Kumar, R. Challenges in Applying Machine Learning Models for Hydrological Inference: A Case Study for Flooding Events Across Germany. *Water Resour. Res.* **2020**, *56*, e2019WR025924. [CrossRef]

51. Badrzadeh, H.; Sarukkalige, R.; Jayawardena, A.W. Hourly Runoff Forecasting for Flood Risk Management: Application of Various Computational Intelligence Models. *J. Hydrol.* **2015**, *529*, 1633–1643. [CrossRef]

52. Schoppa, L.; Disse, M.; Bachmair, S. Evaluating the Performance of Random Forest for Large-Scale Flood Discharge Simulation. *J. Hydrol.* **2020**, *590*, 125531. [CrossRef]

53. Zhang, X.; Peng, Y.; Zhang, C.; Wang, B. Are Hybrid Models Integrated with Data Preprocessing Techniques Suitable for Monthly Streamflow Forecasting? Some Experiment Evidences. *J. Hydrol.* **2015**, *530*, 137–152. [CrossRef]

54. Iwashita, F.; Friedel, M.J.; Ferreira, F.J.F. A Self-Organizing Map Approach to Characterize Hydrogeology of the Fractured Serra-Geral Transboundary Aquifer. *Hydrol. Res.* **2017**, *49*, 794–814. [CrossRef]

55. Kimura, N.; Yoshinaga, I.; Sekijima, K.; Azechi, I.; Baba, D. Convolutional Neural Network Coupled with a Transfer-Learning Approach for Time-Series Flood Predictions. *Water* **2020**, *12*, 96. [CrossRef]

56. Kim, H.I.; Han, K.Y. Urban Flood Prediction Using Deep Neural Network with Data Augmentation. *Water* **2020**, *12*, 899. [CrossRef]

57. Sun, A.Y.; Scanlon, B.R.; Zhang, Z.; Walling, D.; Bhanja, S.N.; Mukherjee, A.; Zhong, Z. Combining Physically Based Modeling and Deep Learning for Fusing GRACE Satellite Data: Can We Learn From Mismatch? *Water Resour. Res.* **2019**, *55*, 1179–1195. [CrossRef]

58. Xiang, Z.; Yan, J.; Demir, I. A Rainfall-Runoff Model With LSTM-Based Sequence-to-Sequence Learning. *Water Resour. Res.* **2020**, *56*, e2019WR025326. [CrossRef]

59. Tennant, C.; Larsen, L.; Bellugi, D.; Moges, E.; Zhang, L.; Ma, H. The Utility of Information Flow in Formulating Discharge Forecast Models: A Case Study From an Arid Snow-Dominated Catchment. *Water Resour. Res.* **2020**, *56*, e2019WR024908. [CrossRef]

60. Kan, G.; Liang, K.; Yu, H.; Sun, B.; Ding, L.; Li, J.; He, X.; Shen, C. Hybrid Machine Learning Hydrological Model for Flood Forecast Purpose. *Open Geosci.* **2020**, *12*, 813–820. [CrossRef]

61. Boucher, M.-A.; Quilty, J.; Adamowski, J. Data Assimilation for Streamflow Forecasting Using Extreme Learning Machines and Multilayer Perceptrons. *Water Resour. Res.* **2020**, *56*, e2019WR026226. [CrossRef]

62. Hernández, F.; Liang, X. Hybridizing Bayesian and Variational Data Assimilation for High-Resolution Hydrologic Forecasting. *Hydrol. Earth Syst. Sci.* **2018**, *22*, 5759–5779. [CrossRef]

63. Aytaç, E. Unsupervised Learning Approach in Defining the Similarity of Catchments: Hydrological Response Unit Based k-Means Clustering, a Demonstration on Western Black Sea Region of Turkey. *Int. Soil Water Conserv. Res.* **2020**, *8*, 321–331. [CrossRef]

64. Ley, R.; Casper, M.C.; Hellebrand, H.; Merz, R. Catchment Classification by Runoff Behaviour with Self-Organizing Maps (SOM). *Hydrol. Earth Syst. Sci.* **2011**, *15*, 2947–2962. [CrossRef]

65. Ley, R.; Hellebrand, H.; Casper, M.C.; Fenicia, F. Is Catchment Classification Possible by Means of Multiple Model Structures? A Case Study Based on 99 Catchments in Germany. *Hydrology* **2016**, *3*, 22. [CrossRef]

66. Swain, J.B.; Sahoo, M.M.; Patra, K.C. Homogeneous Region Determination Using Linear and Nonlinear Techniques. *Phys. Geogr.* **2016**, *37*, 361–384. [CrossRef]

67. Chen, I.-T.; Chang, L.-C.; Chang, F.-J. Exploring the Spatio-Temporal Interrelation between Groundwater and Surface Water by Using the Self-Organizing Maps. *J. Hydrol.* **2018**, *556*, 131–142. [CrossRef]

68. Chang, L.-C.; Chang, F.-J.; Yang, S.-N.; Tsai, F.-H.; Chang, T.-H.; Herricks, E.E. Self-Organizing Maps of Typhoon Tracks Allow for Flood Forecasts up to Two Days in Advance. *Nat. Commun.* **2020**, *11*, 1983. [CrossRef] [PubMed]

69. Prinzio, M.; Castellarin, A.; Toth, E. Data-Driven Catchment Classification: Combining Linear and Non-Linear Approaches to Improve Hydrological Predictions in Ungauged Catchments. In Proceedings of the EGU General Assembly Conference Abstracts, Vienna, Austria, 22–27 April 2012; p. 11717.

70. Brown, S.C.; Lester, R.E.; Versace, V.L.; Fawcett, J.; Laurenson, L. Hydrologic Landscape Regionalisation Using Deductive Classification and Random Forests. *PLoS ONE* **2014**, *9*, e112856. [CrossRef] [PubMed]

71. Li, Y.; Hu, T.; Zheng, G.; Shen, L.; Fan, J.; Zhang, D. An Improved Simplified Urban Storm Inundation Model Based on Urban Terrain and Catchment Modification. *Water* **2019**, *11*, 2335. [CrossRef]

72. Ebtehaj, A.M.; Foufoula-Georgiou, E. On Variational Downscaling, Fusion, and Assimilation of Hydrometeorological States: A Unified Framework via Regularization. *Water Resour. Res.* **2013**, *49*, 5944–5963. [CrossRef]

73. See, L. Data Fusion Methods for Integrating Data-Driven Hydrological Models. In *Quantitative Information Fusion for Hydrological Sciences*; Cai, X., Yeh, T.-C.J., Eds.; Studies in Computational Intelligence; Springer: Berlin/Heidelberg, Germany, 2008; pp. 1–18, ISBN 978-3-540-75384-1.

74. Meng, T.; Jing, X.; Yan, Z.; Pedrycz, W. A Survey on Machine Learning for Data Fusion. *Inf. Fusion* **2020**, *57*, 115–129. [CrossRef]

75. Qian, S.S.; Reckhow, K.H.; Zhai, J.; McMahon, G. Nonlinear Regression Modeling of Nutrient Loads in Streams: A Bayesian Approach. *Water Resour. Res.* **2005**, *41*, 1–10. [CrossRef]

76. Rawat, S.; Rawat, S. Multi-Sensor Data Fusion by a Hybrid Methodology—A Comparative Study. *Comput. Ind.* **2016**, *75*, 27–34. [CrossRef]

77. Yuan, Q.; Shen, H.; Li, T.; Li, Z.; Li, S.; Jiang, Y.; Xu, H.; Tan, W.; Yang, Q.; Wang, J.; et al. Deep Learning in Environmental Remote Sensing: Achievements and Challenges. *Remote Sens. Environ.* **2020**, *241*, 111716. [CrossRef]

78. Sist, M.; Schiavon, G.; Del Frate, F. A New Data Fusion Neural Network Scheme for Rainfall Retrieval Using Passive Microwave and Visible/Infrared Satellite Data. *Appl. Sci.* **2021**, *11*, 4686. [CrossRef]

79. Zhuo, L.; Han, D. Multi-Source Hydrological Soil Moisture State Estimation Using Data Fusion Optimisation. *Hydrol. Earth Syst. Sci.* **2017**, *21*, 3267–3285. [CrossRef]

80. Fehri, R.; Bogaert, P.; Khlifi, S.; Vanclooster, M. Data Fusion of Citizen-Generated Smartphone Discharge Measurements in Tunisia. *J. Hydrol.* **2020**, *590*, 125518. [CrossRef]

81. Abbaszadeh, P.; Moradkhani, H.; Zhan, X. Downscaling SMAP Radiometer Soil Moisture Over the CONUS Using an Ensemble Learning Method. *Water Resour. Res.* **2019**, *55*, 324–344. [CrossRef]

82. Serifi, A.; Günther, T.; Ban, N. Spatio-Temporal Downscaling of Climate Data Using Convolutional and Error-Predicting Neural Networks. *Front. Clim.* **2021**, *3*, 26. [CrossRef]

83. Sun, A.Y.; Tang, G. Downscaling Satellite and Reanalysis Precipitation Products Using Attention-Based Deep Convolutional Neural Nets. *Front. Water* **2020**, *2*, 56. [CrossRef]

84. Rittger, K.; Krock, M.; Kleiber, W.; Bair, E.H.; Brodzik, M.J.; Stephenson, T.R.; Rajagopalan, B.; Bormann, K.J.; Painter, T.H. Multi-Sensor Fusion Using Random Forests for Daily Fractional Snow Cover at 30 m. *Remote Sens. Environ.* **2021**, *264*, 112608. [CrossRef]

85. Tang, K.; Zhu, H.; Ni, P. Spatial Downscaling of Land Surface Temperature over Heterogeneous Regions Using Random Forest Regression Considering Spatial Features. *Remote Sens.* **2021**, *13*, 3645. [CrossRef]

86. Tran, H.; Nguyen, P.; Ombadi, M.; Hsu, K.; Sorooshian, S.; Qing, X. A Cloud-Free MODIS Snow Cover Dataset for the Contiguous United States from 2000 to 2017. *Sci. Data* **2019**, *6*, 180300. [CrossRef]

87. Miao, Q.; Pan, B.; Wang, H.; Hsu, K.; Sorooshian, S. Improving Monsoon Precipitation Prediction Using Combined Convolutional and Long Short Term Memory Neural Network. *Water* **2019**, *11*, 977. [CrossRef]

88. Lei, X.; Chen, W.; Panahi, M.; Falah, F.; Rahmati, O.; Uuemaa, E.; Kalantari, Z.; Ferreira, C.S.S.; Rezaie, F.; Tiefenbacher, J.P.; et al. Urban Flood Modeling Using Deep-Learning Approaches in Seoul, South Korea. *J. Hydrol.* **2021**, *601*, 126684. [CrossRef]

89. Veettil, A.V.; Konapala, G.; Mishra, A.K.; Li, H.-Y. Sensitivity of Drought Resilience-Vulnerability-Exposure to Hydrologic Ratios in Contiguous United States. *J. Hydrol.* **2018**, *564*, 294–306. [CrossRef]

90. Tran, H.; Nguyen, P.; Ombadi, M.; Hsu, K.; Sorooshian, S.; Andreadis, K. Improving Hydrologic Modeling Using Cloud-Free MODIS Flood Maps. *J. Hydrometeorol.* **2019**, *20*, 2203–2214. [CrossRef]

91. Mosavi, A.; Ozturk, P.; Chau, K. Flood Prediction Using Machine Learning Models: Literature Review. *Water* **2018**, *10*, 1536. [CrossRef]

92. Shahabi, H.; Shirzadi, A.; Ghaderi, K.; Omidvar, E.; Al-Ansari, N.; Clague, J.J.; Geertsema, M.; Khosravi, K.; Amini, A.; Bahrami, S.; et al. Flood Detection and Susceptibility Mapping Using Sentinel-1 Remote Sensing Data and a Machine Learning Approach: Hybrid Intelligence of Bagging Ensemble Based on K-Nearest Neighbor Classifier. *Remote Sens.* **2020**, *12*, 266. [CrossRef]

93. Lin, Q.; Leandro, J.; Wu, W.; Bhola, P.; Disse, M. Prediction of Maximum Flood Inundation Extents With Resilient Backpropagation Neural Network: Case Study of Kulmbach. *Front. Earth Sci.* **2020**, *8*, 332. [CrossRef]

94. Löwe, R.; Böhm, J.; Jensen, D.G.; Leandro, J.; Rasmussen, S.H. U-FLOOD—Topographic Deep Learning for Predicting Urban Pluvial Flood Water Depth. *J. Hydrol.* **2021**, *603*, 126898. [CrossRef]

95. Sahoo, S.; Russo, T.A.; Elliott, J.; Foster, I. Machine Learning Algorithms for Modeling Groundwater Level Changes in Agricultural Regions of the U.S. *Water Resour. Res.* **2017**, *53*, 3878–3895. [CrossRef]

96. Bui, D.T.; Panahi, M.; Shahabi, H.; Singh, V.P.; Shirzadi, A.; Chapi, K.; Khosravi, K.; Chen, W.; Panahi, S.; Li, S.; et al. Novel Hybrid Evolutionary Algorithms for Spatial Prediction of Floods. *Sci. Rep.* **2018**, *8*, 15364. [CrossRef]

97. Zhou, Y.; Guo, S.; Chang, F.-J. Explore an Evolutionary Recurrent ANFIS for Modelling Multi-Step-Ahead Flood Forecasts. *J. Hydrol.* **2019**, *570*, 343–355. [CrossRef]

98. Chang, F.-J.; Chiang, Y.-M.; Tsai, M.-J.; Shieh, M.-C.; Hsu, K.-L.; Sorooshian, S. Watershed Rainfall Forecasting Using Neuro-Fuzzy Networks with the Assimilation of Multi-Sensor Information. *J. Hydrol.* **2014**, *508*, 374–384. [CrossRef]

99. Araya, S.N.; Ghezzehei, T.A. Using Machine Learning for Prediction of Saturated Hydraulic Conductivity and Its Sensitivity to Soil Structural Perturbations. *Water Resour. Res.* **2019**, *55*, 5715–5737. [CrossRef]

100. Jian, J.; Shiklomanov, A.; Shuster, W.D.; Stewart, R.D. Predicting Near-Saturated Hydraulic Conductivity in Urban Soils. *J. Hydrol.* **2021**, *595*, 126051. [CrossRef]

101. Watt-Meyer, O.; Brenowitz, N.D.; Clark, S.K.; Henn, B.; Kwa, A.; McGibbon, J.; Perkins, W.A.; Bretherton, C.S. Correcting Weather and Climate Models by Machine Learning Nudged Historical Simulations. *Geophys. Res. Lett.* **2021**, *48*, e2021GL092555. [CrossRef]

102. Xu, T.; Liang, F. Machine Learning for Hydrologic Sciences: An Introductory Overview. *WIREs Water* **2021**, *8*, e1533. [CrossRef]

103. Koch, J.; Gotfredsen, J.; Schneider, R.; Troldborg, L.; Stisen, S.; Henriksen, H.J. High Resolution Water Table Modeling of the Shallow Groundwater Using a Knowledge-Guided Gradient Boosting Decision Tree Model. *Front. Water* **2021**, *3*. [CrossRef]

104. Wunsch, A.; Liesch, T.; Broda, S. Groundwater Level Forecasting with Artificial Neural Networks: A Comparison of Long Short-Term Memory (LSTM), Convolutional Neural Networks (CNNs), and Non-Linear Autoregressive Networks with Exogenous Input (NARX). *Hydrol. Earth Syst. Sci.* **2021**, *25*, 1671–1687. [CrossRef]

105. Kratzert, F.; Klotz, D.; Brenner, C.; Schulz, K.; Herrnegger, M. Rainfall–Runoff Modelling Using Long Short-Term Memory (LSTM) Networks. *Hydrol. Earth Syst. Sci.* **2018**, *22*, 6005–6022. [CrossRef]

106. Xu, W.; Jiang, Y.; Zhang, X.; Li, Y.; Zhang, R.; Fu, G. Using Long Short-Term Memory Networks for River Flow Prediction. *Hydrol. Res.* **2020**, *51*, 1358–1376. [CrossRef]

107. Li, P.; Zhang, J.; Krebs, P. Prediction of Flow Based on a CNN-LSTM Combined Deep Learning Approach. *Water* **2022**, *14*, 993. [CrossRef]

108. Lees, T.; Reece, S.; Kratzert, F.; Klotz, D.; Gauch, M.; De Bruijn, J.; Kumar Sahu, R.; Greve, P.; Slater, L.; Dadson, S.J. Hydrological Concept Formation inside Long Short-Term Memory (LSTM) Networks. *Hydrol. Earth Syst. Sci.* **2022**, *26*, 3079–3101. [CrossRef]

109. Hussain, D.; Hussain, T.; Khan, A.A.; Naqvi, S.A.A.; Jamil, A. A Deep Learning Approach for Hydrological Time-Series Prediction: A Case Study of Gilgit River Basin. *Earth Sci. Inform.* **2020**, *13*, 915–927. [CrossRef]

110. Anderson, S.; Radić, V. Evaluation and Interpretation of Convolutional Long Short-Term Memory Networks for Regional Hydrological Modelling. *Hydrol. Earth Syst. Sci.* **2022**, *26*, 795–825. [CrossRef]

111. Ni, L.; Wang, D.; Singh, V.P.; Wu, J.; Wang, Y.; Tao, Y.; Zhang, J. Streamflow and Rainfall Forecasting by Two Long Short-Term Memory-Based Models. *J. Hydrol.* **2020**, *583*, 124296. [CrossRef]

112. Young, C.-C.; Liu, W.-C.; Wu, M.-C. A Physically Based and Machine Learning Hybrid Approach for Accurate Rainfall-Runoff Modeling during Extreme Typhoon Events. *Appl. Soft Comput.* **2017**, *53*, 205–216. [CrossRef]

113. Tamiru, H.; Dinka, M.O. Application of ANN and HEC-RAS Model for Flood Inundation Mapping in Lower Baro Akobo River Basin, Ethiopia. *J. Hydrol. Reg. Stud.* **2021**, *36*, 100855. [CrossRef]

114. Wagenaar, D.; Curran, A.; Balbi, M.; Bhardwaj, A.; Soden, R.; Hartato, E.; Mestav Sarica, G.; Ruangpan, L.; Molinario, G.; Lallemant, D. Invited Perspectives: How Machine Learning Will Change Flood Risk and Impact Assessment. *Nat. Hazards Earth Syst. Sci.* **2020**, *20*, 1149–1161. [CrossRef]

115. Williams, R.D.; Brasington, J.; Hicks, D.M. Numerical Modelling of Braided River Morphodynamics: Review and Future Challenges. *Geogr. Compass* **2016**, *10*, 102–127. [CrossRef]

116. Flora, K.; Khosronejad, A. On the Impact of Bed-Bathymetry Resolution and Bank Vegetation on the Flood Flow Field of the American River, California: Insights Gained Using Data-Driven Large-Eddy Simulation. *J. Irrig. Drain. Eng.* **2021**, *147*, 04021036. [CrossRef]

117. Williams, R.D.; Measures, R.; Hicks, D.M.; Brasington, J. Assessment of a Numerical Model to Reproduce Event-Scale Erosion and Deposition Distributions in a Braided River. *Water Resour. Res.* **2016**, *52*, 6621–6642. [CrossRef]

118. Perrin, O.; Christophe, S.; Jacquinod, F.; Payrastre, O. Visual Analysis of Inconsistencies in Hydraulic Simulation Data. In Proceedings of the International Archives of the Photogrammetry, Remote Sensing and Spatial Information Sciences, Hannover, Germany, 25 August 2020; Copernicus GmbH: Goettingen, Germany, 2020; Volume XLIII-B4-2020, pp. 795–801.

119. Mirza Alipour, S.; Leal, J. Emulation of 2D Hydrodynamic Flood Simulations at Catchment Scale Using ANN and SVR. *Water* **2021**, *13*, 2858. [CrossRef]

120. Ardourel, V.; Jebeile, J. Numerical Instability and Dynamical Systems. *Eur. J. Philos. Sci.* **2021**, *11*, 49. [CrossRef]

121. Hosseiny, H. A Deep Learning Model for Predicting River Flood Depth and Extent. *Environ. Model. Softw.* **2021**, *145*, 105186. [CrossRef]

122. Forghani, M.; Qian, Y.; Lee, J.; Farthing, M.W.; Hesser, T.; Kitanidis, P.K.; Darve, E.F. Application of Deep Learning to Large Scale Riverine Flow Velocity Estimation. *Stoch. Environ. Res. Risk Assess.* **2021**, *35*, 1069–1088. [CrossRef]

123. Cheng, C.; Zhang, G.-T. Deep Learning Method Based on Physics Informed Neural Network with Resnet Block for Solving Fluid Flow Problems. *Water* **2021**, *13*, 423. [CrossRef]

124. Fukami, K.; Fukagata, K.; Taira, K. Machine-Learning-Based Spatio-Temporal Super Resolution Reconstruction of Turbulent Flows. *J. Fluid Mech.* **2021**, *909*, 1–24. [CrossRef]

125. Buzzicotti, M.; Bonaccorso, F.; Di Leoni, P.C.; Biferale, L. Reconstruction of Turbulent Data with Deep Generative Models for Semantic Inpainting from TURB-Rot Database. *Phys. Rev. Fluids* **2021**, *6*, 050503. [CrossRef]
126. Baracchini, T.; Chu, P.Y.; Šukys, J.; Lieberherr, G.; Wunderle, S.; Wüest, A.; Bouffard, D. Data Assimilation of in Situ and Satellite Remote Sensing Data to 3D Hydrodynamic Lake Models: A Case Study Using Delft3D-FLOW v4.03 and OpenDA v2.4. *Geosci. Model Dev.* **2020**, *13*, 1267–1284. [CrossRef]
127. Carrassi, A.; Bocquet, M.; Bertino, L.; Evensen, G. Data Assimilation in the Geosciences: An Overview of Methods, Issues, and Perspectives. *WIREs Clim. Chang.* **2018**, *9*, e535. [CrossRef]
128. Houser, P.; Lannoy, G.; Walker, J. Hydrologic Data Assimilation. In *Advances in Water Science Methodologies*; CRC Press: Boca Raton, FL, USA, 2012; ISBN 978-953-51-0294-6.
129. Nearing, G.S.; Kratzert, F.; Sampson, A.K.; Pelissier, C.S.; Klotz, D.; Frame, J.M.; Prieto, C.; Gupta, H.V. What Role Does Hydrological Science Play in the Age of Machine Learning? *Water Resour. Res.* **2021**, *57*, e2020WR028091. [CrossRef]
130. Xiong, Z.; Sang, J.; Sun, X.; Zhang, B.; Li, J. Comparisons of Performance Using Data Assimilation and Data Fusion Approaches in Acquiring Precipitable Water Vapor: A Case Study of a Western United States of America Area. *Water* **2020**, *12*, 2943. [CrossRef]
131. Farchi, A.; Laloyaux, P.; Bonavita, M.; Bocquet, M. Using Machine Learning to Correct Model Error in Data Assimilation and Forecast Applications. *Q. J. R. Meteorol. Soc.* **2021**, *147*, 3067–3084. [CrossRef]
132. Arcucci, R.; Zhu, J.; Hu, S.; Guo, Y.-K. Deep Data Assimilation: Integrating Deep Learning with Data Assimilation. *Appl. Sci.* **2021**, *11*, 1114. [CrossRef]
133. Geer, A. Learning Earth System Models from Observations: Machine Learning or Data Assimilation? *Philos. Trans. R. Soc. A* **2021**, *379*, 20200089. [CrossRef] [PubMed]
134. Wagener, T.; Boyle, D.P.; Lees, M.J.; Wheater, H.S.; Gupta, H.V.; Sorooshian, S. A Framework for Development and Application of Hydrological Models. *Hydrol. Earth Syst. Sci.* **2001**, *5*, 13–26. [CrossRef]
135. Vrugt, J.A.; Stauffer, P.H.; Wöhling, T.; Robinson, B.A.; Vesselinov, V.V. Inverse Modeling of Subsurface Flow and Transport Properties: A Review with New Developments. *Vadose Zone J.* **2008**, *7*, 843–864. [CrossRef]
136. Carrera, J.; Alcolea, A.; Medina, A.; Hidalgo, J.; Slooten, L.J. Inverse Problem in Hydrogeology. *Hydrogeol. J.* **2005**, *13*, 206–222. [CrossRef]
137. Beven, K. A Manifesto for the Equifinality Thesis. *J. Hydrol.* **2006**, *320*, 18–36. [CrossRef]
138. Mostafaie, A.; Forootan, E.; Safari, A.; Schumacher, M. Comparing Multi-Objective Optimization Techniques to Calibrate a Conceptual Hydrological Model Using In Situ Runoff and Daily GRACE Data. *Comput. Geosci.* **2018**, *22*, 789–814. [CrossRef]
139. Guo, J.; Zhou, J.; Lu, J.; Zou, Q.; Zhang, H.; Bi, S. Multi-Objective Optimization of Empirical Hydrological Model for Streamflow Prediction. *J. Hydrol.* **2014**, *511*, 242–253. [CrossRef]
140. Huang, X.; Lei, X.; Jiang, Y. Comparison of Three Multi-Objective Optimization Algorithms for Hydrological Model. In Proceedings of the International Symposium on Intelligence Computation and Applications, Wuhan, China, 27–28 October 2012; Li, Z., Li, X., Liu, Y., Cai, Z., Eds.; Springer: Berlin/Heidelberg, Germany, 2012; pp. 209–216.
141. Vrugt, J.A.; Gupta, H.V.; Bastidas, L.A.; Bouten, W.; Sorooshian, S. Effective and Efficient Algorithm for Multiobjective Optimization of Hydrologic Models. *Water Resour. Res.* **2003**, *39*, 1–19. [CrossRef]
142. Essenfelder, A.H.; Giupponi, C. A Coupled Hydrologic-Machine Learning Modelling Framework to Support Hydrologic Modelling in River Basins under Interbasin Water Transfer Regimes. *Environ. Model. Softw.* **2020**, *131*, 104779. [CrossRef]
143. Barnhart, B.L.; Sawicz, K.A.; Ficklin, D.L.; Whittaker, G.W. MOESHA: A Genetic Algorithm for Automatic Calibration and Estimation of Parameter Uncertainty and Sensitivity of Hydrologic Models. *Trans. ASABE* **2017**, *60*, 1259–1269. [CrossRef] [PubMed]
144. Zhang, X.; Srinivasan, R.; Liew, M.V. On the Use of Multi-Algorithm, Genetically Adaptive Multi-Objective Method for Multi-Site Calibration of the SWAT Model. *Hydrol. Process.* **2010**, *24*, 955–969. [CrossRef]
145. Kamali, B.; Mousavi, S.J.; Abbaspour, K.C. Automatic Calibration of HEC-HMS Using Single-Objective and Multi-Objective PSO Algorithms. *Hydrol. Process.* **2013**, *27*, 4028–4042. [CrossRef]
146. Krapu, C.; Borsuk, M.; Kumar, M. Gradient-Based Inverse Estimation for a Rainfall-Runoff Model. *Water Resour. Res.* **2019**, *55*, 6625–6639. [CrossRef]
147. Navarro-Hellín, H.; Martínez-del-Rincon, J.; Domingo-Miguel, R.; Soto-Valles, F.; Torres-Sánchez, R. A Decision Support System for Managing Irrigation in Agriculture. *Comput. Electron. Agric.* **2016**, *124*, 121–131. [CrossRef]
148. Mullapudi, A.; Lewis, M.J.; Gruden, C.L.; Kerkez, B. Deep Reinforcement Learning for the Real Time Control of Stormwater Systems. *Adv. Water Resour.* **2020**, *140*, 103600. [CrossRef]
149. Wolpert, D.H.; Macready, W.G. No Free Lunch Theorems for Optimization. *IEEE Trans. Evol. Comput.* **1997**, *1*, 67–82. [CrossRef]
150. Yusoff, Y.; Ngadiman, M.S.; Zain, A.M. Overview of NSGA-II for Optimizing Machining Process Parameters. *Procedia Eng.* **2011**, *15*, 3978–3983. [CrossRef]
151. Bianchi, L.; Dorigo, M.; Gambardella, L.M.; Gutjahr, W.J. A Survey on Metaheuristics for Stochastic Combinatorial Optimization. *Nat. Comput.* **2009**, *8*, 239–287. [CrossRef]
152. Vrugt, J.A.; Robinson, B.A. Improved Evolutionary Optimization from Genetically Adaptive Multimethod Search. *Proc. Natl. Acad. Sci. USA* **2007**, *104*, 708–711. [CrossRef] [PubMed]
153. Confesor, R.B., Jr.; Whittaker, G.W. Automatic Calibration of Hydrologic Models With Multi-Objective Evolutionary Algorithm and Pareto Optimization1. *JAWRA J. Am. Water Resour. Assoc.* **2007**, *43*, 981–989. [CrossRef]

154. Ercan, M.B.; Goodall, J.L. Design and Implementation of a General Software Library for Using NSGA-II with SWAT for Multi-Objective Model Calibration. *Environ. Model. Softw.* **2016**, *84*, 112–120. [CrossRef]

155. Saravanan, V.S.; McDonald, G.T.; Mollinga, P.P. Critical Review of Integrated Water Resources Management: Moving beyond Polarised Discourse. *Nat. Resour. Forum* **2009**, *33*, 76–86. [CrossRef]

156. Sun, R.; Hernández, F.; Liang, X.; Yuan, H. A Calibration Framework for High-Resolution Hydrological Models Using a Multiresolution and Heterogeneous Strategy. *Water Resour. Res.* **2020**, *56*, e2019WR026541. [CrossRef]

157. Ahmad, A.; El-Shafie, A.; Razali, S.F.M.; Mohamad, Z.S. Reservoir Optimization in Water Resources: A Review. *Water Resour. Manag.* **2014**, *28*, 3391–3405. [CrossRef]

158. Xiang, X.; Li, Q.; Khan, S.; Khalaf, O.I. Urban Water Resource Management for Sustainable Environment Planning Using Artificial Intelligence Techniques. *Environ. Impact Assess. Rev.* **2021**, *86*, 106515. [CrossRef]

159. Chaves, P.; Chang, F.-J. Intelligent Reservoir Operation System Based on Evolving Artificial Neural Networks. *Adv. Water Resour.* **2008**, *31*, 926–936. [CrossRef]

160. Cai, Y.; Wang, H.; Yue, W.; Xie, Y.; Liang, Q. An Integrated Approach for Reducing Spatially Coupled Water-Shortage Risks of Beijing-Tianjin-Hebei Urban Agglomeration in China. *J. Hydrol.* **2021**, *603*, 127123. [CrossRef]

161. Zhang, D.; Peng, Q.; Lin, J.; Wang, D.; Liu, X.; Zhuang, J. Simulating Reservoir Operation Using a Recurrent Neural Network Algorithm. *Water* **2019**, *11*, 865. [CrossRef]

162. Zarei, M.; Bozorg-Haddad, O.; Baghban, S.; Delpasand, M.; Goharian, E.; Loáiciga, H.A. Machine-Learning Algorithms for Forecast-Informed Reservoir Operation (FIRO) to Reduce Flood Damages. *Sci. Rep.* **2021**, *11*, 24295. [CrossRef] [PubMed]

163. Munawar, H.S.; Hammad, A.W.A.; Waller, S.T.; Thaheem, M.J.; Shrestha, A. An Integrated Approach for Post-Disaster Flood Management Via the Use of Cutting-Edge Technologies and UAVs: A Review. *Sustainability* **2021**, *13*, 7925. [CrossRef]

164. Abraham, A. Adaptation of Fuzzy Inference System Using Neural Learning. In *Fuzzy Systems Engineering: Theory and Practice*; Nedjah, N., de Macedo Mourelle, L., Eds.; Studies in Fuzziness and Soft Computing; Springer: Berlin/Heidelberg, Germany, 2005; pp. 53–83, ISBN 978-3-540-32397-6.

165. Ighalo, J.O.; Adeniyi, A.G.; Marques, G. Internet of Things for Water Quality Monitoring and Assessment: A Comprehensive Review. In *Artificial Intelligence for Sustainable Development: Theory, Practice and Future Applications*; Hassanien, A.E., Bhatnagar, R., Darwish, A., Eds.; Studies in Computational Intelligence; Springer International Publishing: Cham, Switzerland, 2021; pp. 245–259, ISBN 978-3-030-51920-9.

166. Jan, F.; Min-Allah, N.; Düştegör, D. IoT Based Smart Water Quality Monitoring: Recent Techniques, Trends and Challenges for Domestic Applications. *Water* **2021**, *13*, 1729. [CrossRef]

167. Leigh, C.; Kandanaarachchi, S.; McGree, J.M.; Hyndman, R.J.; Alsibai, O.; Mengersen, K.; Peterson, E.E. Predicting Sediment and Nutrient Concentrations from High-Frequency Water-Quality Data. *PLoS ONE* **2019**, *14*, e0215503. [CrossRef] [PubMed]

168. Mathany, T.M.; Saraceno, J.F.; Kulongoski, J.T. *Guidelines and Standard Procedures for High-Frequency Groundwater-Quality Monitoring Stations—Design, Operation, and Record Computation*; Techniques and Methods; U.S. Geological Survey: Reston, VA, USA, 2019; Volume 1-D7, p. 66.

169. Seifert-Dähnn, I.; Furuseth, I.S.; Vondolia, G.K.; Gal, G.; de Eyto, E.; Jennings, E.; Pierson, D. Costs and Benefits of Automated High-Frequency Environmental Monitoring—The Case of Lake Water Management. *J. Environ. Manag.* **2021**, *285*, 112108. [CrossRef] [PubMed]

170. Jung, H.; Senf, C.; Jordan, P.; Krueger, T. Benchmarking Inference Methods for Water Quality Monitoring and Status Classification. *Environ. Monit. Assess.* **2020**, *192*, 261. [CrossRef]

171. Jesus, G.; Casimiro, A.; Oliveira, A. A Survey on Data Quality for Dependable Monitoring in Wireless Sensor Networks. *Sensors* **2017**, *17*, 2010. [CrossRef]

172. Khatri, P.; Gupta, K.K.; Gupta, R.K. Drift Compensation of Commercial Water Quality Sensors Using Machine Learning to Extend the Calibration Lifetime. *J. Ambient Intell. Humaniz. Comput.* **2021**, *12*, 3091–3099. [CrossRef]

173. Kahiluoto, J.; Hirvonen, J.; Näykki, T. Automatic Real-Time Uncertainty Estimation for Online Measurements: A Case Study on Water Turbidity. *Environ. Monit. Assess.* **2019**, *191*, 259. [CrossRef]

174. Rodriguez-Perez, J.; Leigh, C.; Liquet, B.; Kermorvant, C.; Peterson, E.; Sous, D.; Mengersen, K. Detecting Technical Anomalies in High-Frequency Water-Quality Data Using Artificial Neural Networks. *Environ. Sci. Technol.* **2020**, *54*, 13719–13730. [CrossRef]

175. Jiang, J.; Tang, S.; Han, D.; Fu, G.; Solomatine, D.; Zheng, Y. A Comprehensive Review on the Design and Optimization of Surface Water Quality Monitoring Networks. *Environ. Model. Softw.* **2020**, *132*, 104792. [CrossRef]

176. Chachuła, K.; Słojewski, T.M.; Nowak, R. Multisensor Data Fusion for Localization of Pollution Sources in Wastewater Networks. *Sensors* **2022**, *22*, 387. [CrossRef] [PubMed]

177. Cao, H.; Guo, Z.; Wang, S.; Cheng, H.; Zhan, C. Intelligent Wide-Area Water Quality Monitoring and Analysis System Exploiting Unmanned Surface Vehicles and Ensemble Learning. *Water* **2020**, *12*, 681. [CrossRef]

178. Huang, W.; Yang, Y. Water Quality Sensor Model Based on an Optimization Method of RBF Neural Network. *Comput. Water Energy Environ. Eng.* **2019**, *9*, 1. [CrossRef]

179. Jafari, H.; Rajaee, T.; Kisi, O. Improved Water Quality Prediction with Hybrid Wavelet-Genetic Programming Model and Shannon Entropy. *Nat. Resour. Res.* **2020**, *29*, 3819–3840. [CrossRef]

180. Jiang, Y.; Li, C.; Sun, L.; Guo, D.; Zhang, Y.; Wang, W. A Deep Learning Algorithm for Multi-Source Data Fusion to Predict Water Quality of Urban Sewer Networks. *J. Clean. Prod.* **2021**, *318*, 128533. [CrossRef]

181. Pattnaik, B.S.; Pattanayak, A.S.; Udgata, S.K.; Panda, A.K. Machine Learning Based Soft Sensor Model for BOD Estimation Using Intelligence at Edge. *Complex Intell. Syst.* **2021**, *7*, 961–976. [CrossRef]
182. Lu, H.; Ma, X. Hybrid Decision Tree-Based Machine Learning Models for Short-Term Water Quality Prediction. *Chemosphere* **2020**, *249*, 126169. [CrossRef]
183. Chen, K.; Chen, H.; Zhou, C.; Huang, Y.; Qi, X.; Shen, R.; Liu, F.; Zuo, M.; Zou, X.; Wang, J.; et al. Comparative Analysis of Surface Water Quality Prediction Performance and Identification of Key Water Parameters Using Different Machine Learning Models Based on Big Data. *Water Res.* **2020**, *171*, 115454. [CrossRef]
184. Castrillo, M.; García, Á.L. Estimation of High Frequency Nutrient Concentrations from Water Quality Surrogates Using Machine Learning Methods. *Water Res.* **2020**, *172*, 115490. [CrossRef]
185. Wang, P.; Yao, J.; Wang, G.; Hao, F.; Shrestha, S.; Xue, B.; Xie, G.; Peng, Y. Exploring the Application of Artificial Intelligence Technology for Identification of Water Pollution Characteristics and Tracing the Source of Water Quality Pollutants. *Sci. Total Environ.* **2019**, *693*, 133440. [CrossRef] [PubMed]
186. Holmberg, M.; Forsius, M.; Starr, M.; Huttunen, M. An Application of Artificial Neural Networks to Carbon, Nitrogen and Phosphorus Concentrations in Three Boreal Streams and Impacts of Climate Change. *Ecol. Model.* **2006**, *195*, 51–60. [CrossRef]
187. Najafzadeh, M.; Ghaemi, A. Prediction of the Five-Day Biochemical Oxygen Demand and Chemical Oxygen Demand in Natural Streams Using Machine Learning Methods. *Environ. Monit. Assess.* **2019**, *191*, 380. [CrossRef] [PubMed]
188. Noori, N.; Kalin, L.; Isik, S. Water Quality Prediction Using SWAT-ANN Coupled Approach. *J. Hydrol.* **2020**, *590*, 125220. [CrossRef]
189. Haghiabi, A.H.; Nasrolahi, A.H.; Parsaie, A. Water Quality Prediction Using Machine Learning Methods. *Water Qual. Res. J.* **2018**, *53*, 3–13. [CrossRef]
190. Bui, D.T.; Khosravi, K.; Tiefenbacher, J.; Nguyen, H.; Kazakis, N. Improving Prediction of Water Quality Indices Using Novel Hybrid Machine-Learning Algorithms. *Sci. Total Environ.* **2020**, *721*, 137612. [CrossRef]
191. Green, M.B.; Pardo, L.H.; Bailey, S.W.; Campbell, J.L.; McDowell, W.H.; Bernhardt, E.S.; Rosi, E.J. Predicting High-Frequency Variation in Stream Solute Concentrations with Water Quality Sensors and Machine Learning. *Hydrol. Process.* **2021**, *35*, e14000. [CrossRef]
192. Chang, N.-B.; Bai, K.; Chen, C.-F. Integrating Multisensor Satellite Data Merging and Image Reconstruction in Support of Machine Learning for Better Water Quality Management. *J. Environ. Manag.* **2017**, *201*, 227–240. [CrossRef]
193. Hamshaw, S.D.; Dewoolkar, M.M.; Schroth, A.W.; Wemple, B.C.; Rizzo, D.M. A New Machine-Learning Approach for Classifying Hysteresis in Suspended-Sediment Discharge Relationships Using High-Frequency Monitoring Data. *Water Resour. Res.* **2018**, *54*, 4040–4058. [CrossRef]
194. Chen, Y.; Song, L.; Liu, Y.; Yang, L.; Li, D. A Review of the Artificial Neural Network Models for Water Quality Prediction. *Appl. Sci.* **2020**, *10*, 5776. [CrossRef]
195. Shi, P.; Li, G.; Yuan, Y.; Kuang, L. Data Fusion Using Improved Support Degree Function in Aquaculture Wireless Sensor Networks. *Sensors* **2018**, *18*, 3851. [CrossRef] [PubMed]
196. Shen, L.Q.; Amatulli, G.; Sethi, T.; Raymond, P.; Domisch, S. Estimating Nitrogen and Phosphorus Concentrations in Streams and Rivers, within a Machine Learning Framework. *Sci. Data* **2020**, *7*, 161. [CrossRef] [PubMed]
197. Sperotto, A.; Molina, J.L.; Torresan, S.; Critto, A.; Pulido-Velazquez, M.; Marcomini, A. A Bayesian Networks Approach for the Assessment of Climate Change Impacts on Nutrients Loading. *Environ. Sci. Policy* **2019**, *100*, 21–36. [CrossRef]
198. Song, C.; Yao, L.; Hua, C.; Ni, Q. A Novel Hybrid Model for Water Quality Prediction Based on Synchrosqueezed Wavelet Transform Technique and Improved Long Short-Term Memory. *J. Hydrol.* **2021**, *603*, 126879. [CrossRef]
199. Zhou, J.; Chu, F.; Li, X.; Ma, H.; Xiao, F.; Sun, L. Water Quality Prediction Approach Based on T-SNE and SA-BiLSTM. In Proceedings of the 2020 IEEE 22nd International Conference on High Performance Computing and Communications; IEEE 18th International Conference on Smart City; IEEE 6th International Conference on Data Science and Systems (HPCC/SmartCity/DSS), Yanuca Island, Fiji, 14–16 December 2020; pp. 708–714.
200. Milošević, D.; Medeiros, A.S.; Stojković Piperac, M.; Cvijanović, D.; Soininen, J.; Milosavljević, A.; Predić, B. The Application of Uniform Manifold Approximation and Projection (UMAP) for Unconstrained Ordination and Classification of Biological Indicators in Aquatic Ecology. *Sci. Total Environ.* **2022**, *815*, 152365. [CrossRef]
201. Baker, V.R. The Modern Evolution of Geomorphology—Binghamton and Personal Perspectives, 1970–2019 and Beyond. *Geomorphology* **2020**, *366*, 106684. [CrossRef]
202. Flener, C.; Vaaja, M.; Jaakkola, A.; Krooks, A.; Kaartinen, H.; Kukko, A.; Kasvi, E.; Hyyppä, H.; Hyyppä, J.; Alho, P. Seamless Mapping of River Channels at High Resolution Using Mobile LiDAR and UAV-Photography. *Remote Sens.* **2013**, *5*, 6382–6407. [CrossRef]
203. Tonina, D.; McKean, J.A.; Benjankar, R.M.; Wright, C.W.; Goode, J.R.; Chen, Q.; Reeder, W.J.; Carmichael, R.A.; Edmondson, M.R. Mapping River Bathymetries: Evaluating Topobathymetric LiDAR Survey. *Earth Surf. Process. Landf.* **2019**, *44*, 507–520. [CrossRef]
204. Emanuele, P.; Nives, G.; Andrea, C.; Carlo, C.; Paolo, D.; Andrea Maria, L. Bathymetric Detection of Fluvial Environments through UASs and Machine Learning Systems. *Remote Sens.* **2020**, *12*, 4148. [CrossRef]
205. Carbonneau, P.E.; Dugdale, S.J.; Breckon, T.P.; Dietrich, J.T.; Fonstad, M.A.; Miyamoto, H.; Woodget, A.S. Adopting Deep Learning Methods for Airborne RGB Fluvial Scene Classification. *Remote Sens. Environ.* **2020**, *251*, 112107. [CrossRef]

206. Carbonneau, P.E.; Belletti, B.; Micotti, M.; Lastoria, B.; Casaioli, M.; Mariani, S.; Marchetti, G.; Bizzi, S. UAV-Based Training for Fully Fuzzy Classification of Sentinel-2 Fluvial Scenes. *Earth Surf. Process. Landf.* **2020**, *45*, 3120–3140. [CrossRef]

207. Rivas Casado, M.; Ballesteros Gonzalez, R.; Kriechbaumer, T.; Veal, A. Automated Identification of River Hydromorphological Features Using UAV High Resolution Aerial Imagery. *Sensors* **2015**, *15*, 27969–27989. [CrossRef] [PubMed]

208. Ling, F.; Boyd, D.; Ge, Y.; Foody, G.M.; Li, X.; Wang, L.; Zhang, Y.; Shi, L.; Shang, C.; Li, X.; et al. Measuring River Wetted Width From Remotely Sensed Imagery at the Subpixel Scale With a Deep Convolutional Neural Network. *Water Resour. Res.* **2019**, *55*, 5631–5649. [CrossRef]

209. Demarchi, L.; van de Bund, W.; Pistocchi, A. Object-Based Ensemble Learning for Pan-European Riverscape Units Mapping Based on Copernicus VHR and EU-DEM Data Fusion. *Remote Sens.* **2020**, *12*, 1222. [CrossRef]

210. Szabo, Z.C.; Mikita, T.; Negyesi, G.; Varga, O.G.; Burai, P.; Takacs-Szilagyi, L.; Szabo, S. Uncertainty and Overfitting in Fluvial Landform Classification Using Laser Scanned Data and Machine Learning: A Comparison of Pixel and Object-Based Approaches. *Remote Sens.* **2020**, *12*, 3652. [CrossRef]

211. Demarchi, L.; Bizzi, S.; Piegay, H. Hierarchical Object-Based Mapping of Riverscape Units and in-Stream Mesohabitats Using LiDAR and VHR Imagery. *Remote Sens.* **2016**, *8*, 97. [CrossRef]

212. Khan, S.; Fryirs, K.A. Application of Globally Available, Coarse-Resolution Digital Elevation Models for Delineating Valley Bottom Segments of Varying Length across a Catchment. *Earth Surf. Process. Landf.* **2020**, *45*, 2788–2803. [CrossRef]

213. Moody, D.I.; Brumby, S.P.; Rowland, J.C.; Gangodagamage, C. Undercomplete Learned Dictionaries for Land Cover Classification in Multispectral Imagery of Arctic Landscapes Using CoSA: Clustering of Sparse Approximations. In *Algorithms and Technologies for Multispectral, Hyperspectral, and Ultraspectral Imagery Xix*; Shen, S.S., Lewis, P.E., Eds.; SPIE: Baltimore, MD, USA, 2013; Volume 8743, ISBN 978-0-8194-9534-1.

214. Guillon, H.; Byrne, C.F.; Lane, B.A.; Solis, S.S.; Pasternack, G.B. Machine Learning Predicts Reach-Scale Channel Types From Coarse-Scale Geospatial Data in a Large River Basin. *Water Resour. Res.* **2020**, *56*, e2019WR026691. [CrossRef]

215. De Souza, C.M.P.; de Figueredo, N.A.; da Costa, L.M.; Veloso, G.V.; de Almeida, M.I.S.; Ferreira, E.J. Machine Learning Algorithm In The Prediction Of Geomorphic Indices For Appraisal The Influence Of Landscape Structure On Fluvial Systems, Southeastern-Brazil. *Rev. Bras. Geomorfol.* **2020**, *21*, 365–380. [CrossRef]

216. Guneralp, I.; Filippi, A.M.; Hales, B. Influence of River Channel Morphology and Bank Characteristics on Water Surface Boundary Delineation Using High-Resolution Passive Remote Sensing and Template Matching. *Earth Surf. Process. Landf.* **2014**, *39*, 977–986. [CrossRef]

217. Heasley, E.L.; Millington, J.D.A.; Clifford, N.J.; Chadwick, M.A. A Waterbody Typology Derived from Catchment Controls Using Self-Organising Maps. *Water* **2020**, *12*, 78. [CrossRef]

218. Mandlburger, G.; Kölle, M.; Nübel, H.; Soergel, U. BathyNet: A Deep Neural Network for Water Depth Mapping from Multispectral Aerial Images. *PFG—J. Photogramm. Remote Sens. Geoinf. Sci.* **2021**, *89*, 71–89. [CrossRef]

219. Corenblit, D.; Vautier, F.; González, E.; Steiger, J. Formation and Dynamics of Vegetated Fluvial Landforms Follow the Biogeomorphological Succession Model in a Channelized River. *Earth Surf. Process. Landf.* **2020**, *45*, 2020–2035. [CrossRef]

220. Díaz Gómez, R.; Pasternack, G.B.; Guillon, H.; Byrne, C.F.; Schwindt, S.; Larrieu, K.G.; Solis, S.S. Mapping Subaerial Sand-Gravel-Cobble Fluvial Sediment Facies Using Airborne Lidar and Machine Learning. *Geomorphology* **2022**, *401*, 108106. [CrossRef]

221. Clubb, F.J.; Bookhagen, B.; Rheinwalt, A. Clustering River Profiles to Classify Geomorphic Domains. *J. Geophys. Res. Earth Surf.* **2019**, *124*, 1417–1439. [CrossRef]

222. Harrison, L.R.; Legleiter, C.J.; Overstreet, B.T.; Bell, T.W.; Hannon, J. Assessing the Potential for Spectrally Based Remote Sensing of Salmon Spawning Locations. *River Res. Appl.* **2020**, *36*, 1618–1632. [CrossRef]

223. Alfredsen, K.; Dalsgård, A.; Shamsaliei, S.; Halleraker, J.H.; Gundersen, O.E. Towards an Automatic Characterization of Riverscape Development by Deep Learning. *River Res. Appl.* **2022**, *38*, 810–816. [CrossRef]

224. González-Castro, J.A.; Muste, M. Framework for Estimating Uncertainty of ADCP Measurements from a Moving Boat by Standardized Uncertainty Analysis. *J. Hydraul. Eng.* **2007**, *133*, 1390–1410. [CrossRef]

225. Pan, Z.; Glennie, C.; Hartzell, P.; Fernandez-Diaz, J.C.; Legleiter, C.; Overstreet, B. Performance Assessment of High Resolution Airborne Full Waveform LiDAR for Shallow River Bathymetry. *Remote Sens.* **2015**, *7*, 5133–5159. [CrossRef]

226. Bandini, F.; Olesen, D.; Jakobsen, J.; Kittel, C.M.M.; Wang, S.; Garcia, M.; Bauer-Gottwein, P. Technical Note: Bathymetry Observations of Inland Water Bodies Using a Tethered Single-Beam Sonar Controlled by an Unmanned Aerial Vehicle. *Hydrol. Earth Syst. Sci.* **2018**, *22*, 4165–4181. [CrossRef]

227. Alvarez, L.V.; Moreno, H.A.; Segales, A.R.; Pham, T.G.; Pillar-Little, E.A.; Chilson, P.B. Merging Unmanned Aerial Systems (UAS) Imagery and Echo Soundings with an Adaptive Sampling Technique for Bathymetric Surveys. *Remote Sens.* **2018**, *10*, 1362. [CrossRef]

228. Van Iersel, W.; Straatsma, M.; Middelkoop, H.; Addink, E. Multitemporal Classification of River Floodplain Vegetation Using Time Series of UAV Images. *Remote Sens.* **2018**, *10*, 1144. [CrossRef]

229. Hemmelder, S.; Marra, W.; Markies, H.; De Jong, S.M. Monitoring River Morphology & Bank Erosion Using UAV Imagery—A Case Study of the River Buëch, Hautes-Alpes, France. *Int. J. Appl. Earth Obs. Geoinf.* **2018**, *73*, 428–437. [CrossRef]

230. Boothroyd, R.J.; Williams, R.D.; Hoey, T.B.; Barrett, B.; Prasojo, O.A. Applications of Google Earth Engine in Fluvial Geomorphology for Detecting River Channel Change. *WIREs Water* **2021**, *8*, e21496. [CrossRef]

231. Shi, W.; Zhang, M.; Zhang, R.; Chen, S.; Zhan, Z. Change Detection Based on Artificial Intelligence: State-of-the-Art and Challenges. *Remote Sens.* **2020**, *12*, 1688. [CrossRef]
232. Wheaton, J.M.; Fryirs, K.A.; Brierley, G.; Bangen, S.G.; Bouwes, N.; O'Brien, G. Geomorphic Mapping and Taxonomy of Fluvial Landforms. *Geomorphology* **2015**, *248*, 273–295. [CrossRef]
233. Polidori, L.; El Hage, M. Digital Elevation Model Quality Assessment Methods: A Critical Review. *Remote Sens.* **2020**, *12*, 3522. [CrossRef]
234. Paniconi, C.; Putti, M. Physically Based Modeling in Catchment Hydrology at 50: Survey and Outlook. *Water Resour. Res.* **2015**, *51*, 7090–7129. [CrossRef]
235. Marçais, J.; de Dreuzy, J.-R. Prospective Interest of Deep Learning for Hydrological Inference. *Groundwater* **2017**, *55*, 688–692. [CrossRef]
236. Murdoch, W.J.; Singh, C.; Kumbier, K.; Abbasi-Asl, R.; Yu, B. Definitions, Methods, and Applications in Interpretable Machine Learning. *Proc. Natl. Acad. Sci. USA* **2019**, *116*, 22071–22080. [CrossRef]
237. Pearl, J. The Seven Tools of Causal Inference, with Reflections on Machine Learning. *Commun. ACM* **2019**, *62*, 54–60. [CrossRef]
238. Goldstein, E.B.; Coco, G.; Plant, N.G. A Review of Machine Learning Applications to Coastal Sediment Transport and Morphodynamics. *Earth-Sci. Rev.* **2019**, *194*, 97–108. [CrossRef]
239. Tharwat, A. Parameter Investigation of Support Vector Machine Classifier with Kernel Functions. *Knowl. Inf. Syst.* **2019**, *61*, 1269–1302. [CrossRef]
240. Shen, C.; Laloy, E.; Elshorbagy, A.; Albert, A.; Bales, J.; Chang, F.-J.; Ganguly, S.; Hsu, K.-L.; Kifer, D.; Fang, Z.; et al. HESS Opinions: Incubating Deep-Learning-Powered Hydrologic Science Advances as a Community. *Hydrol. Earth Syst. Sci.* **2018**, *22*, 5639–5656. [CrossRef]
241. Lindersson, S.; Brandimarte, L.; Mård, J.; Baldassarre, G.D. A Review of Freely Accessible Global Datasets for the Study of Floods, Droughts and Their Interactions with Human Societies. *WIREs Water* **2020**, *7*, e1424. [CrossRef]
242. Pitts, J.; Gopal, S.; Ma, Y.; Koch, M.; Boumans, R.M.; Kaufman, L. Leveraging Big Data and Analytics to Improve Food, Energy, and Water System Sustainability. *Front. Big Data* **2020**, *3*, 13. [CrossRef]
243. Kratzert, F.; Klotz, D.; Shalev, G.; Klambauer, G.; Hochreiter, S.; Nearing, G. Towards Learning Universal, Regional, and Local Hydrological Behaviors via Machine Learning Applied to Large-Sample Datasets. *Hydrol. Earth Syst. Sci.* **2019**, *23*, 5089–5110. [CrossRef]
244. Hagen, J.S.; Leblois, E.; Lawrence, D.; Solomatine, D.; Sorteberg, A. Identifying Major Drivers of Daily Streamflow from Large-Scale Atmospheric Circulation with Machine Learning. *J. Hydrol.* **2021**, *596*, 126086. [CrossRef]
245. Sahraei, A.; Chamorro, A.; Kraft, P.; Breuer, L. Application of Machine Learning Models to Predict Maximum Event Water Fractions in Streamflow. *Front. Water* **2021**, *3*, 652100. [CrossRef]
246. Sudmanns, M.; Tiede, D.; Lang, S.; Bergstedt, H.; Trost, G.; Augustin, H.; Baraldi, A.; Blaschke, T. Big Earth Data: Disruptive Changes in Earth Observation Data Management and Analysis? *Int. J. Digit. Earth* **2020**, *13*, 832–850. [CrossRef]
247. Le Coz, J.; Patalano, A.; Collins, D.; Guillén, N.F.; García, C.M.; Smart, G.M.; Bind, J.; Chiaverini, A.; Le Boursicaud, R.; Dramais, G.; et al. Crowdsourced Data for Flood Hydrology: Feedback from Recent Citizen Science Projects in Argentina, France and New Zealand. *J. Hydrol.* **2016**, *541*, 766–777. [CrossRef]
248. Lowry, C.S.; Fienen, M.N.; Hall, D.M.; Stepenuck, K.F. Growing Pains of Crowdsourced Stream Stage Monitoring Using Mobile Phones: The Development of CrowdHydrology. *Front. Earth Sci.* **2019**, *7*, 128. [CrossRef]
249. Etter, S.; Strobl, B.; van Meerveld, I.; Seibert, J. Quality and Timing of Crowd-Based Water Level Class Observations. *Hydrol. Process.* **2020**, *34*, 4365–4378. [CrossRef]
250. Njue, N.; Stenfert Kroese, J.; Gräf, J.; Jacobs, S.R.; Weeser, B.; Breuer, L.; Rufino, M.C. Citizen Science in Hydrological Monitoring and Ecosystem Services Management: State of the Art and Future Prospects. *Sci. Total Environ.* **2019**, *693*, 133531. [CrossRef] [PubMed]
251. Yang, S.; Yang, D.; Chen, J.; Santisirisomboon, J.; Lu, W.; Zhao, B. A Physical Process and Machine Learning Combined Hydrological Model for Daily Streamflow Simulations of Large Watersheds with Limited Observation Data. *J. Hydrol.* **2020**, *590*, 125206. [CrossRef]
252. Yang, T.; Sun, F.; Gentine, P.; Liu, W.; Wang, H.; Yin, J.; Du, M.; Liu, C. Evaluation and Machine Learning Improvement of Global Hydrological Model-Based Flood Simulations. *Environ. Res. Lett.* **2019**, *14*, 114027. [CrossRef]
253. Souffront Alcantara, M.A.; Nelson, E.J.; Shakya, K.; Edwards, C.; Roberts, W.; Krewson, C.; Ames, D.P.; Jones, N.L.; Gutierrez, A. Hydrologic Modeling as a Service (HMaaS): A New Approach to Address Hydroinformatic Challenges in Developing Countries. *Front. Environ. Sci.* **2019**, *7*, 158. [CrossRef]
254. Kraft, B.; Jung, M.; Körner, M.; Koirala, S.; Reichstein, M. Towards Hybrid Modeling of the Global Hydrological Cycle. *Hydrol. Earth Syst. Sci. Discuss.* **2021**, *26*, 1579–1614. [CrossRef]
255. Gomes, V.C.F.; Queiroz, G.R.; Ferreira, K.R. An Overview of Platforms for Big Earth Observation Data Management and Analysis. *Remote Sens.* **2020**, *12*, 1253. [CrossRef]
256. Sperna Weiland, F.C.; Vrugt, J.A.; van Beek, R.P.H.; Weerts, A.H.; Bierkens, M.F.P. Significant Uncertainty in Global Scale Hydrological Modeling from Precipitation Data Errors. *J. Hydrol.* **2015**, *529*, 1095–1115. [CrossRef]
257. Marthews, T.R.; Blyth, E.M.; Martínez-de la Torre, A.; Veldkamp, T.I.E. A Global-Scale Evaluation of Extreme Event Uncertainty in the *EartH2Observe* Project. *Hydrol. Earth Syst. Sci.* **2020**, *24*, 75–92. [CrossRef]

258. Pagano, T.C.; Pappenberger, F.; Wood, A.W.; Ramos, M.-H.; Persson, A.; Anderson, B. Automation and Human Expertise in Operational River Forecasting. *WIREs Water* **2016**, *3*, 692–705. [CrossRef]

259. Mulligan, M. WaterWorld: A Self-Parameterising, Physically Based Model for Application in Data-Poor but Problem-Rich Environments Globally. *Hydrol. Res.* **2012**, *44*, 748–769. [CrossRef]

260. Jiang, Y.; Li, X.; Huang, C. Automatic Calibration a Hydrological Model Using a Master–Slave Swarms Shuffling Evolution Algorithm Based on Self-Adaptive Particle Swarm Optimization. *Expert Syst. Appl.* **2013**, *40*, 752–757. [CrossRef]

261. Nandi, S.; Janga Reddy, M. Comparative Performance Evaluation of Self-Adaptive Differential Evolution with GA, SCE and DE Algorithms for the Automatic Calibration of a Computationally Intensive Distributed Hydrological Model. *H2Open J.* **2020**, *3*, 306–327. [CrossRef]

262. Koch, P.; Golovidov, O.; Gardner, S.; Wujek, B.; Griffin, J.; Xu, Y. Autotune: A Derivative-Free Optimization Framework for Hyperparameter Tuning. In Proceedings of the 24th ACM SIGKDD International Conference on Knowledge Discovery & Data Mining, London, UK, 19–23 August 2018; Association for Computing Machinery: New York, NY, USA, 2018; pp. 443–452.

263. Feurer, M.; Klein, A.; Eggensperger, K.; Springenberg, J.T.; Blum, M.; Hutter, F. Auto-Sklearn: Efficient and Robust Automated Machine Learning. In *Automated Machine Learning: Methods, Systems, Challenges*; Hutter, F., Kotthoff, L., Vanschoren, J., Eds.; The Springer Series on Challenges in Machine Learning; Springer International Publishing: Cham, Switzerland, 2019; pp. 113–134, ISBN 978-3-030-05318-5.

264. Zürn, J.; Burgard, W.; Valada, A. Self-Supervised Visual Terrain Classification From Unsupervised Acoustic Feature Learning. *IEEE Trans. Robot.* **2021**, *37*, 466–481. [CrossRef]

265. Wittscher, L.; Diers, J.; Pigorsch, C. Improving Image Classification Robustness Using Self-Supervision. *Stat* **2022**, *11*, e455. [CrossRef]

266. Pintelas, E.; Livieris, I.E.; Pintelas, P.E. A Convolutional Autoencoder Topology for Classification in High-Dimensional Noisy Image Datasets. *Sensors* **2021**, *21*, 7731. [CrossRef]

267. Feng, J.; Feng, X.; Chen, J.; Cao, X.; Zhang, X.; Jiao, L.; Yu, T. Generative Adversarial Networks Based on Collaborative Learning and Attention Mechanism for Hyperspectral Image Classification. *Remote Sens.* **2020**, *12*, 1149. [CrossRef]

268. He, Z.; He, D.; Mei, X.; Hu, S. Wetland Classification Based on a New Efficient Generative Adversarial Network and Jilin-1 Satellite Image. *Remote Sens.* **2019**, *11*, 2455. [CrossRef]

269. Hofmann, J.; Schüttrumpf, H. FloodGAN: Using Deep Adversarial Learning to Predict Pluvial Flooding in Real Time. *Water* **2021**, *13*, 2255. [CrossRef]

270. Dong, H.; Ma, W.; Wu, Y.; Zhang, J.; Jiao, L. Self-Supervised Representation Learning for Remote Sensing Image Change Detection Based on Temporal Prediction. *Remote Sens.* **2020**, *12*, 1868. [CrossRef]

271. Xu, Z.; Du, J.; Wang, J.; Jiang, C.; Ren, Y. Satellite Image Prediction Relying on GAN and LSTM Neural Networks. In Proceedings of the ICC 2019—2019 IEEE International Conference on Communications (ICC), Shanghai, China, 20–24 May 2019; pp. 1–6.

272. Kao, I.-F.; Liou, J.-Y.; Lee, M.-H.; Chang, F.-J. Fusing Stacked Autoencoder and Long Short-Term Memory for Regional Multistep-Ahead Flood Inundation Forecasts. *J. Hydrol.* **2021**, *598*, 126371. [CrossRef]

273. Kao, I.-F.; Zhou, Y.; Chang, L.-C.; Chang, F.-J. Exploring a Long Short-Term Memory Based Encoder-Decoder Framework for Multi-Step-Ahead Flood Forecasting. *J. Hydrol.* **2020**, *583*, 124631. [CrossRef]

274. Li, X.; Song, G.; Du, Z. Hybrid Model of Generative Adversarial Network and Takagi-Sugeno for Multidimensional Incomplete Hydrological Big Data Prediction. *Concurr. Comput. Pract. Exp.* **2021**, *33*, e5713. [CrossRef]

275. Mazher, A. Visualization Framework for High-Dimensional Spatio-Temporal Hydrological Gridded Datasets Using Machine-Learning Techniques. *Water* **2020**, *12*, 590. [CrossRef]

276. Van Oosterom, P.; Stoter, J. 5D Data Modelling: Full Integration of 2D/3D Space, Time and Scale Dimensions. In Proceedings of the International Conference on Geographic Information Science, Zurich, Switzerland, 14–17 September 2010; Fabrikant, S.I., Reichenbacher, T., van Kreveld, M., Schlieder, C., Eds.; Springer: Berlin/Heidelberg, Germany, 2010; pp. 310–324.

277. Bauer, P.; Stevens, B.; Hazeleger, W. A Digital Twin of Earth for the Green Transition. *Nat. Clim. Chang.* **2021**, *11*, 80–83. [CrossRef]

278. Bartos, M.; Kerkez, B. Pipedream: An Interactive Digital Twin Model for Natural and Urban Drainage Systems. *Environ. Model. Softw.* **2021**, *144*, 105120. [CrossRef]

279. Veenendaal, B.; Brovelli, M.A.; Li, S. Review of Web Mapping: Eras, Trends and Directions. *ISPRS Int. J. Geo-Inf.* **2017**, *6*, 317. [CrossRef]

Article

Generating Continuous Rainfall Time Series with High Temporal Resolution by Using a Stochastic Rainfall Generator with a Copula and Modified Huff Rainfall Curves

Dinh Ty Nguyen [1] and Shien-Tsung Chen [2,*]

1   Ph.D. Program for Civil Engineering, Water Resources Engineering, and Infrastructure Planning, Feng Chia University, Taichung City 407, Taiwan; dinhtybkdn@gmail.com
2   Department of Hydraulic and Ocean Engineering, National Cheng Kung University, Tainan City 701, Taiwan
*   Correspondence: chen@gs.ncku.edu.tw

Abstract: In this study, a stochastic rainfall generator was developed to create continuous rainfall time series with a high temporal resolution of 10 min. The rainfall-generation process involved Monte Carlo simulation for stochastically generating rainfall parameters such as rainfall quantity, duration, inter-event time, and type. A bivariate copula was used to preserve the correlation between rainfall quantity and rainfall duration in the generated rainfall series. A modified Huff curve method was used to overcome the drawbacks of rainfall type classification by using the conventional Huff curve method. The number of discarded rainfall events was lower in the modified Huff curve method than in the conventional Huff curve method. Moreover, the modified method includes a new rainfall type that better represents rainfall events with a relatively uniform temporal pattern. The developed rainfall generator was used to reproduce rainfall series for the Yilan River Basin in Taiwan. The statistical indices of the generated rainfall series were close to those of the observed rainfall series. The results obtained for rainfall type classification indicated the necessity and suitability of the proposed new rainfall type. Overall, the developed stochastic rainfall generator can suitably reproduce continuous rainfall time series with a resolution of 10 min.

Keywords: stochastic rainfall generator; Huff rainfall curve; copula

Citation: Nguyen, D.T.; Chen, S.-T. Generating Continuous Rainfall Time Series with High Temporal Resolution by Using a Stochastic Rainfall Generator with a Copula and Modified Huff Rainfall Curves. *Water* **2022**, *14*, 2123. https://doi.org/10.3390/w14132123

Academic Editors: Fi-John Chang, Li-Chiu Chang and Jui-Fa Chen

Received: 19 May 2022
Accepted: 1 July 2022
Published: 3 July 2022

Publisher's Note: MDPI stays neutral with regard to jurisdictional claims in published maps and institutional affiliations.

1. Introduction

A stochastic rainfall generator is a statistical model that produces synthetic rainfall time series with desired statistical properties. Synthetic rainfall time series can be used for various purposes, such as rainfall–runoff modeling [1,2], design flood estimation [3–5], rainfall projection under climate change scenarios [6–8], and prediction in ungauged basins [9,10]. The Richardson-type rainfall generator [11,12], a popular stochastic rainfall generator, can reproduce long-term continuous daily precipitation time series. This generator uses Markov chains to determine the occurrence of wet or dry days and then simulates the rainfall quantity on wet days through Monte Carlo techniques. Although the aforementioned generator has been proven to be successful in reproducing daily precipitation time series, it cannot be used to obtain sub-daily or high-temporal-resolution rainfall time series.

High-temporal-resolution rainfall data can be used for different purposes, such as analyzing short-duration extreme rainfall events and simulating floods in small catchment areas. To generate rainfall time series with high temporal resolution [13,14], the temporal characteristics of a rainfall event must be determined. High-temporal-resolution rainfall data can be generated using two models: the profile- and pulse-based models. The profile-based model combines the total rainfall quantity and rainfall profile (rainfall type) to obtain a rainfall hyetograph. Typical rainfall types include rainfall described by the Chicago curve [15], Huff curve [16], and triangular curve [17]. The pulse-based model considers a rainfall event to consist of a cluster of rain cells whose occurrences follow a Poisson distribution. Rodriguez-Iturbe et al. [18,19] examined two popular pulse-based models in detail: the Neyman–Scott cluster model [20–25] and Bartlett–Lewis cluster model [26–31].

In the present study, the profile-based model with Huff rainfall curves, proposed by Huff [16] and modified by Huff and Vogel [32] and Huff [33], was adopted. Huff rainfall curves are presented using four types of dimensionless cumulative hyetographs according to the quarter in which the peak rainfall intensity occurs. Huff curves are popularly used in design storm, runoff simulation, design flood, and rainfall predictions [34–40]. However, Huff curves are not representative of rainfall events with uniform temporal distribution, and the rainfall type cannot be determined when a rainfall event has multiple peak intensities. Therefore, this paper proposes modified Huff rainfall curves for solving the limitations of the original Huff curves. The modified Huff rainfall curves have an additional rainfall type to account for rainfall events with uniform temporal distribution and can be used to classify rainfall events with multiple peak intensities. The proposed rainfall generator uses Monte Carlo simulation for stochastically generating modified Huff rainfall curves and other rainfall parameters (rainfall quantity, duration, and inter-event time) to form continuous rainfall time series with a temporal resolution of 10 min.

A suitable rainfall generator should reproduce rainfall data with anticipated statistical properties. Usually, a single rainfall parameter can be suitably generated through Monte Carlo simulation with an appropriate marginal distribution. However, various rainfall parameters can be interrelated. For example, a rainfall event with a longer duration is typically associated with higher cumulative rainfall. Therefore, the generation of rainfall parameters individually may result in the correlation between rainfall variables being lost and distorted rainfall data being obtained. Hence, the present study used a copula [41] to account for the correlation between parameters during the rainfall generation process. Copulas are mathematical functions that model the dependence among interrelated variables. An advantage of a copula is that it allows the dependence structure of variables to be modeled without the selection of marginal distributions. Therefore, copulas are widely employed in frequency analysis in hydrology [42–48]. The present study examined the correlation between rainfall parameters and constructed copulas for rainfall quantity and duration to generate rainfall data with appropriate correlation properties.

The remainder of this paper is structured as follows. The basic structure of the proposed continuous rainfall generator, the copula theory, and the modified Huff rainfall curves are described in Section 2. Section 3 presents information on the study area, the Yilan River Basin in Taiwan, and the collected 10 min rainfall data. Section 4 details the development of the proposed stochastic rainfall generator, with a focus on the modified Huff rainfall curves and adopted copula functions. The rainfall-generation results are presented in Section 5, and the conclusions of this study are provided in Section 6.

2. Methodology

2.1. Continuous Rainfall Time Series Generation

In this study, a stochastic rainfall generator was developed to produce a continuous rainfall time series with a high temporal resolution of 10 min. A continuous rainfall time series contains data for alternate wet and dry periods. The wet period indicates a rainfall event, and the dry period is called the inter-event time. A rainfall event can be characterized by rainfall duration, quantity, and type. Therefore, a continuous rainfall time series contains data related to four parameters: total rainfall quantity (R), rainfall duration (D), rainfall type, and rainfall inter-event time (T) (Figure 1).

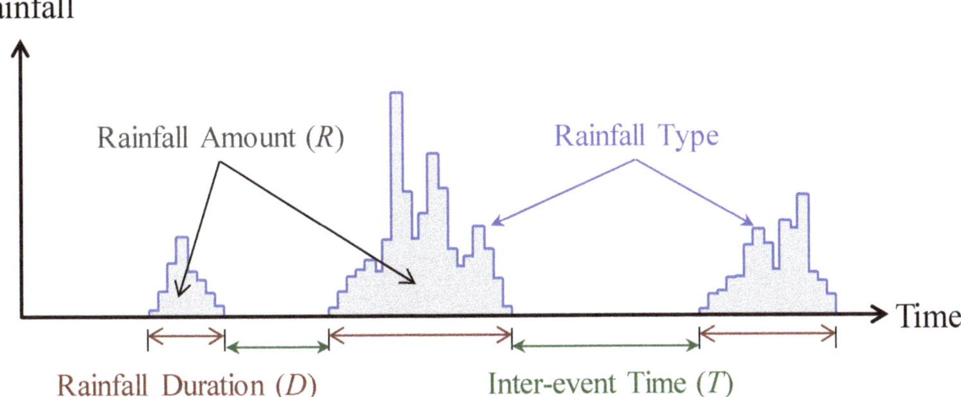

Figure 1. Example of rainfall time series and parameters.

In this study, the generation of continuous rainfall time series basically involved Monte Carlo simulation. First, the statistical properties and probability distribution of the four rainfall parameters were obtained and analyzed. Subsequently, on the time coordinate, an alternating sequence of rainfall duration and rainfall inter-event time was randomly generated according to their statistical properties and probability distributions. Total rainfall quantity, rainfall type and rainfall duration were generated simultaneously to construct a rainfall event. Because of the statistical correlation between rainfall quantity and duration, the copula method (Section 2.2) was used for simultaneously producing rainfall quantity and duration data. Moreover, modified Huff rainfall curves (Section 2.3) were used for better describing the temporal distribution of rainfall data. By using a repetitive generation process based on Monte Carlo simulation, continuous rainfall time series with the desired length were constructed.

2.2. Bivariate Copula

A copula is a multivariate distribution function that links the univariate distribution functions of each random variable. Copulas, originally introduced in the theorem proposed by Sklar [41], can model the correlation among variables without the assumptions made about the marginal distributions. According to this theorem, the joint cumulative distribution function F_{XY} of random variables X and Y with respect to the marginal cumulative distribution functions F_X and F_Y can be expressed as follows:

$$F_{XY}(x,y) = C(F_X(x), F_Y(y)) = C(u,v) \tag{1}$$

where C is a bivariate copula, and u and v are the cumulative probabilities of x and y, respectively. Let $\mathbf{I} = [0,\ 1]$, and let the bivariate copula C be a mapping function defined on a unit square, where $C : [0,\ 1]^2 \rightarrow \mathbf{I}$. When F_X and F_Y are continuous, a unique copula representation exists. Thus, F_{XY} defines a joint distribution function with the marginal distributions F_X and F_Y [49]. For any u and v in $\mathbf{I} = [0,\ 1]$, the copula is bounded as follows:

$$C(u,0) = 0, C(0,v) = 0, \ C(u,1) = u, \ C(1,v) = v \tag{2}$$

The aforementioned copula satisfies the two-increasing property; thus, for all $u_1 \leq u_2$ and $v_1 \leq v_2$, the following equation is satisfied:

$$C(u_2,v_2) - C(u_2,v_1) - C(u_1,v_2) + C(u_1,v_1) \geq 0 \tag{3}$$

Copulas are categorized into various families. The Archimedean copula family is popular in hydrology because of its simplicity and practicality in application [3,50–52]. This family includes various copulas with different numbers of parameters. Salvadori and De Michele [53,54] and Genest and Favre [55] suggested the use of one-parameter copulas in hydrology. The present study adopted three commonly used one-parameter (θ) copulas from the Archimedean copula family: the Frank, Clayton, and Gumbel copulas. The parameter θ can be calculated from the relationship between θ and Kendall's tau (τ), which is the rank correlation coefficient. Table 1 lists the copula functions adopted in this study and the relationships between θ and τ for these functions [56].

Table 1. Three copula functions used in this study and the relationships between θ and τ for these functions.

Copula	Function	Parameter Space	Relationship between θ and τ
Clayton	$C(u,v) = \left(u^{-\theta} + v^{-\theta} - 1\right)^{-1/\theta}$	$\theta > 0$	$\tau = \frac{\theta}{\theta+2}$
Frank	$C(u,v) = -\frac{1}{\theta}\ln\left\{1 + \frac{[\exp(-\theta u)-1]\cdot[\exp(-\theta v)-1]}{\exp(-\theta)-1}\right\}$	$\theta \neq 0$	$\tau = 1 + \frac{4}{\theta}\left[\frac{1}{\theta}\int_0^\theta \frac{t}{\exp(t)-1}dt - 1\right]$
Gumbel	$C(u,v) = \exp\left\{-\left[(-\ln u)^\theta + (-\ln v)^\theta\right]^{1/\theta}\right\}$	$\theta \geq 1$	$\tau = 1 - \frac{1}{\theta}$

2.3. Modified Huff Rainfall Curves

Profile-based models typically use Huff rainfall curves as the basis for defining the temporal distribution of rainfall events. These curves are empirical and dimensionless, probabilistic representations of cumulative hyetographs. The temporal rainfall patterns were classified into four types (Type 1 to Type 4) according to the time when the peak intensity occurred (i.e., in the first, second, third, or fourth quarter of rainfall duration). Figure 2 presents the temporal patterns, median (solid line), and 10% and 90% cumulative probabilities (lower and upper dashed lines, respectively) of the four types of Huff curves. The cumulative rainfall depth and cumulative rainfall duration were standardized by the total rainfall depth and total rainfall duration, respectively, and are presented in dimensionless form within the interval from 0% to 100%.

Although Huff rainfall curves are popular because of their ease of use, they have certain limitations. First, if a rainfall event has multiple peak intensities, then the rainfall type cannot be determined, and this event is omitted or randomly classified into one of the four rainfall types [57]. Second, a rainfall event with a relatively uniform temporal distribution cannot be well-represented by the four Huff curves. This paper proposes the solution described in the following text to the problem caused by the existence of multiple peak rainfall intensities. When a rainfall event has multiple peak intensities, the maximum total rainfall in a quarter instead of the peak intensity should be used to determine the rainfall type. However, a rainfall event with multiple peak intensities and the same maximum total rainfall in two or more quarters cannot be classified using the aforementioned solution. Nevertheless, this type of rare event occupies a very small portion of rainfall time series. Because Huff curves cannot be used to designate a rainfall event with a relatively uniform temporal distribution, this paper proposes an additional rainfall type (Type 5) for labeling such a rainfall event. The Schutz index (explained in the following paragraph) was used to distinguish Type 5 rainfall events from the other types of rainfall events.

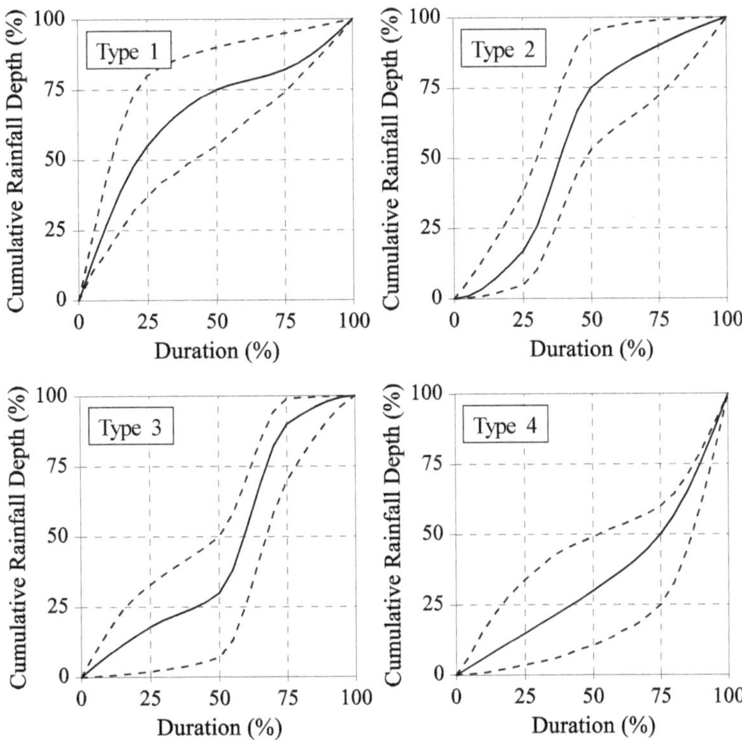

Figure 2. Huff's four rainfall types with the corresponding median (solid line) and 10% and 90% cumulative probabilities (lower and upper dashed lines, respectively).

The Schutz index [58] was originally proposed as an equity measure for income metrics in economics. Therefore, the Schutz index is an appropriate measure for assessing the uniformity of a distribution. Schutz [58] proposed the aforementioned index on the basis of the Lorenz curve [59], which is a probability plot with respect to a variable accumulated in a nondecreasing order. Figure 3 illustrates a Lorenz curve for income distribution among a population. Figure 3a depicts the histogram of the income of each 10% of the population versus the cumulative population in a nondecreasing order (the unit of income and population is percentage in this graph). By accumulating the income with respect to the population, the Lorenz curve (the red curve in Figure 3b) can be obtained. The 45° diagonal in Figure 3b indicates the perfect-equity income. When the Lorenz curve is close to the 45° diagonal, a uniform income distribution is identified. The Schutz index is an objective measure of the closeness of the Lorenz curve to the line of perfect equity. Thus, the Schutz index quantifies the total deviation of the income of each category (y_i) from the mean income (y_{mean}). Figure 4 presents an example of a rainfall event to describe the calculation of the Schutz index. The rainfall in each time step (y_i) of an original event is sorted and rearranged in a nondecreasing order, where $i = 1, 2, \ldots, n$, and n is the last time step. The Schutz index (S) is calculated using the following equation:

$$S = \frac{1}{2} \cdot \frac{\sum_{i=1}^{n} |y_i - y_{mean}|}{\sum_{i=1}^{n} y_i} \tag{4}$$

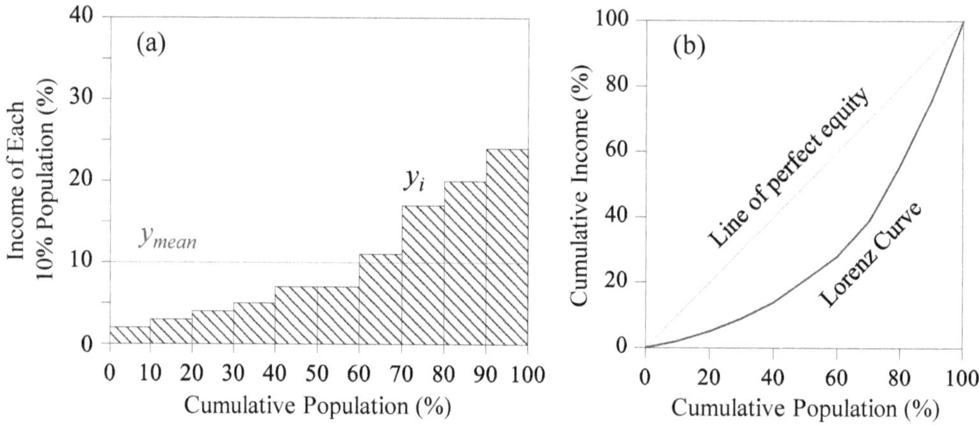

Figure 3. Example of the derivation of a Lorenz curve: (**a**) income of each 10% of the population arranged in a nondecreasing order and (**b**) comparison of the Lorenz curve and the line of perfect equity.

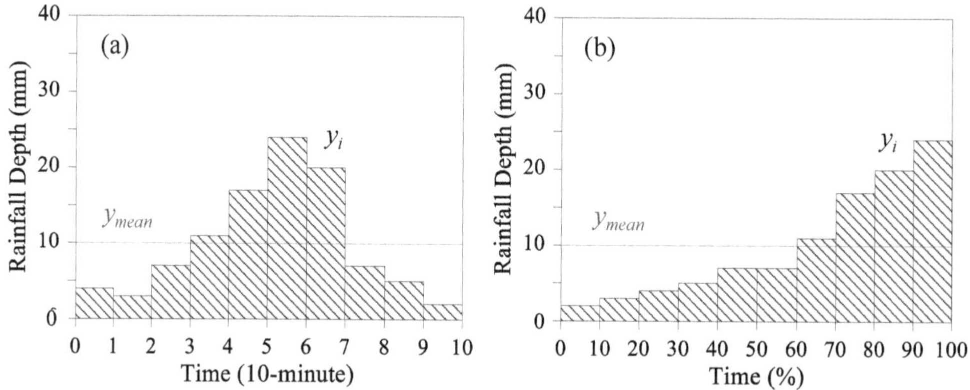

Figure 4. Example of the calculation of the Schutz index by using rainfall data: (**a**) original rainfall event and (**b**) sorted nondecreasing rainfall depth.

When S is 0, the rainfall in each step (y_i) is the same as the mean rainfall (y_{mean}), and the rainfall distribution is perfectly uniform. Thus, in the aforementioned scenario, the cumulative rainfall pattern corresponds to the diagonal in Figure 3b. When S approaches 1, the rainfall distribution is far from uniform. Thus, a small S value indicates that the rainfall distribution can be categorized as Type 5 (uniform rainfall distribution), and a large S value suggests that the rainfall distribution belongs to one category among Types 1 to 4. The modified Huff curve method is illustrated in Figure 5. First, the Schutz index (S) of a rainfall event is calculated to determine whether the rainfall distribution is uniform. If S is smaller than a threshold (which is determined in the following section), the rainfall distribution is considered to be uniform and categorized as Type 5. If S is larger than the threshold, the rainfall distribution is not close to Type 5 and belongs to one category among Types 1 to 4. In this circumstance, the existence of multiple peak rainfall intensities is checked. If only one peak intensity exists, the conventional Huff method is used to identify the rainfall type. If multiple peak intensities exist, the maximum total rainfall in a quarter is used to determine the rainfall type.

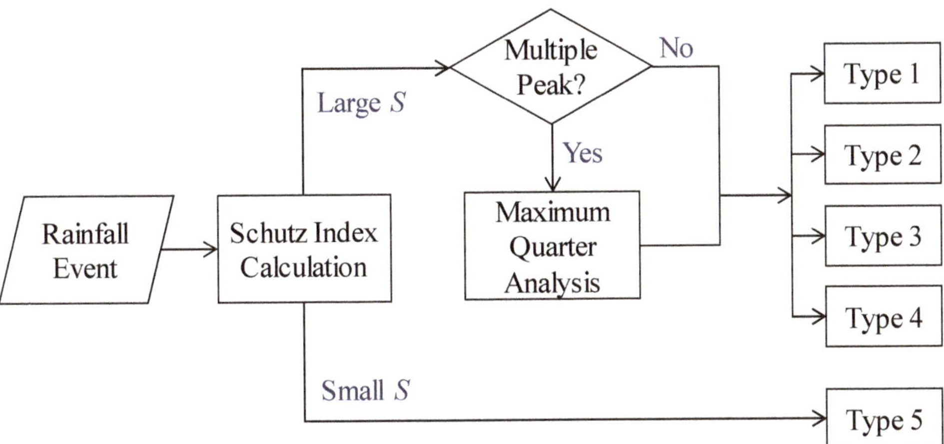

Figure 5. Process of the modified Huff method for determining the rainfall type.

3. Study Area and Rainfall Data

In this study, 10 min rainfall data from 2012 to 2018 were collected for the Yilan River Basin in Taiwan. The Yilan River's watershed is located in northeastern Taiwan (Figure 6), where the typical climate is humid and rainy. Nine rain gauges collect 10 min rainfall data for the Yilan River Basin. As displayed in Figure 6, seven of these rain gauges are located inside the basin, whereas two are located outside the basin.

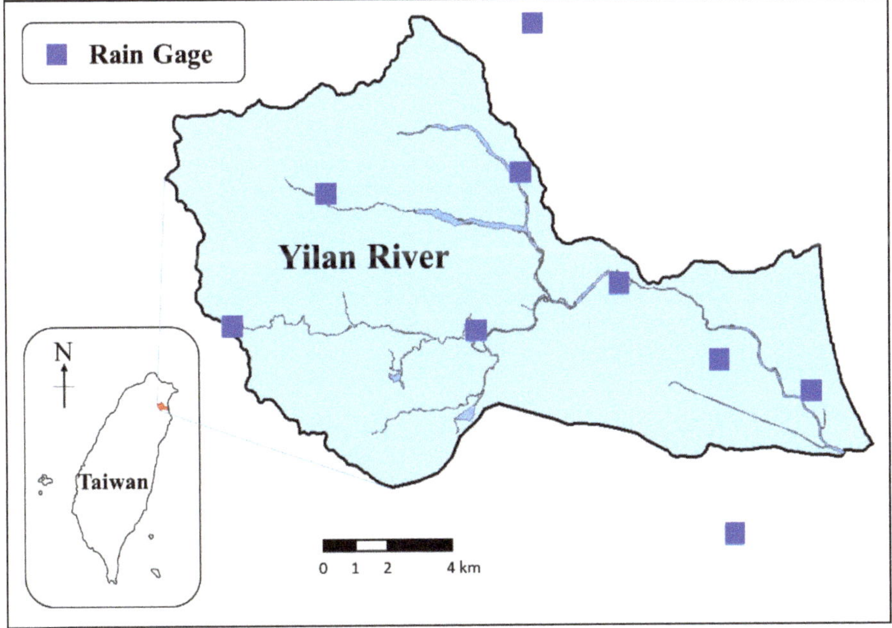

Figure 6. Locations of the study area and rain gauges.

Rainfall time series must be analyzed to obtain their statistical properties. The primary task is to define and distinguish a rainfall event. Researchers have used various methods to distinguish rainfall events and determine the rainfall inter-event time [60,61]. The selection of minimum rainfall duration and minimum rainfall inter-event time is dependent on the temporal resolution of the available data. Because the time resolution of the data collected in this study was 10 min, and the rainfall types were distinguished according to Huff rainfall curves, the minimum duration of a rainfall event was set as 40 min. Thus, rainfall events shorter than 40 min were not included in the rainfall event database. Moreover, the minimum inter-event time was set as 1 h. Thus, when a dry period was shorter than 1 h, this period and the wet periods preceding and following it were regarded as a rainfall event [62,63]. This approach considerably reduced the number of discarded rainfall events with durations less than 40 min.

Summer and winter monsoons occurred in the study area. Therefore, the rainfall events in this study were divided into those occurring in the summer (from May to October) and winter (from November to April of next year) seasons. A total of 4317 summer rainfall events and 3246 winter rainfall events were identified in this study.

4. Stochastic Rainfall-Generation Model Development

4.1. Rainfall Type

A threshold value of the Schutz index (S) was determined to identify Type 5 rainfall events. No a priori criterion can be used to set the threshold value; however, Bonta and Shahalam [64] suggested using the data of more than 120 storms to obtain stable Huff curves. The present study assumed that the number of Type 5 rainfall events was not greater than the numbers of Type 1 to Type 4 rainfall events. This assumption is rational because the number of rainfall events with a uniform distribution (Type 5) is usually less than those with a nonuniform distribution (Types 1 to 4). Therefore, a grid search method was used for S under the condition of increasing the number of Type 5 rainfall events; however, the number of Type 5 rainfall events was constrained by the minimum number of Type 1 to Type 4 rainfall events. Consequently, the number of Type 5 rainfall events could not exceed the minimum number of Type 1 to Type 4 rainfall events. Table 2 lists the threshold values of S for the summer and winter seasons as well as the numbers and percentages of Type 1 to Type 5 rainfall events. The threshold values for the summer and winter seasons were 0.29 and 0.30, respectively. By using the derived thresholds and the process described in the previous section (Figure 5), the identified rainfall events were categorized into different types. Figure 7 shows the classification of rainfall types for the summer season by using modified Huff rainfall curves. The bold black curves connecting squares, circles, and triangles indicate the 10%, 50%, and 90% percentiles of the rainfall categories, respectively. The colored curves represent the observed rainfall distributions. The rainfall classification curves obtained for the winter season were analogous to those obtained for the summer season. In Figure 7, the Type 5 rainfall events exhibit the characteristic of uniform distribution, which is different from the characteristics of Type 1 to Type 4 rainfall events. The results displayed in Figure 7 support the rationale that the Type 5 category is essential for better representing a uniform rainfall distribution.

Table 2. Threshold values of the Schutz index as well as the numbers and percentages of Type 1 to Type 5 rainfall events.

	Summer Season		Winter Season	
	Number of Events	Percentage (%)	Number of Events	Percentage (%)
Type 1	1069	24.76	869	26.77
Type 2	1003	23.24	721	22.21
Type 3	857	19.85	610	18.79
Type 4	704	16.31	534	16.45
Type 5	684	15.84	512	15.78
Schutz index	0.29		0.30	

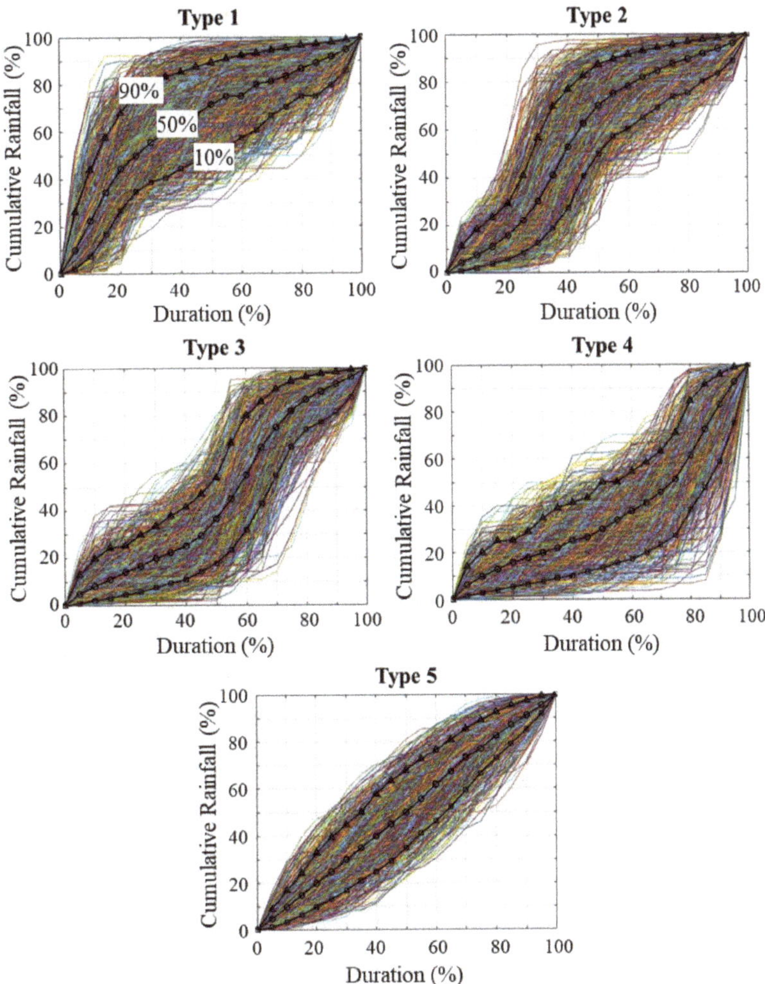

Figure 7. Summer rainfall events categorized using the modified Huff method.

In the traditional Huff method, approximately 40% of the identified rainfall events would need to be ignored because they are associated with multiple peak rainfall intensities. However, in the proposed method, only 5% of the identified rainfall events were ignored.

4.2. Copula Function

This study examined the correlation between each pair of the rainfall parameters by using normalized rank scatter plots [48]. The rainfall quantity, duration, and inter-event time were normalized between 0 and 1 and sorted in ascending order. These normalized and sorted data were then used to draw scatter plots. Figure 8 displays the normalized rank scatter plots for different pairs of rainfall parameters in the summer season. The rows in this figure indicate different rainfall types (Types 1 to 5), and the three columns denote three pairs of parameters. The patterns in the first column in Figure 8 indicate that the rainfall quantity and duration (R, D) are correlated, especially for large values. However, the patterns in the second and third columns suggest that no correlation exists between rainfall duration and inter-event time (D, T) and between rainfall quantity and inter-event time (R, T), respectively. The normalized rank scatter plots for the winter season are similar to those for the summer season and thus are not shown in this paper. Pearson correlation coefficients were calculated using the data in the rank-normalized plots to quantify the correlation between the rainfall parameters. The average correlation coefficients for (R, D) in the summer and winter seasons were 0.70 and 0.85, respectively. However, the correlation coefficients for (D, T) and (R, T) were close to 0. Therefore, the rainfall quantity and duration (R, D) were adopted to construct a bivariate copula for rainfall generation.

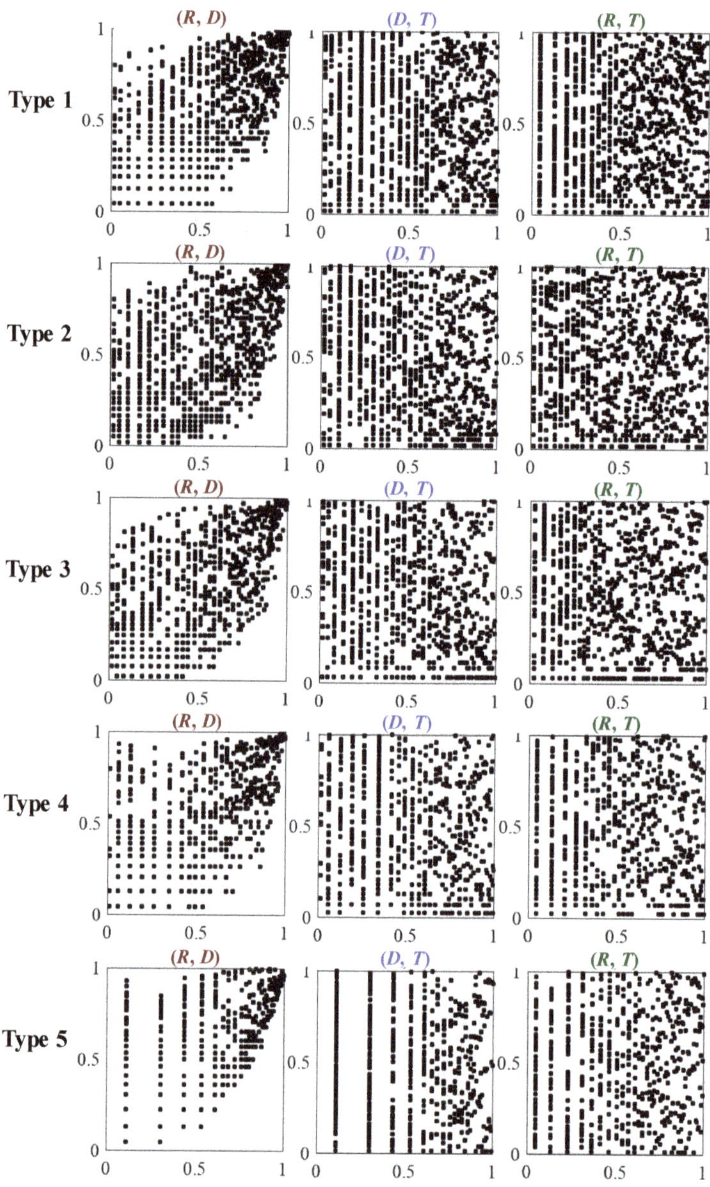

Figure 8. Normalized-rank scatter plots for different pairs of rainfall parameters in summer.

The next step involved determining which bivariate distribution function (from among the Frank, Clayton, and Gumbel copulas) was best suited for describing the correlation between the rainfall quantity and duration in the study area. Kendall's tau (τ) was calculated and used to determine the θ value of the three copulas adopted in this study (Table 1). The τ and θ values derived for the three copulas are listed in Table 3. After deriving the aforementioned values, the three copulas were used to model the correlation between rainfall quantity and rainfall duration. The cumulative probability functions of the three copulas for the summer season are illustrated in Figure 9. The blue, red, and green curves indicate the cumulative probability functions of the Clayton, Frank, and Gumbel copulas, respectively. The black dotted line represents the cumulative probability function of the empirical copula, which was constructed using the observed rainfall quantity and duration. The Clayton, Frank, Gumbel, and empirical copulas exhibited similar patterns to each other except for the Type 5 rainfall data under low cumulative probabilities. In general, the Clayton, Frank, and Gumbel copulas can suitably model the correlation between rainfall quantity and rainfall duration. The present study used the root mean square error (RMSE) for objectively determining the copula with the best fit to the empirical copula. Table 4 lists the RMSEs in probability (the vertical axis in Figure 9) for different copulas and rainfall types. The minimum RMSEs are highlighted in bold although some values are the same after being rounded off. The results indicate that the Frank copula was the most appropriate copula for modeling the correlation between the rainfall quantity and duration in the study area.

Table 3. Values of τ and θ for the three copulas in summer and winter.

| | Summer Season | | | | Winter Season | | | |
| | Kendall's Tau τ | Parameter θ | | | Kendall's Tau τ | Parameter θ | | |
		Clayton	Frank	Gumbel		Clayton	Frank	Gumbel
Type 1	0.487	1.901	5.510	1.950	0.602	3.021	7.975	2.511
Type 2	0.417	1.428	4.394	1.714	0.633	3.455	8.893	2.727
Type 3	0.465	1.739	5.136	1.870	0.613	3.161	8.273	2.581
Type 4	0.436	1.546	4.680	1.773	0.623	3.307	8.581	2.653
Type 5	0.490	1.920	5.553	1.960	0.743	5.785	13.70	3.892

Table 4. RMSEs in probability values for different copulas and rainfall types.

| | Summer Season | | | Winter Season | | |
	Clayton	Frank	Gumbel	Clayton	Frank	Gumbel
Type 1	0.030	**0.027**	0.028	0.038	**0.036**	0.037
Type 2	0.024	0.019	**0.018**	0.026	**0.024**	0.025
Type 3	0.026	**0.020**	0.020	0.023	**0.022**	0.022
Type 4	0.034	**0.030**	0.031	0.042	**0.040**	0.041
Type 5	0.059	**0.059**	0.060	0.052	**0.052**	0.052

Because the rainfall quantity and duration were fitted using the Frank copula, the rainfall inter-event time was modeled using a univariate probability distribution. In this study, the observation data was directly applied to construct the empirical distribution for the rainfall inter-event time.

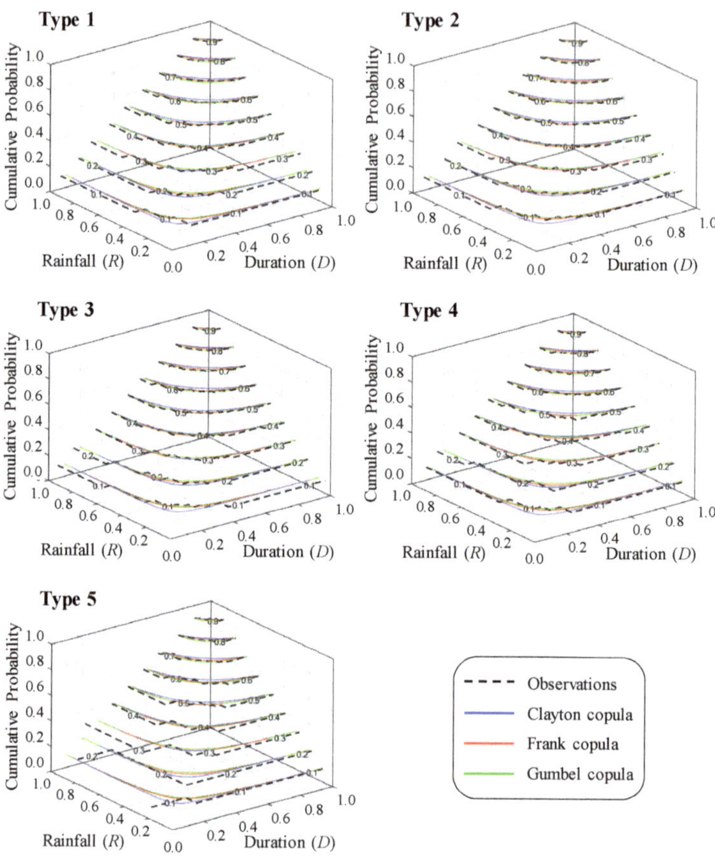

Figure 9. Copula functions for different rainfall types in summer.

4.3. Procedure for Stochastic Rainfall Generation

In this study, rainfall data generation was conducted through Monte Carlo simulation, which involves repeated random sampling. First, a 50% probability is used to determine whether a rainfall time series begins with a rainfall event or rainfall inter-event time. Rainfall events and inter-event times are then alternately produced. To produce a rainfall inter-event time, a number in the range of (0, 1) is randomly generated and substituted into the empirical probability distribution. To produce a rainfall event, three rainfall parameters, namely rainfall quantity, duration, and type, are generated.

The rainfall quantity and duration are generated simultaneously by using the Frank copula. First, a random number is generated as the cumulative probability of the copula, which is represented by a contour curve in Figure 9. Next, a point on the contour curve is randomly selected with equal probability. The values of rainfall quantity and duration can then be obtained from the location of the selected point. For the generation of rainfall type, one out of the five rainfall types (Types 1 to 5) is randomly selected according to their probabilities of occurrence (Table 2). After the rainfall type is determined, a random number is generated as the percentile for that rainfall type (Figure 7). Thus, a rainfall curve can be retrieved for the identified rainfall type with the generated percentile value.

By using the aforementioned process to produce rainfall parameters repetitively, synthetic continuous rainfall time series of any desired length can be generated. This study repeated the aforementioned procedure 10,000 times to generate a continuous rainfall time series with 10,000 sets of rainfall events and inter-event times. The results of this study are discussed in the following section.

5. Results and Discussion

This paper proposes a methodology for generating a continuous rainfall time series with a temporal resolution of 10 min. The statistical properties of the generated rainfall time series should correspond to those of the observed rainfall time series. The observed rainfall time series from 2012 to 2018 in the study area was used as the database to generate a synthetic rainfall time series. The study area has two main seasons: summer (from May to October) and winter (from November to April of next year). The summer season includes the "plum rain" season (May and June) and typhoon rain season (July to October). The typical rainfall types occurring in the summer season are stationary frontal rainfall, convective rainfall, and typhoon rainfall. The winter season is dominated by the northeast monsoon, which constantly brings moist ocean air into the study area. Short-duration, heavy rainfall is typically received in summer, whereas long-duration, moderate rainfall is typically received in winter.

The statistics of four rainfall parameters were calculated to assess the performance of the proposed rainfall generator. Figure 10 presents the average rainfall quantity and the standard deviation of the rainfall quantity in the summer (red circle) and winter (blue square) seasons. The average and standard deviation were calculated with respect to rainfall type (Types 1 to 5); therefore, five points were obtained for each of the aforementioned parameters in each season. These points lie close to the 45° diagonal, which indicates that the average generated rainfall quantity and the standard deviation of the generated rainfall quantity are analogous to the corresponding observation data. As displayed in Figure 10, the average rainfall quantity and the standard deviation of the rainfall quantity were larger in the summer than in the winter season for all types of rainfall except for Type 5 rainfall (the smallest value represented by a red circle). Type 5 rainfall has a relatively uniform temporal distribution, and Type 5 rainfall events are usually short-duration events with low rainfall quantities. Therefore, in this study, the average rainfall quantity and standard deviation of the rainfall quantity were low for Type 5 rainfall. In general, the aforementioned parameters were larger for Type 2 and Type 3 rainfall than for the other types of rainfall.

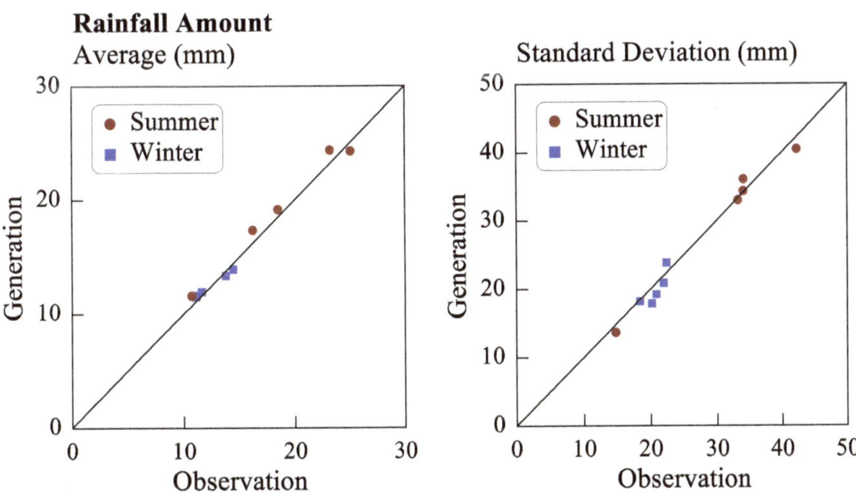

Figure 10. Comparison of the average and standard deviation values obtained for the observed and generated rainfall quantities.

Figure 11 displays the average rainfall duration and the standard deviation of the rainfall duration, and the duration unit in this graph is 10 min. This figure reveals that the average rainfall durations were generally longer in the winter than in the summer season. The shortest average rainfall durations in the summer and winter seasons (the lowest red circle and blue square in the left part of Figure 11) were observed for Type 5 rainfall. Moreover, the standard deviations in winter were larger than those in summer. The aforementioned results correspond to the rainfall characteristics in the study area. The points plotted for the average rainfall duration and the standard deviation of the rainfall duration are close to the 45° diagonal, which indicates that the generated values are close to the observed values. The rainfall quantity and duration were generated using the copula method. Figure 12 illustrates the correlation between the rainfall quantity and the rainfall duration in terms of τ. Positive correlation coefficients were obtained between the aforementioned factors, which indicates that in general, the longer the rainfall duration, the higher the rainfall quantity. The higher correlation in winter than in summer suggests that more persistent rainfall was received for a longer duration in the winter rainfall events than in the summer rainfall events. The lower correlation in summer can be attributed to the various rainfall patterns observed during this season (i.e., frontal, convective, and typhoon rainfalls). Overall, the results displayed in Figures 10–12 indicate that the selected copula can accurately reproduce the correlation between rainfall quantity and rainfall duration.

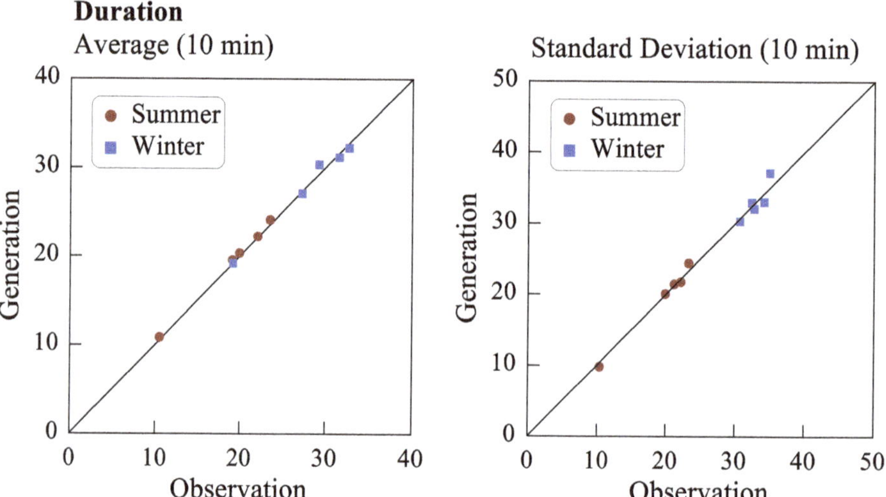

Figure 11. Comparison of the average and standard deviation values obtained for the observed and generated rainfall duration.

Figure 13 presents the average and standard deviation values obtained for the rainfall inter-event time (the time unit in this figure is 10 min). The rainfall inter-event times in the summer and winter seasons were similar (i.e., approximately 3000 min (approximately 2 days)). The developed rainfall generator reproduced the inter-event time on the basis of the empirical probability distribution. The averages of the generated rainfall inter-event times were in line with the corresponding observations; however, the standard deviations were somewhat smaller than the corresponding observations, especially for the winter season. Figure 14 presents the number of rainfall events (in percentage) generated for the different rainfall types. The percentages of rainfall events generated for each rainfall type were close to the corresponding observation data presented in Table 2. In conclusion, the proposed rainfall generation model can suitably reproduce rainfall time series with high temporal resolution.

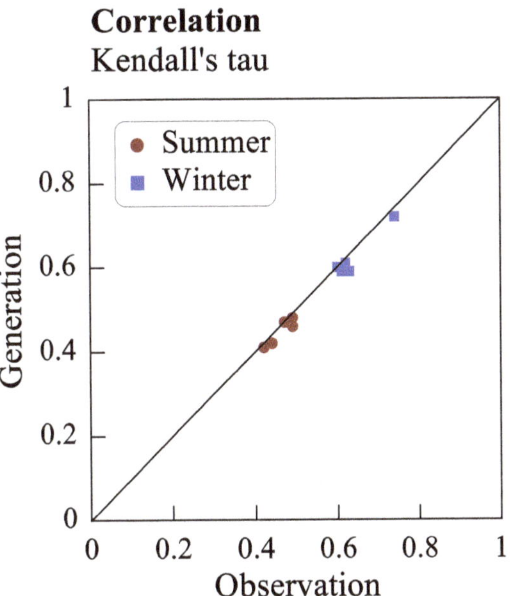

Figure 12. Comparison of the observed and generated correlations between rainfall quantity and duration.

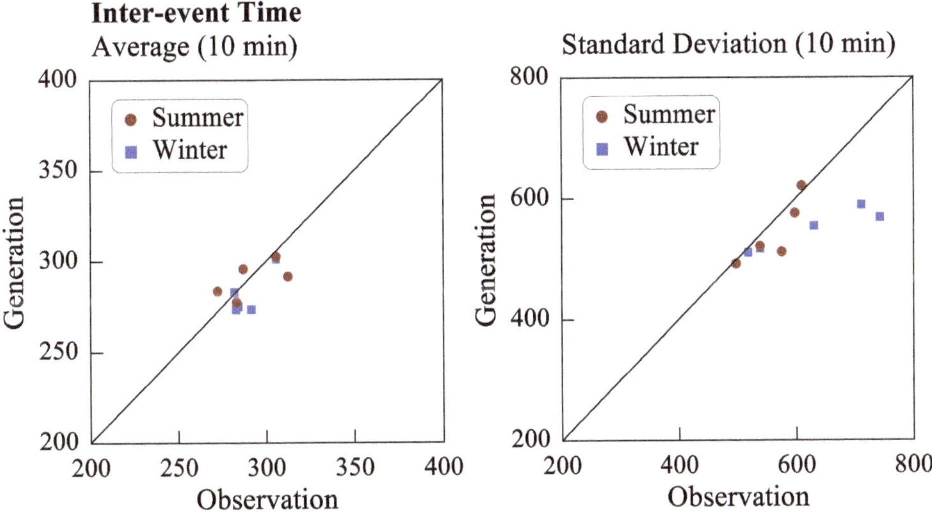

Figure 13. Comparison of the average and standard deviation values obtained for the observed and generated inter-event times.

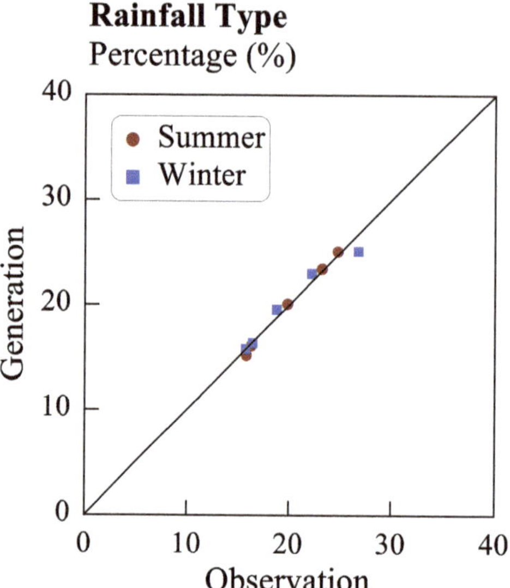

Figure 14. Comparison of the observed and generated numbers of rainfall events for different rainfall types.

6. Conclusions

This paper proposes a stochastic rainfall generator for producing a continuous rainfall time series with high temporal resolution. Observed 10 min rainfall time series from 2012 to 2018 for the Yilan River Basin in Taiwan were collected, and the rainfall events identified in the collected data were divided into summer and winter rainfall events (4317 and 3246 events, respectively). Although the stochastic rainfall-generation process adopted in the proposed generator is based on conventional Monte Carlo simulation, the use of modified Huff curves and a copula enables the generated rainfall series to exhibit appropriate rainfall types and precise correlation between rainfall quantity and rainfall duration.

The modified Huff method not only overcomes the limitation related to classifying rainfall events with multiple peak intensities but also includes a new rainfall type for labeling rainfall events with uniform temporal distribution. The Schutz index was used in this study to distinguish this new rainfall type, which is ignored in the traditional Huff method. The modified Huff method reduced the number of omitted rainfall events in the study area from 40% (with the conventional Huff method) to 5%. Moreover, a copula was used to model the correlation between each pair among three rainfall parameters in the generation process. The results indicated that the rainfall quantity and duration were correlated. Three copulas from the Archimedean family were used in this study, and the Frank copula was found to be the optimal copula for modeling the correlation between rainfall quantity and rainfall duration.

The proposed stochastic rainfall generator was used to generate a continuous rainfall time series with 10,000 sets of alternating rainfall events and inter-event times. The generated rainfall time series was assessed by comparing its statistical indices with those of the observed rainfall data. The results of this comparison indicated that the mean values obtained for the generated and observed rainfall quantity, duration, and inter-event time were similar. The standard deviations of the generated rainfall quantity and duration were close to those of the observed rainfall quantity and duration, respectively. Only the standard deviation of the rainfall inter-event time in winter was marginally underestimated. By using the Frank copula, the correlation between rainfall quantity and rainfall duration can be suitably preserved in the rainfall time series generated using the proposed stochastic rainfall generator. Moreover, the differences between the statistical properties of Type 5 rainfall events and other types of rainfall events support the rationality and necessity of using the modified Huff rainfall curves adopted in this study. In summary, the results of this study indicate that the developed stochastic rainfall generator can accurately reproduce continuous rainfall time series with a temporal resolution of 10 min.

Nonetheless, some issues are discussed in the following as potential improvements for future works. This study used all collected rainfall observations to examine the performance of the rainfall generator. Future work may conduct the cross-validation scheme to check the generation performance with respect to a certain period or a particular site. This study adopted the bivariate copula due to the correlation relationship among the rainfall variables in the study area. The trivariate copula can be tested to model the correlation among multiple variables. This study proposed an additional rainfall type on the basis of Huff rainfall curves, and analysis results demonstrated the success of the modified Huff method. However, the temporal distribution of rainfall events are not certainly restricted to the five types of modified Huff model in this study. Alternative rainfall type methods can be adopted in generating the rainfall time series. The proposed rainfall generator focused on reproducing the rainfall time series with correct statistical characteristics. The spatial rainfall feature was not considered herein. Future works can focus on developing a rainfall generator accounting for the spatial and temporal characteristics simultaneously.

Author Contributions: Conceptualization, S.-T.C.; formal analysis, D.T.N. and S.-T.C.; methodology, D.T.N. and S.-T.C.; visualization, S.-T.C.; writing—original draft, D.T.N. and S.-T.C.; writing—review and editing, S.-T.C. All authors have read and agreed to the published version of the manuscript.

Funding: This research received no external funding.

Institutional Review Board Statement: Not applicable.

Informed Consent Statement: Not applicable.

Conflicts of Interest: The authors declare no conflict of interest.

References

1. Breinl, K. Driving a lumped hydrological model with precipitation output from weather generators of different complexity. *Hydrol. Sci. J.* **2016**, *61*, 1395–1414. [CrossRef]
2. Chlumecký, M.; Buchtele, J.; Richta, K. Application of random number generators in genetic algorithms to improve rainfall-runoff modelling. *J. Hydrol.* **2017**, *553*, 350–355. [CrossRef]
3. Candela, A.; Brigandì, G.; Aronica, G.T. Estimation of synthetic flood design hydrographs using a distributed rainfall–runoff model coupled with a copula-based single storm rainfall generator. *Nat. Hazards Earth Syst. Sci.* **2014**, *14*, 1819–1833. [CrossRef]
4. Winter, B.; Schneeberger, K.; Dung, N.V.; Huttenlau, M.; Achleitner, S.; Stötter, J.; Merz, B.; Vorogushyn, S. A continuous modelling approach for design flood estimation on sub-daily time scale. *Hydrol. Sci. J.* **2019**, *64*, 539–554. [CrossRef]
5. Chimene, C.A.; Campos, J.N.B. The design flood under two approaches: Synthetic storm hyetograph and observed storm hyetograph. *J. Appl. Water Eng. Res.* **2020**, *8*, 171–182. [CrossRef]
6. Chen, S.T.; Yu, P.S.; Tang, Y.H. Statistical downscaling of daily precipitation using support vector machines and multivariate analysis. *J. Hydrol.* **2010**, *385*, 13–22. [CrossRef]
7. Yang, T.C.; Yu, P.S.; Wei, C.M.; Chen, S.T. Projection of climate change for daily precipitation: A case study in Shih-Men Reservoir catchment in Taiwan. *Hydrol. Processes* **2011**, *25*, 1342–1354. [CrossRef]
8. Khazaei, M.R.; Zahabiyoun, B.; Saghafian, B. Assessment of climate change impact on floods using weather generator and continuous rainfall-runoff model. *Int. J. Climatol.* **2012**, *32*, 1997–2006. [CrossRef]
9. Tukimat, N.N.A.; Syukri, N.A.; Malek, M.A. Projection the long-term ungauged rainfall using integrated Statistical Downscaling Model and Geographic Information System (SDSM-GIS) model. *Heliyon* **2019**, *5*, e02456. [CrossRef]
10. Grimaldi, S.; Nardi, F.; Piscopia, R.; Petroselli, A.; Apollonio, C. Continuous hydrologic modelling for design simulation in small and ungauged basins: A step forward and some tests for its practical use. *J. Hydrol.* **2021**, *595*, 125664. [CrossRef]
11. Katz, R.W. Precipitation as a chain-dependent process. *J. Appl. Meteorol.* **1977**, *16*, 671–676. [CrossRef]
12. Richardson, C.W. Stochastic simulation of daily precipitation, temperature, and solar radiation. *Water Resour. Res.* **1981**, *17*, 182–190. [CrossRef]
13. Westra, S.; Mehrotra, R.; Sharma, A.; Srikanthan, R. Continuous rainfall simulation: 1. A regionalized subdaily disaggregation approach. *Water Resour. Res.* **2012**, *48*, W01535. [CrossRef]

14. De Luca, D.L.; Petroselli, A. STORAGE (STOchastic RAinfall GEnerator): A user-friendly software for generating long and high-resolution rainfall time series. *Hydrology* **2021**, *8*, 76. [CrossRef]

15. Keifer, C.J.; Chu, H.H. Synthetic storm pattern for drainage design. *J. Hydraul. Div.* **1957**, *83*, 1–25. [CrossRef]

16. Huff, F.A. Time distribution of rainfall in heavy storms. *Water Resour. Res.* **1967**, *3*, 1007–1019. [CrossRef]

17. Yen, B.C.; Chow, V.T. Design hyetographs for small drainage structures. *J. Hydraul. Div.* **1980**, *106*, 1055–1076. [CrossRef]

18. Rodriguez-Iturbe, I.; Cox, D.R.; Isham, V. Some models for rainfall based on stochastic point processes. *Proc. R. Soc. London. A. Math. Phys. Sci.* **1987**, *410*, 269–288.

19. Rodriguez-Iturbe, I.; De Power, B.F.; Valdes, J.B. Rectangular pulses point process models for rainfall: Analysis of empirical data. *J. Geophys. Res. Atmos.* **1987**, *92*, 9645–9656. [CrossRef]

20. Entekhabi, D.; Rodriguez-Iturbe, I.; Eagleson, P.S. Probabilistic representation of the temporal rainfall process by a modified Neyman-Scott rectangular pulses model: Parameter estimation and validation. *Water Resour. Res.* **1989**, *25*, 295–302. [CrossRef]

21. Cowpertwait, P.S. Further developments of the Neyman-Scott clustered point process for modeling rainfall. *Water Resour. Res.* **1991**, *27*, 1431–1438. [CrossRef]

22. Sørup, H.J.D.; Christensen, O.B.; Arnbjerg-Nielsen, K.; Mikkelsen, P.S. Downscaling future precipitation extremes to urban hydrology scales using a spatio-temporal Neyman–Scott weather generator. *Hydrol. Earth Syst. Sci.* **2016**, *20*, 1387–1403. [CrossRef]

23. Cowpertwait, P.S.P.; Kilsby, C.G.; O'Connell, P.E. A space-time Neyman-Scott model of rainfall: Empirical analysis of extremes. *Water Resour. Res.* **2022**, *38*, 1131. [CrossRef]

24. Lee, J.; Kim, U.; Kim, S.; Kim, J. Development and application of a rainfall temporal disaggregation method to project design rainfalls. *Water* **2022**, *14*, 1401. [CrossRef]

25. De Luca, D.L.; Apollonio, C.; Petroselli, A. The benefit of continuous hydrological modelling for drought hazard assessment in small and coastal ungauged basins: A case study in Southern Italy. *Climate* **2022**, *10*, 34. [CrossRef]

26. Islam, S.; Entekhabi, D.; Bras, R.L.; Rodriguez-Iturbe, I. Parameter estimation and sensitivity analysis for the modified Bartlett-Lewis rectangular pulses model of rainfall. *J. Geophys. Res. Atmos.* **1990**, *95*, 2093–2100. [CrossRef]

27. Onof, C.; Wheater, H.S. Modelling of British rainfall using a random parameter Bartlett-Lewis rectangular pulse model. *J. Hydrol.* **1993**, *149*, 67–95. [CrossRef]

28. Khaliq, M.N.; Cunnane, C. Modelling point rainfall occurrences with the modified Bartlett-Lewis rectangular pulses model. *J. Hydrol.* **1996**, *180*, 109–138. [CrossRef]

29. Onof, C.; Wang, L.P. Modelling rainfall with a Bartlett–Lewis process: New developments. *Hydrol. Earth Syst. Sci.* **2020**, *24*, 2791–2815. [CrossRef]

30. Islam, M.A.; Yu, B.; Cartwright, N. Coupling of satellite-derived precipitation products with Bartlett-Lewis model to estimate intensity-frequency-duration curves for remote areas. *J. Hydrol.* **2022**, *609*, 127743. [CrossRef]

31. Rafatnejad, A.; Tavakolifar, H.; Nazif, S. Evaluation of the climate change impact on the extreme rainfall amounts using modified method of fragments for sub-daily rainfall disaggregation. *Int. J. Climatol.* **2022**, *42*, 908–927. [CrossRef]

32. Huff, F.A.; Vogel, J.L. Hydrometeorology of Heavy Rainstorms in Chicago and Northeastern Illinois, Phase I—Historical Studies, Illinois State Water Survey. 1976. Available online: http://hdl.handle.net/2142/77792 (accessed on 15 May 2022).

33. Huff, F.A. Time Distributions of Heavy Rainstorms in Illinois. Circular No. 173; Department of Energy and Natural Resources, State of Illi-nois. 1990. Available online: https://www.ideals.illinois.edu/bitstream/handle/2142/94492/ISWSC-173.pdf?sequence=1 (accessed on 15 May 2022).

34. Yu, P.S.; Chen, S.T.; Chen, C.J.; Yang, T.C. The potential of fuzzy multi-objective model for rainfall forecasting from typhoons. *Nat. Hazards* **2005**, *34*, 131–150. [CrossRef]

35. Azli, M.; Rao, A.R. Development of Huff curves for peninsular Malaysia. *J. Hydrol.* **2010**, *388*, 77–84. [CrossRef]

36. Golian, S.; Saghafian, B.; Maknoon, R. Derivation of probabilistic thresholds of spatially distributed rainfall for flood forecasting. *Water Resour. Manag.* **2010**, *24*, 3547–3559. [CrossRef]

37. Chen, S.T. Multiclass support vector classification to estimate typhoon rainfall distribution. *Disaster Adv.* **2013**, *6*, 110–121.

38. Dolšak, D.; Bezak, N.; Šraj, M. Temporal characteristics of rainfall events under three climate types in Slovenia. *J. Hydrol.* **2016**, *541*, 1395–1405. [CrossRef]

39. Wartalska, K.; Kaźmierczak, B.; Nowakowska, M.; Kotowski, A. Analysis of hyetographs for drainage system modeling. *Water* **2020**, *12*, 149. [CrossRef]

40. Dunkerley, D. Regional rainfall regimes affect the sensitivity of the Huff quartile classification to the method of event delineation. *Water* **2022**, *14*, 1047. [CrossRef]

41. Sklar, M. Fonctions de repartition an dimensions et leurs marges. *Publ. Inst. Statist. Univ. Paris* **1959**, *8*, 229–231.

42. De Michele, C.; Salvadori, G. A generalized Pareto intensity-duration model of storm rainfall exploiting 2-copulas. *J. Geophys. Res. Atmos.* **2003**, *108*, 4067. [CrossRef]

43. Favre, A.C.; El Adlouni, S.; Perreault, L.; Thiémonge, N.; Bobée, B. Multivariate hydrological frequency analysis using copulas. *Water Resour. Res.* **2004**, *40*, W01101. [CrossRef]

44. Kao, S.C.; Chang, N.B. Copula-based flood frequency analysis at ungauged basin confluences: Nashville, Tennessee. *J. Hydrol. Eng.* **2012**, *17*, 790–799. [CrossRef]

45. Dodangeh, E.; Singh, V.P.; Pham, B.T.; Yin, J.; Yang, G.; Mosavi, A. Flood frequency analysis of interconnected Rivers by copulas. *Water Resour. Manag.* **2020**, *34*, 3533–3549. [CrossRef]
46. Pathak, A.A.; Dodamani, B.M. Connection between meteorological and groundwater drought with copula-based bivariate frequency analysis. *J. Hydrol. Eng.* **2021**, *26*, 05021015. [CrossRef]
47. Razmkhah, H.; Fararouie, A.; Ravari, A.R. Multivariate flood frequency analysis using bivariate copula functions. *Water Resour. Manag.* **2022**, *36*, 729–743. [CrossRef]
48. Jang, J.H.; Chang, T.H. Flood risk estimation under the compound influence of rainfall and tide. *J. Hydrol.* **2022**, *606*, 127446. [CrossRef]
49. Genest, C.; Rivest, L.P. On the multivariate probability integral transformation. *Stat. Probab. Lett.* **2001**, *53*, 391–399. [CrossRef]
50. Kao, S.C.; Govindaraju, R.S. Trivariate statistical analysis of extreme rainfall events via the Plackett family of copulas. *Water Resour. Res.* **2008**, *44*, W02415. [CrossRef]
51. Kao, S.C.; Govindaraju, R.S. A copula-based joint deficit index for droughts. *J. Hydrol.* **2010**, *380*, 121–134. [CrossRef]
52. Amirataee, B.; Montaseri, M.; Rezaie, H. An advanced data collection procedure in bivariate drought frequency analysis. *Hydrol. Processes* **2020**, *34*, 4067–4082. [CrossRef]
53. Salvadori, G.; De Michele, C. Frequency analysis via copulas: Theoretical aspects and applications to hydrological events. *Water Resour. Res.* **2004**, *40*, W12511. [CrossRef]
54. Salvadori, G.; De Michele, C. On the use of copulas in hydrology: Theory and practice. *J. Hydrol. Eng.* **2007**, *12*, 369–380. [CrossRef]
55. Genest, C.; Favre, A.C. Everything you always wanted to know about copula modeling but were afraid to ask. *J. Hydrol. Eng.* **2007**, *12*, 347–368. [CrossRef]
56. Gao, C.; Xu, Y.P.; Zhu, Q.; Bai, Z.; Liu, L. Stochastic generation of daily rainfall events: A single-site rainfall model with copula-based joint simulation of rainfall characteristics and classification and simulation of rainfall patterns. *J. Hydrol.* **2018**, *564*, 41–58. [CrossRef]
57. Vandenberghe, S.; Verhoest, N.E.C.; De Baets, B. Properties and performance of a copula-based design storm generator. In Proceedings of the International workshop: Advances in Statistical Hydrology, Taormina, Italy, 23–25 May 2010.
58. Schutz, R.R. On the measurement of income inequality. *Am. Econ. Rev.* **1951**, *41*, 107–122.
59. Lorenz, M.O. Methods of measuring the concentration of wealth. *Publ. Am. Stat. Assoc.* **1905**, *9*, 209–219. [CrossRef]
60. Cameron, D.; Beven, K.; Naden, P. Flood frequency estimation by continuous simulation under climate change (with uncertainty). *Hydrol. Earth Syst. Sci.* **2020**, *4*, 393–405. [CrossRef]
61. Sharafati, A.; Zahabiyoun, B. Stochastic generation of storm pattern. *Life Sci. J.* **2013**, *10*, 1575–1583.
62. Bonta, J.V.; Rao, A.R. Factors affecting the identification of independent storm events. *J. Hydrol.* **1988**, *98*, 275–293. [CrossRef]
63. Vernieuwe, H.; Vandenberghe, S.; De Baets, B.; Verhoest, N. A continuous rainfall model based on vine copulas. *Hydrol. Earth Syst. Sci.* **2015**, *19*, 2685–2699. [CrossRef]
64. Bonta, J.V.; Shahalam, A. Cumulative storm rainfall distributions: Comparison of Huff curves. *J. Hydrol.* **2003**, *42*, 65–74.

Article

Estimation of Threshold Rainfall in Ungauged Areas Using Machine Learning

Kyung-Su Chu, Cheong-Hyeon Oh, Jung-Ryel Choi and Byung-Sik Kim *

Department of Urban and Environmental and Disaster Management, Graduate School of Disaster Prevention, Kangwon National University, Samcheok 25913, Korea; chu_93@kangwon.ac.kr (K.-S.C.); och@kangwon.ac.kr (C.-H.O.); lovekurt82@gmail.com (J.-R.C.)
* Correspondence: hydrokbs@kangwon.ac.kr; Tel.: +82-33-570-6819

Abstract: In recent years, Korea has seen abnormal changes in precipitation and temperature driven by climate change. These changes highlight the increased risks of climate disasters and rainfall damage. Even with weather forecasts providing quantitative rainfall estimates, it is still difficult to estimate the damage caused by rainfall. Damaged by rainfalls differently for inch watershed, but there is a limit to the analysis coherent to the characteristic factors of the inch watershed. It is time-consuming to analyze rainfall and runoff using hydrological models every time it rains. Therefore, in fact, many analyses rely on simple rainfall data, and in coastal basins, hydrological analysis and physical model analysis are often difficult. To address the issue in this study, watershed characteristic factors such as drainage area (A), mean drainage elevation (H), mean drainage slope (S), drainage density (D), runoff curve number (CN), watershed parameter (L_p), and form factor (R_s) etc. and hydrologic factors were collected and calculated as independent variables, and the threshold rainfall calculated by the Ministry of Land, Infrastructure and Transport (MOLIT) was calculated as a dependent variable and used in the machine learning technique. As for machine learning techniques, this study uses the support vector machine method (SVM), the random forest method, and eXtreme Gradient Boosting (XGBoost). As a result, XGBoost showed good results in performance evaluation with RMSE 20, MAE 14, and RMSLE 0.28, and the threshold rainfall of the ungauged watersheds was calculated using the XGBoost technique and verified through past rainfall events and damage cases. As a result of the verification, it was confirmed that there were cases of damage in the basin where the threshold rainfall was low. If the application results of this study are used, it is judged that it is possible to accurately predict flooding-induced rainfall by calculating the threshold rainfall in the ungauged watersheds where rainfall-outflow analysis is difficult, and through this result, it is possible to prepare for areas vulnerable to flooding.

Keywords: machine learning; random forest; regression analysis; support vector machine; threshold rainfall; threshold runoff; XGBoost

Citation: Chu, K.-S.; Oh, C.-H.; Choi, J.-R.; Kim, B.-S. Estimation of Threshold Rainfall in Ungauged Areas Using Machine Learning. *Water* **2022**, *14*, 859. https://doi.org/10.3390/w14060859

Academic Editors: Fi-John Chang, Li-Chiu Chang and Jui-Fa Chen

Received: 22 January 2022
Accepted: 6 March 2022
Published: 10 March 2022

Publisher's Note: MDPI stays neutral with regard to jurisdictional claims in published maps and institutional affiliations.

1. Introduction

Climate change has increased rainfall in Korea, resulting in various natural disasters that cause rapidly increasing social and economic loss [1]. However, Korean weather forecasts only provide rainfall information in absolute terms, and the same heavy rain warnings and special reports apply to all areas in Korea, which means a failure to reflect regional differences. For this reason, even with accurate forecasts, the forecast system fails to provide specific information on how different areas are affected and damaged by weather events. Forecasts focused on physical aspects of weather events do not provide sufficient information on how people's properties and safety are affected by them.

It is for this reason that the World Meteorological Organization (WMO) emphasizes the need for 'impact forecasts' that consider the socioeconomic effects that may be caused by weather events [2]. In Korea, different organizations provide different definitions of impact forecasting. However, they can be summarized as follows: forecast that scientifically estimates the socioeconomic impact of weather at different times and places and delivers the estimates along with detailed weather information [3,4]. Outside of Korea, the WMO defines impact forecast as forecast that provides information on expected risks along with weather forecasts when disaster-causing high impact weather is expected. According to the Met Office of the United Kingdom, it is defined as a forecast that estimates the socioeconomic impact of a climate disaster at the time and place of its occurrence by considering meteorological disasters, level of exposure to disasters, and regional vulnerabilities. The National Weather Service of the United States defines it as a service aimed at providing the people with information on the social, economic, and environmental impact of weather, hydrological, and climate events [5]. Leading countries in the field of meteorology already provide information on socioeconomic impact of weather events along with high-resolution weather information. In the United Kingdom, the Flood Forecasting Centre (FFC) provides Flood Guidance Statements (FGS) that assess the risks of all flood types over five days and publish the findings daily [6,7]. The FFC uses the information to publish a table of flood risks which divides flood impact into four stages.

Impact forecasting requires threshold rainfall. Threshold rainfall means the rainfall amount that causes inundation. Accurate impact assessment requires calculation of the precise inundation-causing rainfall in each area. However, in Korea, research on threshold rainfall has been lacking. Most researchers use simplified analysis methods rather than refined hydraulic and hydrological analyses. Hydrological analyses of coastal areas are too complex to conduct properly.

As for previous literature on threshold rainfall calculation, ref [8] developed a flash food monitoring and prediction (F2MAP) model to calculate the flash flood-threshold runoff from rainfall. Ref [9] analyzed the relationship between flash flood index and runoff number characteristics to develop an equation between the two. Ref [10] proposed a threshold runoff calculation method using the flash flood guidance (FFG) model, which is more suitable for Korea rather than those used in the United States. The researchers presented the method as a way to acquire basic data for a flash flood forecast system. Ref [11] analyzed runoff in Jeju using the SWAT-K model that combines DEM, landcover, soil map methods, and developed a threshold runoff simulation method (TRSM) specifically for the island. Ref [12] used ArcGIS and HEC-GEOHMS to divide the Nakdonggang River watershed into 2268 sectors, drew rainfall-peak flow curves for different initial loss scenarios and antecedent moisture conditions, and calculated the threshold rainfalls. Ref [13] estimated threshold rainfalls for different durations using events with damage caused by past rainfall in urban areas and others without such damage. Ref [14] stressed the need for impact forecast and estimated threshold rainfalls using the SWMM model. Ref [7] also linked the grid base inundation analysis model (GIAM) for grid-based inundation analysis. Using the Huff distribution [14], the researcher converted the data into time-series rainfall data to simulate inundation depths, and inversely estimated the threshold rainfall based on the inundation depths. Ref [15] collected data on rainfall and typhoon damage over the last five years where inundation was caused, analyzed the relationship between rainfall and the damages, and developed an equation for threshold rainfall ($y = ax^b$). As can be seen from the literature cited above, Korean studies on threshold rainfall mostly used hydrological models. Few researchers studied threshold rainfall by considering hydrological characteristics.

More recently, a number of researchers used machine learning to improve the accuracy of threshold rainfall analysis [16–18]. Additionally growing is the body of literature that study rainfall-runoff, rainfall damage, and flood estimation with machine learning and deep learning rather than hydrological models [19–23]. However, few studies were identified in Korea that used machine learning to calculate threshold rainfall. Ref [19] sought to predict river water levels using observation data and deep learning algorithms. To that end, the researchers used tensor flow to predict water levels at the Okcheon Observatory location along the upper stream sectionof the Daecheong Dam within the Geumgang watershed and used TensorFlow to develop a multiple regression model and a long short-term memory (LSTM) artificial neural network model. Ref [24] used three machine learning techniques (support vector machine, decision-making tree, and random forest) to develop a function for predicting rainfall damage in the Seoul Metropolitan Area (SMA) and found that support vector machine analysis using meteorological observation data from two days before yields the highest prediction performance. Ref [25] used the machine learning method on Gyeonggi-do, the province that suffers the worst rainfall damage each year. Choi used the data on rainfall damage of public facilities from the 2006–2015 Disaster Yearbooks published by the Ministry of the Interior and Safety (MIST) as the dependent variable. Ref [26] used machine learning methods such as ESN and DeepESN to predict rainfall using rainfall, pressure, and humidity from 2004 to 2014 as mediating variables. The correlation factors calculated using DeepESN yielded better results. Ref [27] performed hydrological rainfall adjustment using Light GBM and XGBoost. They found clear adjustment effects across all rainfall events after Light GBM and XGBoost learning, despite the fact that rainfall is adjusted 5 to 20 mm less.

Much of the literature cited above only used a single hydrological model. However, in this study, two models were coupled and used to calculate the marginal rainfall [8] and it is considered an advantage of this paper to apply the results to machine learning. Figure 1 shows the flow chart of this study. It was analyzed through machine learning using threshold rainfall and topographic factors of standard watershed units. The threshold rainfall was used as a dependent variable, and the topographic factor of the standard watershed unit was designated as an independent variable. In addition, the model with the smallest error was selected using error performance analysis to calculate the threshold rainfall in the ungauged basin where hydrological analysis was difficult. The ungauged basin means a coastal area where it is difficult to calculate the threshold rainfall.

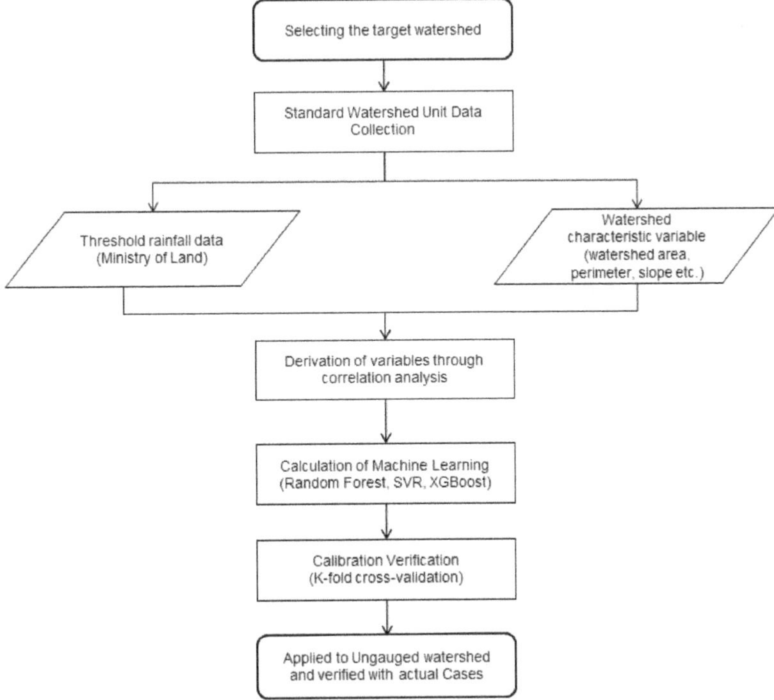

Figure 1. Flowchart of study.

2. Theoretical Background

2.1. Definition of Threshold Rainfall

In this study, threshold rainfall is calculated using the method used by the MOLIT in 2007, according to which threshold rainfall means the three-hour rainfall causing inundation depth at which the flow overflows the river embankment [8]. Threshold rainfall can be calculated by determining the rainfall of the rainfall-runoff curve corresponding to the threshold rainfall. In general, the runoff calculation equation for a rainfall-runoff model can be expressed as follows [15,28].

$$R_t = R_i + R_p \tag{1}$$

where R_t is the total runoff, R_i is the runoff at the impermeable layer, and R_p is the flow at the permeable area.

$$R_t = \text{FFG} \times I + f(\text{FFG}) \times (I - 1) \tag{2}$$

In a rainfall-runoff model, rainfall and soil moisture constitute the inputs. However, the opposite is true with the flash flood threshold; calculation of flash flood threshold requires current soil moisture and required flow as inputs. As such, the equation on upper stream water and small-sized rivers is converted for FFG using the repetitive calculation method, as shown in Figure 2, to calculate the rainfall that causes threshold runoff. The FFG is the rainfall corresponding to the threshold runoff in the relationship of the rainfall and runoff curve. If there is no impervious area, the relationship between R and FFG can be expressed in Figure 2 and Equation (2). R means the threshold runoff (mm), FFG means the flash flow guidance (mm), and $f()$ means the fall-runoff process. Moreover, I means rainfall intensity.

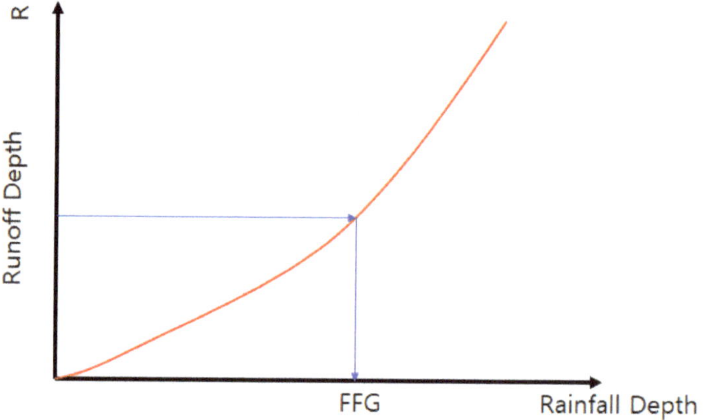

Figure 2. Concept of flash flood concept.

2.2. Machine Learning Method

Machine learning is an area of artificial intelligence where numerical models, algorithms, and programs are used to have a machine learn from given data as humans do, and new information is derived, or decisions are made based on what it learns [29]. In other words, machine learning means a system that uses accumulated empirical data to build models and improve performance. The amount of data matters in machine learning, and higher-quality data leads to higher-performing results. As for machine learning methods, this study used random forest, support vector machine, and XGBoost.

(1) Random Forest

The random forest method uses bootstraps to create several samples and applies them to a decision tree model to compile the results [30]. A decision-making trees produces estimates by creating and learning one-time training data from a given dataset. On the other hand, a random forest creates multiple training data from a given data set and creates and combines multiple decision-making threes for improved prediction [31]. The observations not used by individual decision-making trees are out-of-bagging (OOB) data and are used for estimating prediction probability and identifying variables. The prediction probability of OOB observations for each observation k within the x_i category (0 or 10) [32].

This study used Python and the random forest method to calculate threshold rainfalls. Figure 3 shows the conceptual diagram of a random forest.

$$\hat{p}_k(x_i) = \frac{\sum_{j \in OOB_i} I\hat{y}(x_i, t_j) = k}{|OOB_i|} \text{ , for } k = 0, 1 \tag{3}$$

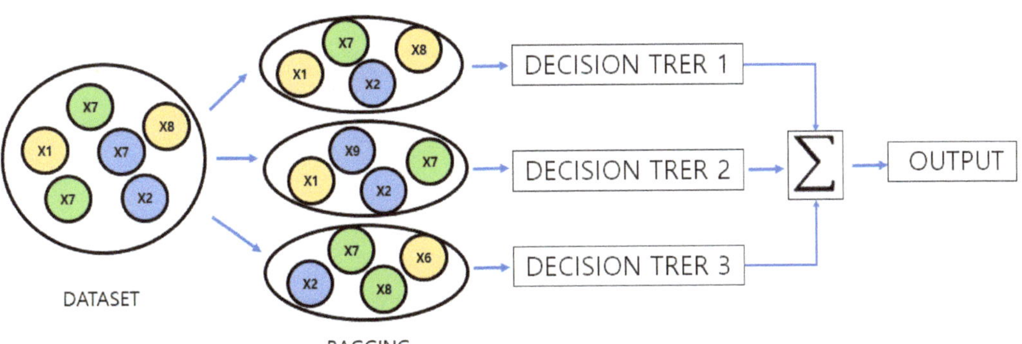

Figure 3. Concept of Random Forest.

(2) Support Vector Machine

Support vector machine (SVM) is a supervised learning algorithm used for both linear and non-linear classification issues. The purpose of SVM algorithms is to determine the lines or boundaries dividing an n-dimensional space into separate groups, so that they can be classified as their proper categories when new data are given. There may be multiple lines or boundaries for dividing an n-dimensional space into classes. However, the optimal boundary should be identified to determine categories. This optimal determination boundary is called the hyperplane. A support vector is the vector closest to the hyperplane and affects its position. The support vector machine is an algorithm that determines the optimal hyperplane that maximizes the margin, which means the distance between different data points.

The support vector regression (SVR) model has a small number of support vectors, and thus is known to be less sensitive to outliers. Ref [33] developed support vector regression that adopts a ε-insensitive loss function into the support vector machine. Support vector regression is estimated using the function shown in Equation (4) [32].

$$f(x) = \omega^t x + b \tag{4}$$

Equation (5) shows the constraints for calculating the optimal hyperplane function while calculating the error that minimizes Equation (4). $\frac{1}{2}\|\omega\|^2$ describes the degree of flattening of the function. If the data cannot be completely linearly separated, a slack variable $\xi(i = 1, \ldots I)$ is introduced to process it. ξ means the distance between the margin and the data outside the boundary between the margins. The main superparameters of the support vector regression are C (cost) and γ, and C adjusts the complexity of the estimation model and the degree of error tolerance. An increase in C means imposing a high penalty on errors within the margin. ϵ is not considered in the calculation process if the error is less than ϵ due to the maximum deviation between the actual value and the estimated value.

In this study, SVM was converted into SVR to predict arbitrary real values and used, and a Gaussian kernel (RBF) known for its excellent performance was applied. Figure 4 shows the conceptual diagram of the support vector machine.

$$\text{Minimuze}: \frac{1}{2}\parallel w \parallel^2 + C\sum_{i=1}^{l}(\xi_i + \xi_i^*)$$
$$\text{Subject to}: y_i - wx_i - b \leq \epsilon + \xi_i$$
$$wx_i + b - y_i \leq \epsilon + \xi_i^* \tag{5}$$
$$\xi_i \xi_i^* \geq 0$$

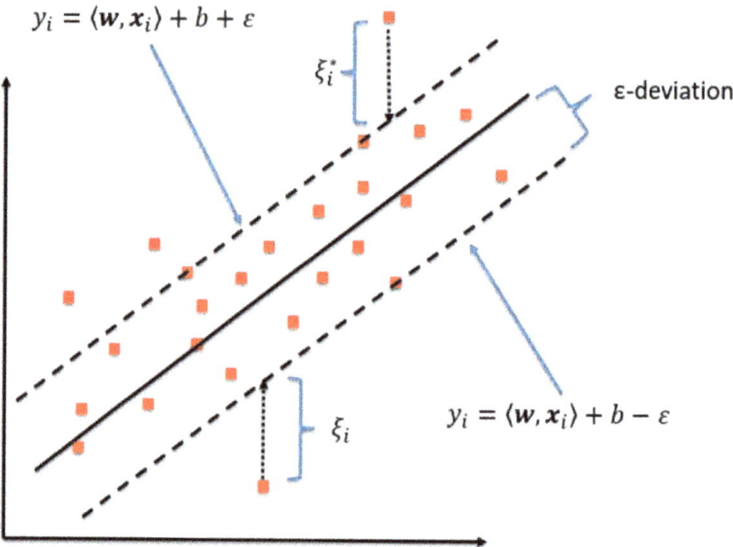

Figure 4. Concept of Support Vector Machine [34].

(3) eXtreme Gradient Boosting

Similarly to random forests, XGBoost is an ensemble algorithm that addresses the errors of multiple decision trees [35]. It offers improved prediction performance over gradient boosting machine (GBM) through distribution and parallel processing. In general, it is ten times faster than GBM. The efficiency and scalability of this method has been validated in multiple previous studies [36,37]. This boosting method lowers errors by grouping multiple classification and regression trees (CARTs).

$$\hat{y}_i = \sum_{k=1}^{k} f_k(x_i), f_k \in F \tag{6}$$

Equation (6) shows an ensemble model of trees, where K is the number of trees and F represents the set of CARTs. f_k corresponds to the weight of each independent tree and leaf. The scores of the leaves are summed up and compared for final prediction.

$$Obj = \sum_{i}^{n} = l(y_i, \hat{y}_i) + \sum_{k=1}^{K} \Omega(f_k) \tag{7}$$

Equation (7) represents an XGBoost model. The first $l(y_i, \hat{y}_i)$ is a loss function that represents the difference between a prediction and an actual observation. The second $\Omega(f_k)$ is the normalization term that controls the complexity of the model to prevent overfitting.

2.3. Performance Assessment Using K-Fold Cross Validation

This study uses MAE, RMSE, and RMSLE to compare the performance of different models. Most studies use the above three indicators a lot for data comparison [38–40]. They are widely used to objectively assess the accuracy of a regression equation by analyzing differences between observations and estimates. MAE and RMSE are statistical indicators for confirming the degree of errors included in an estimate calculated using an equation, when compared with an observation. A value closer to 0 represents better fit. RMSLE represents the average ratio of observations to predictions.

$$\text{MAE} = \frac{1}{N} \sum_{i=1}^{N} |y_i - \hat{y}_i| \tag{8}$$

$$\text{RMSE} = \sqrt{\frac{1}{n} \sum_{i=1}^{n} (y_i - \hat{y}_i)^2} \tag{9}$$

$$\text{RMSLE} = \sqrt{\frac{1}{n} \sum_{i=1}^{n} (\log(y_i + 1) - \log(\hat{y}_i + 1))^2} \tag{10}$$

Ideally, these errors need to be tested by applying them to actual ungauged watersheds. However, due to data and time constraint, the prediction models were validated using five-fold cross validation. K-fold cross validation is a model assessment method that uses a part of the overall data as a validation set. It ensures that all data are used as dataset at least once. Figure 5 shows dividing the data into five datasets and validating the models with a different dataset each time. An average cross-validation uses five datasets. This study selects the optimal parameters following cross validation to calculate threshold rainfalls.

Figure 5. Concept of cross vaildation.

3. Selection of Target Watersheds and Variables

3.1. Selection of Target Watersheds

This study chose the Han River watershed as its target, as the area includes the highest number of standard watersheds according to the water resource unit map. 290 of Korea's 850 standard watersheds are included in the Han River watershed. 237 of the 290 watersheds are inland, and the other 53 are coastal watersheds, as shown in Figure 6.

The (a) section of Figure 7 shows the learning watersheds; 80% of the learning watersheds were used for machine learning, and the other 20% were used for validation. High-performing models were selected with (a), and predictions were performed for the watersheds highlighted yellow in (b). The data set of the basin used for machine learning was randomly selected and proceeded.

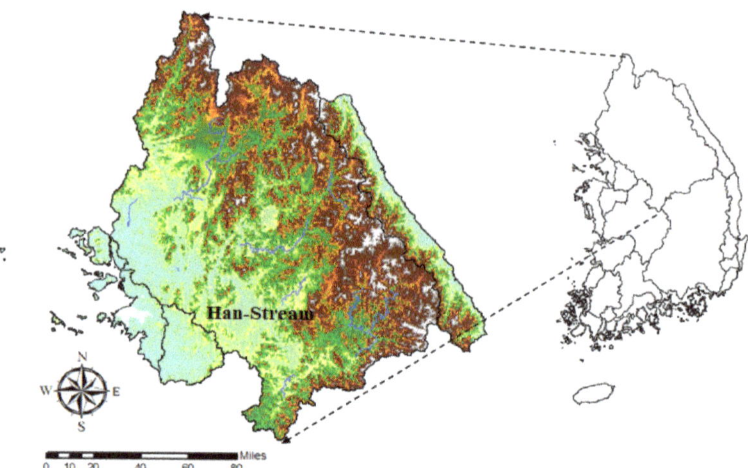

Figure 6. Study Areas.

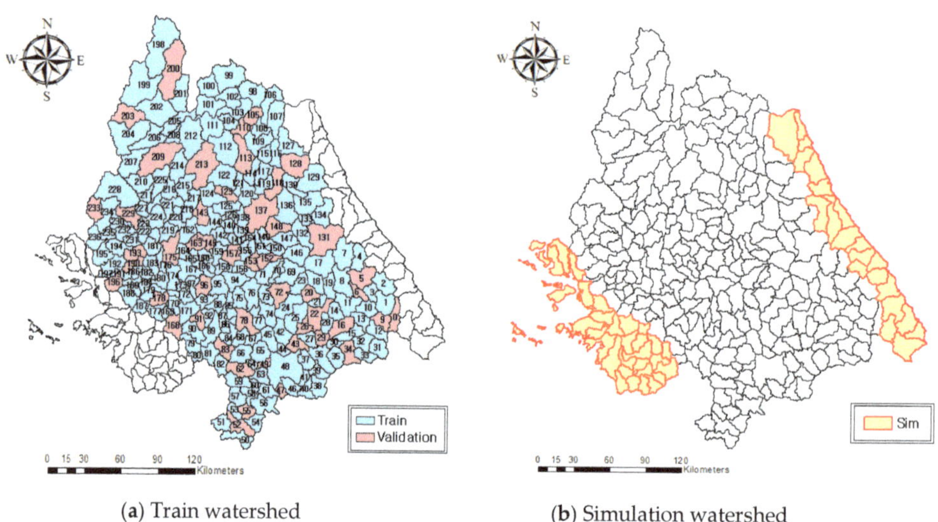

(**a**) Train watershed   (**b**) Simulation watershed

Figure 7. Training and Prediction Watersheds for Machine Learning.

3.2. Dependent and Independent Variables

This study used the threshold rainfalls calculated using the MOLIT method [8,15,41] as dependent variables. Figure 8 shows the calculated threshold rainfalls on the map.

Characteristic factors of the watersheds were used as independent variables. The analysis only considered topographical factors and hydrological factors. The watershed characteristic factors used in were collected from the Water Resources Management Information System (www.wamis.go.kr, accessed on 31 December 2011) and the geographic information system (GIS). Data on 15 characteristic factors were collected, including: drainage area (km^2), mean drainage elevation (m), mean drainage slope (%), highest drainage elevation (m), drainage density, runoff curve number, river length (km), drainage perimeter (km), form factor, circularity ratio, stream frequency, channel maintenance constant, relative relief, number of reliefs, and river length ratio.

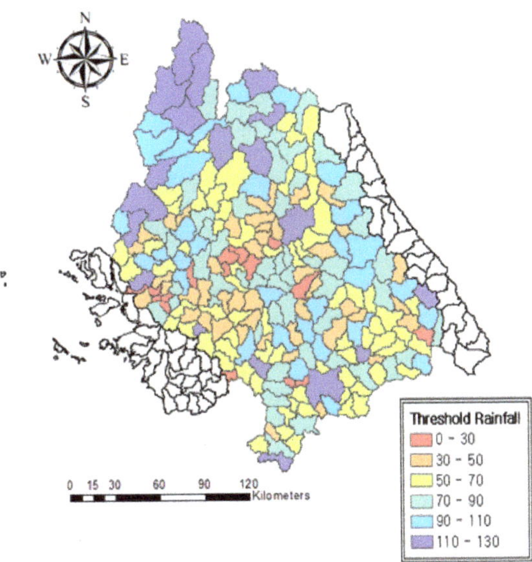

Figure 8. Standard watershed Threshold rainfall.

The drain area refers to the area on the plane of the basin and refers to the plane area within the closed curve, which is usually made up of a watershed. The basin average elevation is calculated by arithmetically averaging the elevation values corresponding to each cell of the DEM (Digital Elevation Model). The mean drain slope is calculated by arithmetically averaging the slope corresponding to each cell of the DEM in degrees. Highest drain evaluation means the highest elevation in the basin, and drain density means the length of rivers per unit area. It means that the degree of outflow of the basin is quantified by the Soil Conservation Service (SCS) using the runoff curve number land use and soil map. River length is the total length of all rivers in a given drainage basin. The drain perimeter is defined as the length measured along the boundary of the watershed of a given order projected on the horizontal plane of the map, and the form factor is defined as the ratio of the main river length of the watershed to the diameter of the circle having the same area as the watershed area. The circularity ratio is a dimensionless parameter defined as the ratio of the basin area to the area of a circle with the same length as the basin circumference. Stream frequency is defined as the ratio of river water in the basin to the basin area, and the channel maintenance constant is the reciprocal of the aqueous density. Relative relief is defined as the ratio of watershed undulations to watershed circumference, number of reliefs is defined as the product of watershed undulations and water density, and river length ratio is defined as the ratio of river length w to average river length $w - 1$. Table 1 shows a summary of the watershed characteristics factor.

Table 1. Summary of independent variables.

	A	H	S	Em	H	CN	L_w	L_p	R_s	R_c	Cf	C	R_p	R_n	RL
Count	290	290	290	290	290	290	290	290	290	290	290	290	290	290	290
mean	144.6	324.0	35.4	253.3	1.7	58.7	12.9	67.3	1.0	0.4	2.4	0.7	13.3	1376	1.8
Max	571.6	930.3	65.1	302.7	4.0	87.9	63.3	262.3	3.6	0.7	12.6	9.5	36.8	3921.4	4.4
min	39.0	4.9	4.0	103.7	0.1	33.7	0.0	32.7	0.0	0.0	0.1	0.3	0.9	32.8	0.7

A correlation analysis was performed to select statistically correlated independent variables, as independent variables not correlated to dependent variables may lower the prediction performance. The correlation analysis was performed as shown in Figure 9. Given the fact that the threshold runoffs required for calculating threshold rainfalls were calculated from peak flood volumes and overflowing runoffs, the following variables were determined to be significantly correlated: drainage area, river length, drainage perimeter, relative relief, and river length ratio. Among those factors, river length, river length ratio, relative relief, and drainage perimeter were determined to be more highly correlated with the independent variables. The correlation coefficient was 0.65 for threshold rainfall and drainage area, 0.64 for drainage perimeter, and 0.31 for river length. As such, drainage area, drainage perimeter, and river length were finally selected as independent variables. In most data analyses, principal component analysis (PCA) should be used to reduce initial independent variables [42], but in this study, a principal component analysis was omitted because the amount of data for each independent variable was not large.

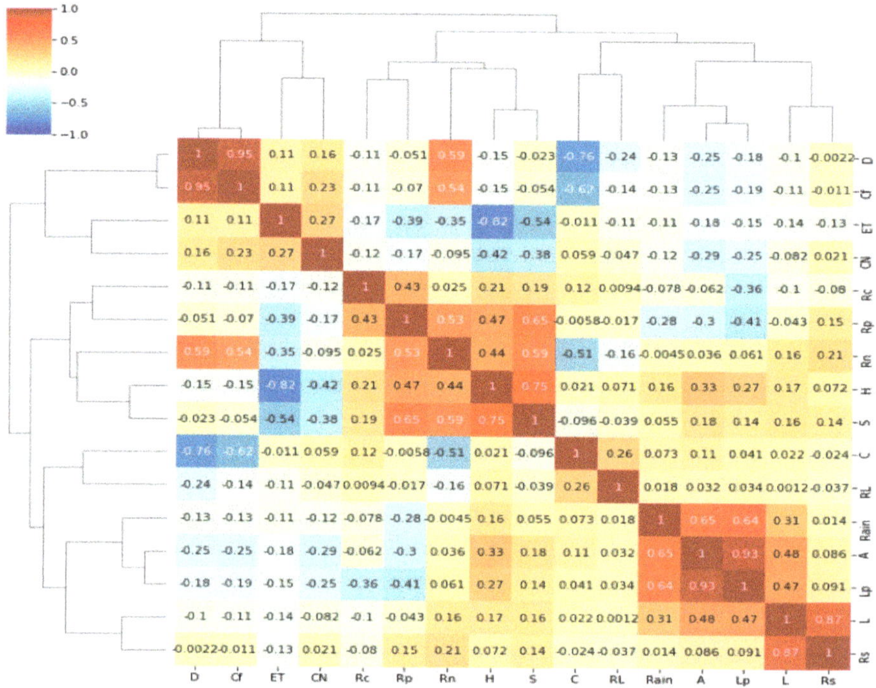

Figure 9. Variable correlation analysis.

4. Machine Learning Application and Results

This study used SVM, random forest, and XGBoost. Excel 2010 and Python ver. 3.6 were used to record and statistically analyze the collected data and generate graphs. This study also used the model packages provided by Python-based Scikit-learn.

Effective machine learning requires pre-processing of the data to be used. The independent variable data went through data scaling and missing values were removed. As data scales vary depending on the variable, the data were standardized to render them more suitable for machine learning. Independent variables were analyzed using RobustScaler, which is less affected by outliers. A higher accuracy can be expected by removing outliers. However, the small number of inputs in this study means possible overfitting. Therefore, this study addressed outliers through pre-processing rather than outlier removal.

4.1. Validation of Prediction Models

The optimal parameters for each model were selected through K-fold cross-validation. An error closer to 0 indicates a better result. Table 2 shows the MAE, RMSLE, and RMSLE values of each of the five datasets created by dividing the datasets through the k-fold cross-validation. All three performance assessments found that XGBoost produces the results closest to actual observations compared with the other models.

Table 2. Comparison of model performance evaluation.

Model		MAE	RMSE	RMSLE
Support Vector	Fold 1	15	19	0.26
	Fold 2	23	38	0.4
	Fold 3	19	26	0.29
	Fold 4	21	26	0.47
	Fold 5	28	38	0.34
Random Forest	Fold 1	12	19	0.28
	Fold 2	20	32	0.46
	Fold 3	16	20	0.34
	Fold 4	22	27	0.47
	Fold 5	21	26	0.45
XGBoost	Fold 1	14	20	0.28
	Fold 2	20	33	0.38
	Fold 3	16	20	0.29
	Fold 4	21	27	0.46
	Fold 5	25	37	0.35

Parameters were applied to increase the accuracy of machine learning, and n_estimator represents variables that adjust the number of trees to generate. max_depth means the number of tree depths. min_samples_split represents the minimum number of sample data to split nodes, and min_samples_leaf means the minimum number of sample data required for a leaf node. learning_rate means a parameter that, in machine learning and statistics, moves toward the min loss function and determines the size of each stage of repetition. The calculated parameter values area n_estimators: 100, learning_rate = 0.04, min_samples_leaf = 3, min_samples_split = 2, max_depth = 4.

Figure 10 compares the existing threshold rainfalls with those calculated using XGBoost. Most threshold rainfalls are distributed between 40 and 60 mm and between 60 and 80 mm and are close to actual observations.

Figure 11 is a map representing threshold rainfall values. The watersheds with low threshold rainfalls in (a) are reflected in (b) as well.

4.2. Calculation of Threshold Rainfalls in Ungauged Watersheds

This study used XGBoost, which produced good results in error performance assessment, to calculate the threshold rainfalls of ungauged watersheds. Figure 12 shows the distribution of the threshold rainfalls calculated for the ungauged basins. The majority of watersheds show threshold rainfalls between 40 mm and 80 mm. Figure 13 is a map showing the threshold rainfalls of the ungauged basins other than the those in the inland areas.

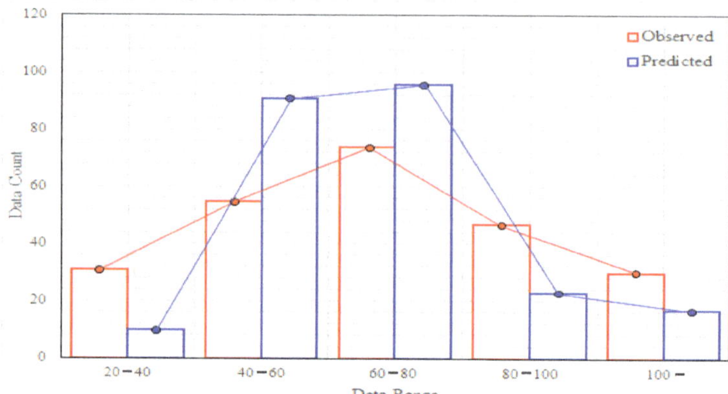

Figure 10. Threshold rainfall distribution.

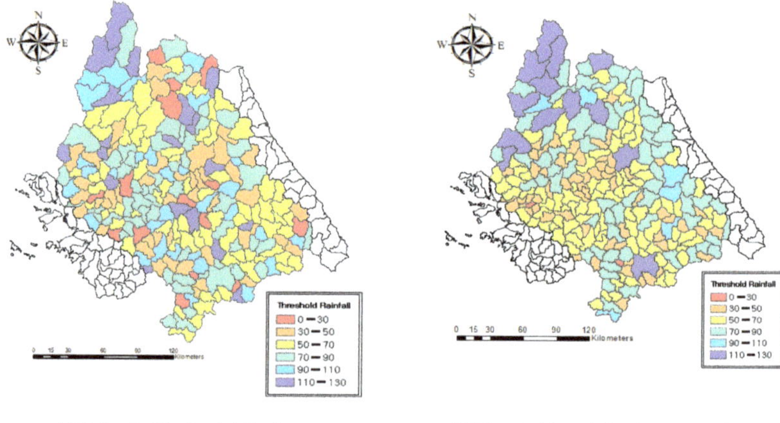

(**a**) Estimated limit rainfall value   (**b**) Threshold rainfall value using XGBoost

Figure 11. Comparison of training data against actual values.

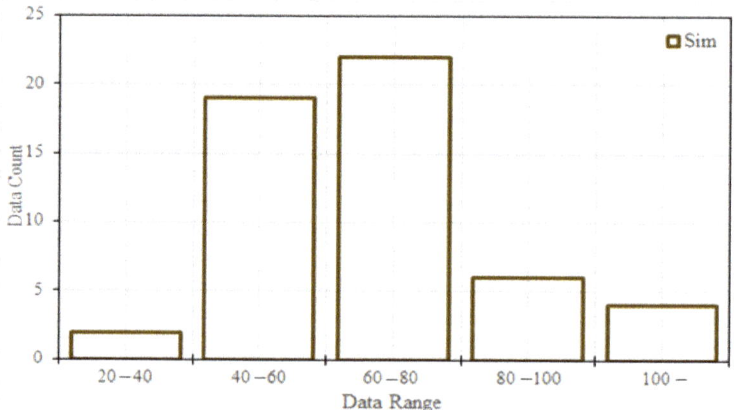

Figure 12. Ungauged watershed threshold rainfall distribution.

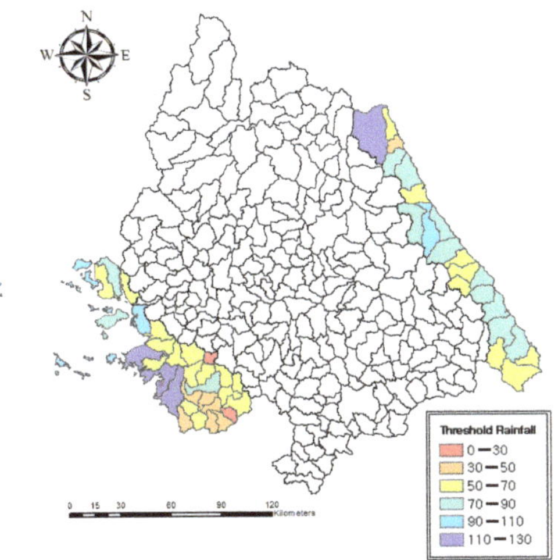

Figure 13. Threshold rainfall values for ungauged watersheds.

4.3. Validation Using Real World Cases and Assessment

As shown in Figure 14, based on the rainfall events in 2017, 2020, and 2021 of each affected watershed, among the ungauged watersheds outside the purple lines around the inland part of the Han River, Yongin, Cheonan, Samcheok, Gangneung, and Sokcho watersheds were found to be vulnerable against heavy rain. An application to actual rainfall events showed that damage was caused when the rainfall exceeds the specified rainfall in the legends. However, the researchers' ability to verify damages in other areas was restricted by the fact that damage was verified from news articles and social network posts.

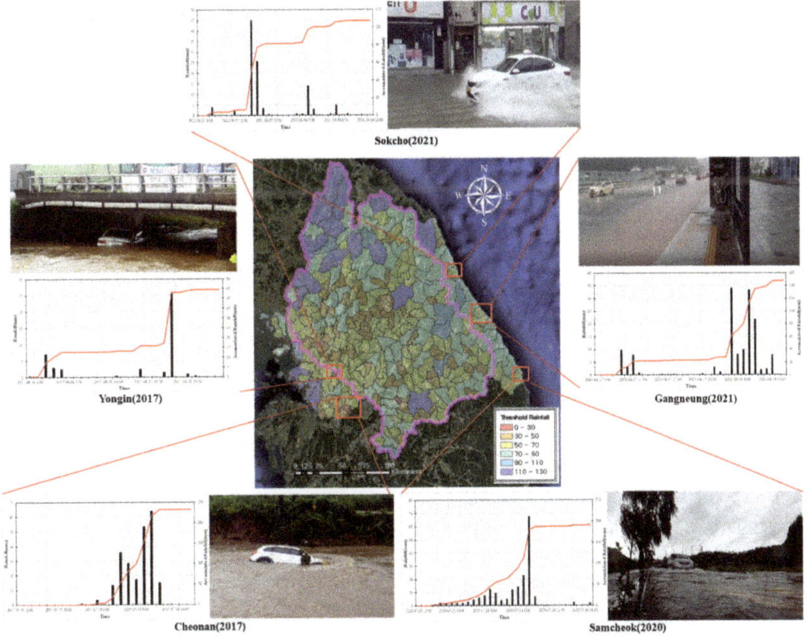

Figure 14. Threshold rainfall values for ungauged watersheds.

5. Conclusions

Damage caused by localized heavy rain continues to increase in Korea. However, research on inundation-causing threshold rainfall is still largely absent in the country. More researchers need to study technologies for predicting and responding to inundation in advance. This study can be summarized as follows.

The purpose of this study was to identify threshold rainfalls in areas not readily available for hydrological analysis, using the calculation method and characteristic factors used by the MOLIT. Three machine learning methods (SVM, random forest, and XGBoost) were compared in terms of accuracy using MAE, RMSE, and RMSLE, and XGBoost was selected as the best-performing method. Watershed characteristics, hydrological factors, and XGBoost were used to calculate the threshold rainfalls of the ungauged coastal watersheds. In this study, it is judged that what can reflect actual topographic and hydrological factors can be differentiated from other machine learning and marginal rainfall papers. In addition, distinct from conventional simple data, data using physical models were used in machine learning techniques, so high accuracy could be secured through a small number of data, and anyone could use it by using widely known machine learning techniques.

However, this study has its limitations. First, outliers were found while calculating the threshold runoffs of the hydrological models. More sophisticated hydrological models and more accurate data may be needed for analysis. In addition, threshold rainfall calculation based on the runoff-rainfall curve simply used polynomials. However, higher accuracy may be achieved by applying a machine learning method to threshold runoff and runoff-rainfall curve calculation.

This study compared the calculated threshold rainfalls with real world cases identified from news reports and social network posts, which was found to pose limitations to quantitative assessment.

The researcher plans to conduct a similar study nation-wide. Watershed data with more diverse hydrological models and outliers will improve the accuracy of the findings. Although not included in this study, quantitative validation using real world events will yield meaningful results. The current water forecast system provides only quantitative figures without considering the damage caused, which some regard as insufficient for supporting effective decision-making to prevent and prepare for damage caused by natural disasters. Threshold rainfall prediction suggested in this study may, if implemented on a continued basis, provide accurate information on rainfall damage in advance and help decisionmakers make better decisions on disaster control.

Author Contributions: K.-S.C. and C.-H.O., J.-R.C. carried out the survey of previous study and wrote the graph of the data. B.-S.K. suggested idea of study and contributed to the writing of the paper. All authors have read and agreed to the published version of the manuscript.

Funding: This work was funded by the Korea Meteorological Administration Research and Development Program under Grant KMI [2021—00312]. This work was financially supported by Ministry of the Interior and Safety as Human Resource Development Project in Disaster Management(C2001777-01-01).

Data Availability Statement: Not applicable.

Conflicts of Interest: The authors declare no conflict of interest.

References

1. Lee, J.Y.; Choi, C.H.; Kang, D.S.; Kim, B.S.; Kim, T.W. Estimating Design Floods at Ungauged Watersheds in South Korea Using Machine Learning Models. *Water* **2020**, *12*, 3022. [CrossRef]
2. WMO. *Guideline on Multi-Hazard Impact-Based Forecast and Warning System*; WMO-No. 1150; World Meteorological Organization: Geneva, Switzerland, 2015; p. 22.
3. Jeong, K.Y. Impact Forecast Vision and Direction. *Meteorol. Technol. Policy* **2016**, *9*, 6–22.
4. Go, Y.H. Development and Activation of Impact Forecasting Service. Available online: https://www.kma.go.kr/down/t_policy/t_policy_201706.pdf (accessed on 2 June 2021).
5. Lee, D.H.; Kim, G.Y. Weather Disaster Impact Forecast. *KISTEP Technol. Trend Brief* **2019**, *10*, p1–p4.
6. FFC (Flood Forecasting Centre). *Flood Guidance Statement User Guide (Version 4)*; Flood Forecasting Centre: Exeter, UK, 2017; p. 6.

7. Lee, B.J. Analysis on Inundation Characteristics for Flood Impact Forecasting in Gangnam Drainage Basin. *Korean Meteorol. Soc.* **2017**, *27*, 189–197.

8. Kim, B.S.; Hong, J.B.; Kim, H.S.; Yoon, S.Y. Development of Flash Flood model Using Digital Terrain Analysis Model and Rainfall RADAR: I. Methodology and Model Development. *J. Korean Soc. Civ. Eng. B* **2007**, *27*, 151–159.

9. Kim, B.S.; Kim, H.S. Estimation of the Flash Flood Severity using Runoff hydrograph and Flash flood index. *J. Korea Water Resour. Assoc.* **2008**, *41*, 185–196. [CrossRef]

10. Cho, B.G.; Ji, H.S.; Bae, D.H. Development and Evaluation of Computational Method for Korean Threshold Runoff. *J. Korea Water Resour. Assoc.* **2011**, *44*, 875–887. [CrossRef]

11. Chung, I.M.; Lee, C.W.; Kim, J.T.; Na, H.N.; Kim, N.W. Development of Threshold Runoff Simulation Method for Runoff Analysis of Jeju Island. *J. Environ. Sci. Int.* **2011**, *20*, 1347–1355. [CrossRef]

12. Park, J.B.; Kang, D.G.; Park, M.J. Rainfall Thresholds Estimation to Develop Flood Forecasting and Warning System for Nakdong Small River Basins. *J. Korean Soc. Hazard Mitig.* **2013**, *13*, 311–317. [CrossRef]

13. Kang, H.S.; Choi, C.W.; Lee, H.S.; Hwang, J.G.; Moon, H.J. Development of an ANN-Based Urban Flood Alert Criteria Prediction Model and the Impact of Training Data Augmentation. *J. Korean Soc. Hazard Mitig.* **2021**, *21*, 257–264. [CrossRef]

14. Huff, F.A. Time Distribution of Rainfall in Heavy Storms. *Water Resour. Res.* **1967**, *3*, 1007–1019. [CrossRef]

15. Lee, S.H.; Kang, D.H.; Kim, B.S. A Study on the Method of Calculating the Threshold Rainfall for Rainfall Impact Forecasting. *J. Korean Soc. Hazard Mitig.* **2018**, *18*, 93–102. [CrossRef]

16. Zagorecki, A.T.; Johnson, D.E.; Ristvej, J. Data mining and machine learning in the context of disaster and crisis management. *Int. J. Emerg. Manag.* **2013**, *9*, 351–365. [CrossRef]

17. Chang, F.-J.; Guo, S. Advances in hydrologic forecasts and water resources management. *Water* **2020**, *12*, 1819. [CrossRef]

18. Park, J.; Park, J.-H.; Choi, J.-S.; Joo, J.C.; Park, K.; Yoon, H.C.; Park, C.Y.; Lee, W.H.; Heo, T.-Y. Ensemble Model Development for the Prediction of a Disaster Index in Water Treatment Systems. *Water* **2020**, *12*, 3195. [CrossRef]

19. Jung, S.H.; Lee, D.E.; Lee, K.S. Prediction of River Water Level Using Deep-Learning Open Library. *J. Korean Soc. Hazard Mitig.* **2018**, *18*, 1–11. [CrossRef]

20. Kao, I.F.; Zhou, Y.; Chang, L.C.; Chang, F.J. Exploring a Long Short-Term Memory based Encoder-Decoder framework for multi-step-ahead flood forecasting. *J. Hydrol.* **2020**, *583*, 124631. [CrossRef]

21. Chang, L.C.; Chang, F.J.; Yang, S.N.; Tsai, F.H.; Chang, T.H.; Herricks, E.E. Self-organizing maps of typhoon tracks allow for flood forecasts up to two days in advance. *Nat. Commun.* **2020**, *11*, 1983. [CrossRef] [PubMed]

22. Chang, F.J.; Tsai, M.J. A nonlinear spatio-temporal lumping of radar rainfall for modeling multi-step-ahead inflow forecasts by data-driven techniques. *J. Hydrol.* **2016**, *535*, 256–269. [CrossRef]

23. Zhou, Y.; Guo, S.; Chang, F.J. Explore an evolutionary recurrent ANFIS for modelling multi-step-ahead flood forecasts. *J. Hydrol.* **2019**, *570*, 343–355. [CrossRef]

24. Choi, C.H.; Kim, J.S.; Kim, D.H.; Lee, J.H.; Kim, D.H.; Kim, H.S. Development of Heavy Rain Damage Prediction Functions in the Seoul Capital Area Using Machine Learning Techniques. *J. Korean Soc. Hazard Mitig.* **2018**, *18*, 435–447. [CrossRef]

25. Choi, C.H. Development of Combined Heavy Rain Damage Prediction Models Using Machine Learning and Effectiveness of Disaster Prevention Projects. Ph.D. Thesis, Inha University, Incheon, Korea, 2019.

26. Yen, M.H.; Liu, D.W.; Hsin, Y.C.; Lin, C.E.; Chen, C.C. Application of the deep learning for the prediction of rainfall in Southern Taiwan. *Sci. Rep.* **2019**, *9*, 12774. [CrossRef] [PubMed]

27. Lee, Y.M.; Ko, C.M.; Shin, S.C.; Kim, B.S. The Development of a Rainfall Correction Technique based on Machine Learning for Hydrological Applications. *J. Environ. Sci. Int.* **2019**, *28*, 125–135. [CrossRef]

28. Yu, B.I. Study on Flash Floods and Debris Flow Guidance of Gangwon-do. Master's Thesis, Kangwon University, Chuncheon-si, Korea, 2016.

29. Kim, D.W.; Hong, S.J.; Kim, J.W.; Han, D.G.; Hong, I.P.; Kim, H.S. Water Quality Analysis of Hongcheon River Basin under Climate Change. *J. Wetl. Res.* **2015**, *17*, 348–358. [CrossRef]

30. Suh, J.D. Foreign Exchange Rate Forecasting Using the GARCH extended Random Forest Model. *J. Ind. Econ. Bus.* **2016**, *29*, 1607–1628.

31. Choi, C.H.; Park, K.H.; Park, H.K.; Lee, M.J.; Kim, J.S.; Kim, H.S. Development of Heavy Rain Damage Prediction Function for Public Facility Using Machine Learning. *J. Korean Soc. Hazard Mitig.* **2017**, *17*, 443–450. [CrossRef]

32. Jang, D.R. A Study on the Art Price Prediction Model Using Machine Learning. Master's Thesis, Hongik University, Seoul, Korea, 2020.

33. Vapnik, V. Support-Vector Networks. *Mach. Learn.* **1995**, *20*, 273–297.

34. Kleynhans, T.; Montanaro, M.; Gerace, A.; Kanan, C. Predicting Top-of-Atmosphere Thermal Radiance Using MERRA-2 Atmospheric Data with Deep Learning. *Remote Sens.* **2017**, *9*, 1133. [CrossRef]

35. Kim, I.H.; Lee, K.S. Tree based ensemble model for developing and evaluating automated valuation models: The case of Seoul residential apartment. *J. Korean Data Inf. Sci. Soc.* **2020**, *31*, 375–389. [CrossRef]

36. Chen, T.; Guestrin, C. XGBoost: A scalable tree boosting system. In Proceedings of the 22nd ACM SIGKDD International Conference on Knowledge Discovery and Data Mining, San Francisco, CA, USA, 13–17 August 2016; pp. 785–794.

37. Ke, G.; Meng, Q.; Finley, T.; Wang, T.; Chen, W.; Ma, W.; Ye, Q.; Tie, Y.L. LightGBM: A Highly efficient gradient boosting decision tree. In Proceedings of the 31st Conference on Neural Information Processing Systems, Long Beach, CA, USA, 4–9 December 2017; pp. 3149–3157.

38. Ioannou, K.; Karampatzakis, D.; Amanatidis, P.; Aggelopoulos, V.; Karmiris, I. Low-Cost Automatic Weather Stations in the Internet of Things. *Information* **2021**, *12*, 146. [CrossRef]
39. Stefanidis, S.; Dafis, S.; Stathis, D. Evaluation of Regional Climate Models (RCMs) Performance in Simulating Seasonal Precipitation over Mountainous Central Pindus (Greece). *Water* **2020**, *12*, 2750. [CrossRef]
40. Ioannou, K.; Myronidis, D.; Lefakis, P.; Stathis, D. The use of artificial neural networks (anns) for the forecast of precipitation levels of lake doirani(N. greece). *Fresenius Environ. Bull.* **2010**, *19*, 1921–1927.
41. Choo, K.S.; Kang, D.H.; Kim, B.S. A Study on the Estimation of the Threshold Rainfall in Standard Watershed Units. *J. Korean Soc. Disaster Secur.* **2021**, *14*, 1–11.
42. Myronidis, D.; Ivanova, E. Generating Regional Models for Estimating the Peak Flows and Environmental Flows Magnitude for the Bulgarian-Greek Rhodope Mountain Range Torrential Watersheds. *Water* **2020**, *12*, 784. [CrossRef]

Article

Using a Self-Organizing Map to Explore Local Weather Features for Smart Urban Agriculture in Northern Taiwan

Angela Huang and Fi-John Chang *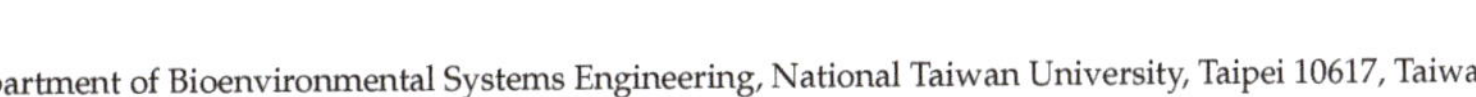

Department of Bioenvironmental Systems Engineering, National Taiwan University, Taipei 10617, Taiwan; d05622005@ntu.edu.tw
* Correspondence: changfj@ntu.edu.tw

Abstract: Weather plays a critical role in outdoor agricultural production; therefore, climate information can help farmers to arrange planting and production schedules, especially for urban agriculture (UA), providing fresh vegetables to partially fulfill city residents' dietary needs. General weather information in the form of timely forecasts is insufficient to anticipate potential occurrences of weather types and features during the designated time windows for precise cultivation planning. In this research, we intended to use a self-organizing map (SOM), which is a clustering technique with powerful feature extraction ability to reveal hidden patterns of datasets, to explore the represented spatiotemporal weather features of Taipei city based on the observed data of six key weather factors that were collected at five weather stations in northern Taiwan during 2014 and 2018. The weather types and features of duration and distribution for Taipei on a 10-day basis were specifically examined, indicating that weather types #2, #4, and #7 featured to manifest the dominant seasonal patterns in a year. The results can serve as practical references to anticipate upcoming weather types/features within designated time frames, arrange potential/further measures of cultivation tasks and/or adjustments in response, and use water/energy resources efficiently for the sustainable production of smart urban agriculture.

Keywords: weather types and features; meteorological feature extraction; artificial neural network; self-organizing map (SOM); urban agriculture; resource utilization efficiency; urban northern Taiwan

Citation: Huang, A.; Chang, F.-J. Using a Self-Organizing Map to Explore Local Weather Features for Smart Urban Agriculture in Northern Taiwan. *Water* **2021**, *13*, 3457. https://doi.org/10.3390/w13233457

Academic Editor: Kwok-wing Chau

Received: 2 November 2021
Accepted: 2 December 2021
Published: 6 December 2021

Publisher's Note: MDPI stays neutral with regard to jurisdictional claims in published maps and institutional affiliations.

1. Introduction

Urban agriculture (UA), which is defined by the Food and Agriculture Organization (FAO) as "the small areas within the city for growing crops and raising small livestock or milk cows for own-consumption subsistence or small-scale sale in local/neighborhood markets" [1,2], takes advantage of vacant rooftops, balconies, and community spaces to plant vegetables for neighborhoods' fresh diet in urban areas. In East Asia, Europe, and North America, many urban farmers produce potential high-quality food at an affordable cost [3]. Compared with large-scale commercial cultivation on rural farms, UA usually occupies smaller land areas and is operated by community volunteers/seniors (non-professionals) for leisure, social interaction, and partial self-sufficiency purposes [4,5]. Leafy vegetables are often the primary products of UA since they are highly valued for their nutritional content, with dietary diversity with shorter growing periods [6,7]. Therefore, the planting activities of UA generally produce short-term vegetable crops with more diversified species but less yield quantity and crop rotation is often carried out after each harvest upon the growers' interests.

Weather plays a critical role in outdoor agricultural production by affecting the optimal growth, development, and yields of crops, as well as the incidence and spread of pests/diseases, water needs, and fertilizer requirements for cultivation. The spatiotemporal (short-term and annual) variations of weather factors (i.e., temperature, rainfall, humidity, sunshine, etc.) of a particular place over the selected time interval during the cultivation season should be considered for assessing their influences on crop growth. In addition, climate change has also caused impacts on agricultural operation and management [8,9].

Decision making in agriculture is based on knowledge of the crop behavior and cultivation information, which may also include characterization of the local growing conditions, management practices, and the response of the crop to these variables at any given time. In particular, the timing of planting, cultivation, and harvest based on cultural tasks are often determined based upon weather forecasts; therefore, a good climate-based strategy for agronomic planning can help to reduce the stresses of crop growth and increase the effectiveness of the timing of preventive measures and cultural operations, as well as engage farmers to organize and use appropriate cultural practices to cope with or take advantage of weather forecasts in various ways. On the other hand, once the crop season starts and resources demands and technology are committed, only certain cultivation operations can be adopted in response to weather phenomena by relying on advanced notice of the occurrence of erratic weather for minimizing the hazardous effects during mid-season [10].

Despite the agricultural weather forecast being available for upcoming days, many farmers in East Asia today still follow the traditional solar terms of the Lunar Calendar (also known as the Farmer Calendar or the Yellow Calendar in Chinese) from rural proverbs as important rules of thumb for the timing of cultivation practices in light of anticipating local weather in each season. Based on the experiences of ancestors, farmers are suggested to use specific agricultural operations for better efficiency due to the expectation for the occurrence of certain seasonal weather phenomenon on specific dates in a year.

In addition to nowcasting, short-term forecasts or monthly climate projections for agricultural producers are usually preferred by farmers when making agricultural decisions. The use of non-forecast climate information on seasonal pattern analysis, i.e., historical climate information, long-term climate outlooks, and decision calendars, can also be valuable as a practically useful reference for cultivation tasks and agricultural risk management [11,12]. The seasonal pattern of weather types requires probing into years of historical data of various weather factors, and feature patterns can be extracted through effective and efficient data-mining techniques to explore more information that might not otherwise be disclosed. Various machine learning methods were employed in precedent studies to classify weather features and make forecasts [13,14], for example, a deep neural network (DNN) for weather forecasting [15]; a recurrent neural network (RNN) and long short-term memory (LSTM) for air temperature forecasting [16]; an RNN for hourly rainfall forecasting during typhoon periods [17]; a multilayer perceptron neural network (MLP) for air temperature prediction inside greenhouse [18]; a convolutional neural network (CNN) for wind speed prediction [19] and weather pattern clustering [20]; a deep convolutional neural network (DCNN) for weather phenomenon classification based on images [21]; a backpropagation neural network (BPNN) for weather system prediction [22–24]; a self-organizing map (SOM) for estimating meteorological variables of evaporation [25], a method to train an SOM for clustering high-dimensional flood inundation maps [26]; an adaptive model of the enhanced multiple linear regression model (EMLRM) for rainfall forecasting [27]; a combined modular models comparison using moving average (MA), MLP, and support vector regression (SVR) for daily and monthly prediction on rainfall time series [28]; an artificial neural network (ANN)-based lower upper bound estimation (LUBE) and multi-objective fully informed particle swarm (MOFIPS) for interval forecasting for streamflow discharge [29]; and a comparison of BPNN, group method of data handing (GMDH), and autoregressive integrated moving average (ARIMA) for monthly rainfall forecasting [30].

As one of the powerful methods of exploratory data analysis for data mining and visualization interpreting [31,32], SOMs have been usually used to discover intrinsic patterns by downscaling the complex weather data sets from a high-dimensional space to a low-dimensional one through clustering similar data patterns into neighboring SOM units for easy comprehension [33,34]. For the feature extraction from large datasets, dimensionality reduction is a critical step to eliminate redundant information for simplifying the subsequent processes of classification and the search for information while retaining meaningful properties of the original datasets. Compared with traditional methods, such as principal component analysis (PCA) and wavelet decomposition methods, SOM may better classify datasets and present results in a two-dimensional topology well [35].

An increasing number of SOM applications were also adopted to assist with agricultural decision making when considering the weather information from recent years [36]. For agricultural applications, some data-driven models were developed to identify land covers for agricultural control and management and provide information for production systems to better manage their crops according to the specific conditions on farms [37–40]. Time series of climatic and agro-climatic indices were used to examine the signs of climate changes in rainfall, temperature, and agricultural drought to identify potential impacts on the agricultural water balance [41]. In addition, SOM was also applied to data clustering and pattern recognition for many types of climatic and meteorological data to analyze synoptic climatology at various spatial and temporal scales [42–44], forecasting and nowcasting [45–47], and investigation of extreme climate events [48–50], as well as variations of meteorological variables, such as evaporation and rainfall pattern analysis [25,51,52], cloud classification [53], and climate change analysis [54,55].

Generally, monthly/annual meteorological statistics provide very rough information about respective weather factors rather than substantially reflect the overall phenomenon of weather features during a certain period and give no characteristics on the temporal and sequential distribution of repetition and alteration trends. Weather forecasts are provided based on the surrounding atmospheric circulation condition. Both types of information are often used as reference guidelines for crop cultivation for farmers, but it is challenging for farmers to anticipate and grasp precisely what weather phenomena and features would potentially occur during a specific time window. Therefore, this research aimed to use data-mining techniques to discover/induce the annual pattern and distribution of weather types and features of a region based on historical meteorological data, including their occurrence time, frequency, continuity, and intensity, so that agricultural operationists can anticipate the occurrences and trends of weather types and features during each specified period for engaging in appropriate cultivation tasks in advance.

In this regard, this research adopted an SOM to cluster large and complex historical meteorological data into several categories (types) of similar weather features while exploring each type's characteristics on temporal distribution, sequential continuity, occurrence time, and frequency so that crops can adapt to the weather in northern urban Taiwan owing to the measures that were taken beforehand (when necessary). Therefore, the eventual findings are expected to provide practical weather reference to help with sustainable production in terms of species selection, planting schedules, precautions arrangement, and further efficiency enhancement of water and energy resources that are used for the planning and design of urban agriculture/farming for promotion purposes.

2. Materials and Methodology

Various weather factors for agricultural forecasting are intertwined and affect farm planning and operations from place to place and from season to season. This research aimed to explore the representative spatiotemporal weather features by collecting and analyzing the observed data of 6 key weather factors at 5 weather stations in northern Taiwan. An SOM, as one of the effective artificial neural network approaches, was adopted to cluster the meteorological data to reveal the hidden weather features. Subsequently, the temporal patterns of such features at the Taipei Station were specifically examined so that potential measures in response to certain weather phenomena at certain periods for the smart and efficient utilization of resources (water/energy) in urban agriculture can be provided.

To cross-reference to the results of weather types and features, the Da-an rooftop farm, which has been a successful urban agriculture (UA) site since 2014 and had a complete work log and harvest records in Taipei City in northern Taiwan, was chosen as the real case in operation for reference. In addition, due to the special characteristics of UA, i.e., it usually involves smaller planting areas, lower harvest weight, higher crop rotation frequency, more species diversity, and short-term crops (with growing periods of a few months), the data collected during 2014–2018 were selected to form the weather types and features in this study.

2.1. Materials of Weather Data Collection

Taipei City, New Taipei City, Taoyuan City, and Keelung City, which constitute the main metropolitan area in northern Taiwan, were selected for this research. The four cities cover an area of 3675 km^2 (accounting for 10.2% of the total area of Taiwan) and have a population of 8.82 million (about 38.3% of the total population of Taiwan). The daily meteorological data were collected from five Central Bureau of Weather (CBW) ground weather stations in northern Taiwan, including the Banqiao, Tamshui, Taipei, Keelung, and Xinwu Weather Stations (locations are shown in Figure 1) from 1 January 2014 to 31 December 2018 [56]. Being located in downtown areas in Taipei Metropolitan, the five weather stations were selected because they are governed directly by the Central Weather Bureau in Taiwan and can provide the most comprehensive, complete, and extensive monitoring weather data at urban areas in northern Taiwan. With Taipei City being located in the Taipei Basin and the Taipei Weather Station being situated in the main urban area in the city, the Taipei Weather Station was specifically examined to explore the weather types and features in this research because the Da-an rooftop farms, with its harvest logs for reference, was located there. With a similar approach, the weather types and patterns at the other four weather stations are also valuable and deserve further in-depth exploration in future research.

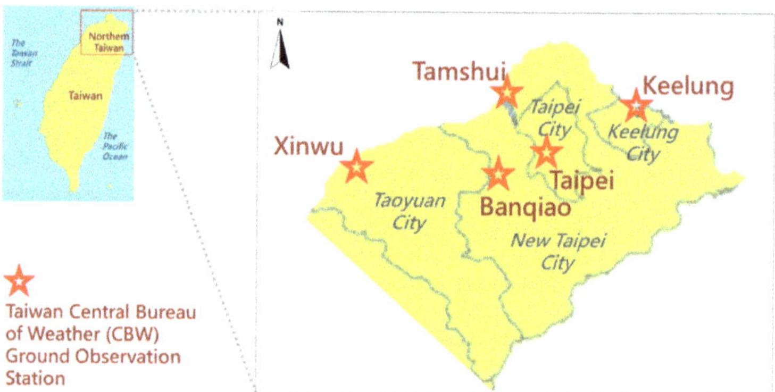

Figure 1. The location of five Central Weather Bureau stations in northern Taiwan.

The meteorological factors (variables) for agricultural weather forecasts that immediately affect farm planning or operations vary from place to place and from season to season. According to the Köppen climate classification, Taiwan is an island that is classified as having a "warm oceanic climate/humid subtropical climate," while Taipei City is classified as "temperate, no dry season, hot summer" (Köppen: Cfa) [57]. Temperature (for heat succession), relative humidity (for transpiration), precipitation (for irrigation), sunshine hours (insolation duration), global radiation (light and thermal condition for plant physiology), and total cloud cover (character of prevailing clouds that reduce the global solar radiation) are the core weather factors that influence crop-growing processes; therefore, each dataset comprised the daily logs of the 6 weather factors. Wind speed and direction were excluded from this research because the wind blowing around UA sites with a relatively small scale is often affected by the surrounding buildings. It is noted that these heterogeneous datasets were normalized (within a value range of 0–1.0) for preprocessing. Then, the normalized datasets were input into the SOM for clustering the weather features, and the clustering results were displayed in a honeycomb arrangement.

A total of 9130 datasets were collected from the 5 weather stations in northern Taiwan over the 5 years, and their basic statistics are given in Table 1.

Table 1. Basic statistics of the meteorological data collected at five weather stations in northern Taiwan during 2014–2018.

Station Name		Temperature (°C)	Relative Humidity (%)	Precipitation (mm)	Sunshine Duration (Total h)	Global Radiation (MJ/m²)	Total Cloud Cover (Level 0–10)
Banqiao	Max	32.3	96.0	246.0	12.7	29.8	10.0
	Min	5.4	46.0	0.0	0.0	0.0	0.0
	Avg *	23.5	72.9	6.1	3.7	11.4	7.0
	Std **	5.5	7.9	17.0	3.5	7.3	2.6
Tamshui	Max	32.3	100.0	379.5	12.5	29.1	6.7
	Min	5.1	1.8	0.0	0.0	0.0	0.0
	Avg	23.0	76.7	5.4	4.5	13.2	5.0
	Std	5.5	9.4	19.7	4.0	7.8	1.6
Taipei	Max	33.2	97.0	306.7	11.6	28.9	10.0
	Min	5.6	5.6	0.0	0.0	0.0	0.0
	Avg	23.8	23.8	6.1	3.6	12.0	7.2
	Std	5.6	5.6	18.2	3.5	7.2	2.5
Xinwu	Max	31.7	100.0	246.5	12.9	30.6	10.0
	Min	5.7	50.0	0.0	0.0	0.0	0.0
	Avg	23.0	79.8	3.9	5.1	14.9	6.2
	Std	5.5	8.5	14.6	4.2	8.5	2.9
Keelung	Max	32.4	96.0	337.6	12.6	31.1	10.0
	Min	5.4	49.0	0.0	0.0	0.0	0.0
	Avg	23.1	76.1	9.2	3.8	11.7	7.6
	Std	5.3	8.9	22.6	4.0	9.0	2.6
All Stations	Max	33.2	100.0	379.5	12.9	31.1	10.0
	Min	5.1	1.8	0.0	0.0	0.0	0.0
	Avg	23.3	75.6	6.1	4.2	12.6	6.6
	Std	5.5	9.1	18.7	3.9	8.1	2.7

* Average, ** standard deviation.

2.2. Self-Organizing Map (SOM)

The SOM proposed by Kohonen in 1982 [58,59] is an artificial neural network that is configured with an unsupervised learning algorithm. It consists of repeatedly learning processes to gradually update the data nodes in the output map until converging to a stable and representative solution of the input space. Each of its learning steps starts with randomly selecting an input weight vector. A node in the input layer is searched for by competing with each other in the output map (the topological layer) to find the most "similar" one (also called the "winning" node or the "best matching unit" (BMU)) to best match the input vector. Next, the training continues to make the BMU and its neighbors closer to the input vector in a manner that is governed by the learning rate and the neighborhood function [34,42]. The map is then reconfigured to adaptively transform high-dimensional input patterns into two-dimensional arrays of neurons in a topologically ordered fashion, which facilitates the detection of the inherent structure and the interrelationships between data [25]. Thus, the patterns of a large number of clusters and the transitional nodes between patterns can be more readily understood and discerned [60]. The SOM technique preserves the neighborhood relations of the input data to form a meaningful topological map [30] so that a large amount of information can be stored in the weight values of the SOM's neurons with similar characteristics in input vectors [61,62]. An SOM is capable of conserving the space continuum between daily meteorological datum so that there is some resemblance in the neighboring clusters. Therefore, an SOM can cover the overall data characters to offer a more detailed presentation of particular features [63], and the result of the clustering can provide a means to visualize the complex distribution and reveal weather patterns and attributes of the temporal sequence over a region of interest [42].

Figure 2 illustrates the concept of the SOM approach in this research. The datasets were input into the SOM network for clustering according to weather similarities by calculating the differences, namely, the distances, between every two inputs on the multi-dimensional topology. The shortest distance (i.e., the minimal difference) between any two inputs indicates that they share more resembling weather characters than others; therefore, they are grouped into one category (i.e., one neuron). Weather features may be clustered into various categories (neurons) depending on the research purposes. Given relatively limited differences in terms of geographical distance among the five weather stations and environmental conditions in northern Taiwan, the variation in weather features may not be too drastic; therefore, this study determined the network size of the SOM to be 3×3 (= 9 categories of weather types in total).

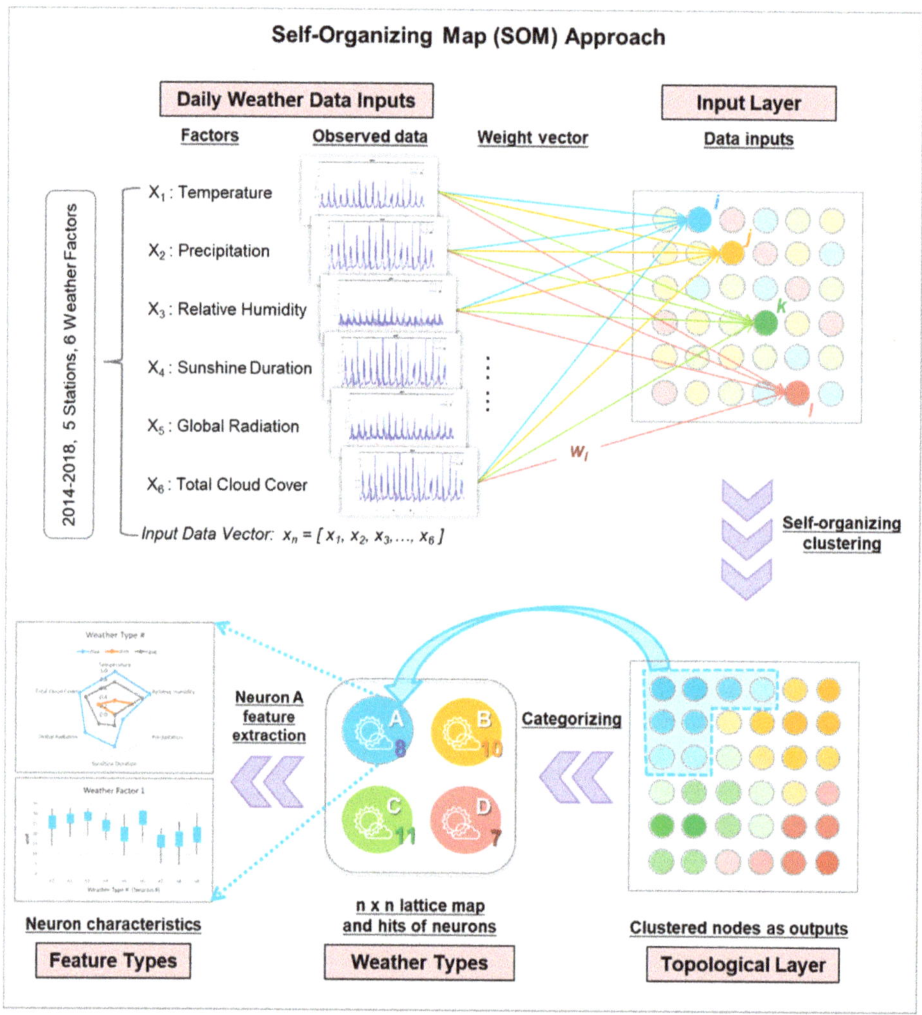

Figure 2. The SOM network structure for clustering weather features in this research.

In reference to the traditional Chinese Farmers' Calendar, the concept of a "10-day period", also called the "xun" in Chinese, has been commonly used as a temporal cycle for planning and engaging cultivation tasks in Chinese society. There are basically three 10-day periods in a month. With variations of week numbers and start/end dates in a month every year, as well as month/seasons, it is relatively complicated to specify the same period in a year for the analysis and comparison of weather types/features and generalizing weather features over very long temporal periods. Therefore, following the agricultural tradition in Taiwan, the 10-day period was adopted to be the conveniently unified and appropriate temporal scale in this research. Therefore, this research considered the 10-day period to be the temporal scale. The software that was used to run the SOM in this research was MATLAB version 2019b.

3. Results

The strength of an SOM is the ability to directly use uncompressed data rather than only using traditional statistics with diluted weather attributes or converting them into certain performance indicators [63]. SOM also effectively provides a means to visualize the complex distribution of weather features to classify and reveal the synoptic weather patterns and attributes of the temporal sequence over a region of interest. In this research, the SOM led to two outcomes for further analysis: first, the result of nine weather feature types (neurons) at each weather station in northern Taiwan, and second, the distribution of weather types throughout the years at the specific weather station indicated the temporal trend/pattern of local weather features. These two results are delineated as follows.

3.1. Types of Weather Features

Figure 3 illustrates the SOM results. Figure 3a shows the number of datasets in each SOM neuron that represented similar features. Figure 3b shows the numbering labels that were associated with the neurons of the SOM topological map. It is evident that neurons #2, #3, #4, and #7 accounted for 64% of the total inputs; therefore, these four neurons were the most representative types to depict the overall weather characteristics in northern Taiwan.

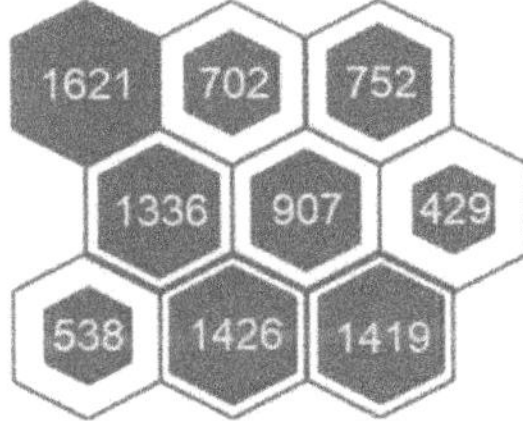

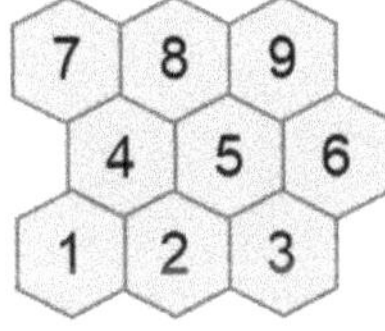

(a) The hits of datasets clustered in each neuron   *(b) Numbering label for each neuron*

Figure 3. The counts of datasets and numbering labels that were associated with the neurons of the SOM topological map. (**a**) The hits of datasets clustered in each neuron. (**b**) Numbering label for each neuron.

Figure 4 presents the heatmap of the SOM results, indicating the weather-type configuration that can be considered to be the six stratified layers of weather factors with the corresponding characteristics. Every neuron was held at the same position across all weather factor layers so that similar heatmaps in different factor planes represent a high correlation of features. On each weather factor layer, every neuron is shaded in a specific color with a spectrum from light to dark to indicate the weight significance from the highest (in yellow) to the lowest (in black). Therefore, each heatmap represents the feature intensity of a weather factor learned by the SOM. The four corners of the SOM can thus be taken as the most extreme nodes in terms of climate variability, with a smooth continuum in between [60]. All data inputs that were assigned in each neuron for each weather factor layer were extracted to further explore their typical characteristics in common via statistics methods, where the characteristics of each weather factor are illustrated in the form of a stock chart. They are visually presented to show the significance and variation of each specific weather factor in each neuron in this research. With the neuron labels (#1–#9) on the horizontal axis, each blue rectangle comprises the ranges of average value plus/minus one standard deviation of all data in that specific neuron. The top and bottom tips of black vertical lines indicate the maximum and minimum values of all data in that specific neuron, respectively. It is noted that in the precipitation layer, the enormous values of the maximum rainfalls are labeled directly on top so that the precipitation variation is still visually explicit.

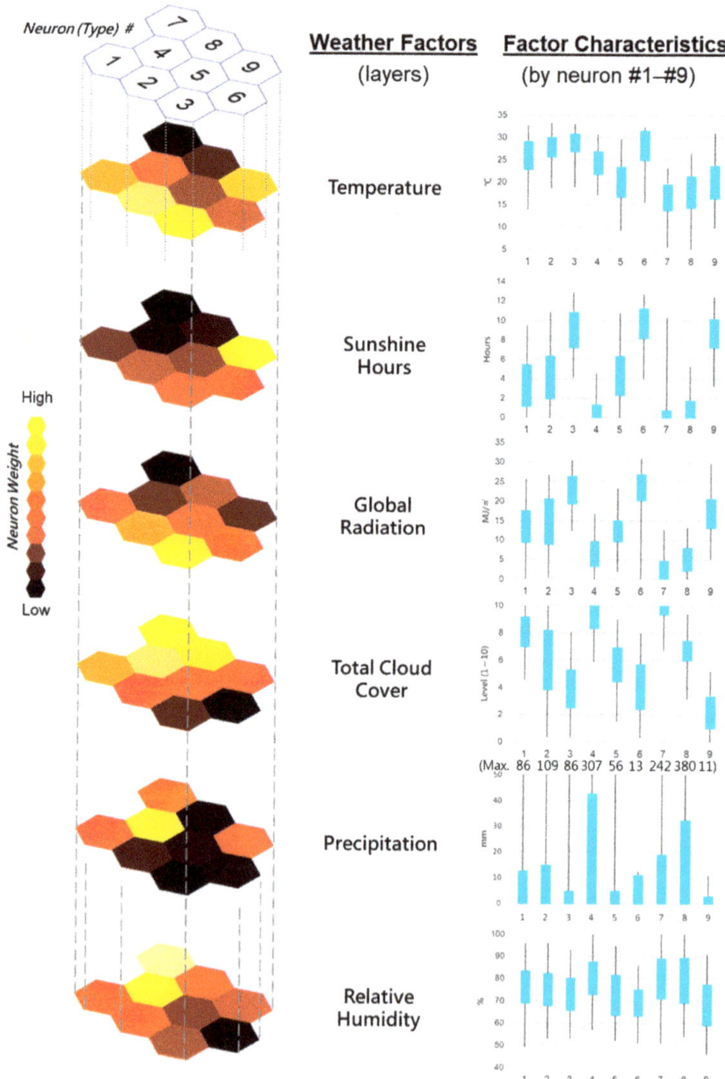

Figure 4. The heatmap of weather factors and the corresponding characteristics.

Among all nine neurons, the temperature weighed most significantly (the highest value) in neuron #3, then followed by neurons #2, #9, #1, and #4 accordingly with gradually darker colors, and lastly, neuron #7 in black denoted the lowest value. As for the total cloud cover layer, neurons #6 and #3 on the lower-right presented the least cloud cover (i.e., the sunniest condition), then neurons #2, #5, and #9 followed with gradually higher temperatures toward the upper-left, and neuron #7 in yellow denoted the most cloudy condition (i.e., least sunny).

Therefore, through integrating the characteristics of the six weather factors with various significances allocated, each neuron designated one type of weather feature. Since neurons #7 and #2 comprised the two most significant inputs, they were taken to illustrate how the six weather factors are transformed into a radar diagram of each weather type and what their overall features were, as shown in Figure 5.

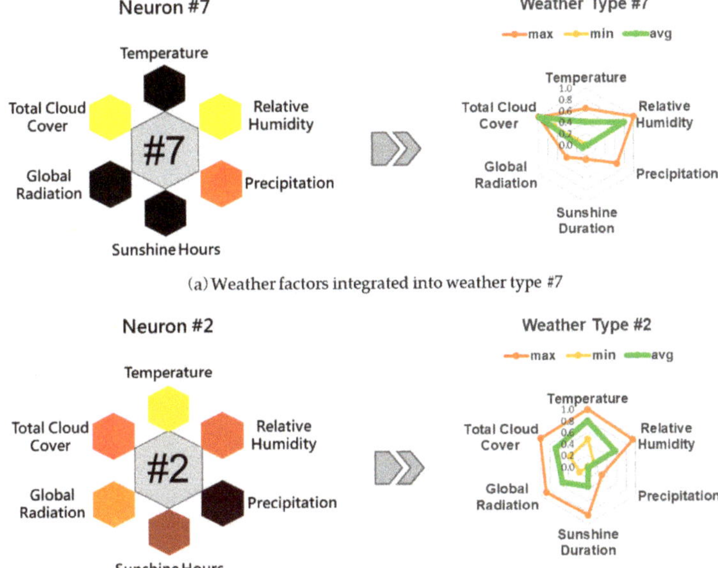

Figure 5. The significance of weather factors in each weather type are transformed into radar diagrams (with examples of weather types #7 and #2). (**a**) Weather factors integrated into weather type #7. (**b**) Weather factors integrated into weather type #2.

In neuron #7, the temperature, sunshine hours, and global radiation factors were all colored black, indicating these three factors displayed the lowest levels. In contrast, relatively high humidity and total cloud cover appear (in yellow), suggesting that these factors displayed the highest levels. Precipitation was orange, meaning it displayed a medium level. As a consequence, neuron #7, which was also denoted as weather type #7, integrated the weather characteristics of the lowest temperature, the highest humidity, medium rainfall, the least sunshine hours, the lowest global radiation, and the highest total cloud cover, which complied with the general weather phenomena in winter from our observation (Figure 5a). With the same rationale, neuron #2 exemplified the weather characteristics of high temperature, medium humidity, and low precipitation, along with lower sunshine hours, high global radiation, and lower total cloud cover (Figure 5b).

When converting the above results of neuron #2 to the radar diagram shown in Figure 5b, the green line linking the green points (average values) formed the "shape of type #2 in terms of average values". The same rationale applied to the orange points and line, denoting the "shape of type #2 in terms of the maximum values", as well as to the yellow ones, denoting the "shape of type #2 in terms of the minimum values". Therefore, the daily average, maximum, and minimum records for each weather type were delineated explicitly. Overall, type #2 indicated that the maxima of temperature, relative humidity, and total cloud cover reached the highest level of 1.0 at the outmost edge on the hexagon, while the sunshine hours and global radiation positions were at 0.8 (the second-highest level) and precipitation positions were at 0.2 (a very low level). The "shape of the average values" profiled the dominant weather factors, which were temperature, humidity, and total cloud cover positioning in the range 0.6–0.8. The other three factors played much less significant roles, where precipitation, sunshine hours, and global radiation positioning were at around 0.0, 0.4, and 0.6, respectively.

3.2. Weather Pattern at Five Stations in Northern Taiwan

This research originally explored the weather features from five stations in northern Taiwan, where Table 2 illustrates the integrated results. Each station presented its own weather characteristics, but overall, types #2, #3, #4, #7, and #9 were the most typical ones at the Banqiao, Taipei, and Xinwu stations, while Tamshui and Keelung presented their own types of weather features. The seasonal characteristics of weather features at the other four stations also deserve more elaboration in the form of the comprehensive analysis that was given for the Taipei Station in this research.

Table 2. Features of weather types at 5 weather stations in northern Taiwan.

Station	Main Types	Occupying Percentage	Absent Types	Summer Types	Concentration in Summer Months	Winter Types	Concentration in Winter Months
Banqiao	4, 7 (1~9)	40%	0	1, 2, 3, 4, 6	April–September	4, 5, 7, 8, 9	October–March
Tamshui	3, 8	55%	1, 6, 7	2, 3, 4	April–October	5, 8	October–May
Taipei	2, 4, 7	70%	1, 6	2, 3, 4	April–October	5, 7, 8, 9	October–April
Keelung	1, 5, 6	83%	2, 3, 5	4, 6	April–October	7, 8	October–April
Xinwu	2, 3, 4, 7	72%	1, 6	2, 3, 4	April–October	5, 7, 8, 9	October–April

3.3. Temporal Pattern of Weather Types

The result of the Taipei weather station is taken as the main focus for elaboration in this research. Figure 6 shows the occurrences (counts) of the nine weather types on a 10-day basis from 2014 to 2018. Each cell is visually shaded with a color spectrum indicating the frequency intensification. The higher the occurrence frequency is, the darker the cell color is. Therefore, the occurrence time, frequency, and duration of each weather type are presented explicitly on the pattern of weather types that occurred at specific weather stations throughout the years between 2014 and 2018.

Figure 7 shows the percentage of each weather type that occurred on an annual basis. Figures 6 and 7 explicitly display that there was a stable trend and similar occurrence percentages for the variation of weather types over the five years. Though the annual occurrences of each weather type might be slightly different and/or offset, the timings of the sequential trend and the duration were generally consistent. In this regard, it is essential to examine the high-frequency occurrence distribution, duration, continuity, and extent profiles rather than the forecasts of weather features on specific dates. Therefore, it was reasonable to aggregate the 5-year occurrences under the same 10-day basis into overall seasonal features by summing all the values in each cell to strengthen the significance and distribution of feature types, as illustrated in Figure 8a.

In addition, when the specific weather type lasted continuously (especially with high occurrence frequency) and multiple weather types took place over the same time frame, these weather types could be considered altogether to be the "representing weather features" of the sequential regularity and pattern. In this regard, from the distribution of concentration and continuity, it is explicit that the nine weather types that were distributed throughout the thirty-six 10-day periods formed three distinct sections of temporal groups annually, referring to sections A, B, and C marked with dashed rectangles in Figure 8a.

Section A (marked with a dashed orange rectangle) represented the period from "early spring through late fall" (mainly full summer). Section B (marked with a dashed blue rectangle) represented the period from "mid-fall through next mid-spring" (mainly full winter), with two sub-sections, namely, section B2 from early autumn to winter at the year-end and section B1 from winter at the beginning of the year (1 January) to early summer. Section C (marked with a dashed yellow rectangle) referred to the absent or barely occurring types all year round. There were some overlaps of 10-day periods between sections A and B1, similarly between sections B2 and A. This explained the seasonal alternation with diverse weather types and complex variations during spring and fall.

Figure 8b also delineates the two leading weather types with the associated occurrence percentages summed from major (in pink shades) and secondary (in yellow shades) weather types in each 10-day period. With shaded cells indicating that the percentage exceeded the designated one listed in the sum column, this demonstrates how significant and representative the weather type was in each 10-day period. In addition, the last row shows the subtotal percentage of the two leading types, together with over 60% (in green shades) as a high concentration and continuum of occurrences. The 10-day periods without green shaded cells represent the percentage <60% as non-leading major and secondary occurrences, which also refers to more diverse and variant weather types taking place in springs and falls as the seasonal transition.

Year	Weather Type #	JAN 1	JAN 2	JAN 3	FEB 1	FEB 2	FEB 3	MAR 1	MAR 2	MAR 3	APR 1	APR 2	APR 3	MAY 1	MAY 2	MAY 3	JUN 1	JUN 2	JUN 3	JUL 1	JUL 2	JUL 3	AUG 1	AUG 2	AUG 3	SEP 1	SEP 2	SEP 3	OCT 1	OCT 2	OCT 3	NOV 1	NOV 2	NOV 3	DEC 1	DEC 2	DEC 3	Sum (days)
2014	1																																						0
2014	2								1	5	1	3	2	1	3	3	2	5	6	5	4	6	5	6	3	6	4	2		1	1		1					76
2014	3								1	4		1		1						5	6	4	4	2	8	4	6	4	1	1	1	1						54
2014	4	1					1	1		1	3	5		6	6	7	8	4	4						1	1	2	4	6	2	3	2	2	2	2			74
2014	5	2	3	1	2	2	4	3	2		3			1															3	1	3	1	2	2	1		4	40
2014	6																																					0
2014	7	3	3	2	7	7		9	4	1	4		2	3		1																		3	6	5	8	5	73
2014	8	1	1			1					1																							2		1	2		9
2014	9	3	3	8	1		3	1	2		1																				5	3	1		5	1		2	39
2015	1																																						0
2015	2								3	2	3	1	3	3	5	1	5	8	7	5	4	7	2	6	1	4	6	2	3	1	5	3	2	1			1	94
2015	3										3	1	1		2	2	2	2	3	4	5	4	6	2	2	3	4	2	3	1								52
2015	4	1					2			1	1	3	3	4	3	8	3			1	1		2	2	8	2		6	4	4	5	4	4	4	1	2	2	81
2015	5	4	4	3	3	7	4	2	3			1	2	1														1			1		2	4	2		1	44
2015	6																																					0
2015	7	2	5	5	6		2	8	2	5	3	2	1	2																1			1		4	5	7	61
2015	8	1	1		1	1										1																			3	2		10
2015	9	2		3		2					2	2		2														1		3		2	2	1		1		23
2016	1																																						0
2016	2					2							4	5	6	6	4	3	8	5	6	3	7	5	1	4	1	5	2	5	7	2	1				1	97
2016	3										1	2	1	2	1	1	2	3	1	2	3	8	3	3	8	1	2	2	4	1	2							53
2016	4	2					1	1		2	5	5	5	3	3	4	4	4	1	3	1			2	2	5	7	3	4	4	2	4	2	6	5		1	91
2016	5	1		2		2		1		2	2		3	1																	4				1	2	4	22
2016	6																																						0
2016	7	6	9	8	6	7	7	2	9	5		2																	3				2		3	3	3	75
2016	8		1	1							1																						2	1	1			7
2016	9	1			4	2		3		1																					3	1	3		4	2		21
2017	1																						2																2
2017	2											1	2	5	6	4	6	2	10	9	5	7	3	8	5	3	1	6	6	2		1	2					94
2017	3									1	3		2	2		1					5	3	2	2	4	1	6	4	2			1						39
2017	4	2		1	2	1					4	2		2	5	5	3	3	6	4	8		1						2	8	5	6	4	3	1		1	93
2017	5	2	3	3	3	3	1	2			2			1																	4	1			1	2	4	32
2017	6																										1											1
2017	7	3	7	6	5	2	6	6	5	4				3	2		1														2	1	4	5	8	7	2	79
2017	8						1	2																											1	2		6
2017	9	3								4	1	2	3	1																			2				2	19
2018	1																																						0
2018	2												1	4	4	3	2	6	9	7	3	4	7	2	8	7	5	3	4	4	4	1	1	1		2		92
2018	3											2	1	2	3	1	1	2		1	7	2	3				1	4		1	4							30
2018	4	1	1					5	3						3	6	6	1	1	3	6	4	2	1	1		5	8	5	2	5	6	5	2	2	1	1	96
2018	5	1	4	2		4	1					2	5	1	1																1		2	4	1	2		33
2018	6																																						0
2018	7	6	2	8	10	2	7	3	1	1	2	2																		4	1	2	1	5	4	4	8	73
2018	8						1	2																															3
2018	9	2	3							2	2	3	3	3														1		2	5		5	1	2	1	3	3	2	38

Figure 6. The occurrence distribution on a 10-day basis for each weather type at Taipei Station.

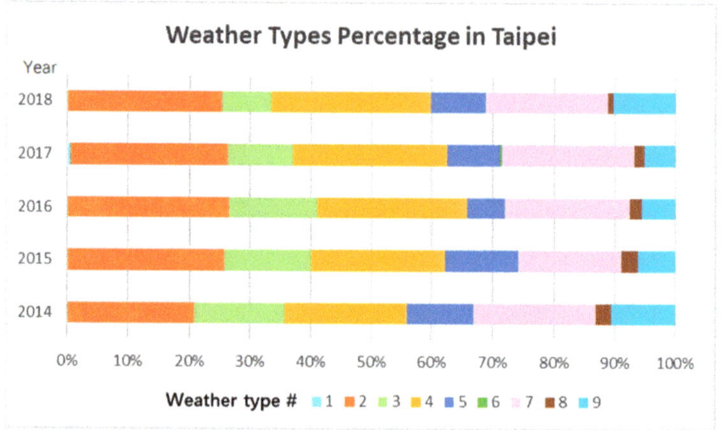

Figure 7. The occurrence percentages of the nine weather types at Taipei Station.

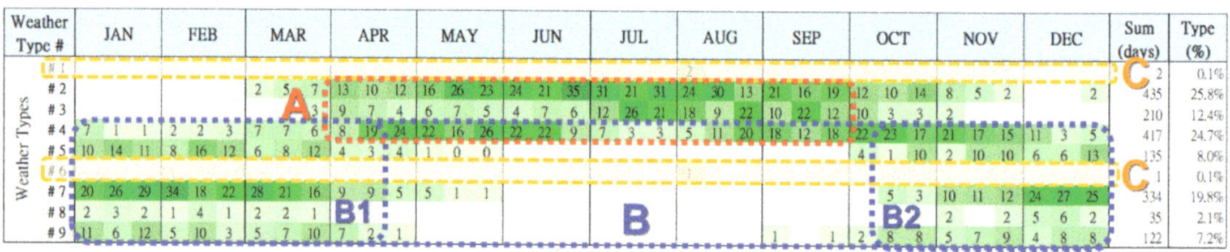

Weather Type #	JAN			FEB			MAR			APR			MAY			JUN		
#1																		
#2							2	5	7	13	10	12	16	26	23	24	21	35
#3									3	9	7	4	6	7	5	4	7	6
#4	7	1	1	2	2	3	7	7	6	8	19	24	22	16	26	22	22	9
#5	10	14	11	8	16	12	6	8	12	4	3	4	1	0	0			
#6																		
#7	20	26	29	34	18	22	28	21	16	9	9	5	5	1	1			
#8	2	3	2	1	4	1	2	2	1									
#9	11	6	12	5	10	3	5	7	10	7	2	1						

Weather Type #	JUL			AUG			SEP			OCT			NOV			DEC			Sum (days)	Type (%)
#1																			2	0.1%
#2	31	21	31	24	30	13	21	16	19	12	10	14	8	5	2			2	435	25.8%
#3	12	26	21	18	9	22	10	22	12	10	3	3	2						210	12.4%
#4	7	3	3	5	11	20	18	12	18	22	23	17	21	17	15	11	3	5	417	24.7%
#5										4	1	10	2	10	10	6	6	13	135	8.0%
#6																			1	0.1%
#7											5	3	10	11	12	24	27	25	334	19.8%
#8													2		2	5	6	2	35	2.1%
#9						1			1	2	8	8	5	7	9	4	8	8	122	7.2%

(a) Seasonal weather features of sections A, B and C on aggregated occurrences

Weather Type #	JAN			FEB			MAR			APR			MAY			JUN		
Major type #	7	7	7	7	7	7	7	7	7	2	4	4	4	2	4	2	4	2
In each 10-day (%)	40	52	53	68	36	54	56	42	29	26	38	48	44	52	47	48	44	70
Secondary type #	9	5	9	5	5	5	5	5	5	3,7	2	2	2	4	2	4	2	4
In each 10-day (%)	22	28	22	16	32	29	12	16	22	18	20	24	32	32	42	44	42	18
Subtotals (%)	62	80	75	84	68	83	68	58	51	44	58	72	76	84	89	92	86	88

Weather Type #	JUL			AUG			SEP			OCT			NOV			DEC			Shaded cells with % over
Major type #	2	3	2	2	2	3	2	3	2	4	4	4	4	4	4	7	7	7	>40%
In each 10-day (%)	62	52	56	48	60	40	42	44	38	44	46	31	42	34	30	48	54	45	
Secondary type #	3	2	3	3	4	4	4	2	4	2	2	2	7	7	7	4	9	5	>20%
In each 10-day (%)	24	42	38	36	22	36	36	32	36	24	20	25	20	22	24	22	16	24	
Subtotals (%)	86	94	95	84	82	76	78	76	74	68	66	56	62	56	54	70	70	69	>60%

(b) Leading weather types analysis on a 10-day basis

Figure 8. The occurrences distribution of nine weather types on a 10-day basis and the leading types at Taipei Station. (**a**) Seasonal weather features of sections A, B and C on aggregated occurrences. (**b**) Leading weather types analysis on a 10-day basis.

3.4. Frequency, Duration, and Distributions of Type Occurrences

The collective occurrences of the nine weather types at Taipei Station were analyzed from two perspectives: the features and distribution of occurring types and the comprehensive outcome during designated timespans (season, month, and 10-day periods), which are elaborated in the following.

Figure 8a illustrates that among all nine feature types, types #2 (25.8%), #4 (24.7%), and #7 (19.8%) were the main weather types at the Taipei Station and altogether accounted for 70.3% of the occurrences out of the occurrences of all types over the five years. There were barely any type #1 or #6 weather patterns each year.

Figure 9 illustrates the occurrence distribution and duration of types in sections A, B, and C. Section A in Figure 9a basically spanned from April to early October, where the highest frequency weather types were types #2 (red bar), #4 (green bar), and #3 (yellow bar), implying they are the representative weather types during the season from mid-spring to early fall. These types tended to appear sporadically or were absent before April and disappeared gradually after mid-October till the year-end. The occurrence timespans and high frequencies of types #2 and #4 seemed complementary to each other, with one rising (appearing) and the other falling (disappearing), while type #3 peaked shortly in the hottest days of mid-summer (mid-July to mid-September). Section B in Figure 9b, on the other hand, shows the duration starting from mid-October to the next late April (covering late fall, winter, and the following spring), with a gradual transition through types #4, #7, #5, and #9. There were barely any type #1 or #6 weather patterns, with only 1–2 occurrences in early August in Section C.

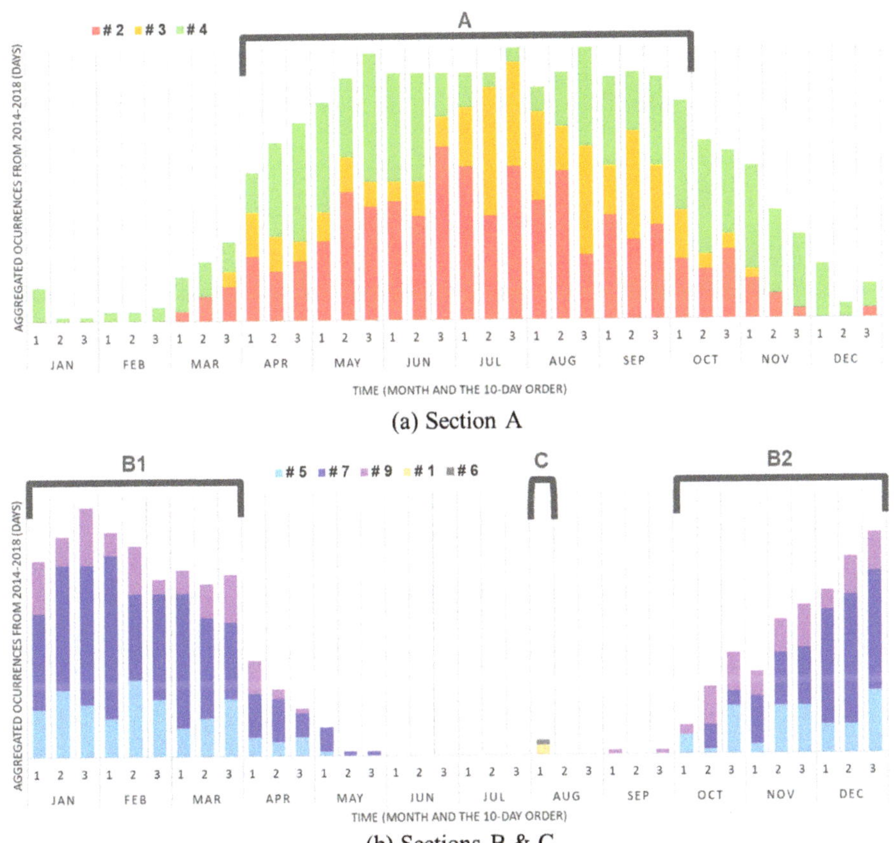

Figure 9. The occurrence distribution of the main weather types in sections A, B & C at Taipei Station from 2014 and 2018. (**a**) Section A. (**b**) Section B & C.

3.5. Main Features of Weather Types in Taipei City

To elaborate the weather types and features in the Taipei area for further application to UA, Figure 10 highlights the features of the two leading weather types for sections A, B, and C to visually present the variations of the six weather factors during the designated period.

Comprehensively, with cross-referencing to Figures 8–10, the annual weather pattern in the Taipei area presented the following features in the sequential timing of a year are elaborated as follows.

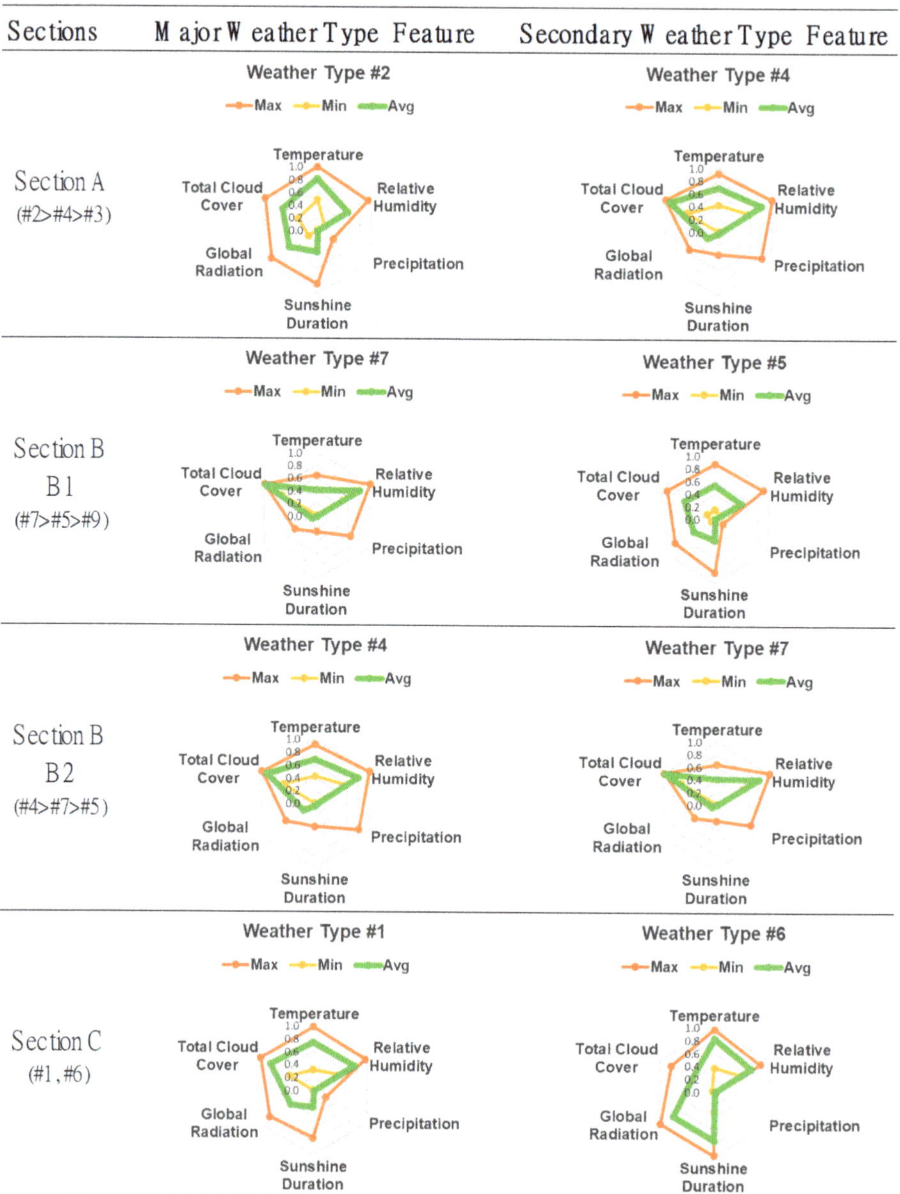

Figure 10. The leading weather types of Sections A, B, and C in radar diagrams.

Though mainly concentrating from April to mid-June and from mid-August to November (before and after summer), type #4 weather patterns occurred in almost every 10-day period throughout 2014–2018. This was taken as the basic annual weather feature in Taipei.

Intensifying from January to March and in December, type #7 weather patterns dominated the weather features in winter, with the lowest temperature range (14–20 °C), little rainfall, high humidity (80%), the least global radiation (0–5 MJ/m^2) and sunshine hours, the highest level of cloud cover of the year, and sometimes accompanied by the second most frequent weather pattern, namely, type #5, with its relatively mild/less "winter touch" features.

During the periods of March–April (spring) and October–November (fall), all the weather types, except for types #1 and #6, appeared alternatively, indicating greater variations and less stable features, for instance, temperature (14–31 °C), precipitation (15–43 mm), global radiation (0–26 MJ/m^2), sunshine duration (0.5–11 h), and cloud cover (levels 1–10). Types #7 (winter) and #5 played the most significant roles and frequently occurred in early March and then gradually disappeared, followed by types #4 and #2, mainly from April (as the summer began to emerge). Such weather features proceeded in a mirrored way from October to November as winter approached.

From May to September (late spring through summer to early fall), types #2 and #4 occur were dominant and alternated, with their high temperature (25–30 °C), little precipitation, and large variation in global radiation, sunshine hours, and cloud cover. Furthermore, accompanied with a slight trace of type #3, the summer feature reached its climax occurrences, with the highest temperature (27–31 °C), very limited rainfall, the most abundant global radiation (19–26 MJ/m^2) and sunshine hours (7–11 h), the least rainfall, and lower cloud cover (levels 2.5–5) in the summer peak from mid-July to September. Then, from October, the weather feature returned gradually to the basic type (type #4) and transformed into fall–winter patterns.

4. Discussion

Weather classification of typical local weather features from past experiences and historical data can greatly enhance agricultural decision making by identifying the timings for effective cultivation activities and the corresponding efficiency of resource use. For example, the statistics of the historical average, maximum, and minimum of certain weather factors (e.g., temperature and precipitation) within a time interval (month or year) provides a very rough idea of the local weather condition with variation ranges. In addition, some specific daily phenomena with the presence of various weather factors occurred frequently and repetitively during a certain period in a year; therefore, such alternation of seasonal features can be considered to be the "typical weather patterns" and is expected to continue in the following years. This section discusses and extends the use of SOM, weather types, and features and particularly focuses on the potential application of the feature results.

4.1. SOM Approach Contributions to the Weather Typing

The results of the weather features from the SOM approach can effectively excavate out more hidden details of characteristics on the key weather types. It explicitly visualizes the weather patterns and trends in terms of their occurring time, frequency, intensity, distribution, duration, and transition nexus with various types. Furthermore, it is rational to elucidate/interpret what weather types are likely to occur during designated periods of interest and how meteorological factors are expected to appear substantially. Therefore, it is practical to grasp the temporal distribution and seasonal changes of weather features to plan for appropriate strategies and measures when necessary.

4.2. Potential Applications

The investigation of weather features and historical meteorological data throughout the year can be a reference for agricultural decisions in various cultivation activities, such as species selection, planting, harvesting, transplanting, defoliation, fertilization, and irrigation [11]. In addition, necessary precautions can be further adopted to preempt crop stress control, sheltering, disease risk reduction, and pest control, as well as to explore sustainable resource mechanisms in terms of collection, storage, and utilization.

The symbolic representation of temporal weather features is the key to effective crop planting plans that suggest farmers take necessary actions and measures for optimal crop growing and harvesting, with more efficient resource utilization and protection from potential weather damage. For example, the timings of meteorological conditions of temperature, precipitation, and sunshine hours affect crop-growing schedules regarding seeding, seedling, flowering, fruiting, and harvesting, particularly through the necessary heat for succession, water supply for irrigation, and sunlight exposure for photosynthesis. Therefore, the understanding of the temporal distribution and intensity of local weather types would contribute to enhancing the cropping (such as the species selection) under favorable weather, with adequate supporting facilities being installed when necessary (such as the provision of shade for crops from sunscald and increasing ventilation to prevent against potential insects/disease due to high temperature and humidity).

The features of weather types can also enhance the prediction of rainwater and solar power output by calculating the related green resources input based on meteorology for cultivation during the process of species selection, maintenance, and resources regulation regarding collection, storage, and provision. Particularly for urban agriculture via rooftop farming and community gardening, the water and electricity that are needed for irrigating farms often come from public utilities. Hence, drawing practical planting plans and operation strategies by taking advantage of favorable weather features during designated periods for specific measurements/adaptations can be sustainable. For instance, installing an external rainwater tank and solar power facility, as well as prioritizing the use of rainfall and sunlight as green/renewable resources for efficient collection, storage, and reuse for farming irrigation, reduces the dependence on tap water and municipal electricity.

The Da-an rooftop farm with elevated planters extensively laid out on its rooftop can be taken as a study site for the application of weather types and features that are relevant to UA activities. The selection of crop species and some auxiliary measurements are proposed as application examples in the following.

Sweet potato leaves (SPL) (*Ipomoea batatas* (L.) Lam.) has been one of the commonly planted species at the Da-an site over the years. As popular leafy vegetables, SP is a subtropical herbaceous trailing vine with a relatively short growing cycle to produce frequent harvests during growing seasons; it requires continuous heat, abundant daily watering (with good drainage), long sunlight exposure, and is very sensitive to chilly temperatures [64]. SPL prefers cooler and drier seasons as seedlings, then it can mature fast in hot and humid months (with a favorable optimal temperature range between 20–30 °C). SP leaves can be picked for harvest within as short as 18 days in hot summers but as long as 30 days in cold winters [65]. According to the analysis of major weather types that were developed for Taipei in this research, SPL can serve as a very good species that is suitable to grow under the weather conditions in Taipei, and its most favorable growing seasons would fall within section A. By referring to Figures 4, 8 and 10, with SPL seedlings planted in April, they can grow prosperously from May to September when weather types #2 and #4 are dominant (with characteristics of high temperature (25–30 °C); constantly high humidity; limited daily precipitation but sometimes showers up to over 300 mm; and large variations in global radiation, sunshine hours, and cloud cover). It is noted that type #3 is concentrated from mid-July to mid-September (with characteristics of very hot days, reaching the highest temperature (27–31 °C), sunshine hours (7–11 h daily), and radiation (19–26 MJ/m^2). SPL grows slowly from November to next February, during which, weather types transform gradually from types #4 to #7 with winter characteristics. In other words, as seen in Figures 4 and 8b, the total occurrence frequency of weather types #2, #3, and #4 accounted for 72–95% in section A starting from late April to the end of September. Therefore, SPL would enjoy the high temperature and humidity, as well as the long sunshine hours and strong radiation, with some fluctuation between these three weather types.

With types #2, #3, and #4 occurring alternately during section A, cross-referencing between Figures 4, 8, and 9, types #2 and #3 featured low daily precipitation (12–15 mm) whereas type #4 featured high daily precipitation (42 mm). Furthermore, all three types had occasional extremely high daily precipitation (with a maximum of up to 86–307 mm). These showed that there were potential demands to set up provisional rainwater tanks, such as additional water containers and large inclined planes to increase rainwater collection areas and storage volume as watershed during intensive showers, which could be set aside for later irrigation in a sustainable way. Therefore, by way of providing an SPL irrigation plan to satisfy SPL's abundant water needs in hot summers, the daily irrigation water supply can be managed to come from an additional setup of a rainwater tank to collect and store torrential rain as a supplement to irrigation water from May to September, specifically during section A. The potential size of the suggested rainwater tank depends on the SPL planting area, the number of plants, and the loading capacity of the rooftop of the building.

The sunshine duration can affect plants' growth due to sunlight exposure for photosynthesis; therefore, some crop species prefer longer sunlight exposure to grow prosperously than others. The features of sunshine duration, global radiation, and total cloud cover level can also impact the efficiency of solar power utilization. Therefore, seasons with long sunshine hours and radiation (type #3) showed higher suitability for heliophyte species or require some shading facility for sciophyte species to prevent sunscald. On the other hand, this also suggests the great potential to establish a solar power facility for providing green energy to the automatic irrigation system in the farm.

5. Conclusions

This study proposed an SOM neural network to cluster and identify the specific features of weather types based on six meteorological variables at five weather stations in northern metropolitan Taiwan. The daily meteorological datasets, comprising temperature, precipitation, relative humidity, sunshine duration, global radiation, and total cloud cover from 2014–2018, were collected as inputs to the SOM network and then were classified into a topological map based on the similarities of weather features to investigate their multi-collinear relationships for spatiotemporal distribution analysis.

The results of the weather features from the SOM classification not only corresponded to what we know about the general weather features in the five metropolitan areas but also provided detailed and integrated information of weather features on the occurrence period, duration, frequency, intensity, and variation range from historical data sets on a 10-day basis. This study contributes to the practicability for urban agriculture planning by exploring the in-depth weather features in northern Taiwan to conduct urban farming in planting arrangement, installing equipment, and managing the crop-growing process (seeding, seedling, growth rate, maturing, flowering, fruiting, harvesting, etc.) in response to local weather features before launching planting in the five study areas.

The results of this study can also be applied to selecting appropriate species for planting in favorable seasons with appropriate weather features. The size of the rainwater tank and the scale of solar power equipment can also be identified to comply with weather features to achieve the optimal resource utilization efficiency to grow and irrigate vegetables so as to reduce the municipal water and electricity use during farming operations.

Given the arising phenomenon of climate change, the chances of unprecedented weather with continuous hot/chilly days, rainstorms, and droughts tend to occur more frequently. However, such weather conditions were not included in the weather typing in this research because outlier data (especially for extremely high precipitation, which is mostly caused by typhoons or sudden torrential rains) were removed at the data preprocessing stage. It is suggested that the extreme weather events in cities of northern Taiwan require further investigation regarding their occurrence time and frequency, as well to help with providing advice as a further reference/strategy for urban agriculture operations if possible so that more risks may be anticipated and stronger adaptation measures could be prepared.

Author Contributions: Conceptualization, A.H. and F.-J.C.; methodology, A.H. and F.-J.C.; software, A.H.; validation, A.H. and F.-J.C.; formal analysis, A.H.; investigation, A.H.; resources, A.H. and F.-J.C.; data curation, A.H.; writing—original draft preparation, A.H.; writing—review and editing, A.H. and F.-J.C.; visualization, A.H.; supervision, F.-J.C.; project administration, F.-J.C.; funding acquisition, F.-J.C. All authors have read and agreed to the published version of the manuscript.

Funding: This study was supported by the Ministry of Science and Technology, Taiwan (grant number: 107-2621-M-002-004-MY3), and the Chi-Seng Water Management Research and Development Foundation, Taipei.

Institutional Review Board Statement: Not applicable.

Informed Consent Statement: Not applicable.

Data Availability Statement: The data presented in this study are available from the Taiwan Central Weather Bureau website.

Acknowledgments: The datasets provided by the CWB, Taiwan, are acknowledged. The authors would also like to thank the editors and anonymous reviewers for their constructive comments that greatly contributed to enriching the manuscript.

Conflicts of Interest: The authors declare no conflict of interest.

References

1. FAO. Urban and Peri-Urban Agriculture, IV. Characteristics of Urban and Peri-Urban Agriculture. 2020. Available online: http://www.fao.org/unfao/bodies/coag/coag15/x0076e.htm (accessed on 21 September 2021).
2. Da Cruz, S.M.S.; Steglich, A.; Cruz, P.V.; Vieira, A.C.M. Investigating the challenges and opportunities of urban agriculture in global north and global south countries. In *Frontiers of Science and Technology: Reports on Technologies for Sustainability–Selected Extended Papers from the Brazilian-German Conference on Frontiers of Science and Technology Symposium (BRAGFOST)*; Kanoun, O., Viehweger, C., Eds.; Walter de Gruyter GmbH & Co. KG: Berlin, Germany, 2017; pp. 95–110.
3. Badami, M.G.; Ramankutty, N. Urban agriculture and food security: A critique based on an assessment of urban land constraints. *Glob. Food Sec.* **2015**, *4*, 8–15. [CrossRef]
4. Pearson, L.J.; Pearson, L.; Pearson, C.J. Sustainable urban agriculture: Stocktake and opportunities. *Int. J. Sustain. Agric. Res.* **2010**, *8*, 7–19. [CrossRef]
5. Specht, K.; Schimichowski, J.; Fox-Kämper, R. Multifunctional Urban Landscapes: The Potential Role of Urban Agriculture as an Element of Sustainable Land Management. In *Sustainable Land Management in a European Context. Human-Environment Interactions*; Weith, T., Barkmann, T., Gaasch, N., Rogga, S., Strauß, C., Zscheischler, J., Eds.; Springer: Basel, Switzerland, 2021; Volume 8, pp. 291–303.
6. Ellis, F.; Sumberg, J. Food production, urban areas and policy responses. *World Dev.* **1998**, *26*, 213–225. [CrossRef]
7. Zezza, A.; Tasciotti, L. Urban agriculture, poverty, and food security: Empirical evidence from a sample of developing countries. *Food Policy* **2010**, *35*, 265–273. [CrossRef]
8. Barrios, S.; Ouattara, B.; Strobl, E. The impact of climatic change on agricultural production: Is it different for Africa? *Food Policy* **2008**, *33*, 287–298. [CrossRef]
9. Agnolucci, P.; Rapti, C.; Alexander, P.; De Lipsis, V.; Holland, R.A.; Eigenbrod, F.; Ekins, P. Impacts of rising temperatures and farm management practices on global yields of 18 crops. *Nat. Food* **2020**, *1*, 562–571. [CrossRef]
10. Das, H.P.; Doblas-Reyes, F.J.; Garcia, A.; Hansen, J.; Mariani, L.; Nain, A.S.; Ramesh, K.; Rathore, L.S.; Venkataraman, R. Chapter 5: Weather and climate forecasts for agriculture. Guide to agricultural, meteorological practices. In *Guide to Agricultural Meteorological Practices*, 2010 ed.; Gommes, R., Challinor, A., Das, H., Dawod, M.A., Mariani, L., Tychon, B., Trampf, W., Eds.; WMO: Geneva, Switzerland, 2012. Available online: https://library.wmo.int/doc_num.php?explnum_id=3996 (accessed on 22 October 2021).
11. Frisvold, G.B.; Murugesan, A. Use of weather information for agricultural decision making. *Weather Clim. Soc.* **2013**, *5*, 55–69. [CrossRef]
12. Haigh, T.; Takle, E.; Andresen, J.; Widhalm, M.; Carlton, J.S.; Angel, J. Mapping the decision points and climate information use of agricultural producers across the US Corn Belt. *Clim. Risk Manag.* **2015**, *7*, 20–30. [CrossRef]
13. Abhishek, K.; Singh, M.P.; Ghosh, S.; Anand, A. Weather forecasting model using artificial neural network. *Procedia Technol.* **2012**, *4*, 311–318. [CrossRef]
14. Narvekar, M.; Fargose, P. Daily Weather Forecasting using Artificial Neural Network. *Int. J. Comput. Appl.* **2015**, *121*, 9–13. [CrossRef]
15. Liu, J.N.; Hu, Y.; You, J.J.; Chan, P.W. Deep neural network based feature representation for weather forecasting. In *Proceedings on the International Conference on Artificial Intelligence (ICAI)*; The Steering Committee of the World Congress in Computer Science, Computer Engineering and Applied Computing (WorldComp): Las Vegas, NV, USA, 2014; p. 1.
16. Tran, T.T.K.; Bateni, S.M.; Ki, S.J.; Vosoughifar, H. A Review of Neural Networks for Air Temperature Forecasting. *Water* **2021**, *13*, 1294. [CrossRef]
17. Chiang, Y.M.; Chang, F.J.; Jou, B.J.D.; Lin, P.F. Dynamic ANN for precipitation estimation and forecasting from radar observations. *J. Hydrol.* **2007**, *334*, 250–261. [CrossRef]
18. Francik, S.; Kurpaska, S. The use of artificial neural networks for forecasting of air temperature inside a heated foil tunnel. *Sensors* **2020**, *20*, 652. [CrossRef]

19. Trebing, K.; Mehrkanoon, S. Wind speed prediction using multidimensional convolutional neural networks. In Proceedings of the 2020 IEEE Symposium Series on Computational Intelligence (SSCI), Canberra, Australia, 1–4 December 2020; pp. 713–720.

20. Chattopadhyay, A.; Hassanzadeh, P.; Pasha, S. Predicting clustered weather patterns: A test case for applications of convolutional neural networks to spatio-temporal climate data. *Sci. Rep.* **2020**, *10*, 1317. [CrossRef] [PubMed]

21. Xiao, H.; Zhang, F.; Shen, Z.; Wu, K.; Zhang, J. Classification of weather phenomenon from images by using deep convolutional neural network. *Earth Space Sci.* **2021**, *8*, e2020EA001604. [CrossRef]

22. Riyazuddin, Y.M.; Basha, S.M.; Reddy, K.K. An approach for prediction of weather system by using back propagation neural network. *Int. J. Sci. Dev. Res.* **2017**, *2*, 117–121.

23. Kakar, S.A.; Sheikh, N.; Naseem, A.; Iqbal, S.; Rehman, A.; Kakar, A.U.; Kakar, B.A.; Kakar, H.A.; Khan, B. Artificial neural network based weather prediction using Back Propagation Technique. *Int. J. Adv. Comput. Sci. Appl.* **2018**, *9*, 462–470. [CrossRef]

24. Wica, M.; Witkowsk, M.; Szumiec, A.; Ziebura, T. Weather forecasting system with the use of neural network and backpropagation algorithm. In *Proceedings of the International Conference on Data Engineering and Communication Technology*; Springer: Singapore, 2019; Volume 2468, pp. 37–41.

25. Chang, F.J.; Chang, L.C.; Kao, H.S.; Wu, G.R. Assessing the effort of meteorological variables for evaporation estimation by self-organizing map neural network. *J. Hydrol.* **2010**, *384*, 118–129. [CrossRef]

26. Chang, L.C.; Wang, W.H.; Chang, F.J. Explore training self-organizing map methods for clustering high-dimensional flood inundation maps. *J. Hydrol.* **2021**, *595*, 125655. [CrossRef]

27. Reddy, P.C.; Sureshbabu, A. An adaptive model for forecasting seasonal rainfall using predictive analytics. *Int. J. Intell. Syst.* **2019**, *12*, 22–32. [CrossRef]

28. Wu, C.L.; Chau, K.W. Prediction of rainfall time series using modular soft computing methods. *Eng. Appl. Artif. Intell.* **2013**, *26*, 997–1007. [CrossRef]

29. Taormina, R.; Chau, K.W. ANN-based interval forecasting of streamflow discharges using the LUBE method and MOFIPS. *Eng. Appl. Artif. Intell.* **2015**, *45*, 429–440. [CrossRef]

30. Wang, W.; Du, Y.; Chau, K.; Chen, H.; Liu, C.; Ma, Q. A Comparison of BPNN, GMDH, and ARIMA for Monthly Rainfall Forecasting Based on Wavelet Packet Decomposition. *Water* **2021**, *13*, 2871. [CrossRef]

31. Ruß, G.; Kruse, R.; Schneider, M.; Wagner, P. Visualization of agriculture data using self-organizing maps. In *Applications and Innovations in Intelligent Systems XVI. SGAI 2008*; Allen, T., Ellis, R., Petridis, M., Eds.; Springer: London, UK, 2008; pp. 47–60.

32. Ponmalai, R.; Kamath, C. *Self-Organizing Maps and Their Applications to Data Analysis*; (No. LLNL-TR-791165); Lawrence Livermore National Lab.: Livermore, CA, USA, 2019.

33. Chang, F.J.; Chang, L.C.; Kang, C.C.; Wang, Y.S.; Huang, A. Explore spatio-temporal PM2. 5 features in northern Taiwan using machine learning techniques. *Sci. Total Environ.* **2020**, *736*, 139656. [CrossRef] [PubMed]

34. Doan, Q.V.; Kusaka, H.; Sato, T.; Chen, F. A structural self-organizing map algorithm for weather typing. *Geosci. Model Dev.* **2021**, *14*, 2097–2111. [CrossRef]

35. Hidalgo, D.R.; Cortés, B.B.; Bravo, E.C. Dimensionality reduction of hyperspectral images of vegetation and crops based on self-organized maps. *Inf. Process. Agric.* **2021**, *8*, 310–327. [CrossRef]

36. Liu, Y.; Weisberg, R.H. A review of self-organizing map applications in meteorology and oceanography. In *Self-Organizing Maps: Applications and Novel Algorithm Design*; Mwasiagi, J.I., Ed.; IntechOpen: Rijeka, Croatia, 2011; Volume 1, pp. 253–272. Available online: https://www.intechopen.com/chapters/13302 (accessed on 23 February 2020). [CrossRef]

37. Taşdemir, K.; Wirnhardt, C. Neural network-based clustering for agriculture management. *EURASIP J. Adv. Signal Process* **2012**, *1*, 1–13. [CrossRef]

38. Satizábal, H.; Barreto-Sanz, M.; Jiménez, D.; Pérez-Uribe, A.; Cock, J. Enhancing decision-making processes of small farmers in tropical crops by means of machine learning models. In *Technologies and Innovations for Development*; Springer: Paris, France, 2012; pp. 265–277.

39. Mohan, P.; Patil, K.K. Deep learning based weighted SOM to forecast weather and crop prediction for agriculture application. *Int. J. Intell. Eng. Syst.* **2018**, *11*, 167–176. [CrossRef]

40. Rubem, A.P.S.; de Moura, A.L.; de Oliveira, E.; de Mello, J.S.; Alves, L.A.; Tavares, R.S. Productive performance of small peri-urban farms using self-organizing maps and data envelopment analysis. *WIT Trans. Ecol. Environ.* **2015**, *192*, 133–145.

41. Vergni, L.; Todisco, F. Spatio-temporal variability of precipitation, temperature and agricultural drought indices in Central Italy. *Agric. For. Meteorol.* **2011**, *151*, 301–313. [CrossRef]

42. Hewitson, B.C.; Crane, R. Self-organizing maps: Applications to synoptic climatology. *Clim. Res.* **2002**, *22*, 13–26. [CrossRef]

43. Alexander, L.V.; Uotila, P.; Nicholls, N.; Lynch, A. A new daily pressure dataset for Australia and its application to the assessment of changes in synoptic patterns during the last century. *J. Clim.* **2010**, *23*, 1111–1126. [CrossRef]

44. Loikith, P.C.; Lintner, B.R.; Sweeney, A. Characterizing large-scale meteorological patterns and associated temperature and precipitation extremes over the northwestern United States using self-organizing maps. *J. Clim.* **2017**, *30*, 2829–2847. [CrossRef]

45. Borah, N.; Sahai, A.; Chattopadhyay, R.; Joseph, S.; Abhilash, S.; Goswami, B. A self-organizing map–based ensemble forecast system for extended range prediction of active/break cycles of Indian summer monsoon. *J. Geophys. Res. Atmos.* **2013**, *118*, 9022–9034. [CrossRef]

46. Chang, L.C.; Shen, H.Y.; Chang, F.J. Regional flood inundation nowcast using hybrid SOM and dynamic neural networks. *J. Hydrol.* **2014**, *519*, 476–489. [CrossRef]

47. Chang, L.C.; Chang, F.J.; Yang, S.N.; Tsai, F.H.; Chang, T.H.; Herricks, E.E. Self-organizing maps of typhoon tracks allow for flood forecasts up to two days in advance. *Nat. Commun.* **2020**, *11*, 1983. [CrossRef]

48. Cavazos, T. Using self-organizing maps to investigate extreme climate events: An application to wintertime precipitation in the Balkans. *J. Clim.* **2000**, *13*, 1718–1732. [CrossRef]

49. Cassano, E.N.; Glisan, J.M.; Cassano, J.J.; Gutowski, W.J., Jr.; Seefeldt, M.W. Self-organizing map analysis of widespread temperature extremes in Alaska and Canada. *Clim. Res.* **2015**, *62*, 199–218. [CrossRef]

50. Gibson, P.B.; Perkins-Kirkpatrick, S.E.; Uotila, P.; Pepler, A.S.; Alexander, L.V. On the use of self-organizing maps for studying climate extremes. *J. Geophys. Res. Atmos.* **2017**, *122*, 3891–3903. [CrossRef]

51. Fassnacht, S.R.; Derry, J.E. Defining similar regions of snow in the Colorado River Basin using self-organizing maps. *Water Resour. Res.* **2010**, *46*, W04507. [CrossRef]

52. Hsu, K.C.; Li, S.T. Clustering spatial–temporal precipitation data using wavelet transform and self-organizing map neural network. *Adv. Water Resour.* **2010**, *33*, 190–200. [CrossRef]

53. Zhang, W.; Wang, J.; Jin, D.; Oreopoulos, L.; Zhang, Z. A deterministic self-organizing map approach and its application on satellite data based cloud type classification. In Proceedings of the 2018 IEEE International Conference on Big Data, Seattle, WA, USA, 10–13 December 2018; pp. 2027–2034.

54. Nguyen-Le, D.; Yamada, T.J. Using weather pattern recognition to classify and predict summertime heavy rainfall occurrence over the Upper Nan river basin, northwestern Thailand. *Weather Forecast* **2019**, *34*, 345–360. [CrossRef]

55. Ohba, M.; Sugimoto, S. Differences in climate change impacts between weather patterns: Possible effects on spatial heterogeneous changes in future extreme rainfall. *Clim. Dynam.* **2019**, *52*, 4177–4191. [CrossRef]

56. CWB Observation Data Inquire System. Available online: https://e-service.cwb.gov.tw/HistoryDataQuery/index.jsp (accessed on 1 August 2019).

57. World Maps of Köppen–Geiger Climate Classification. Available online: http://koeppen-geiger.vu-wien.ac.at (accessed on 12 July 2021).

58. Kohonen, T. A simple paradigm for the self-organized formation of structured feature maps. In *Competition and Cooperation in Neural Nets*; Springer: Berlin, Heidelberg, 1982; pp. 248–266.

59. Kohonen, T. *Self-Organizing Maps*, 3rd ed.; Springer: Berlin, Germany, 2001.

60. Sheridan, S.C.; Lee, C.C. The self-organizing map in synoptic climatological research. *Prog. Phys. Geogr.* **2011**, *35*, 109–119. [CrossRef]

61. Heikkinen, M.; Poutiainen, H.; Liukkonen, M.; Heikkinen, T.; Hiltunen, Y. Subtraction analysis based on self-organizing maps for an industrial wastewater treatment process. *Math. Comput. Simul.* **2011**, *82*, 450–459. [CrossRef]

62. Lakshminarayanan, S. Application of self-organizing maps on time series data for identifying interpretable driving manoeuvres. *Eur. Transp. Res. Rev.* **2020**, *12*, 25. [CrossRef]

63. Skific, N.; Francis, J. Self-Organizing Maps: A Powerful Tool for the Atmospheric Sciences. In *Applications of Self-Organizing Maps*; Johnsson, M., Ed.; IntechOpen: Rijeka, Croatia, 2012. Available online: https://www.intechopen.com/chapters/40865#B7 (accessed on 23 February 2020). [CrossRef]

64. Welbaum, G.E. *Vegetable Production and Practices*, 1st ed.; CABI: Boston, MA, USA, 2015.

65. Huang, A.; Chang, F.J. Prospects for Rooftop Farming System Dynamics: An Action to Stimulate Water-Energy-Food Nexus Synergies toward Green Cities of Tomorrow. *Sustainability* **2021**, *13*, 9042. [CrossRef]

Article

Stochastic Modeling for Estimating Real-Time Inundation Depths at Roadside IoT Sensors Using the ANN-Derived Model

Shiang-Jen Wu [1,*], Chih-Tsu Hsu [2] and Che-Hao Chang [3]

1   Department of Civil and Disaster Prevention Engineering, National United University, Miaoli 36063, Taiwan
2   National Center for High-Performance Computing, Hsinchu 30076, Taiwan; 1003135@narlabs.org.tw
3   Department of Civil Engineering, National Taipei University of Technology, Taipei 10608, Taiwan;
     chchang@ntut.edu.tw
*   Correspondence: sjwu@nuu.edu.tw

Abstract: This paper aims to develop a stochastic model (SM_EID_IOT) for estimating the inundation depths and associated 95% confidence intervals at the specific locations of the roadside water-level gauges, i.e., Internet of Things (IoT) sensors under the observed water levels/rainfalls and the precipitation forecasts given. The proposed SM_EID_IOT model is an ANN-derived one, a modified artificial neural network model (i.e., the ANN_GA-SA_MTF) in which the associated ANN weights are calibrated via a modified genetic algorithm with a variety of transfer functions considered. To enhance the reliability and accuracy of the proposed SM_EID_IOT model in the estimations of the inundation depths at the IoT sensors, a great number of the rainfall induced flood events as the training and validation datasets are simulated by the 2D hydraulic dynamic (SOBEK) model with the simulated rain fields via the stochastic generation model for the short-term gridded rainstorms. According to the results of model demonstration, Nankon catchment, located in northern Taiwan, the proposed SM_EID_IOT model can estimate the inundation depths at the various lead times with high reliability in capturing the validation datasets. Moreover, through the integrated real-time error correction method integrated with the proposed SM_EID_IOT model, the resulting corrected inundation-depth estimates exhibit a good agreement with the validated ones in time under an acceptable bias.

Keywords: ANN; roadside IoT sensors; simulations of the gridded rainstorms; 2D inundation simulation and real-time error correction

Citation: Wu, S.-J.; Hsu, C.-T.; Chang, C.-H. Stochastic Modeling for Estimating Real-Time Inundation Depths at Roadside IoT Sensors Using the ANN-Derived Model. *Water* **2021**, *13*, 3128. https://doi.org/10.3390/w13213128

Academic Editors: Fi-John Chang, Li-Chiu Chang and Jui-Fa Chen

Received: 22 September 2021
Accepted: 25 October 2021
Published: 5 November 2021

Publisher's Note: MDPI stays neutral with regard to jurisdictional claims in published maps and institutional affiliations.

1. Introduction

Owing to climate change and the occurrence of extreme rainstorm events, rainfall-induced flood frequently takes place, causing severe damage to people's lives and properties. Hence, flood early warning operation plays an important role in the prevention and mitigation of flood-induced hazards. Recently, with the establishment of the dike system, flooding is triggered merely as a result of overtopping from the embankments; in contrast, inundation frequently occurs in the urban and drainage zone owing to the failure of draining the runoff through the sewer systems [1]. In the past, the flood early warning operation was executed based on specific thresholds (e.g., rainfall or inundation depth) in accordance with real-time measurements; however, the real-time practical inundation depths, especially in urban areas, are hardly measured owing to the limitation of measurement equipment or hindrance in data acquisition, processing, and analysis [1–3].

To achieve the goal of immediately capturing and transferring the temporal changes in the inundation depths on the roads, the IoT is commonly utilized to set up the roadside sensors in order to measure the flooding/inundation depths, especially on the roads where the water levels result from the rainstorms of which the corresponding strength is perhaps greater than the draining capability with respect to the sewers. Moreover, to achieve the goals of flood early warning and flood-induced hazard mitigation, receiving and estimating the inundation information is an essential task. Among the flooding information, the potential inundation region and associated area are supposed to be known in advance. In spite of the difficulty in obtaining real-time measured observations (e.g., water level), they could be established through the hydraulic numerical models under consideration of the design rainfall events regarding the various return periods [4–6]. For example, Chen et al. [4] established a potential inundation-map database by means of the hydraulic numerical model (HEC_RAS) with the design rainfall events of the various return periods. In referring to the above inundation-map database, the possible flooding area under conditions of rainfall characteristics could be quantified for the flood warning systems and emergency. In addition to the hydraulic/hydrological numerical modeling with the given precipitations, another commonly used data-derived method is to roughly and rapidly perform the flooding mapping in accordance with the at-site observations [7], such as the water-level gauges [8] or the observed inundation depths recorded at the roadside IoT sensors [9,10]. Furthermore, the observations related to the water levels/inundation depths can be generally incorporated with a GIS model with the digital elevation map (DEM) to estimate the area of the floodplain [11,12]. For illustration, Shastry and Durand [12] proposed the two-step algorithm for effectively regulating the more accurate floodplain topography by combining the results from the flood model associated with the DEM and inundation-related observations. In conclusion, the at-site inundation-depth estimates/forecasts should be advantageous to flooding prevention and mitigation.

Generally speaking, the well-known flood simulation models applied in the inundation simulation can be classified into two types: deterministic models (i.e., physical-based models) and statistical-related models (i.e., data-driven models) [13,14]. Deterministic-based flooding simulation models have been proposed to forecast the water levels/inundation depths within the specific zone under the given precipitation of high resolution in time and space, leading to a possible problem where a long computation time might affect the effectiveness and performance [15–18] attributed to the uncertainties in the complicated model structure and insufficient direct measurements regarding physical signification [18–20]. Recently, artificial intelligent (AI) modeling has been comprehensively applied in the prediction of the flood-related hydrological variates (e.g., precipitation, discharge, and water level) [18,21–26]. Of the relevant AI models, the ANN-based model can be more efficiently applied in modeling difficult and complicated phenomena described in terms of nonlinear mathematic relationships by constructing the linear multi-layer network using all possible predictor variables through the multiple training algorithm, especially for hydrological forecasts, such as the precipitation, discharge, and water level [9,10,27,28]. For example, Campolo et al. [10] utilized the logistic function as the transfer function, namely, the activation function, to train an ANN model that describes the spatial relationship between rainfall and water levels to issue forecasting information on the distributed water levels. Shamseldin [27] proposed an ANN-derived rainfall-runoff model based on the structure of the multi-layer perceptron with a specific transfer function (i.e., the logistic/sigmoid function) to provide the river-runoff forecasts using the weighted average of rainfall and expectation of the rainfall index as well as the observed discharge as model inputs. Furthermore, Tamiru and Dinka [28] combine the results from the ANN model and the hydraulic numerical model (HEC-RAS) to carry out the flood-triggered inundation simulation; in detail, the inundation simulation is implemented by the HEC-RAS model under the boundary condition of runoff hydrographs at the up-stream and lateral branches estimated by the ANN model. However, the performance of the resulting forecasts from the ANN models is possibly impacted by uncertainties in the network structure, as well as selection of the transfer functions and associated parameters (i.e., connection weights between different layers, ANN weights). Thus, Wu et al. [18] presented a modified ANN (called ANN_GA-SA_MTF) model by adopting a variety of transfer functions in which the ANN weights are calibrated using the genetic algorithm based on the parameter sensitivity (GA-SA) [29]. Particularly, within the ANN_GA-SA_MTF model, a real-time error correction model for the water-level forecasts derived using the time series and Kalman filtering approach RTEC_TS&KF [30] are combined in order to boost the accuracy of the estimates.

Therefore, this study intends to develop a stochastic model for estimating the inundation depths at the roadside water-level IoT sensors by training an ANN-derived models, named the SM_EID_IOT model. With training the proposed SM_EID_IOT model, to enhance the reliability of its results, a great number of rainfall-induced inundation simulations are adopted as the training dataset; in particular, the relevant concepts regarding the real-time error correction technique on the basis of the different estimations and observations at the previous times during the rainfall-induced flood event are applied in the development of the proposed SM_EID_IOT model in order to obtain more accurate model outputs. It is expected that the proposed SM_EID_IOT can not only provide the inundation depths at the roadside IoT sensors with high accuracy, but also quantify their corresponding reliability, which is advantageous to the decision-making regarding early flood warning operation and infrastructure-planning of a water-proofing system as a reference.

2. Methodology

2.1. Model Concept

As mentioned in Section 1, an ANN-based inundation-depth estimation model at the IoT sensors of interest, called the SM_EID_IOT model, is developed herein; the framework of the model development can be generally classified into the three parts: (1) generation of the gridded rainstorm events in the study area; (2) 2D inundation simulations by means of the well-known hydraulic dynamic numerical modeling; and (3) establishment of an ANN-derived model for estimating inundation depths at the IoT sites.

At first, to facilitate the accuracy and reliability of the results from the proposed SM_FIDEP_IOT, a great number of the regional rainfall events are simulated via the stochastic modeling for generating the gridded short-term rainstorms (i.e., SM_GSTR model) [31]. Afterwards, they are used in the two-dimensional (2D) inundation simulation by the hydraulic dynamic numerical model (i.e., SOBEK) [32] to reproduce the big data involving the rainfall-induced inundation simulations, including the gridded inundations and corresponding floodplain area treated as the training datasets. Within the development of the proposed SM_EID_IOT model, this study adopts the ANN-based model, ANN_GA-SA_MTF, proposed by Wu et al. [18] for describing the relationship between the at-site inundation depths and the related rainfall and water levels. The associated connection weights of the neurons at various layers are calibrated through the genetic algorithm based on the sensitivities of model parameters (named the SA-GA method) [29] under consideration of the multiple transfer functions.

Unlike the well-known ANN-based models, in order to reduce the effect of uncertainties in the observations and model parameters, the resulting inundation-depth estimates from the proposed SM_EID_IOT model need to be immediately corrected in accordance with the difference between the observed inundation depths and forecasted ones at the forward time steps through the real-time error correction method, RTEC_TS&KF [30]. The aforementioned relevant methods and concepts are addressed below.

2.2. Generation of Gridded Rainstorm Events

It is well-known that a large training dataset is desired for training the ANN model. Therefore, in this study, the stochastic modeling of gridded short-term rainstorms developed by Wu et al. (2021) (named the SM_GSTR model) is employed to simulate a great number of rainstorms at all grids within the study area. Within the SM_GSTR model, the event-based rainstorm is basically grouped into three rainfall characteristics, including the event-based rainfall duration, gridded rainfall depths, and gridded storm depths composed of the dimensionless rainfalls at the various dimensionless times; with respect to the gridded storm pattern, it can be grouped into two components, the areal average of the dimensionless rainfalls (i.e., the storm pattern) and the associated deviations at the various dimensionless times. Of these, the gridded rainfall characteristics, gridded rainfall depths, and deviations regarding the areal average of the storm patterns are treated as the spatially correlated variates and the areal average storm patterns are regarded as the temporally correlated variates [31]. Figure 1 shows the process of characterizing the gridded rainstorms into five features (i.e., the gridded rainfall characteristics).

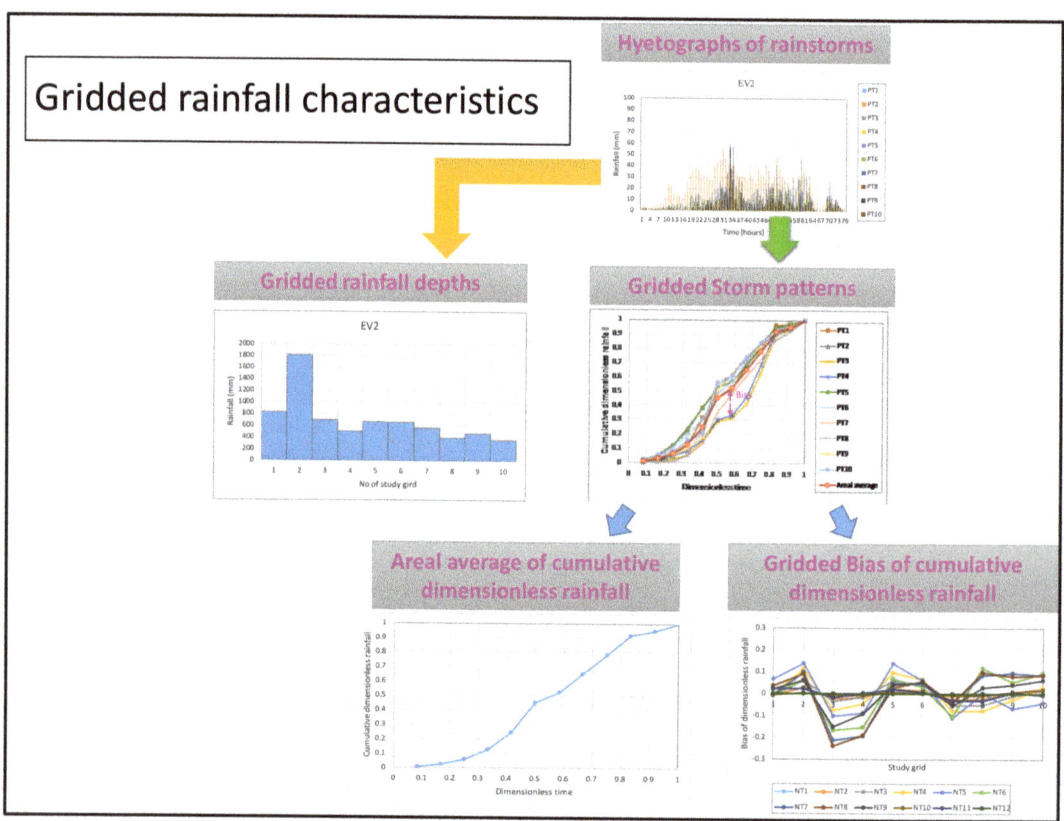

Figure 1. Graphical process of extracting the gridded rainfall characteristics from observed hyetographs of rainstorm events [31]).

Upon obtaining the gridded rainfall characteristic, within the SM_GSTR model, the non-normal correlated multivariate Monte Carlo simulation approach (Chang et al., 1996) based on the correlation structures of gridded rainfall characteristics in time and space is adopted to generate a desired number of event-based rainfall events through the transform algorithms. The transform algorithms could be employed via the Nataf bivariate distribution model [32], including the transformation to standard normal space, orthogonal transform, and inverse transformation, based on the following correlation relationship:

$$\rho_{ij} = \int_{-\infty}^{\infty} \int_{-\infty}^{\infty} \left[\frac{x_i - \mu_i}{\sigma_i} \right] \left[\frac{x_j - \mu_j}{\sigma_j} \right] \varnothing_{ij}(\langle z_i, z_j | \rho_{ij}^* \rangle) \mathrm{d}z_i \mathrm{d}z_j \tag{1}$$

$$z_i = \frac{x_i - \mu_i}{\sigma_i}; \ z_j = \frac{x_j - \mu_j}{\sigma_j} \tag{2}$$

where X_i and X_j are the correlated variables at the points i and j, respectively, with the means μ_i and μ_j, the standard deviations σ_i and σ_j, and correlation coefficient ρ_{ij}, respectively; i and Z_j are bivariate standard normal variables corresponding to the variables X_i and X_j, with the correlation coefficient ρ_{ij}^* and the joint standard normal density function $\varnothing_{ij}(\cdot)$. To generate a number of variables with high correlation ρ_{ij}, a set of semi-empirical formulae [33] was derived to modify ρ_{ij} in the original space to ρ_{ij}^* in the normal space through a transformation factor T_{ij}, depending on the marginal distributions and correlation of X_i and X_j, as follows:

$$\rho_{ij}^* = T_{ij} \times \rho_{ij} \tag{3}$$

Eventually, the simulated gridded rainstorms can be achieved in accordance with the process of combining the generated gridded rainfall characteristics, as shown Figure 2 [31].

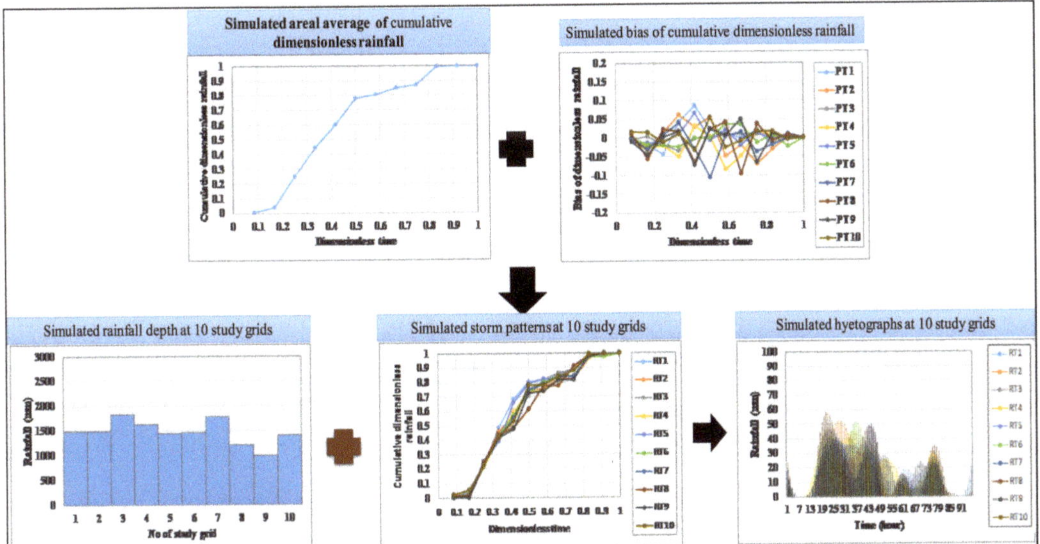

Figure 2. Graphical process of combining the simulated gridded rainfall characteristics as the gridded rainstorms [31].

2.3. 2D Inundation Simulation Modeling

Using the estimated runoff hydrograph from the observed and predicted precipitation and tide levels, the water level hydrographs at various cross-sections along the river and at the computation grids within the region can be calculated through the inundation simulation models developed using the depth-averaged Navier–Stokes equation (NSE), named the Saint-Venant shallow water equations. Several numerical models for simulating 2D inundation have been developed based on the NSE, such as SOBEK 1D-2D [32], MIKE 11-21 [34,35], TrimR2D [36], and TELEAC-2D [37] In general, the above inundation-simulation models can be classified into numerical, statistical, and flood inundation mapping models. The hydraulic numerical SOBEK model is a sophisticated one-dimensional open-channel dynamic flow and two-dimensional overland flow modeling system (named SOBEK 1D-2D hydrodynamic model); it can be used to simulate and tackle problems in river management, flood protection, design of canals, irrigation systems, water quality, navigation, and dredging. Therefore, this study uses the SOBEK model to carry out the inundation simulation with a large number of generated grid-based rainstorms.

2.4. Artificial Neural Network Model Associated with Multiply Transfer Functions

It is well-known that the related artificial neural network (ANN) models are frequently adopted in the forecast/estimation regarding flood-rated variates. In spite of the prediction of the hydrological variates being effectively carried out by the ANN-based models, their reliability and accuracy should be influenced by the uncertainties in the transfer function (i.e., activation functions) and selected and associated neuron weights between two layers (i.e., ANN weights) attributed to the variation in the observations [18,38,39]. Moreover, although the back-propagation (BP) algorithm with the gradient descent method is commonly utilized in training the ANN model, the formula of adjusting the connection weights regarding the neurons is difficult to derive under constraint with the transfer functions (or activation function) used. By doing so, the training performance fails to achieve the local optimal values with high likelihood, contributing to the given inappropriate initial values and leaning rate, which leads to the problem with oscillation, thereby reducing the convergence speed [18,22].

Furthermore, the numbers of neurons at hidden layers are significantly increased owing to the performance of training the ANN model. Thus, if few neurons are considered, the corresponding network structure does not easily emulate the underlying function attributed to the insufficient parameters; in contrast, as a result of a great number of the neurons adopted in the network structure, the overfitting problem might occur [40]. Therefore, several methods are proposed to estimate the number of neurons included in Table 1.

Table 1. Formulae for estimating the number of hidden neurons (Wu et al., 2021) [18].

No of Formulae	Formula	Reference
1	$N_{HN} = \left(\sqrt{1+8 \times N_{IP}} - 1\right)/2$	Li et al. method [41]
2	$N_{HN} = N_{IP} - 1$	Tamura and Tateishi method [28]
3	$N_{HN} = \frac{2 \times N_{IP}}{N_{IP}} + 1$	Zhang et al. method [42]
4	$N_{HN} = \sqrt{N_{IP} \times N_{OP}}$	Shibata and Ikeda method [43])
5	$N_{HN} = 2^{N_{IP}} - 1$	Hunter et al. method [44]
6	$N_{HN} = \frac{\left[4 \times (N_{IP})^2 + 3\right]}{\left[(N_{IP})^2 - 8\right]}$	Sheela and Deepa method [45]

Therefore, the network structure and the types of the transfer functions are supposed to be regarded as the uncertainty factors for training the ANN models; to figure out the problem with the above uncertainties in the training of the ANN model, Wu et al. [18] proposed an ANN-derived model (named the ANN_GA-SA_MTF) by adopting the network structure of three layers with multiple transfer functions (see Table 2) in which the associated ANN weights are calibrated by means of the genetic algorithm based on the sensitivities to model parameters (called the GA-SA algorithm) [29].

Table 2. Transform functions commonly used (Wu et al., 2021, Maca et al., 2014) [18,38].

	Transfer function	Formula	Derivative
TF1	Logistic (soft step, Sigmoid)	$f(x) = \frac{1}{1+e^{-\alpha x}}$	$f'(x) = f(x)(1 - f(x))$
TF2	Tanh	$f(x) = \tan h(x) = \frac{2}{1+e^{-2\alpha x}} - 1$	$f'(x) = 1 - f(x)^2$
TF3	Arctan	$f(x) = \tan^{-1}(\alpha x)$	$f'(x) = \frac{1}{(\alpha x)^2 + 1}$
TF4	Identity	$f(x) = \alpha x$	$f'(x) = \alpha$
TF5	Rectified linear unit (ReLU)	$f(x) = \begin{cases} 0 \; for \; x < 0 \\ 1 \; for \; x \geq 0 \end{cases}$	$f'(x) = \begin{cases} 0 \; for \; x < 0 \\ 1 \; for \; x \geq 0 \end{cases}$
TF6	Parametric rectified linear unit (PReLU, leaky ReLU)	$f(x) = \begin{cases} \alpha \, x \; for \; x < 0 \\ x \; for \; x \geq 0 \end{cases}$	$f'(x) = \begin{cases} \alpha \; for \; x < 0 \\ 1 \; for \; x \geq 0 \end{cases}$
TF7	Exponential linear unit (ELU)	$f(x) = \begin{cases} \alpha \, (e^x - 1) \, for \; x < 0 \\ x \; for \; x \geq 0 \end{cases}$	$f'(x) = \begin{cases} f(x) + \alpha \; for \; x < 0 \\ 1 \; for \; x \geq 0 \end{cases}$
TF8	Inverse abs (IA)	$y(x) = \frac{x}{1+|\alpha x|}$	$y'(x) = \frac{1}{(1+|a\alpha x|)^2}$
TF9	Rootsig (RS)	$y(x) = \frac{\alpha x}{1+\sqrt{1+(\alpha x)^2}}$	$y'(x) = \frac{1}{\left(1+\sqrt{1+(\alpha x)^2}\right)\sqrt{1+a(\alpha x)^2}}$
TF10	Sech function (SF)	$y(x) = \frac{2}{\exp(\alpha x) + \exp(-\alpha x)}$	$y'(x) = -y(x)\tan h(\alpha x)$

To quantify the reliability of the model outputs, the proposed ANN_GA-SA_MTF model is collaborated with a nonparametric method, named the weighted likelihood sample quantile estimator method proposed by Yang and Tung [46], to compute the quantiles of resulting model outputs via the following equation:

$$X_{p,WL} = \sum_{r \in \omega} W_{r,n,WL} X_{(r)} \tag{4}$$

in which $W_{r,n,WL} = \frac{F_{r,n}(p)}{\sum_{s \in \omega} F_{s:n}(p)}$, with ω being the band width that contains a set of order statistics $(X_{(1)} \leq X_{(2)} \leq \ldots \leq X_{(n)})$ that are deemed significant in contributing to the estimation of X_p as a result of the band width being no smaller than 2.0.

In addition to quantifying the reliability of the model estimates using the results from the ANN_GA-SA_MTF model, the weighted averages of model estimates are issued as forecasts using the following equation:

$$\hat{Y}_{WA} = \sum_{i=1}^{N_{TF}} \left[W_{TF}^i \times \hat{Y}\left(\theta_{TF_i}^j\right) \right] \tag{5}$$

$$W_{TF}^i = \frac{\dfrac{1}{E\left(\theta_{TF}^i\right)}}{\sum_{i=1}^{N_{TF}} \dfrac{1}{E\left(\theta_{TF}^i\right)}} \tag{6}$$

in which N_{TF} is the number of transfer functions considered; Y_k and $\hat{Y}_k(\theta_{TF}^i)$ denote the observed hydrological data and estimated ones by the ANN_GA-SA_MTF model with the jth set of the appropriate parameters θ_{TF}^i, respectively; and W_{TF}^i represents the weighted factor of the ith transfer function with the appropriate parameters θ_{TF}^i calculated, with the $E\left(\theta_{TF}^i\right)$ being the objective-function value (i.e., the root mean square error, RMSE).

In particular, to provide more reliable and accurate model outputs, the real-time error correction method established using the time-series approaches and Kalman filtering [30] is adopted within the ANN_GA-SA_MTF to immediately adjust the forecasts ($Y_{corr}^{t_{pred}}$) based on real-time observations through the Internet of Things (IoT) using the ANN_GA-SA_MTF model by means of the following equation:

$$Y_{corr}^{t_{pred}} = Y_{pred}^{t_{pred}} + \varepsilon_{TS}^{t_{pred}} + \varepsilon_{KF}^{t_{pred}} \tag{7}$$

where $Y_{pred}^{t_{pred}}$ stands for the model estimates (i.e., the forecasts); and $\varepsilon_{TS}^{t_{pred}}$ and $\varepsilon_{KF}^{t_{pred}}$ serve as the forecast error estimated by the time series approaches and Kalman filtering method, respectively.

In summary, the framework of developing the ANN_GA-SA_MTF model is generally classified into the four steps (see Figure 3): the parameter calibration using the GA-SA approach, the reliability quantification of model outputs, the estimation of model outputs, and the real-time correction of model outputs; the associated concepts are briefly introduced as follows.

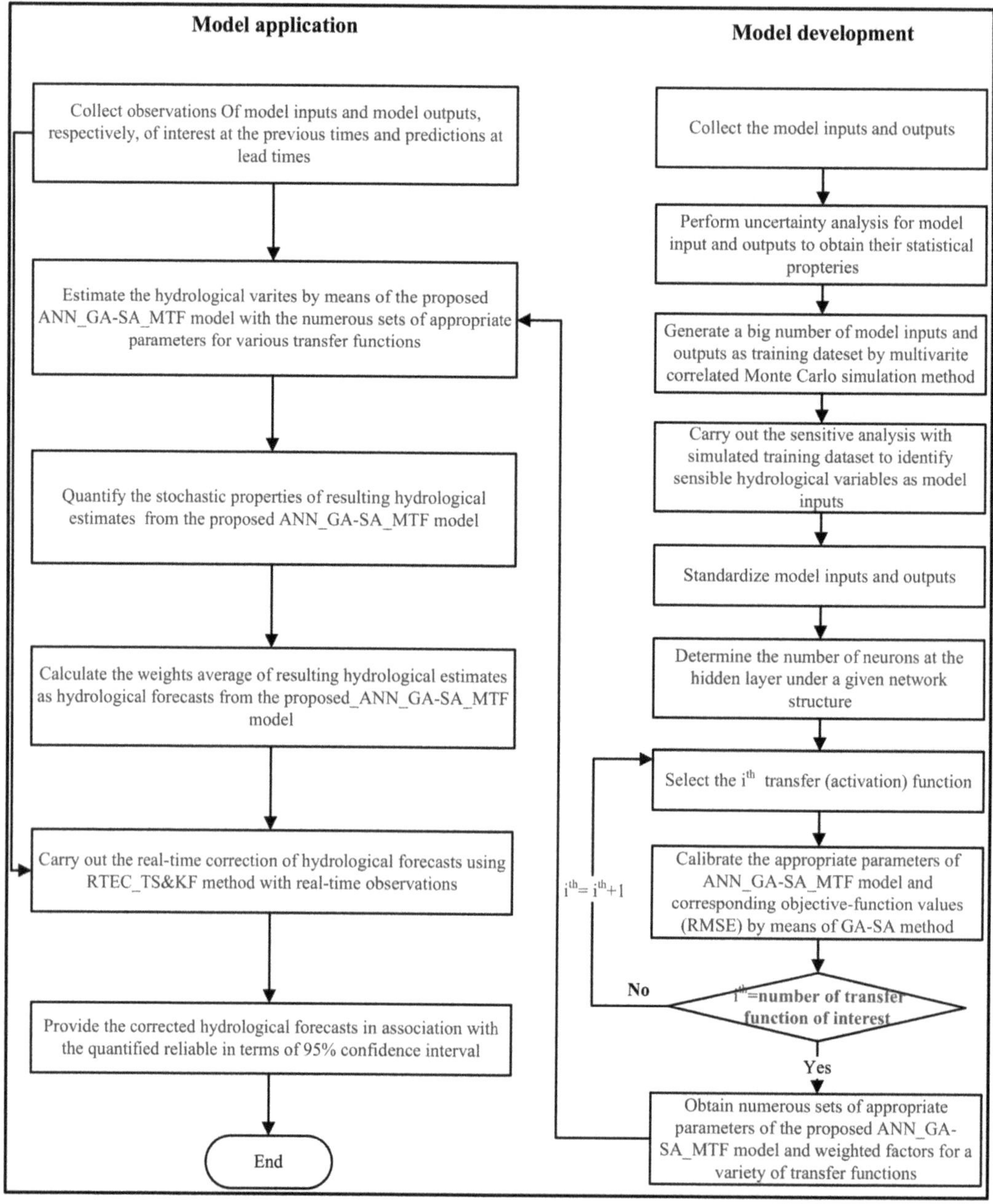

Figure 3. Graphic framework of developing and applying the ANN_GA-SA_MTF model [18].

2.5. Model Formulation

To sum up the aforementioned concepts, this study intends to utilize the ANN_GA-SA_MTF model to develop a smart model for forecasting the inundation depth at the roadside IoT sensors, named the SM_EID_IOT model. As a result of the inundation being significantly increased by the rainstorm, the inundation depths at the specific locations, where the IoT sensors are set up, should be temporally and spatially related to the rainfalls and inundation depths at the previous time steps during an event (Notaro et al., 2013; Lyu et al., 2018). Although the resolution of the rainstorm in space obviously impacts the estimation of the flood/inundation, the areal average rainfalls calculated from a number of the raingauges within the small basin are frequently applied in the hydrological/hydraulic analysis under simplification of the rainfall-runoff simulation [47]. Therefore, in this study, in addition to the uncertainties in the resolutions of the rainfall and inundations in time (i.e., the forward time step from the current time), the distances to the IoT sensor for calculating the areal-average rainfall with the rainfalls at the grids (i.e., gridded rainfall) is treated as the uncertainty factor calculated through the following equation:

$$\overline{R}_{IOT}^t = \frac{1}{N_g} \sum_{i=1}^{N_g} R_i^t \tag{8}$$

where $\overline{R}_{IOT}^t$ accounts for the areal-average rainfall at the IoT sensor; N_g is the number of the grids, the distance of which to the IoT senor is equal to or less than the specific critical distance (i.e., critical spatial resolution); and L_c and R_i^t serve as the gridded rainfalls for the time step t-hour at the ith grid.

Therefore, on the basis of the ANN_GA-SA_MTF model with generated rainfall-inundation depths and associated gridded rainstorms, this study establishes the relationship of the inundation depths at the IoT sensors with the inundation depths and rainfall at the lead times, as well as the previous time steps, and the inundation depths at the forward time step can be written as follows:

$$\hat{h}_{IOT}^{t+1} = f_{\text{ANN}_{GA}-\text{SA\_MTF}}\left(\overline{R}_{IOT}^{t+1}, \overline{R}_{IOT}^t \cdots \overline{R}_{IOT}^{t-T_c-1}, h_{IOT}^t, \cdots h_{IOT}^{t-T_c-1}\right) \tag{9}$$

in which T_c serves as the critical values of the resolutions in time (i.e., critical temporal resolution); $\hat{h}_{IOT}^{t+1}$ is the inundation-depth estimate for the lead time $(t + 1\ \text{h})$; $\overline{R}_{IOT}^{t+1}$ denotes the rainfall forecast at the lead time $(t + 1\ \text{h})$; $\overline{R}_{IOT}^t, \cdots \overline{R}_{IOT}^{t-T_c-1}$ account for the areal average rainfall at the current time $(t\ \text{hour})$ and those from the forward T_R hours calculated from the gridded rainfalls within the specific critical spatial resolution, i.e., the distance L_c to the IoT sensor; and $h_{IOT}^t, \cdots h_{IOT}^{t-T_c-1}$ represent the observed inundation depths from the t hour to the t-T_h hours under consideration of the critical temporal resolution (i.e., the forward time steps T_h.). Note that the critical values of the resolution in time and space T_R and T_h can be determined by evaluating the spatially and temporally varying trend of the at-site inundation depth with the areal average rainfall via the correlation and sensitivity analysis in this study.

2.6. Model Framework

According to the aforementioned concepts, the development and application of the proposed SM_FIDEP_IOT model can be grouped into six parts: (1) generation of the rainstorm events at all grids within the study area; (2) 2D rainfall-induced inundation simulation; (3) extraction of the at-site inundation depths and corresponding rainfall at neighboring grids; (4) identification of critical resolution in time and space; (5) development of the proposed SM_EID_IOT model on the basis of the ANN_GA-SA_MTF model; and (6) integration with the real-time error correction (RTEC_TS&KF) method to adjust the inundation-depth estimates. The detained framework of the model development and application are addressed as follows.

2.6.1. Model Development

Step 1   Collect the historical rainstorm events at all grids within the study area and extract their gridded characteristics, i.e., rainfall duration, gridded rainfall depth, areal average of cumulative dimensionless rainfall, and the associated bias.

Step 2   Reproduce a great number of rainfall fields with high spatiotemporal resolutions comprised of the simulated gridded rainfall characteristics by the SM_GSTR model with the results from uncertainty analysis using the observations obtained in Step [1].

Step 3   Carry out the 2D inundation simulation by means of the SOBEK model with a great number of gridded rainstorms simulated in Step [2] to obtain the simulations of inundations depths at the IoT sensors.

Step 4   Extract the simulated inundation depths at the IoT sensor under consideration of different periods (i.e., critical temporal resolution) and the corresponding simulated rainfalls at the neighboring girds within the specific distances to the IoT sensor (i.e., critical spatial resolution).

Step 5   Perform correlation and sensitivity analysis to determine the appropriate critical resolutions in time and space.

Step 6   Calculate the areal average rainfalls from the simulations of the gridded rainfall under the consideration of critical temporal and spatial resolutions determined in Step [5].

Step 7   Calibrate the parameters of the ANN_GA-SA_MTF model used in the estimation of inundation depths via the proposed SM_EDI_IOT model using the simulations of the inundation depths at the IoT sensor and corresponding areal average rainfalls.

2.6.2. Model Application

Step 1   Collect the observed inundation depths during the rainfall-induced flood event at the IoT sensor and calculate the corresponding areal average rainfalls under the conditions of the appropriate temporal and spatial resolutions.

Step 2   Obtain the resulting inundation-depth estimate at the lead times and associated 95% confidence intervals from the proposed SM_EID_IOI model.

Step 3   Carry out the real-time correction regarding the resulting inundation-depth estimates at the lead times from the proposed SM_EID_IOT model with the RTEC_TS&KF method in accordance with the bias of the inundation-depth estimates in comparison with the observations at the forward time steps during the event.

3. Study Area and Data

The Nankan River—whose length and drainage area and slope are 31 km and 224 km^2, respectively (see Figure 4)—in Taoyuan County is one of the most polluted rivers in northern Taiwan; further, its average slope and mountain area are about 0.0077 and over 900 m, respectively. Additionally, it flows through Guishan, Taoyuan, and Luzhou Districts in Taoyuan City, including six riverside parks and three branches, Dongmon Creek, KengZi Creek, and Kengzi Creek. Of the aforementioned branches, Dongmon Creek is frequently inundated as a result of a reduction in the number of detention ponds and the cross-section area in the river channel. Note that, within the Nankan River watershed, Taiwan Central Weather Bureau (CWB) provides the quantitative precipitation estimation (QPE) with a spatial resolution of 1.5 km × 1.5 km, i.e., the rainfall data of 336 grids (called QPE grids), as shown in Figure 4.

As the purpose of this study is to develop a stochastic ANN-derived model using the training datasets comprising a great number of the rainfall-induced 2D inundation simulations with high resolution in time and space, the hourly rainfall of 20 rainstorm events at 336 grids within the study area (Nankan River watershed) (see Table 3) is adopted as the study data.

Figure 5 shows the hyetographs of the 20 selected radar-based rainstorm events (2005–2017) provided by Taiwan Central Weather Bureau. According to the process of extracting the gridded rainfall characteristics, the gridded rainfall depths and storm patterns of the concerned 20 rainstorm events in the study area can be obtained. Therefore, upon establishing the proposed SM_EID_IOT model, the uncertainties in the gridded rainfall characteristics should be taken into account and quantified for the simulations of the rainstorm events at all grids within the study area. Accordingly, big data regarding the 2D rainfall-induced flood events can be generated at the training and validation datasets.

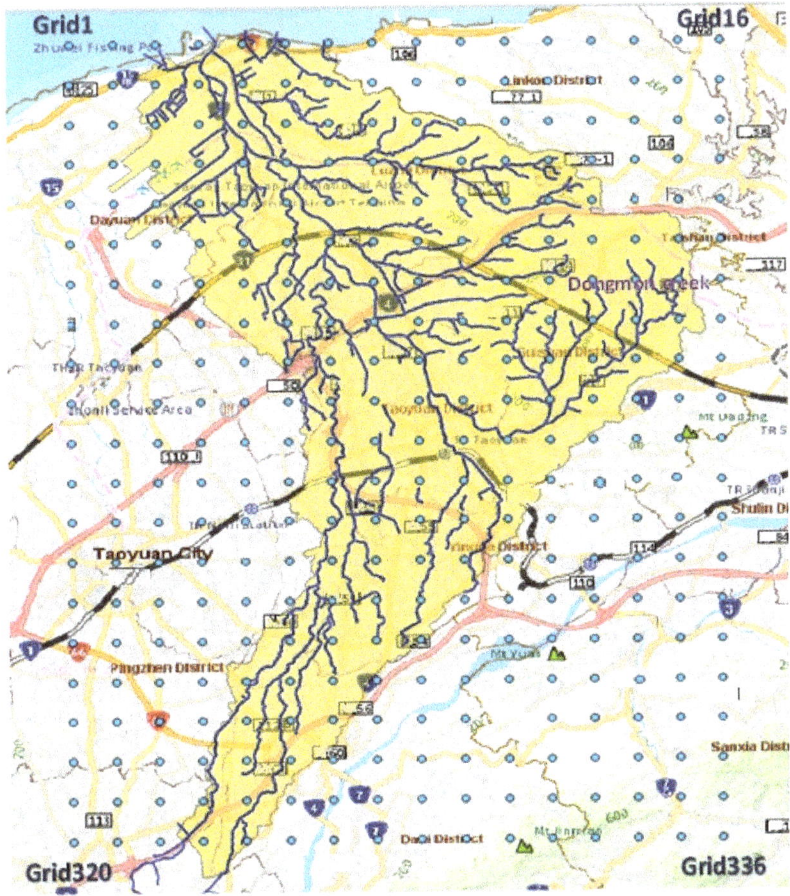

Figure 4. Locations and DEM as well as QPE grids (blue circle) within the study area, Nankan River watershed (yellow region) [18].

Table 3. Summary for hydraulic facilities, hydrological analysis, and topographic features used in the SOBEK model for the Nankan River watershed.

Hydraulic Facility	Number	Hydrologic Analysis	Model
1. Sub-basins	579	Rainfall-runoff modeling	SCS UH
2. Cross-sections	3219	**Topographic feature**	**Measurement date**
3. Gates	18	1. Digital elevation map	2012
4. Bridges	90	2. Map for land-use	2014
5. Sewer	1386	**Hydraulic factors Kn**	**Magnitude**
6. Manholes for sewer system	1382	Roughness coefficient	0.2–0.45

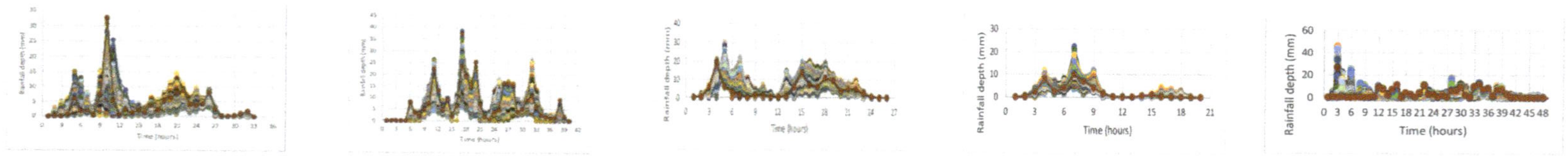

EV1 (Rainfall depth = 140 mm) 2005/7/17 10:00–2005/7/18 18:00
EV2 (Rainfall depth = 231 mm) 2005/8/4 09:00–2005/8/6 00:00
EV3 (Rainfall depth = 126 mm) 2005/8/31 07:00–2005/9/1 08:00
EV4 (Rainfall depth = 55 mm) 2005/10/1 23:00–2005/10/2 18:00
EV5 (Rainfall depth = 106 mm) 2008/7/26 22:00–2008/7/28 22:00
EV6 (Rainfall depth = 403 mm) 2008/9/12 02:00–2008/9/15 00:00 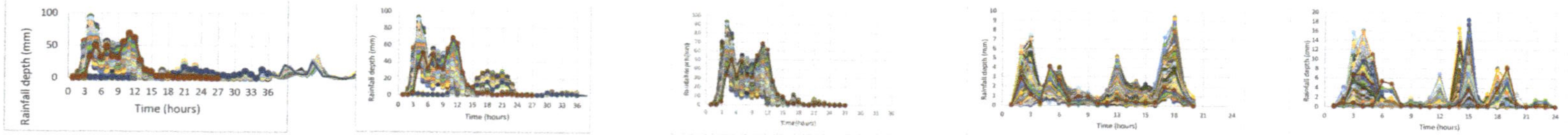
EV7 (Rainfall depth = 121 mm) 2010/10/21 00:00–2010/10/22 13:00
EV8 (Rainfall depth = 367 mm) 2012/6/11 21:00–2012/6/12 23:00
EV9 (Rainfall depth = 20 mm) 2012/6/14 11:00–2012/6/15 06:00
EV10 (Rainfall depth = 30 mm) 2012/8/26 09:00–2012/8/27 08:00
EV11 (Rainfall depth = 185 mm) 2013/5/11 01:00–2013/5/13 01:00 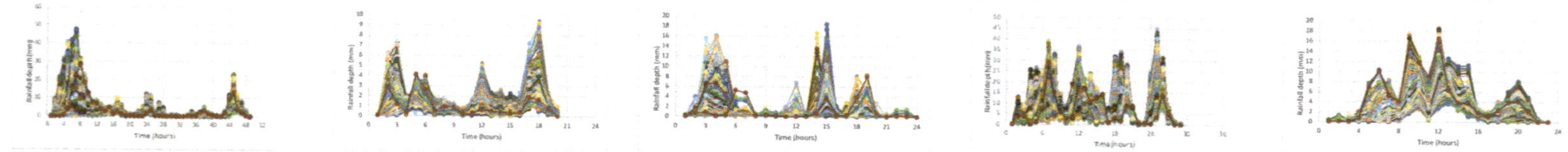
EV12 (Rainfall depth = 75 mm) 2013/7/12 16:00–2013/7/13 14:00
EV13 (Rainfall depth = 56 mm) 2013/10/4 14:00–2013/10/6 23:00
EV14 (Rainfall depth = 196 mm) 2014/5/20 20:00–2014/5/22 00:00
EV15 (Rainfall depth = 43 mm) 2014/7/22 21:00–2014/7/24 03:00
EV16 (Rainfall depth = 84 mm) 2014/9/21 16:00–2014/9/22 12:00 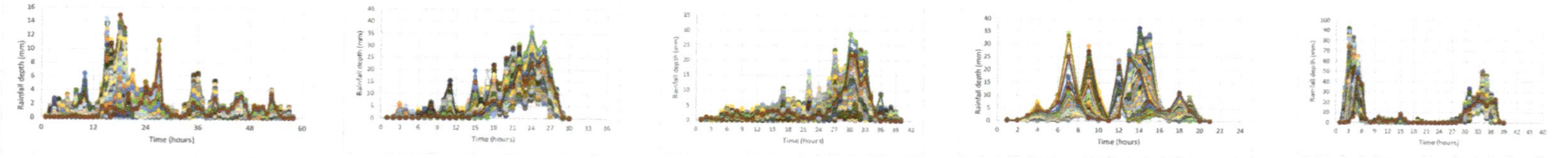
EV17 (Rainfall depth = 185 mm) 2015/8/7 08:00–2015/8/8 13:00
EV18 (Rainfall depth = 138 mm) 2015/9/27 14:00–2015/9/29 05:00
EV19 (Rainfall depth = 106 mm) 2016/9/26 10:00–2016/9/28 04:00
EV20 (Rainfall depth = 271 mm) 2017/6/2 10:00–2017/6/4 04:00

Figure 5. Hyetographs of 20 observed rainstorm events used in the model development and application [18].

4. Results and Discussion

4.1. Simulation of Rainfall-Induced Inundation

Before carrying out the rainfall-induced inundation, a great number of the gridded rainstorms should be reproduced in advance. Thus, in this study, using the SG_GSTR method with the above statistical moments of gridded rainfall characteristics, 1000 simulations can be obtained. After that, the SOBEK 1D-2D hydrodynamic model is employed to carry out the inundation simulation with 20 m $\times$ 20 m DEM of Nankan River watershed provided by Water Resource Agency in Taiwan (see Figure 5). Figure 6 presents the river-channel internet and computation nodes in the SOBEK 1D-2D dynamic model set up according to the geographical and hydrological data in the study area. Within the Nankan River watershed, a mesh composed of 335 $\times$ 520 computation grids whose the spatial resolution is 20 m $\times$ 20 m is adopted. The above topographic, hydrologic, and hydraulic features used in Nankan River SOBEK model are listed in Table 3. Finally, using the SOBEK model for the Nankan River watershed with 1000 simulations of the gridded rainstorm events, the resulting 2D inundation simulations, including the gridded inundation depths and corresponding flooding area, could be accordingly obtained.

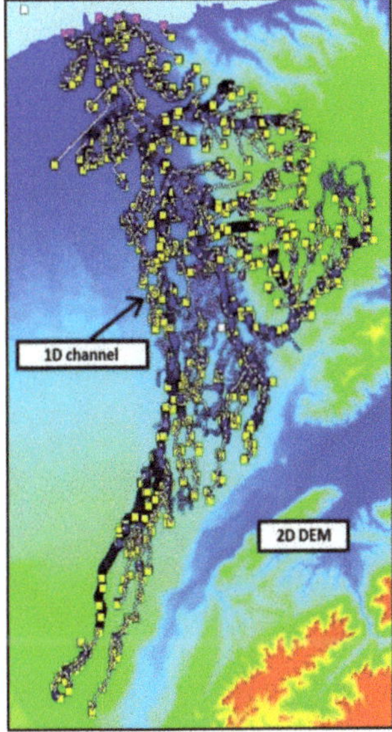

Figure 6. 2D SOBEK model for the study area (Nankan River watershed) (note: the circle is the rainfall-runoff computing node for each sub-basin).

Thereby, this study implements the 2D inundation simulation by taking into account the spatiotemporal uncertainty in the rain field to achieve the goal of providing detailed 2D inundation information with high spatial and temporal resolution as the training datasets for the proposed SM_EID_IOT model.

4.2. Identification of Potential Locations of Roadside IoT Sensors

On the basis of the results from the simulated grid-based inundation depths within the study area, a 2D inundation simulation with high spatial and temporal resolution can be used to determine the locations of the roadside water-level sensors. In detail, the road-side water-level sensors can be set up at the locations in association with high flooding frequency calculated from the above 2D inundation simulations for the correction of flooding forecasts [30]. Under the consideration of IoT quality, the most appropriate locations of roadside water-level sensors can be accordingly determined and named IoT sensors.

To quantify the flooding risk at all grids, this study utilizes the following equation to calculate the flooding probability $\left(P_f\left(h_f > 0\right)\right)$ within the study area, the Nankan River watershed (see Figure 4):

$$P_f\left(h_f > 0\right) = \frac{\sum_{i=1}^{N_{GRS}}\left(I_f\left(h_f^i\right)\right)}{N_{GRS}}$$
$$I_f\left(h_f^i\right) = 1, \; if \; max\left(h_f\right) > 0$$
$$I_f\left(h_f^i\right) = 0, \; if \; max\left(h_f\right) = 0$$

(10)

where $I_f\left(h_f^i\right)$ is the flooding indicator and $max\left(h_f\right)$ stands for the maximum gridded inundation. In Equation (10), if the maximum of gridded inundation depth is greater than zero, the corresponding flooding probability $I_f\left(h_f^i\right)$ is equal to one; otherwise, it is equal to 0. Thus, in reference to Figure 7, it can be seen that inundation mainly takes place in the four regions. Apparently, the first 46 grids associated with high flooding probabilities (approximately 0.7) can be identified in the two locations marked with a red line, near the locations (TWD97X:280200.2, TWD97Y:2765356.7) and (TWD97X:281523.6, TWD97Y:2761435.5), respectively; they can be treated as the potential inundated grids. Therefore, among the aforementioned potential inundated spots, the locations of desired roadside IoT sensors can be determined.

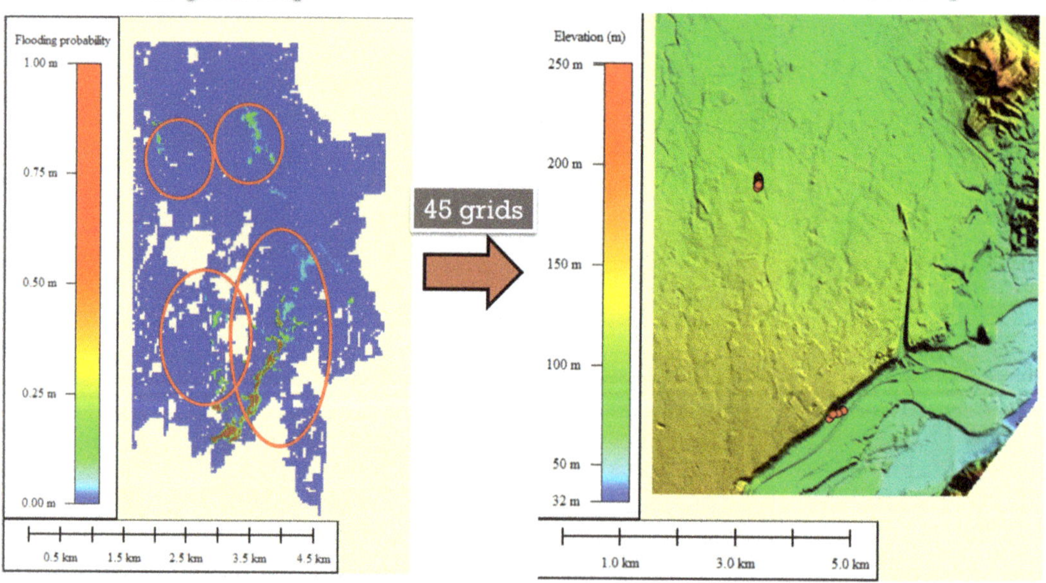

Figure 7. Flooding risk map and locations of potential inundated spots within the study area (the Nankan River watershed).

In conclusion, the quantification of flood risk at all grids within the study area can be carried out using a large number of inundation simulations by the hydraulic numerical model with the generated grid-based rainstorm events through the proposed SM_GSTR model. Additionally, the resulting big data of rainfall-inundation simulations are advantageous to the identification of roadside inundation-depth sensors, which can be used in the real-time error correction of flood forecasts [1].

As the Nankan River watershed lacks practical roadside IoT sensors, the grid with high flooding risk, the TWD97 coordination of which is (TWD97X: 281523.6, TWD97Y:

2761435.5), is selected as the potential IoT sensor. As a result, the corresponding simulations of rainfall-induced inundation, consisting of the inundation depths and associated areal average rainfalls, are employed in the development and validation of the proposed SM_EID_IOT model.

4.3. Determination of Critical Resolutions in Time and Space

In this study, the correlation and sensitivity analysis for the inundation depth at the IoT sensor of interest (TWD97X: 281523.6, TWD97Y: 2761435.5) is carried out to detect the appropriate critical value of the spatial and temporal resolutions. In detail, the critical distance to the IoT sensor for calculating the areal average rainfalls and the forward time steps from the current time for selecting the observed inundation depths and areal average rainfalls are required for deriving Equation (9) within the proposed SM_EID_IOT model.

4.3.1. Critical Resolution in Space

To quantify the fitness of the areal average rainfall for various critical distances to the inundation depth, Pearson correlation coefficients (ρ) are adopted, as shown in Figure 8, presenting that most correlation coefficients gradually increase/decrease with the critical distances, i.e., the critical spatial resolution L_c used in Equation (9). For example, regarding the 5th, 500th, and 1000th simulation cases, the correction coefficient generally declines from the critical distance of 1.5 km to 4 km critical distance; it then remains constant. Contrarily, in the case of the remaining simulation cases, the correlation coefficient remains constant.

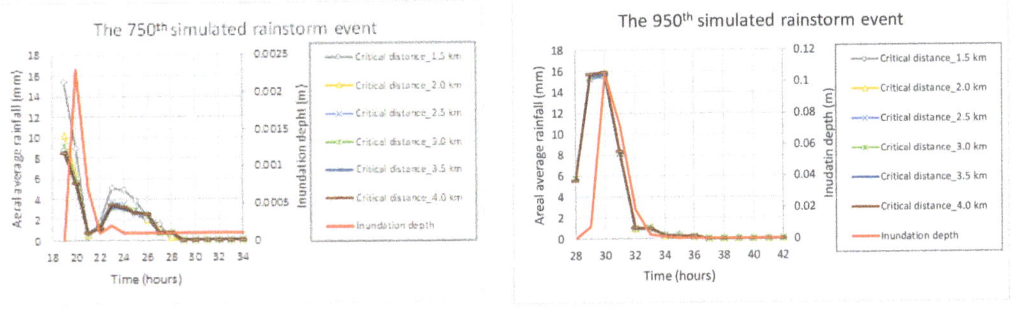

Figure 8. Relationship between the inundation depth and associated areal average of rainfall calculated using the simulated gridded rainfall with the specific distances from the IoT sensor of interest.

To quantify and assess the effect of critical distances regarding the calculation of areal average rainfall on the estimated inundation depth, the statistical properties of Pearson coefficients (i.e., mean value μ and standard deviation σ) calculated from 1000 simulation cases are computed as shown in Figure 9, where the mean value of the correlation coefficient ρ declines with the critical distance from 1.5 km ($\rho = 0.655$) km to 3 km ($\rho = 0.645$), and the correlation coefficient then reaches its constant value ($\rho = 0.645$); further, its standard deviations for various critical distances approximate 0.69, except for the 2 km critical distance ($\sigma = 0.688$).

To sum up, the above results from the correlation analysis reveals that, in spite of the 2 km distance having the smallest variation, the areal average rainfall calculated from the precipitations at the grids, the distance of which to the IoT sensor is less than or equal to 3 km, exhibits a stable variation in terms of the correlation coefficient. In conclusion, the critical resolution in space is assigned as 3 km for the calculation of the areal average rainfalls.

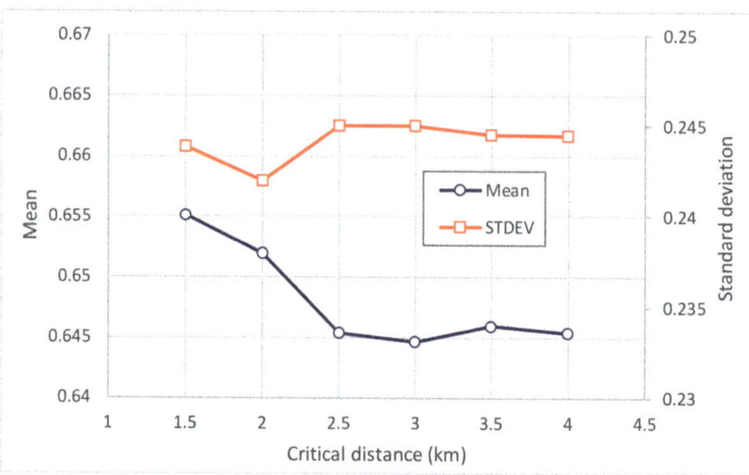

Figure 9. Correlation coefficients and associated statistical properties of inundation depths with the areal average rainfall calculated using the simulated gridded rainfall with specific distances from the IoT sensor of interest.

4.3.2. Critical Resolution in Time

Generally speaking, the water level at the current time is markedly related to the areal average rainfall and inundation depths at the previous time steps. Thus, the temporal resolution of the areal average rainfall and inundation depths in terms of the correlations within the specific forward time steps, i.e., critical temporal resolution T_c hour in the standardized regression equation, possibly affect the estimation of the inundation depths at the current time step for the IoT sensors. The standardized regression equation is as follows (Speed and Yu, 1993):

$$\frac{Y - \overline{Y}}{S_Y} = \sum_{i=1}^{n} \beta_i \frac{X_i - \hat{X}_i}{S_{X_i}} \tag{11}$$

where Y and X are the model output and inputs; $\overline{Y}$ and $\hat{X}$ account for the mean of the model output and inputs, respectively; S_Y and S_X separately represent the corresponding standard deviation; and β_i denotes the regression coefficient that is inversely related to the model outputs; otherwise, the model parameter is proportional to the model output in the case of the associated β_i being positive. Note that the standard regression coefficient β_i accounts for the sensitivities of the ith model parameter to the model outputs, meaning that a larger absolute value indicates that the change in the ith model parameter more significantly impacts the estimation of the model outputs. Moreover, the model parameter is associated with the negative β_i. Accordingly, the above specific forward period k hours should be treated as the sensible factors, which can be determined based on the regression coefficients β (i.e., the sensitivity coefficient) of the standard regression Equation (11).

In this study, using 1000 simulations of the rainfall-induced inundation, the inundation depths at the six particular time steps—0.3, 0.5, 0.6, 0.7,0.8, and 0.9 times the duration—and the inundation depths as well as the areal average rainfall at the forward 6 h are used in the establishment of the standard regression equation; their resulting regression coefficients could be obtained as shown in Figure 10. By observing Figure 10, the average of absolute sensitivity coefficients regarding the areal average rainfall and inundation depths at the forward 1–6 h, it is known that, with respect to the inundation depth, the average of the absolute sensitivity coefficients of the forward period from $T_c = 1$ h to $T_c = 3$ h ranges between 0.563 and 0.43, which are obviously greater than the coefficients regarding the forward $T_c = 4$ h to $T_c = 6$ h (about 0.029–0.127). Furthermore, in the case of the areal average rainfall, the absolute sensitivity coefficients corresponding to the forward 3 h change on average from 0.123 to 0.426, which are significantly greater than the coefficients at the forward 4–6 h (approximately 0.015–0.048).

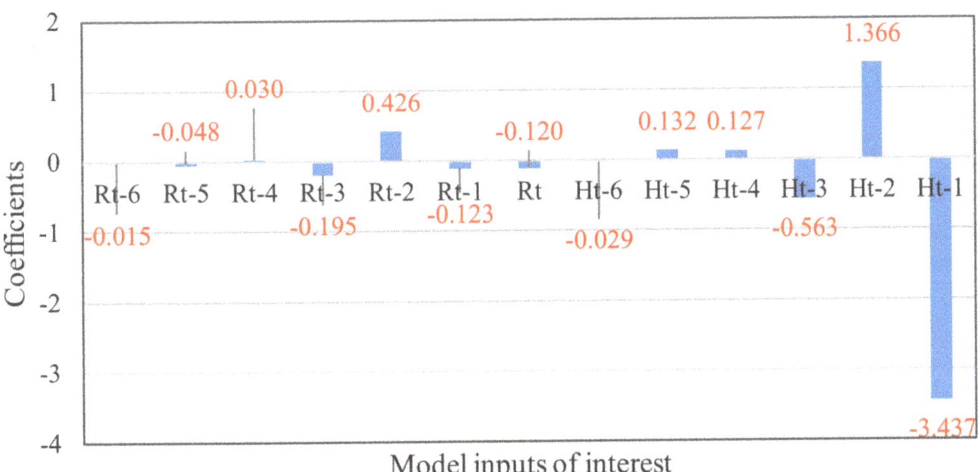

Figure 10. The averages of absolute sensitivity coefficients regarding the areal average rainfall and inundation depths at the various forward time steps.

To sum up the above results, the inundation depths at a particular time step are strongly and significantly related to the areal average of rainfall and inundation depths at the forward $T_c = 3$ h. Therefore, in referring to Equation (9), the relationship of the estimated/forecasted inundation depth regarding the lead time $(t + 1\ \text{h})$ at a specific IoT sensor $\hat{h}_{IOT}^{t+1}$ with the observation of the areal average rainfall at the lead time $(t + 1\ \text{h})$ and current time (t hour) as well as the forward 2 h and the inundation depths at the forward 3 h (i.e., t, $t - 1$ and $t - 2$ h) can be written as follows:

$$\hat{h}_{IOT}^{t+1} = f_{ANN-GA-MTF}\left(\overline{R}_{IOT}^{t+1}, \overline{R}_{IOT}^{t}, \overline{R}_{IOT}^{t-1}, \overline{R}_{IOT}^{t-2}, h_{IOT}^{t}, h_{IOT}^{t-1}, h_{IOT}^{t-2}\right) \tag{12}$$

where $\overline{R}_{IOT}^{t}$ is the rainfall forecast at the lead time $(t + 1\ \text{h})$; $\overline{R}_{IOT}^{t}$, $\overline{R}_{IOT}^{t-1}$, and $\overline{R}_{IOT}^{t-2}$ are the areal average rainfalls calculated using the gridded rainfall at the current time (t hour) and forward 2 h ($t - 1$ and $t - 2$ h) within the distance of 3 km to the location of the IoT sensor; and h_{IOT}^{t}, h_{IOT}^{t-1}, and h_{IOT}^{t-2} represent the observed inundation depths at the current time (t hour) and forward 2 h ($t - 1$ and $t - 2$ h).

4.4. Development of the Proposed SM_EID_IOT Model

In referring to the framework of the model development, the proposed SM_EID_IOT model for estimating the inundation depths at the lead time regarding the IoT sensor is developed based on the ANN-derived ANN_GA-SA_MTF model by taking into account the uncertainty factors, including the areal average rainfall at the lead time and forward 3 h and the inundation depths at the forward 3 h. Furthermore, according to the induction to the ANN_GA-SA_MTF model, the initial conditions regarding the parameters should be given in advance, including the number of the hidden layers, the total number of neurons used, and the candidate transfer functions, as listed in Table 2. It is well-known that the three-layer network structure, comprising one input layer, one output layer, and one hidden layer, is commonly adopted in hydrological/hydraulic modeling (e.g., Wu et al., 2021); thus, the hidden layer used in the derivation of the SM_EID_IOT model is derived on the basis of the three-layer ANN-based model (i.e., the ANN_GA-SA_MTF model).

Moreover, in general, the number of neurons in the entire ANN network structure can be set up by means of a variety of formulae listed in Table 1 in accordance with the number of model inputs and outputs. Figure 11 indicates the resulting number of neurons from the different equations, ranging from 3 neurons to 127 neurons. Accordingly, their average number of 8 neurons is employed in the model development. In addition, the statistical properties, the mean and standard deviation, of the ANN weights are assigned as 0 and 1, respectively. As for the remaining parameters, their initial conditions can be referred to in Table 4.

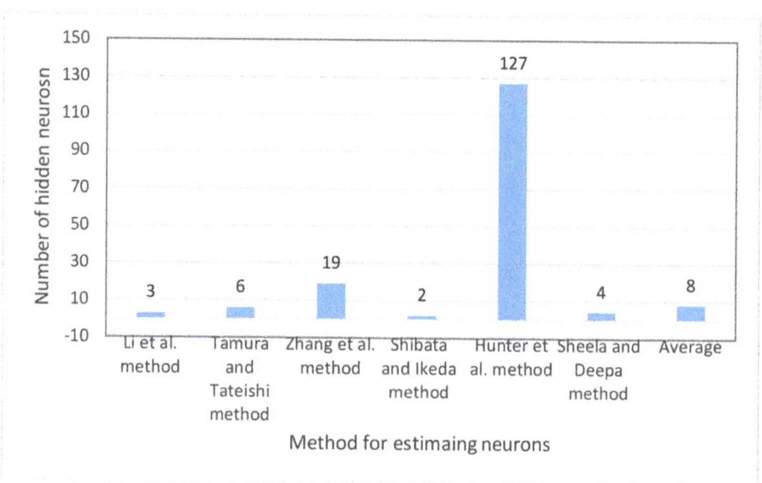

Figure 11. Summary for the estimation of the number of hidden neurons via various methods.

Table 4. Definition of parameters used in the proposed ANN-GA-SA_MTF model.

Parameters	Definition		
Transfer functions used		TF1-TF10	
Input factors	Average rainfall	$\overline{R}_{IOT}^{t+1}, \overline{R}_{IOT}^{t}, \overline{R}_{IOT}^{t-1}, \overline{R}_{IOT}^{t-2}$	
	Inundation depth	$h_{IOT}^{t}, h_{IOT}^{t-1}, h_{IOT}^{t-2}$	
Output factor	Inundation depth	$\hat{h}_{IOT}^{t+1}$	
Number of hidden levels		1	
Number of neurons		8	
Calibration of parameters of transfer function	Number of optimizations	10	
	Weights of neurons (ω_{HL})	Mean	1
		Standard deviation	3
	Bias of function (θ_{TF})	Mean	0
		Standard deviation	1
	Adjusting factor ($\propto_{TF}$)	Mean	1
		Standard deviation	0.005

Using the parameter definition shown in Table 5, the SM_EID_IOT model can be developed by training the ANN_GA-SA_MTF model with 650 simulations of the training datasets, extracted from 1000 rainfall-induced inundation simulations obtained in Sections 4.1 and 4.2. Table 5 summarizes the results from the parameter calibrations under consideration of the transfer function TF1 (Sigmoid function). Furthermore, according to Figure 12, the inundation-depth estimates are the weighted average using the model estimates resulting from a variety of transfer functions with the corresponding weights, as shown in Figure 13, in which TF2 (Tanh function) and TF5 (ReLU function) are associated with the maximum and minimum weights, 0.10045 and 0.0996, respectively, indicating that the various transfer functions significantly contribute to different degrees to the estimation of the inundation depths at the IoT sensors. Consequently, it is necessary to take into account the effect of uncertainty in the formulation of the transfer functions on the estimation of model outputs via the proposed SM_EID_IOT model.

Table 5. Summary for the calibrated parameter of the SM_EID_IOT model in the study area (the Nankan River Watershed).

Transfer Function	Adjust Factor ($\propto_{TF}$)			0.99139							
		First Hidden Layer		**Model Inputs at Input Layer**							
				1	**2**	**3**	**4**	**5**	**6**	**7**	**Bias**
TF1 (Sigmoid)	Connection weights of neurons ω_{HL}	Neuron	1	2.105	−0.146	−3.114	−3.035	1.756	−3.039	−3.355	0.169
			2	1.879	−0.267	2.018	−3.344	−1.904	−0.069	−4.633	−2.275
			2	1.898	3.023	−3.828	−2.855	4.008	−1.024	4.944	−3.015
			4	−0.186	−1.392	2.213	1.737	4.005	6.746	−2.054	−0.052
			5	−1.372	2.638	−1.166	−3.536	1.961	−2.736	−3.007	−0.043
			6	3.729	−3.363	−2.254	−0.530	−4.844	4.774	2.666	5.212
			7	5.364	0.546	5.105	0.169	−4.981	2.625	2.180	3.198
			8	10.950	−3.238	6.896	−1.404	−5.472	1.547	−4.310	−4.908
	Output layer Model output			**Connection neurons at the first hidden layer**							
				1	**2**	**3**	**4**	**5**	**6**	**7**	**8**	**Bias**
		1		−0.264	−0.002	0.357	0.136	−0.560	−1.682	−0.582	0.368	0.085

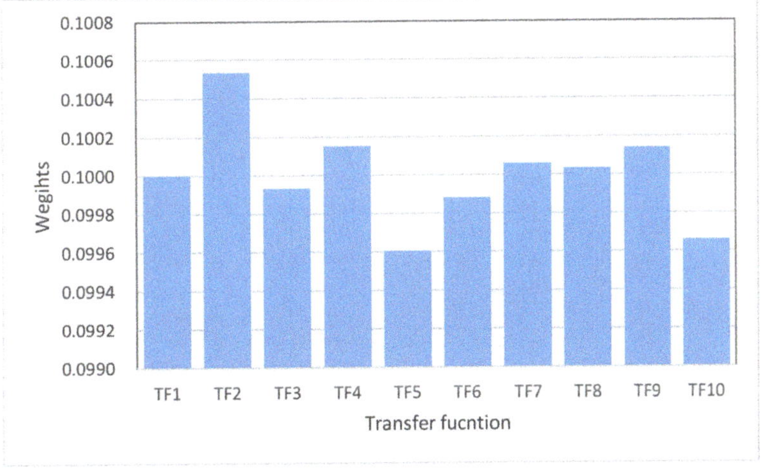

Figure 12. Summary of the weights of the transfer function for calculating the weighted average of inundation-depth estimates.

4.5. Model Validation

To demonstrate the reliability and accuracy of the resulting inundation-depth estimates from the proposed SM_EID_IOT model, a simulated rainfall-induced inundation event, i.e., the 825th simulated rainstorm event (see Figure 13) is adopted as the validated one, where the duration and average rainfall intensity regarding the validated event are 57 h and 3.7 mm/h, respectively; moreover, in the validated water-level hyetograph, the two peaks of the inundation depths are 0.12 m and 0.1 m at the 30th and 40th hours, respectively. As the Center Weather Bureau (CWB) in Taiwan can provide the gridded rainfall forecasts at lead times of 3 h, the model verification focuses on the evaluation in comparison with the inundation-depths estimates at the 1, 2, and 3 h lead times, namely, $t + 1$, $t + 2$, and $t + 3$ h, respectively (t = current time).

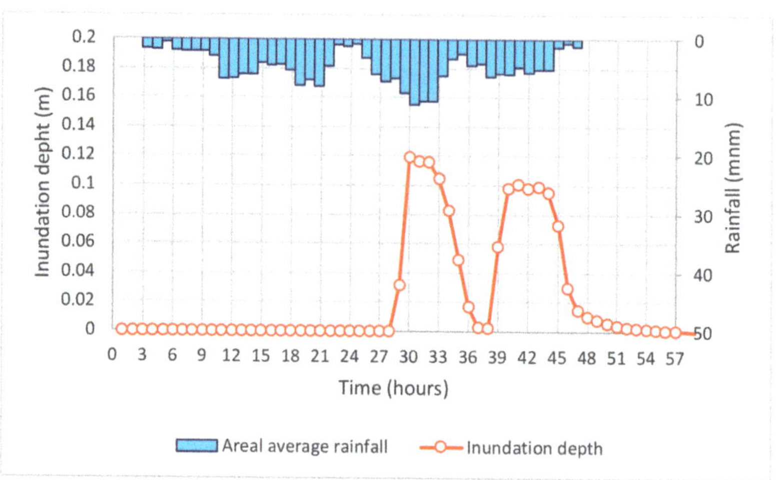

Figure 13. Relationship between the areal average rainfall and inundation depths regarding the 825th validated rainfall-induced flood event.

4.5.1. Reliability Quantification of Inundation-Depth Estimates

Accordingly, as for the 460th simulation case, the resulting inundation-depth estimates and associated 95% confidence intervals at the IoT sensor can be obtained from the SM_EID_IOT model as compared with the validated ones, as shown in Figure 14. It can be seen that the temporal change in the inundation-depth estimates at the three lead times resembles variation regarding the areal average rainfall in accordance with the high correlation. Moreover, at the 1 h lead time, the estimated inundation depths mostly lie around the 95% confidence interval, except for the 30th–31th and 40th–41th hour, where the inundation-depth estimates are about 0.119 m and 0.09 m, exceeding the upper bounds of 0.075 m and 0.085 m, respectively. Similar results can be found for the inundation-depth forecasts at the 2 h and 3 h lead times. The above results imply that the proposed SM_EID_IOT model can produce the inundation-depth estimate with high likelihood of approaching the true values (i.e., observations).

As can be seen in Figure 14, in spite of the proposed SM_EID_IOT model possibly producing reasonable inundation-depth estimates, the inundation-depth estimates at the 1 h lead time are underestimated as compared with the validated data at the 30th–32th hours regarding the 825th simulated event. Hence, to evaluate the accuracy of the inundation-depth forecasts at the various lead times, the performance in comparison with the estimated inundation depths and validated ones at the 31th–51th hours is carried out in terms of the root mean error (RMSE) and correlation coefficients, as shown in Figure 15. According to the results from Figure 15, the RMSE increases with the lead time; however, the correlation coefficient declines with the lead time. For example, although the estimations exhibit an obvious difference from the validation data in association with a large root mean error square (RMSE) (about 0.01 m), the corresponding correlation coefficient approaches 0.3, meaning the change in the average rainfall in time is close to that regarding the inundation depth; similar conclusions are also made based on the results from the 2 h and 3 h lead time. The above difference between the estimation and validations might be caused by the uncertainties in the observation and model parameters.

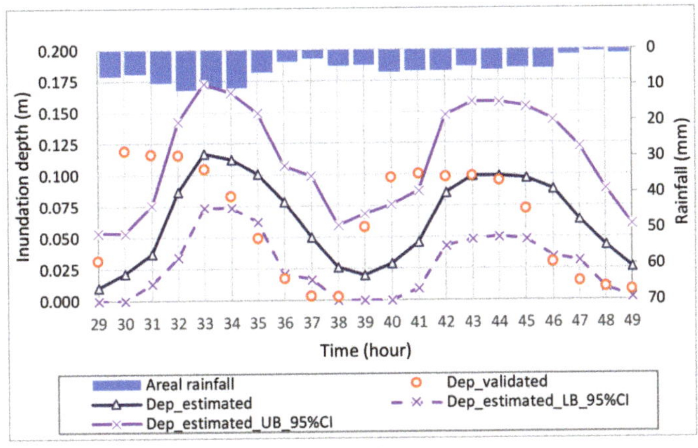

Lead time = 1-hour

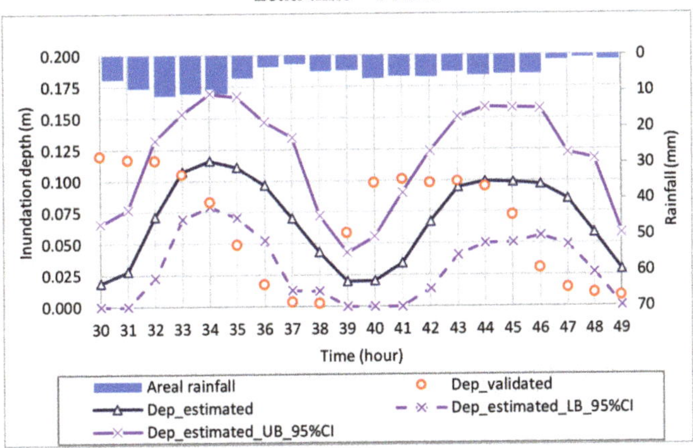

Lead time = 2-hour

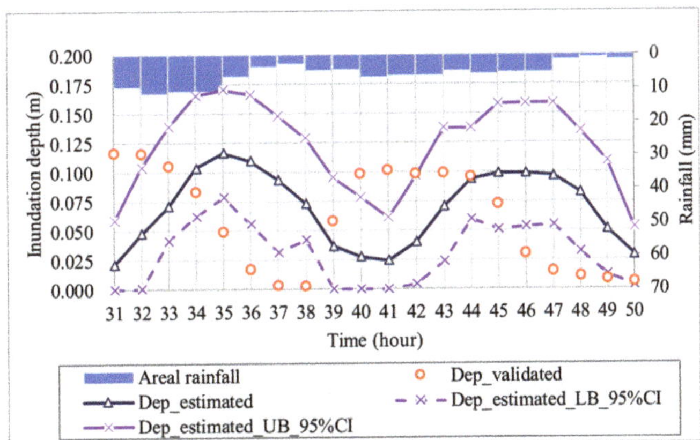

Lead time = 3-hour

Figure 14. Comparison among the validated, estimated, and corrected inundation depths as well as the quantified 95% confidence intervals for the validated rainfall-induced flood event by the proposed SM_EID_IOT model.

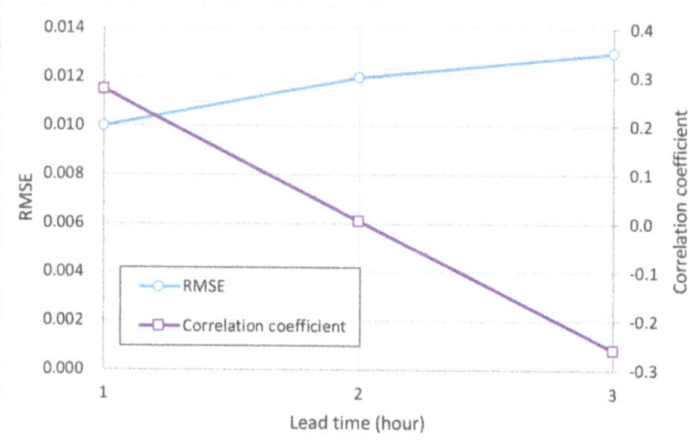

Figure 15. Summary of the performance indices of the inundation-depth estimates and comparison with the validation datasets at various lead times.

4.5.2. Real-Time Correction of Inundation-Depth Estimates

To facilitate the accuracy of the results from the proposed SM_EID_IOT model, the inundation-depth estimates are supposed to be adjusted based on the real-time measured data. The real-time error correction method is developed via the time series approach and Kalmen filtering equation (RTEC_TS&KF) [30]. Figures 16 and 17 show the comparison between the validated, estimated, and corrected inundation depths, respectively, as well as the corresponding performance indices, respectively, indicating that the RMSE value increases with the lead time; in contrast, the correlation coefficient markedly decreases with the lead time. In spite of the RMSE values for the corrected inundation-depth estimates significantly increasing with the lead time, from 0.007 m to 0.014 m, on average, they are less than those for the forecasts (0.012 m). Moreover, with respect to the consistency wof the validations, the correlation coefficients for both inundation-depth estimations and corrections generally decrease with the lead time, 0.28–0.26 (estimations) and 0.74–0.06 (corrections), respectively. For illustration, the correlation coefficients for the corrections at the 1 h lead time approximate 0.7, obviously greater than that from the estimates (about 0.28). In particular, at the 3 h lead time with the worst correlation, −0.25 (estimations) and 0.07 (corrections), the corrections have better consistency with the validated datasets than the estimations. The above results reveal that the corrected inundation-depth forecasts exhibit better agreement with the observations than the underestimated/overestimated inundation-depth forecasts, with the marked errors even for the long lead times being able to be immediately adjusted.

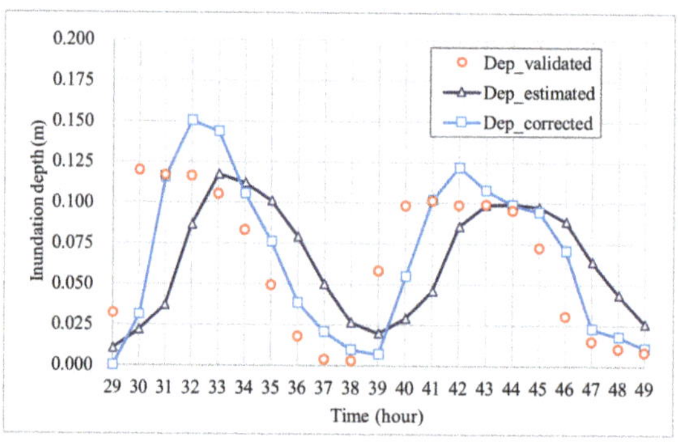

Lead time = 1-hour

Figure 16. *Cont.*

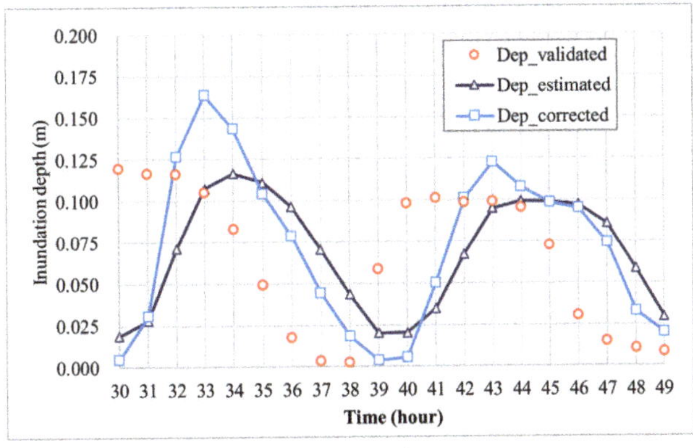

Figure 16. Comparison among the validated inundation depths and the corrected as well as corrected ones by the proposed SM_EID_IOT model during the validated rainfall-induced flood event.

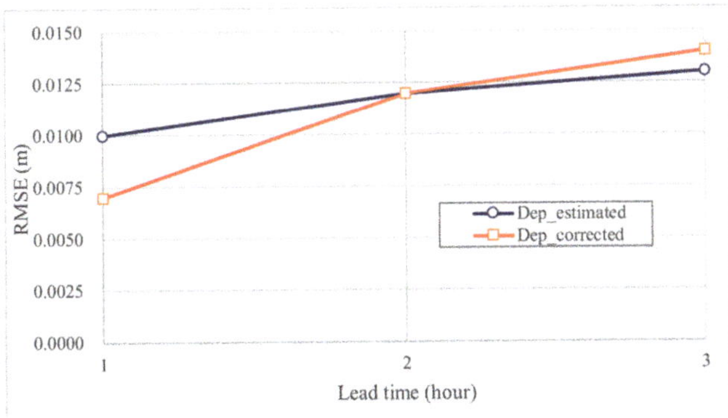

Figure 17. *Cont.*

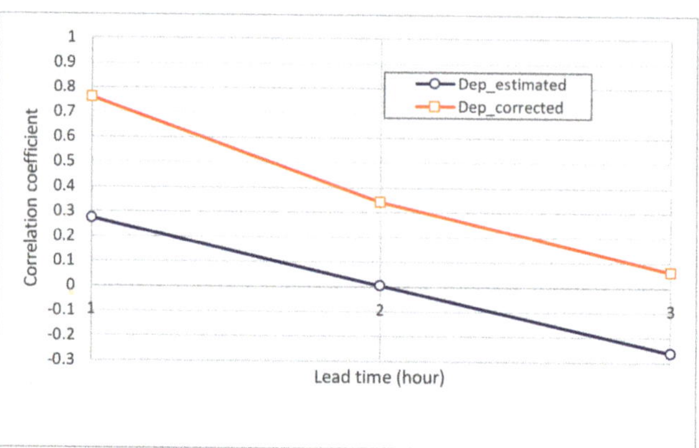

Figure 17. Summary of the performance indices of the inundation-depth estimates and the corresponding corrections for the validated rainfall-induced flood event.

In summary, the accuracy of the resulting inundation-depth estimates at various lead times (hours) from the proposed SM_EID_MTF with a reasonable reliability can be effectively improved based on the difference between the observations and estimates at the previous time steps during a rainfall-induced flooding event.

5. Conclusions

This study intends to propose a stochastic ANN-derived model for the estimation of the inundation depths at the roadside water-level sensors set up through the Internet of Things (IoT) (named the SM_EID_IOT). The proposed SM_EID_IOT is developed on the basis of the artificial neural network model ANN_GS-SA_MTF (Wu et al., 2021), in which the associated parameters are calibrated by means of the modified genetic algorithm (GA-SA) [30] under the consideration of multiple transfer functions. A basin located in northern Taiwan, the Nankan River watershed, is selected as the study area and the associated grid-based precipitation data regarding 20 historical rainstorms provided by Center Weather Bureau in Taiwan are utilized to reproduce 1000 simulations of the rainfall-induced inundations via as the training dataset for the development of the proposed SM_EID_IOT model.

According to the results from the correlation and sensitivity analysis, the inundation depths at the IoT sensor for the forward periods of 3 h (i.e., critical temporal resolution) and the corresponding precipitations at the neighboring grids within the specific distance of 3 km (i.e., critical spatial resolution) to the IoT sensor should be regarded as the uncertainty factors for the resulting inundation-depth estimate (i.e., model inputs) from the proposed SM_EID_IOT model. Additionally, the results from the model demonstration indicate that the validated inundation depths at the lead times of 3 h are almost located within the quantified 95% confidence intervals by the proposed SM_EID_IOT model, revealing that the proposed SM_EID_IOT can provide the inundation-depth estimates at the lead times of 3 h with a high likelihood of approaching the validated datasets. Furthermore, the corrected inundation-depth estimates by the real-time error correction method RTEC_TS&KF method integrated within the proposed SM_EID_IOT model could effectively improve the accuracy of the inundation-depth estimates by 50%; thereby, the estimate exhibits a good match with the validated datasets under a better temporal correlation (i.e., correlation coefficient approaching 0.8). Consequently, the proposed SM_EID_IOT model is capable of estimating more accurate inundation depths at the IoT sensors of interest with high reliability.

In addition to the estimation of the inundation depths at the particular locations at which the roadside water-level IoT sensors are set up, the rainfall-induced flooding map is necessarily delineated in order to primarily estimate the possible inundation area under conditions of the flood-rated indices, such as the flash-food potential index (FFPI) and flooding potential index (FPI) [48], and rainfall-rated variables, such as the radar precipitation [49,50]. As a result of the flooding map being composed of the gridded inundation depths, the inundation depths at the ungauged locations should be quantified; by doing so, future work would improve the application of the framework and detailed concepts of developing the proposed SM_EID_IOT model in the derivation of stochastic ANN-derived modeling (i.e., the ANN_GA-SA MTF model) for the inundation-depth estimates at the ungauged locations within the flood-prone zones.

Author Contributions: S.-J.W. Conceptualization, methodology, investigation, validation, original draft preparation, and writing—review and editing; C.-T.H. Resources and software; C.-H.C. Resources. All authors have read and agreed to the published version of the manuscript.

Funding: This research received no external funding.

Institutional Review Board Statement: Not applicable.

Informed Consent Statement: Not applicable.

Data Availability Statement: This paper was written based on the results from the several researches made by authors which were not mentioned in any reports.

Conflicts of Interest: The authors declare no conflict of interest.

References

1. Wu, S.J.; Chang, C.H.; Hsu, C.T. Real-time error correction of two-dimensional flood-inundation simulations during rainstorm events. *Stoch. Environ. Res. Risk Assess.* **2020**, *34*, 641–667. [CrossRef]
2. Amarnath, G. An algorithm for rapid flood inundation mapping from optical data using a reflectance differencing technique. *J. Risk Manag.* **2014**, *7*, 239–250. [CrossRef]
3. Park, I.; Seong, H.; Ryu, Y.; Rhee, D.S. Measuring inundation depth in a subway station using the laser image analysis method. *Water* **2018**, *10*, 1558. [CrossRef]
4. Chen, A.S.; Hsu, M.H.; Teng, W.H.; Huang, C.J.; Yeh, S.H.; Ien, W.Y. Establishment the Database of Inundation Potential in Tawian. *Nat. Hazards* **2018**, *377*, 107–132.
5. Rebolho, C.; Andreassian, V.; Moine, N.L. Inundation mapping based on reach-scale effective geometric. *Hydrol. Earth Syst. Sci. Discuss.* **2018**, *22*, 5967–5985. [CrossRef]
6. Ongdas, N.; Akiyanova, F.; Karakulov, Y.; Muratbayeva, A.; Zinabdin, N. Application of HEC-RAS (2D) for Flood Hazard Maps Generation for Yesil (Ishim) River in Kazakhstan. *Water* **2020**, *12*, 2672. [CrossRef]
7. Fohringer, J.; Dransch, D.; Kreibich, H.; Schröter, K. Social media as an information source for rapid flood inundation mapping. *Nat. Hazards Earth Syst. Sci.* **2015**, *15*, 2725–2738. [CrossRef]
8. Assumpção, T.H.; Popescu, I.; Jonoski, A.; Solomatine, D.P. Citizen observations contributing to flood modelling: Opportunities and challenges. *Hydrol. Earth Syst. Sci.* **2018**, *22*, 1473–1489. [CrossRef]
9. Chang, L.C.; Mohd Zaki, M.A.; Yang, S.N.; Chang, F.J. Building ANN-Based Regional Multi-Step-Ahead Flood Inundation Forecast Models. *Water* **2018**, *10*, 1283. [CrossRef]
10. Yang, S.N.; Chang, L.C. Regional inundation forecasting using machine learning techniques with the Internet of Thing. *Water* **2020**, *12*, 1578. [CrossRef]
11. Jung, Y.J.; Kim, D.K.; Kim, D.W.; Kim, M.; Lee, S.O. Simplified Flood Inundation Mapping Based on Flood Elevation-Discharge Rating Curves Using Satellite Images in Gauged Watersheds. *Water* **2014**, *6*, 1280–1299. [CrossRef]
12. Shastry, A.; Durand, M. Utilizing Flood Inundation Observations to Obtain Floodplain Topography in Data-Scarce Regions. *Front. Earth Sci.* **2019**, *6*, 243. [CrossRef]
13. Jain, S.K.; Mani, P.M.; Jain, S.K.; Prakash, P.; Singh, V.P.; Tullos, D.; Kumar, S.; Agarwal, S.P.; Dimri, A.P. A brief review of flood forecasting techniques and their applications. *Int. J. River Basin Manag.* **2018**. [CrossRef]
14. Myronidis, D.; Ivanova, E. Generating regional models for estimating the peak flows and environmental flows magnitude for the Bulgarian-Greek Rhodope mountain range torrential watersheds. *Water* **2020**, *12*, 784. [CrossRef]
15. Paschalis, A.; Molnar, P.; Fatichi, S.; Burlando, O. A stochastic model for high-resolution space-time precipitation simulation. *Water Resour. Res.* **2013**, *49*, 8400–8417. [CrossRef]
16. Ran, Q.; Hong, Y.; Li., W.; Gao, J. A modeling study of rainfall-induced shallow landslide mechanisms under different rainfall characteristics. *J. Hydrol.* **2018**, *363*, 790–801. [CrossRef]
17. Kan, G.; He, X.; Li, J.; Ding, L.; Hong, Y. Computer aided numerical methods for hydrological model calibration: An overview and recent development. *Arch. Comput. Methods Eng.* **2019**, *26*, 35–59. [CrossRef]
18. Wu, S.J.; Hsu, C.T.; Chang, C.H. Stochastic modeling of artificial neural networks for real-Time hydrological forecasts based on uncertainties in transfer Functions and ANN weights. *Hydrol. Res.* **2021**. [CrossRef]

19. Gupta, H.V.; Beven, K.J.K.; Wagener, T. Model calibration and uncertainty estimation. In *Encyclopedia of Hydrologic Sciences*; Anderson, M.G., Ed.; Wiley: Chichester, UK, 2005.

20. Melsen, L.; Teuling, A.; Torfs, P.; Zappa, M.; Mizukami, N.; Clark, M.; Uijlenhoet, R. Representation of spatial and temporal variability in large-domain hydrological models: Case study for a mesoscale pre-Alpine basin. *Hydrol. Earth Syst. Sci.* **2016**, *29*, 2207–2226. [CrossRef]

21. Campolo, M.; Andreussi, P.; Soldati, A. River flood forecasting with a neural network model. *Water Resour. Res.* **1999**, *35*, 1191–1197. [CrossRef]

22. Wang, L.; Zeng, Y.; Chen, T. Back propagation neural network with adaptive differential evolution algorithm for time series forecasting. *Expert Syst. Appl.* **2015**, *42*, 855–863. [CrossRef]

23. Ioannou, K.; Myronidis, D.; Lefakis, P.; Stathis, D. The use of artificial neural networks (ANNs) for the forecast of precipitation levels of lake Doirani (N. Greece). *Fresenius Environ. Bull.* **2010**, *19*, 1921–1927.

24. Chang, L.C.; Shen, H.Y.; Chang, F.J. Regional flood inundation nowcast using hybrid SOM and dynamic neural networks. *J. Hydrol.* **2014**, *519*, 476–489. [CrossRef]

25. Sung, J.Y.; Lee, J.; Chung, I.W.; Heo, J.H. Hourly Water Level Forecasting at Tributary Affected by Main River Condition. *Water* **2017**, *6*, 644. [CrossRef]

26. Chang, L.C.; Chang, F.J.; Yang, S.N.; Kao, I.F.; Ku, Y.Y.; Kuo, C.L.; Ir. Mohd Zaki bin Mat Amin; Building an Intelligent Hydroinformatics Integration Platform for Regional Flood Inundation Warning Systems. *Water* **2019**, *11*, 9. [CrossRef]

27. Shamseldin, A.Y. Artificial neural network model for river flow forecasting in a developing country. *J. Hydroinform.* **2010**, *12*, 22–35. [CrossRef]

28. Tamiru, H.; Dinka, M.O. Application of ANN and HEC-RAS model for flood simulation mapping in lower Baro Akobo River Basin, Ethiopia. *J. Hydrol. Reg. Stud.* **2021**, *36*, 100855. [CrossRef]

29. Wu, S.J.; Lien, H.C.; Chang, C.H. Calibration of a conceptual rainfall-runoff model using a genetic algorithm integrated with runoff estimation sensitivity to parameters. *J. Hydroinformatics* **2012**, *14*, 497–511. [CrossRef]

30. Wu, S.J.; Lien, H.C.; Chang, C.H.; Shen, J.C. Real-Time Correction of Water Stage Forecast during Rainstorm Events Using Combination of Forecast Errors. *Stoch. Environ. Res. Risk Assess.* **2012**, *26*, 519–531. [CrossRef]

31. Wu, S.J.; Hsu, C.T.; Chang, C.H. Stochastic modeling of gridded short-term rainstorms. *Hydrol. Res.* **2021**, *52*, 876–904. [CrossRef]

32. Deltares Systems. SOBEK User Manual. Delft, the Netherlands. 2014. Available online: http://content.oss.deltares.nl/delft3d/manuals/SOBEK_User_Manual.pdf (accessed on 22 October 2021).

33. Nataf, A. Determination des distributions don't les marges sontdonnees. *C. R. L'acad. Sci.* **1962**, *225*, 42–43.

34. Liu, P.L.; Der Kiureghian, A. Multivariate distribution models with prescribed marginals covariances. *Probabilistic Eng. Mech.* **1986**, *1*, 105–112. [CrossRef]

35. Hydraulic Institute DHI. *MIKE 11: A Modelling System for Rivers and Channels, Reference Manual*; Danish Hydraulic Institute: Lakewood, CO, USA, 2016.

36. Danish Hydraulic Institute DHI. *MIKE 21: Flow Model FM, Hydrodynamic Module Reference Manual*; Danish Hydraulic Institute: Lakewood, CO, USA, 2019.

37. Horritt, M.D. Establishment the Datatbase of Inundation Potential in Tawian Bates, P.D. Predicting floodplain inundation: Raster-based modelling versus the finite-element approach. *Hydrol. Process.* **2001**, *18*, 825–842. [CrossRef]

38. Imrie, C.; Durucan, S.; Korre, A. River flow prediction using artificial neural networks: Generalization beyond the calibration range. *J. Hydrol.* **2000**, *233*, 138–153. [CrossRef]

39. Maca, P.; Pech, P.; Pavlasek, J. Comparing the Selected Transfer Functions and Local Optimization Methods for Neural Network Flood Runoff Forecast. *Math. Probl. Eng.* **2014**, *2014*, 782351. [CrossRef]

40. Dawson, C.W.; Wilby, R.L. Hydrological modelling using artificial neural networks. *Prog. Phys. Geogr.* **2001**, *25*, 80–108. [CrossRef]

41. Li, J.Y.; Chow, T.W.S.; Yu, Y.L. Estimation theory and optimization algorithm for the number of hidden units in the higher-order feedforward neural network. In Proceedings of the IEEE International Conference on Neural Networks, Perth, WA, Australia, 27 November–1 December 1995; Volume 3, pp. 1229–1233.

42. Zhang, Z.; Ma, X.; Yang, Y. Bounds on the number of hidden neurons in three-layer binary neural networks. *Neural Netw.* **2003**, *16*, 995–1002. [CrossRef]

43. Shibata, K.; Ikeda, Y. Effect of number of hidden neurons on learning in large-scale layered neural networks. In Proceedings of the ICROS-SICE International Joint Conference (ICCAS-SICE '09), Fukuoka, Japan, 18–21 August 2009; pp. 5008–5013.

44. Hunter, D.; Yu, H.; Pukish, M.S.; Kolbusz, J.B.M.; Wilamowski, B.M. Selection of proper neural network sizes and architectures: A comparative study. *IEEE Trans. Ind. Inform.* **2012**, *8*, 228–240. [CrossRef]

45. Sheela, K.; Deepa, S.N. Review on Methods to Fix Number of Hidden Neurons in Neural Networks. *Math. Probl. Eng.* **2013**. [CrossRef]

46. Yang, J.C.; Tung, Y.K. Establishment of flow-duration curve and the assessment of its certainty. In *Final Report of Environment Protection Agency*; Environment Protection Agency: Taipei City, Taiwan, 1996.

47. Terink, W.; Leijnse, H.; Eertwegh, G.V.D.; Uijlenhoet, R. Spatial resolutions in areal rainfall estimation and their impact on hydrological simulations of a lowland catchment. *J. Hydrol.* **2018**, *563*, 319–335. [CrossRef]

48. Zaharia, L.; Costache, R.; Pravalie, R.; Ioana-Toroimac, G. Mapping flood and flooding potential indices: A methodological approach to identifying areas susceptible to flood and flooding risk-Case study: The Prahova catchment (Romania). *Front. Earth Sci.* **2017**, *11*, 229–274. [CrossRef]
49. Shen, X.Y.; Wang, D.C.; Mao, K.B.; Anagnostou, E.; Hong, Y. Inundation extent mapping by synthetic aperture radar: A review. *Remote Sens.* **2019**, *11*, 879. [CrossRef]
50. Try, S.; Tanaka, S.; Tanaka, K.; Sayama, T.; Oeurng, C.; Uk, S. Comparison of gridded precipitation datasets for rainfall-runoff and inundation modeling in the Mekong River Basin. *PLoS ONE* **2020**, *15*, e0226814. [CrossRef]

MDPI

St. Alban-Anlage 66

4052 Basel

Switzerland

Tel. +41 61 683 77 34

Fax +41 61 302 89 18

www.mdpi.com

Water Editorial Office

E-mail: water@mdpi.com

www.mdpi.com/journal/water